Radiation Curing
of Polymeric Materials

ACS SYMPOSIUM SERIES **417**

Radiation Curing of Polymeric Materials

Charles E. Hoyle, EDITOR
University of Southern Mississippi

James F. Kinstle, EDITOR
James River Corporation

Developed from a symposium sponsored
by the Division of Polymeric Materials: Science and Engineering
at the 197th National Meeting
of the American Chemical Society,
Dallas, Texas,
April 9-14, 1989

American Chemical Society, Washington, DC 1990

Library of Congress Cataloging-in-Publication Data

Radiation curing of polymeric materials
  Charles E. Hoyle, editor, James F. Kinstle, editor.

  p.   cm.—(ACS Symposium Series, ISSN 0097–6156; 417).

  "Developed from a symposium sponsored by the Division of Polymeric Materials: Science and Engineering at the 197th National Meeting of the American Chemical Society, Dallas, Texas, April 9–14, 1989."

  Papers from the International Conference on Radiation Curing of Polymeric Materials.

  Includes bibliographical references.

  ISBN 0–8412–1730–0

  1. Radiation curing—Congresses. 2. Radiation—Industrial applications—Congresses. 3. Plastic coatings—Congresses.

  I. Hoyle, Charles E., 1948–   . II. Kinstle, James F. III. American Chemical Society. Division of Polymeric Materials: Science and Engineering. IV. International Conference on Radiation Curing of Polymeric Materials (1989: Dallas, Tex.) V. American Chemical Society. Meeting (197th: 1989: Dallas, Tex.) VI. Series.

TP156.C8R34 1990
668.4—dc20   89–29972
                                                       CIP

The paper used in this publication meets the minimum requirements of American National Standard for Information Sciences—Permanence of Paper for Printed Library Materials, ANSI Z39.48–1984.

∞

Copyright © 1990

American Chemical Society

All Rights Reserved. The appearance of the code at the bottom of the first page of each chapter in this volume indicates the copyright owner's consent that reprographic copies of the chapter may be made for personal or internal use or for the personal or internal use of specific clients. This consent is given on the condition, however, that the copier pay the stated per-copy fee through the Copyright Clearance Center, Inc., 27 Congress Street, Salem, MA 01970, for copying beyond that permitted by Sections 107 or 108 of the U.S. Copyright Law. This consent does not extend to copying or transmission by any means—graphic or electronic—for any other purpose, such as for general distribution, for advertising or promotional purposes, for creating a new collective work, for resale, or for information storage and retrieval systems. The copying fee for each chapter is indicated in the code at the bottom of the first page of the chapter.

The citation of trade names and/or names of manufacturers in this publication is not to be construed as an endorsement or as approval by ACS of the commercial products or services referenced herein; nor should the mere reference herein to any drawing, specification, chemical process, or other data be regarded as a license or as a conveyance of any right or permission to the holder, reader, or any other person or corporation, to manufacture, reproduce, use, or sell any patented invention or copyrighted work that may in any way be related thereto. Registered names, trademarks, etc., used in this publication, even without specific indication thereof, are not to be considered unprotected by law.

PRINTED IN THE UNITED STATES OF AMERICA

ACS Symposium Series

M. Joan Comstock, *Series Editor*

1989 ACS Books Advisory Board

Paul S. Anderson
Merck Sharp & Dohme Research
  Laboratories

Alexis T. Bell
University of California—Berkeley

Harvey W. Blanch
University of California—Berkeley

Malcolm H. Chisholm
Indiana University

Alan Elzerman
Clemson University

John W. Finley
Nabisco Brands, Inc.

Natalie Foster
Lehigh University

Marye Anne Fox
The University of Texas—Austin

G. Wayne Ivie
U.S. Department of Agriculture,
  Agricultural Research Service

Mary A. Kaiser
E. I. du Pont de Nemours and
  Company

Michael R. Ladisch
Purdue University

John L. Massingill
Dow Chemical Company

Daniel M. Quinn
University of Iowa

James C. Randall
Exxon Chemical Company

Elsa Reichmanis
AT&T Bell Laboratories

C. M. Roland
U.S. Naval Research Laboratory

Stephen A. Szabo
Conoco Inc.

Wendy A. Warr
Imperial Chemical Industries

Robert A. Weiss
University of Connecticut

Foreword

The ACS SYMPOSIUM SERIES was founded in 1974 to provide a medium for publishing symposia quickly in book form. The format of the Series parallels that of the continuing ADVANCES IN CHEMISTRY SERIES except that, in order to save time, the papers are not typeset but are reproduced as they are submitted by the authors in camera-ready form. Papers are reviewed under the supervision of the Editors with the assistance of the Series Advisory Board and are selected to maintain the integrity of the symposia; however, verbatim reproductions of previously published papers are not accepted. Both reviews and reports of research are acceptable, because symposia may embrace both types of presentation.

Contents

Preface xiii

1. Photocurable Coatings.. 1
   Charles E. Hoyle

2. Electron-Beam Curing of Polymeric Materials 17
   James F. Kinstle

PHOTOINITIATORS

3. Depth of Cure Profiling of UV-Cured Coatings 27
   Leslie R. Gatechair and Ann M. Tiefenthaler

4. Modified Kubelka–Munk Optical Treatment of Photon Flux
   in a Coating and Its Relationship to UV Curing as Measured
   with a UV N101 Cure Tester 43
   James H. Nobbs and Peter K. T. Oldring

5. Time-Resolved Laser Spectroscopy of Synergistic Processes
   in Photoinitiators of Polymerization.. 59
   Jean-Pierre Fouassier and Daniel-Joseph Lougnot

6. Photochemistry and Photopolymerization Activity
   of Water-Soluble Benzophenone Initiators..................................... 72
   Norman S. Allen, F. Catalina, J. L. Mateo, R. Sastre,
   W. Chen, P. N. Green, and W. A. Green

7. Cationic Photoinitiators: Rearrangement Reactions
   from Direct Irradiation of Diarylhalonium Salts........................... 82
   Nigel P. Hacker and John L. Dektar

8. Novel Photoinitiator for Modern Technology 92
   V. Desobry, K. Dietliker, R. Hüsler, L. Misev, M. Rembold,
   G. Rist, and W. Rutsch

9. Coreactive Photoinitiators for Surface Polymerization 106
   M. Koehler and J. Ohngemach

NOVEL RADIATION PHOTOCURABLE SYSTEMS

10. Effect of Monomer Structure on Concurrent Grafting
    During Radiation Curing 128
    Stephen J. Bett, Paul A. Dworjanyn, Barry A. Fields,
    John L. Garnett, Stan V. Jankiewicz, and David F. Sangster

11. Photoinitiated Cross-Linking of Polyethylenes and Diene
    Copolymers 140
    Bengt Rånby

12. Nonacrylate Curing Mechanisms: Photoinitiated
    Cross-Linking Using the Amine–Ene Reaction.................................151
    G. K. Noren and E. J. Murphy

13. Photoinitiated Cross-Linking of Norbornene Resins
    with Multifunctional Thiols...160
    Anthony F. Jacobine, David M. Glaser, and Steven T. Nakos

14. Acrylated Melamines in UV-Curable Coatings.................................176
    Joel J. Gummeson

15. Coatings Curable with Low-Emission Radiation 194
    Xiaosong Wu, Yan Feng, and Stoil K. Dirlikov

16. Polymeric Coatings Containing Chlorendic Anhydride:
    Decomposition Products of Chlorendic Anhydride.........................210
    John C. Graham and David J. Gloskey

17. Photochemical Approaches to Ordered Polymers..........................220
    Michael A. Meador, Mahmoud Abdulaziz, and
    Mary Ann B. Meador

PROPERTIES OF RADIATION-CURED MATERIALS

18. Comparison of Thermal, Mechanical, and Spectroscopic
    Techniques for Characterization of Radiation-Cured Adhesives...242
    G. M. Allen and K. F. Drain

19. Mechanical Properties of UV-Cured Coatings
    Containing Multifunctional Acrylates 258
    G. K. Noren, J. M. Zimmerman, J. J. Krajewski, and
    T. E. Bishop

20. Structure–Performance Relationships of Urethane Acrylates 272
    JoAnn A. McConnell and F. Kurt Willard

21. Enthalpy Relaxation in UV-Cured Epoxy Coatings 284
    Geoffrey A. Russell and W. Eugene Skiens

22. Temperature Effect on the Phase Transformation
    of UV-Curable Systems 297
    L. Feldman and T. C. Ward

23. Microcolumn Imaging: Simulation of the Microcapsule
    Imaging System 308
    L. Feldman, M. R. Cage, D. J. Shi, T. K. Kiser, and
    R. C. Liang

PHOTODEGRADATION OF RADIATION-CURED FILMS

24. Photolysis Studies of Bisphenol-A-Based Model Compounds:
    Effect of Decomposition Products on the UV Stability
    of Bisphenol-A-Based Epoxy Coatings 325
    John C. Graham, Dennis J. Gaber, Yifang Liu, and
    Pravin K. Kukkala

25. Photooxidative Stability and Photoyellowing of Electron-Beam-
    and UV-Cured Multifunctional Amine-Terminated Diacrylates 346
    Norman S. Allen, Peter J. Robinson, Roy Clancy, and
    Nicholas J. White

RADIATION CURING OF CATIONIC POLYMERIZATION

26. Radiation-Induced Cationic Curing
    of Vinyl-Ether-Functionalized Urethane Oligomers 363
    Stephen C. Lapin

27. UV Cure of Epoxysiloxanes and Epoxysilicones 382
    Richard P. Eckberg and Karen D. Riding

28. UV Cure of Epoxy–Silicone Monomers 398
    J. V. Crivello and J. L. Lee

29. UV-Induced Polymerization of Highly Filled Epoxy Resins
    in Microelectronics 412
    H. Bayer and B. Lehner

LASER-INITIATED POLYMERIZATION

30. Laser-Initiated Polymerization of Multifunctional Acrylates:
    Repetition Rate Effects on Percent Conversion 429
    Charles E. Hoyle and Martin A. Trapp

31. UV-Radiation- and Laser-Induced Polymerization of Acrylic
    Monomers...439
    C. Decker and K. Moussa

HIGH-ENERGY RADIATION CURING

32. High-Energy-Radiation-Induced Cationic Polymerization
    of Vinyl Ethers in the Presence of Onium Salt Initiators...............459
    Per-Erik Sundell, Sonny Jönsson, and Anders Hult

33. Structure–Property Behavior of Caprolactone–Allyl Glycidyl
    Ether Copolymers Cross-Linked by Electron-Beam
    Irradiation 474
    Ha-Chul Kim, Abdel M. El-Naggar, Garth L. Wilkes,
    Youngtai Yoo, and James E. McGrath

34. Electron-Beam Exposure of Organic Materials: Radiation
    Curing of Perfluorinated Acrylates 498
    J. Pacansky and R. J. Waltman

35. Molecular Weight Dependence of Electron-Beam Resist
    Sensitivity 516
    F. Rodriguez, B. C. Dems, A. A. Krasnopoler,
    Y. M. N. Namaste, and S. K. Obendorf

36. Electron-Initiated Graft Modification of Polyolefins........................534
    Sam V. Nablo, I. J. Rangwalla, and John E. Wyman

INDEXES

Author Index 554

Affiliation Index 555

Subject Index 555

Preface

There is no defence or security for any of us except in the highest intelligence and development of all.

—Booker T. Washington

The past two decades have witnessed the birth and subsequent rapid growth of the radiation-curing industry. From a modest beginning, radiation curing has developed into a viable and widely used method for processing polymeric materials that today touches virtually every aspect of the coatings industry. The success of using high-intensity sources of radiation for curing has rested in the sustained efforts of a group of dedicated individuals who have provided the expertise and hard work needed to overcome a number of obstacles along the way. The chapters in this book are a direct result of the guidance and fundamental concepts provided by the many individual scientists who have helped build the industry since its beginnings.

We anticipate that the information and insight included in this collection of papers will generate new interest in the possibilities of radiation curing and thereby lead to expansion of the field in directions that as yet are unimagined. To this end, introductory chapters on both photocuring and electron-beam curing have been incorporated to introduce those not familiar with the concepts of radiation curing to the basic principles involved. The chapters in the book have been divided into sections according to subject, and a brief summary appears at the beginning of each to advise the reader of the contents therein.

We acknowledge the donors of the Petroleum Research Fund, administered by the American Chemical Society, for partial support of the symposium on which this book is based. We specially thank each contributor to this collection for the time expended in submission and correction of manuscripts. Special recognition goes to Patricia Linton and Machell Haynes for their help and patience.

We would like to extend our appreciation to S. P. Pappas, who has provided inspiration over the years through his sustained excellence in investigating numerous aspects of radiation curing. We also acknowledge C. Decker for his friendship and continual encouragement, as well as his unique contributions to the field of radiation curing.

CHARLES E. HOYLE
University of Southern Mississippi
Hattiesburg, MS 39406–0076

JAMES F. KINSTLE
James River Corporation
Neenah, WI 54957–0899

September 29, 1989

Chapter 1

Photocurable Coatings

Charles E. Hoyle

Department of Polymer Science, University of Southern Mississippi, Southern Station Box 10076, Hattiesburg, MS 39406-0076

Since the introduction of photocuring of coatings as a viable industrial process well over a decade ago, the UV curing industry has followed a line of steady growth. This paper deals with the mechanisms and components which characterize photocurable coatings. Free radical and cationic photoinitiated polymerization are discussed in light of key review references from the recent literature. In addition, a section is dedicated to discussion of the lamp sources available for photocuring operations.

About 15-20 years ago, as mandates were issued by the government to reduce the volatile organic components in coating formulations, the UV curing industry was born. Since its infancy the UV curing industry, despite many initial problems with its development, has become one of the most rapidly developing fields in the entire coatings industry. Today the use of UV curable formulations is progressing at a rate which places it far ahead of the general coatings field in annual growth. A number of excellent review articles/books have been recently published on UV curing (1-11).

What are the general characteristics of UV curables, as they are often called in the industry, which account for their present popularity? First, they are comprised of 100% reactive components which provide for environmentally acceptable coatings. Second, they are energy efficient requiring a small fraction of the power normally consumed in thermally cured coatings. Third, they can be easily formulated to meet a variety of applications since functionalized monomers and oligomers (which are the building blocks of UV curables as will be discussed in another section of this paper) are available covering a wide range of properties. Fourth, the UV curing process in itself often imparts desirable properties to the cured coating which cannot be achieved with thermally cured coatings. Fifth, due to the wide range of viscosities inherent in the reactive monomers used in UV curable components, coatings can be readily formulated to meet demanding viscosity requirements dictated by certain coatings applications, i.e., UV curables are often applied by roll coaters, dipping, spraying, brushing, reverse roll coaters, etc. All of the aspects listed above combine together to make UV curables truly unique coatings which can be tailored for any application normally met by thermally cured coatings, but with less complications.

In general, what can be said about the coatings areas in which UV curables are making the highest impact? One of the prime application areas for UV curable coatings is clear coats on substrates ranging from metals and wood to floors and paper. Each of these substrates demands a different set of properties from the UV curable coating. For instance, coatings on metallic substrates must have good adhesion and be able to resist deformation induced by stresses experienced during use. Table I lists several of the current applications for UV curable coatings with a brief comment for each.

This introductory article does not profess to present an all inclusive view of UV curing. This has already been accomplished, and quite successfully so, in a number of books and publications (1-11). It is meant, however, first to introduce the field to the chemist not currently involved heavily in the UV curing industry. The first two sections cover the basics of UV curing with reference to the mechanisms, and traditional components, in the majority of photocurable coatings. Although both cationic and radical UV curing processes are discussed, emphasis is given to the free-radical type photopolymerization process most frequently encountered in current industrial applications. The next sections describe dual free radical/cationic and photo/thermal curable systems. The final three sections deal with oxygen inhibition, lamp sources, and exotherm measurements of photocurable systems.

Photocuring Mechanisms

Before undertaking the selection of the basic components in a UV curable coating system, one must first know something about the general mechanism of UV curing. There are two primary types of UV curable systems. The first, proceeds by a free radical chain process in which low molecular weight monomers and oligomers are converted by absorption of UV radiation into highly-crosslinked, chemically-resistant films. The second type of UV curable system, which shows great promise for the future, polymerizes by a cationic mechanism and has the advantage of oxygen insensitivity compared to free-radical photocurable systems. Scheme I depicts the general scheme for UV curing which is typical of both free-radical and cationic type cure processes. Basically light absorbed by a photoinitiator generates free-radical type initiators or catalysts which induce the crosslinking reactions of functionalized oligomers/monomers to generate a cured film.

The mechanisms for free-radical type photocurable and cationic photocurable systems can be discussed individually.

The basic mechanism for any free-radical photocurable system is shown in Scheme II.

In the first step, a photoinitiator (PI) absorbs UV radiation followed by a subsequent reaction to give a radical initiator (R·). According to the

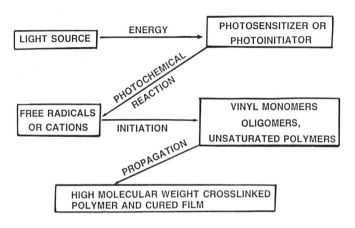

Scheme I

$$PI \xrightarrow{h\nu} (PI)^*$$

$$(PI)^* \rightarrow R\cdot \quad \text{INITIATION}$$

$$R\cdot + nM \xrightarrow{k_p} R(M)_n^\cdot \quad \text{PROPAGATION}$$

$$R(M)_n^\cdot + R(M)_m^\cdot \xrightarrow{k_t} \text{POLYMER}$$

$$\quad \xrightarrow{k_t} \quad \text{TERMINATION}$$

$$R(M)_m^\cdot \rightarrow \text{POLYMER}$$

$$\text{RATE} \propto I^n \quad 0.5 < n < 1.0$$

Scheme II

traditional mechanism, the radical initiator induces a chain-reaction or chain-growth polymerization (rate constant of the propagation step is k_p) which is terminated, in the absence of oxygen effects, by a radical-radical coupling process (rate constant of termination step is k_t). Assuming steady-state kinetics, the rate of the reaction is proportional to the square root of the light intensity (I) and the monomer (M) concentration (see Scheme II). [Recent ESR experimental evidence suggests strongly that termination in highly-viscous photocurable systems comprised of multifunctional acrylate monomers and oligomers may take place in part by a unimolecular process (12); thus the exponent of the light intensity may approach 1.0 in some cases (13,14)]. The constant "K" takes into account the efficiency of the generation of R· in the second step as well as the rate constants for chain propagation and termination. Consequently, the two factors which can be manipulated to change the rate of the curing process for a given set of monomers/oligomers are the light intensity (I) and the monomer concentration (M). Of course, the propagation and termination rate constants are dependent on the particular monomers and oligomers employed in the UV curable formulation and dictated by the photoinitiator(s) used.

The general mechanism for photoinitiated cationic polymerization is given in Scheme III using a triarylsulfonium salt photoinitiator. In the initiation process, the sulfonium salt absorbs a photon of light followed by a set of reactions (summarized by the single step 2) culminating in the generation of the protonic acid $H^+ X^-$, where X^- is usually a tetra- or hexafluoro substituted salt such as PF_6^-, SbF_6^-, BF_4^-, or AsF_6^-. The acid readily attacks the monomer M, which represents a wide range of cationic polymerizable monomers including cyclic ethers, epoxies, cyclicacetals, and certain vinyl monomers. The termination step usually results from reaction of the propagating polymer with an impurity. In the absence of impurities, termination is very slow and for many practical purposes is considered non-existent.

STEP 1 $Ar_3S^+X^- \xrightarrow{h\nu} [Ar_3S^+X^-]^*$

STEP 2 $[Ar_3S^+X^-]^* + RH \longrightarrow Ar_2S + Ar^· + R^· + H^+X^-$

STEP 3 $H^+X^- + M \longrightarrow HM^+X^-$ INITIATION

STEP 4 $HM_nM^+X^- + M \longrightarrow HM_{n+1}M^+X^-$ PROPAGATION

STEP 5 $HM_nM^+X^- + IMPURITY \longrightarrow POLYMER$ TERMINATION

Scheme III

In general, cationic photocure systems (4) are often deemed superior to free-radical photocuring processes since they are:
   (a) Readily sensitized by dyes and/or polycyclic aromatics
   (b) Insensitive to oxygen inhibitors
   (c) Terminated only by impurity quenching
   (d) Characterized by an efficient post thermal cure

The choice to use one photocurable system over another is dictated by the particular requirements of the application and consideration should be given to both free radical and cationic photocurables.

With the basic mechanism of free-radical and cationic photocuring presented in Schemes II and III, we turn to consideration of each of the individual components in UV curable systems. It is this aspect over which the formulator has control in producing a resin system which will cure in an appropriate environment (temperature, air, etc.) at reasonably fast rates to generate the type of film desired.

Traditional Photocurable System Components

All UV curable systems have four basic components which must be included in order to develop a successful coating. They are the photoinitiator(s), oligomer(s), monomer(s), and additive(s). Table II lists the properties of each component which make it essential to the UV curable formulation. In the next few pages, we summarize each of the components and their properties. The photoinitiator section covers both radical and cationic type photoinitiators while the oligomer and monomer sections are restricted to components used in free radical systems.

Photoinitiators. Photoinitiators are the basic link in the UV curable formulation between the lamp source and the resin system. In order for a photoinitiator to be effective, it must have a relatively high extinction coefficient in the region of the electromagnetic spectrum matching the output of the lamp source. This, of course, stems from the basic principal of photochemistry: "in order to have a photochemical process the first necessary step is to absorb the incoming radiation." Having a high extinction coefficient is a necessary but not sufficient requirement for selection of a particular compound as a photoinitiator. The photoinitiator must also be capable of producing reactive species such as radicals (free radical polymerization) or acids (cationic) in an efficient manner with high yield. Additionally, the species generated ($R \cdot$ in Scheme II or $H^+ X^-$ in Scheme III) must be capable of rapidly initiating the curing process while participating minimally in the termination step. For free-radical type polymerizations, this necessitates that the reactive radical must add to the vinyl type monomers even in the presence of dissolved oxygen.

There are two free-radical photoinitiator types currently in use which accomplish this task quite successfully. Both are shown in Scheme IV.

TABLE I. APPLICATIONS FOR PHOTOCURABLE COATINGS

Application	Comments
Plastic Coatings	Rapidly developing area
Metal Coatings	Primarily Epoxy Acrylates
Wire Coatings	Potential growth area
Textile Dressing and Coatings	Potential growth area
Wood Coatings	Traditional area
Paper Coatings	Rapidly developing
Vinyl Floor Coatings	One of the largest world-wide markets for UV curables
Fiberglass Laminates	Opportunity for development of high strength composites
Photoresists	Used for PC board and silicon chip production
Photocurable Inks	Requires curing of highly pigmented resins
Glass Fiber Coatings	Rapid new growth area
Contact Adhesives	Tremendous market potential for growth
Furniture Coatings and Filler Boards	One of the first applications for UV curables--mature market
Conformal Coatings for Electronics	Excellent growth potential

TABLE II. BASIC COMPONENTS OF UV CURABLES

Component	Percentage	Function
Photoinitiator	1-3	Free Radical or Cationic Initiation
Oligomer Resin	25-90	Film Formation and Basic Properties
Monomers	15-60	Film Formation and Viscosity Control
Additives and Fillers	1-50	Surfactants, Pigments, Stabilizers, etc.
Lamp Source	---	Initiate Curing

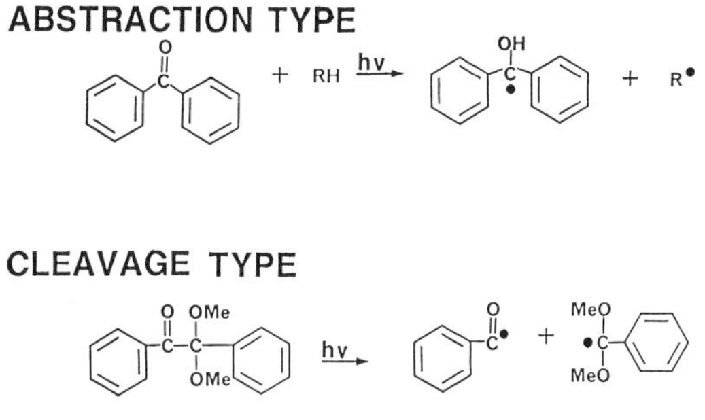

ABSTRACTION TYPE

[Benzophenone] + RH $\xrightarrow{h\nu}$ [semi-pinacol radical] + R•

CLEAVAGE TYPE

[dimethoxy acetophenone derivative] $\xrightarrow{h\nu}$ [benzoyl radical] + [dimethoxy benzyl radical]

Scheme IV

The abstraction type photoinitiator acts by abstraction of a hydrogen atom from a donor compound, the most common of which are aliphatic amines such as triethylamine, methyl diethanol amine (MDEA), or dimethyl ethanol amine. The radical produced (R•) is the initiator of the free radical polymerization process. The semi-pinacol radical is unreactive and simply couples to give the pinacol. The abstraction process is, however, highly dependent on the amine co-initiator. Examples of abstraction type photoinitiators are given below:

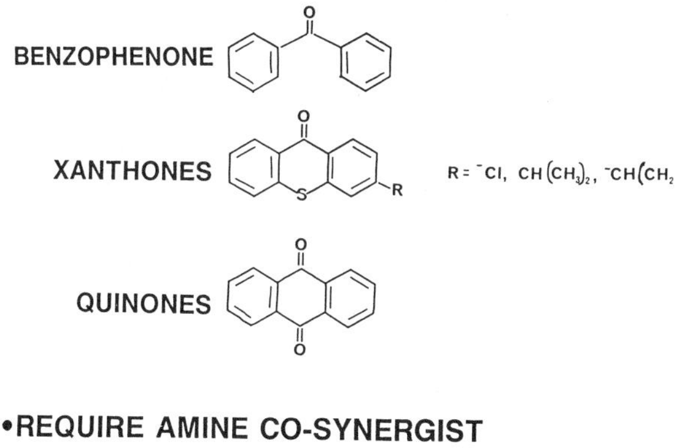

BENZOPHENONE

XANTHONES   R = ‾Cl, $CH(CH_3)_2$, ‾$CH(CH_2$

QUINONES

•**REQUIRE AMINE CO-SYNERGIST**

Cleavage type photoinitiators are best represented by the benzoin acetal compound shown in Scheme IV. Upon absorption of light, cleavage readily occurs (in this case at the carbon alpha to the carbonyl) to give two radical species. The benzoyl radical initiates polymerization by direct reaction with a monomer. The disubstituted benzyl radical, and/or one of its rearrangement products also initiate polymerization effectively. The cleavage type photoinitiator operates without a co-initiator, i.e., an aliphatic amine. This is a critical point since amine synergists, while acting as the species with an available hydrogen for abstraction by benzopenone, are also effective as chain-transfer agents. Chain transfer processes can reduce the chain length of the polymerization and the ultimate crosslink density of the cured film. Examples of four cleavage-type photoinitiators, all of which are capable of producing radicals for initiating free radical polymerization, are shown below:

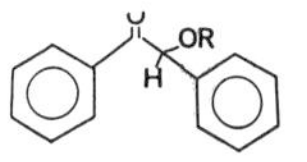

BENZOIN ETHERS

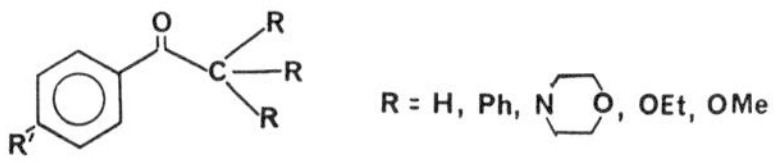

ACETOPHENONES   R = H, Ph, N O, OEt, OMe

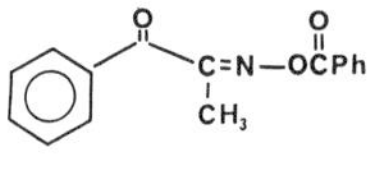

BENZOYL OXIMES

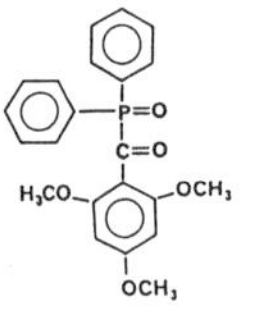

ACYLPHOSPHINES

In contrast to free radical type photoinitiators whose principal function is to generate reactive free radicals, cationic photoinitiators produce either Lewis acids such as BF_3 and PF_5 or strong Bronsted acids such as $H^+ BF_4^-$ and $H^+ PF_6^-$. The aryldiazonium compounds used to generate the Lewis acids, however, are quite thermally unstable and severely limit the potlife of formulated resin systems. In addition, they produce nitrogen as a by-product causing bubbling in thicker coating systems thus restricting practical use to relatively thin films. The photolysis of the diaryliodonium, sulfonium, and selenonium salts (examples shown below) produces strong Bronsted acids. Additionally, they are quite stable in comparison to the diazonium salts.

IODONIUM $\quad Ar_2I^+X^- \xrightarrow[RH]{h\nu} ArI + Ar^\bullet + R^\bullet + H^+X^-$

SULFONIUM $\quad Ar_3S^+X^- \xrightarrow[RH]{h\nu} Ar_2S + R^\bullet + Ar^\bullet + H^+X^-$

SELENONIUM $\quad Ar_3Se^+X^- \xrightarrow[RH]{h\nu} Ar_2Se + R^\bullet + Ar^\bullet + H^+X^-$

$X^- = AsF_6^- \quad PF_6^- \quad BF_4^-$

One of the most useful characteristics of the Bronsted producing photoinitiators is their ability to be sensitized by dyes absorbing well into the visible region. For example, Crivello found (15) that dyes such as benzoflavin or acridine orange efficiently sensitize the photolytic decomposition of diaryliodonium salts using a visible tungsten lamp source. This provides a convenient method for utilizing lamp sources which emit primarily in the visible region to photocure resin systems which may be sensitive to UV radiation.

<u>Oligomers</u>   As indicated in Table II, functionalized oligomers provide the photocured coating with its basic physical properties. A generalized structure for reactive oligomeric resins or pre-polymers is shown below:

$$H_2C=HC-\overset{O}{\underset{\|}{C}}-O\sim\text{OLIGOMER}\sim O-\overset{O}{\underset{\|}{C}}-CH=CH_2$$

- ACRYLATED EPOXY
- ACRYLATED URETHANE
- ACRYLATED POLYETHER
- ACRYLATED POLYESTERS

Oligomers are usually functionalized for UV curable systems by reacting with an acrylate such as hydroxy ethyl acrylate or hydroxy propyl acrylate. The most important types of oligomers are epoxies, urethanes, polyethers, and polyesters. Each of these oligomers is capable of imparting special properties to the final UV cured film in which they are incorporated. This gives the UV curable formulator a wide latitude in developing resin systems for a variety of applications. In the next few paragraphs each of the four basic oligomer systems is covered individually.

The structure of a typical acrylated epoxy oligomer is:

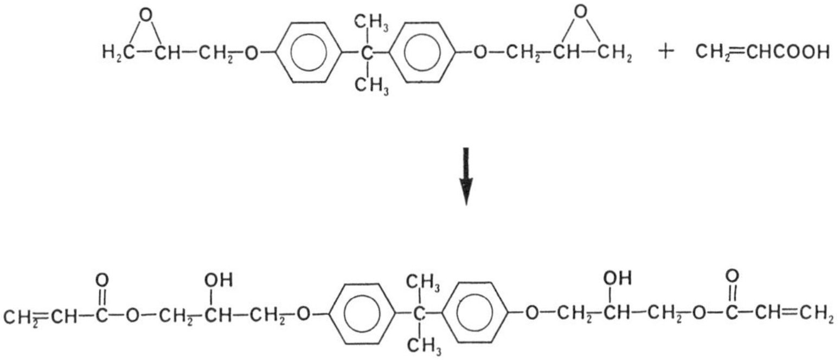

UV resins formulated with acrylated epoxies promote good adhesion, chemical resistance, and flexibility. They are especially useful as coatings on metallic substrates where high gloss, hard coatings which cure rapidly are required. In some cases, it is possible to obtain very low viscosity oligomers with the same final film forming properties as high viscosity oligomers.

The structure of a typical acrylated urethane oligomer is shown below:

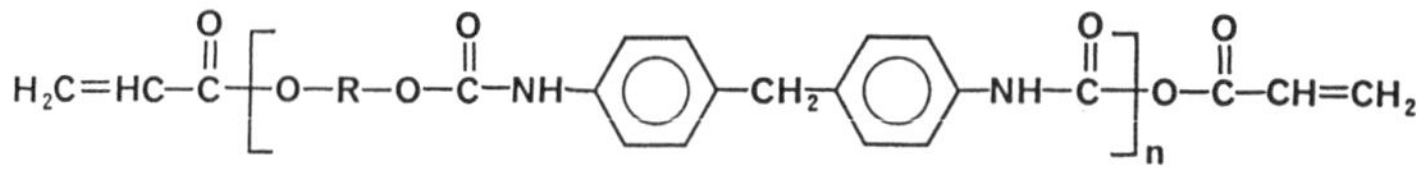

$$H_2C=HC-\overset{O}{\underset{\|}{C}}-\left[O-R-O-\overset{O}{\underset{\|}{C}}-NH--CH_2--NH-\overset{O}{\underset{\|}{C}}-O\right]_n-\overset{O}{\underset{\|}{C}}-CH=CH_2$$

UV cured resins containing acrylated urethane oligomers are usually abrasion resistant, tough, and flexible. These properties are ideal for coatings on floors, paper, printing plates, and in packaging materials.

The structures of an acrylated polyether and polyester are depicted as follows:

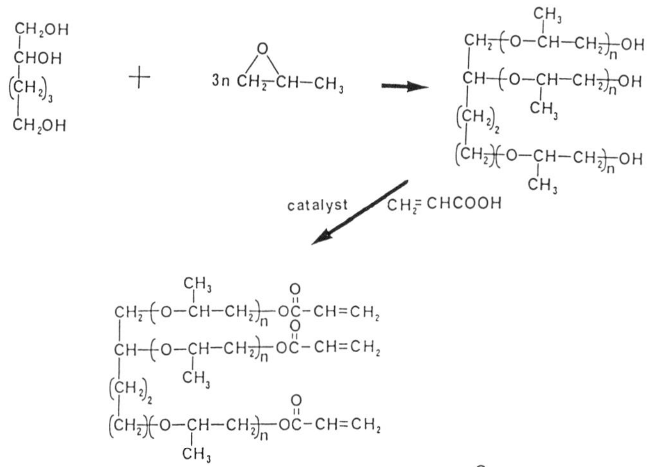

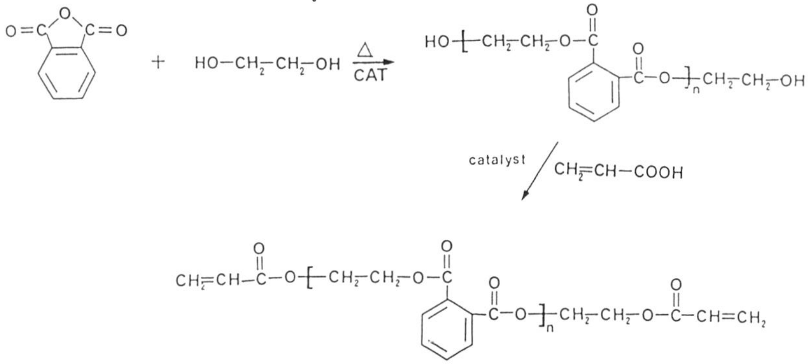

Although acrylated polyethers provide otherwise excellent films when photocured, they are subject to UV degradation when exposed to sunlight and are rarely used in UV curables. Acrylated polyesters, when used in UV curable formulations yield very hard, tough, solvent resistant films which are characterized by very rapid cure rates. A large number of acrylated polyesters are currently available commercially covering a wide range of requirements. In order to achieve high quality cured films, some photocurable formulations combine the acrylated polyesters and acrylated polyurethanes in order to utilize the properties of both components.

Monomers. As shown in Table II, monomers are used in UV curable systems to provide final film properties and viscosity control of the resin. They are also important in determining the speed of cure, crosslink density, and final surface properties of the cured film. Properties which dictate the selection of one monomer over another are listed below:
- (a) Viscosity reduction and control
- (b) Effect on cure speed
- (c) Effect on film properties (flexibility, hardness, adhesion)
- (d) Shrinkage during polymerization
- (e) Cost
- (f) Shelf life
- (g) Volatility, odor, toxicity

In accordance with the above requirements, several monomer types employed in UV curable systems are listed below:
- (a) Styrene (fast cure, low cost)
- (b) N-vinyl pyrrolidone (vinyl pyrol)--highly flexible when cured; low toxicity; often used with acrylates
- (c) Acrylates--versatile; highly reactive

By far the acrylates are the monomers of choice in UV curable systems. Not only do they cure at extremely rapid rates compared to other monomer systems (acrylic > methacrylic > vinyl > allylic), but they are also available in a wide range of structures which are monofunctional, difunctional, trifunctional, and tetrafunctional. Additionally, as shown in the oligomer section, acrylates can be used to derivatize oligomers or pre-polymers. Commonly in UV curable formulations it is necessary to use a number of monomers in order to achieve a balance between speed of cure and properties of the final film. It is not unheard of to use four or five monomers in a single UV curable formulation. For instance, tri- and tetra-functional acrylates result in highly crosslinked films when incorporated into UV curable resins; however, they severely limit the extent and rate of the curing process. Thus, one often combines a tetrafunctional acrylate to increase crosslink density with a mono and/or difunctional acrylate to increase the cure rate.

Additives. As shown in Table II additives are used in UV curable formulations at levels of up to 50% by weight. As in any coatings system, it is often the additives which give the coatings their most important properties. Additives often included in UV curable coatings are surfactants for substrate adhesion, dispersants for pigment dispersion, pigments for color, and stabilizers for both thermal and UV protection.

In general, while additives are often critical in producing the final desired properties of a UV cured coating, they may cause serious problems in the curing process. For instance, one of the early challenges in the UV curing industry was to develop UV curable resins which could be used in highly filled systems. Since filled systems are opaque the problem of light penetration into thick samples can be quite severe. It was not until photoinitiators which absorb light at longer wavelengths approaching the visible region of the spectrum were developed that adequate "through cure" of filled formulations could be accomplished.

Dual Free Radical/Cationic Photocurable Systems

One of the more interesting possibilities of developing truly unique photocurable films lies in the ability of certain onium salts to generate both a reactive radical species and a strong acid upon photolysis. Indeed, several papers describe dual radical/cationic photocurable systems (16-18). The possibilities afforded by dual cure systems provide a unique mechanism for generating cured films with expanded properties for a number of applications.

Dual Photocurable/Thermal Curable Coatings

In the last few years, based on the earlier pioneering work of Crivello, Pappas, and Ledwith (see reference 4 for review) in photocationic polymerization processes, several groups have reported on the development of useful epoxy based UV curable coatings (19-22). Recent developments on this subject (22) have resulted in the development of dual UV curable/thermal curable epoxy coatings based on difunctional epoxies.

These photocured epoxy films have excellent adhesion to a variety of substrates including glass, stainless steel, aluminum, etc. In addition, they show little or no shrinkage on curing and have good mechanical strength and a low dielectric constant making them quite suitable as conformal coatings for a variety of sensitive electrical components. Finally, these films are quite versatile and can be cured to give film thicknesses up to 10 mils.

Oxygen Inhibition In Photopolymerization Processes

In what is probably one of the more revealing papers ever published on the subject of oxygen inhibition in photopolymerization, Decker and Jenkins (23) actually measured the steady-state concentration of oxygen for the photopolymerization of multi-functional acrylate systems (epoxy-acrylate resin) as a function of light intensity. Using a medium pressure mercury lamp source it was found that oxygen was consumed by a chain peroxidation process in which up to as many as eight oxygen molecules were consumed for each initiating radical produced in the system during the initial inhibition period. By adding tertiary amines to the photocurable system, the peroxidation chain process was estimated to consume sixteen oxygen molecules per amine radical generated. Conversely, when a high intensity focused argon-ion laser was used the peroxidation chain process resulted in the consumption of between one and two oxygen atoms for each initiating radical generated. Furthermore, the consumption of oxygen molecules per radical generated was not increased in the presence of added trifunctional amines. A detailed kinetic analysis of the oxygen inhibition and chain peroxidation process for photopolymerization in air (oxygen) saturated systems is given by Decker. This should serve as a guide to the formulator attempting to develop UV curable coatings for processing in the presence of air.

Light Sources

The final factor involved in the UV curing process is the light source. A number of light sources available for photocuring are listed in Table III. By far, the most popular source for UV curing has traditionally been the medium pressure mercury lamp with output at specific lines from the deep UV to the visible superimposed on a low level continuous band. The medium pressure mercury lamps used in UV curing operations are generally operated at 200 Watts/inch power input. This generates an extremely intense UV output providing cure times on the order of a few seconds for even quite thick photocurable coatings on a variety of substrates. This allows line speeds of several hundred feet per second to be met without serious difficulty and with relatively low total energy consumption. There is always the menace of overheating sensitive substrates to which the UV curable coatings are applied. This is a particularly important consideration when applying UV curable coatings to plastics. This problem is minimized by faster cure speeds resulting in shorter exposure of the substrate to the intense heat generated by the mercury lamp sources. In recent years, a number of UV curing operations have turned to the use of the electrodeless lamp systems. Electrodeless lamps have long service lifetimes and extremely short warm-up times.

The pulsed xenon lamp source in Table III merits special attention. It is a well recognized fact that pulsing xenon lamps significantly increases their output below 300 nm. In addition, photon flux densities produced by pulsing xenon lamps are extremely high and can lead, under the correct circumstances, to efficient curing of highly filled, very thick films or composites. Both the output below 300 nm and the high pulse densities make pulsed xenon lamp sources a choice which should be considered for certain critical applications.

The final two entries in Table III are representative of the lamp sources currently available from the laser industry. Although both are extremely important as tools for investigation of basic photopolymerization processes, their widespread use as sources for UV curing is limited by the small cross section of the beam to specialty applications in the imaging and electronics industries.

In summary, one of the most important factors in the success of a UV curing operation is in the correct choice of the lamp system. The interested reader is referred to an excellent book recently written by Roger Phillips (24) for a review of the light sources available for UV curing. Scrutinization of the information provided in this well-written and detailed treatise could well save either the experienced or novice UV processor a good deal of time and money.

TABLE III. LIGHT SOURCES AVAILABLE FOR UV CURING

Lamp Source	Comments
Medium Pressure Mercury Lamp (Undoped)	(1) Standard lamp for UV curing (2) Up to 200 Watts/inch input power (3) Traditional mercury lines at 254 nm, 280 nm, 297 nm, 303 nm, 313 nm, 334 nm, 366 nm, etc. (4) Provides excellent through cure
Low Pressure Mercury Lamp	(1) Very low power output (2) Primary output at 254 nm (3) Has been used to enhance surface cure in conjunction with medium pressure mercury lamps
Electrodeless Lamps	(1) Filled with mercury or other metal/gases (2) UV efficiency (mercury fill) approximately 20% (3) Powered by microwave or RF input
Xenon Lamp	(1) Continuous output from UV to IR
Pulsed Xenon Lamp below	(1) Enhanced output in deep UV 250 nm (2) Ultra-high pulse photon flux can be obtained (3) Excellent for curing highly filled, thick coatings or composites (4) Relatively low heat source
Argon Ion Laser	(1) High coherence length (2) Output available both in visible and UV
Excimer Laser	(1) Output at several lines in deep and near UV (2) Pulsed lamp source (10-20 nsec pulse width) (3) Can be focused to give extremely high photon flux densities

Exotherm Measurements Of Photocuring

With the introduction of highly reactive photocurable coatings came the necessity to have an instrument capable of easily and rapidly assessing such parameters as cure rate and extent of cure for new or modified formulations. This need is compounded by the availability of a large variety of components to the photocurable coatings formulator. For instance, there are well over 20 photoinitiators alone currently available from commercial sources for use in UV curable resins. Add to this the tremendous selection of monomers and oligomers and one is faced with the staggering problem of combining the components into a single efficient fast-curing coating with acceptable physical properties for the desired application. The necessity of quick, efficient, and reliable evaluation of new UV curable formulations in a reproducible and accurate manner cannot be overemphasized. One could imagine using conventional methods for following chemical reactions at low conversions in solution, but the photocuring process demands evaluation of viscous highly crosslinked systems at rather large conversions on a continuous basis while the curing process is in progress. This is beyond the capabilities of most analytical tools. In recent years, thermal techniques (25,26) for measuring the rate of heat evolution during the photopolymerization process have been developed to the point that polymerization rates, extents of polymerization (degree of cure), and even rate constant data can be accurately obtained.

The exotherm apparatus (either DSC or thin-foil) is a highly useful tool for evaluation of the efficiency and speed of UV curable systems. It can be applied as a screening mechanism for UV curable formulations or as a highly sophisticated scientific tool. Its use has led, and will no doubt lead in the future, to the development and improvement of photocurable coatings.

Conclusions

This paper serves to introduce the field of UV curable coatings by considering the polymerization mechanisms inherent to all photocurable systems. From the discussion of the basic components found in most UV curable coatings, the general background necessary to understand the papers included in the first 30 chapters of this book is presented. The field of UV curing is indeed a fascinating and profitable one which promises to proceed on a path of continued growth. It is the expectation that the papers in this book will suggest new applications and lead to the future development and growth of radiation curing.

Acknowledgment

Acknowledgment is gratefully extended to the Southern Society for Coatings Technology for their continued support.

Literature Cited

1. S. P. Pappas (Ed.) U. V. Curing: Science and Technology Vol. 1, Technology Marketing Corporation (1978).
2. C. G. Roffey, Photopolymerization of Surface Coatings, Wiley-Interscience, New York (1982).
3. K. Hashimoto and S. Saraiya, J. Rad. Cur., 8(1), 4 (1981).
4. J. V. Crivello, Adv. Polym. Sci., Vol. 62, pp. 1-48, Springer-Verlag, Berlin (1984).
5. S. Paul, Surface Coatings: Science and Technology, pp. 601-57, Wiley-Interscience, New York (1985).
6. S. P. Pappas (Ed.), UV Curing: Science and Technology Vol. 2, Technology Marketing Corp., Norwalk, CT (1984).
7. G. A. Senich and R. E. Florin, Rev. Macromol. Chem. Phys., C24(2), 239 (1984).
8. G. E. Green, B. P. Stark, and S. A. Zahir, J. Macro. Sci.-Revs. Macro. Chem., C21(2), 187 (1982).
9. V. D. McGinniss, "Radiation Curing," in Kirk-Othmer Encyclopedia of Chemical Technology, 3rd ed., Vol. 19, Wiley-Interscience, New York, pp. 607-624 (1982).
10. E. P. Tripp and J. Weismann, Mod. Paint. and Coat., Mar., 51 (1982).
11. E. Levine, Mod. Paint and Coat., Aug., 26 (1983).
12. J. G. Kloosterboer and G. J. M. Lippits, J. Imag. Sci., 30, 177 (1986).
13. G. R. Tryson and A. R. Schultz, J. Polym. Sci. Polym. Phys. Ed, 17, 2059 (1979).
14. C. Decker, J. Coat. Tech., 56, 29 (1984).
15. J. V. Crivello, J. H. W. Lam, and C. N. Volante, ACS Mtg., Chicago, IL, Aug. 28 -Sept. 2, Ctgs. and Plast. Preprints, 37 (2), 4 (1977).
16. J. V. Crivello and J. H. W. Lam, Polym. Chem. Polym. Lett. Ed., 17 (1979).
17. A.D. Ketley and J-H. Tsao, J. Rad. Cur., 6 (2), 22 (1979).
18. W.C. Perkins, J. Rad. Cur., 8 (1), 16 (1981).
19. J.V. Koleske, O.K. Spurr, and N.J. McCarthy, 14th national SAMPE Tech. Conf., Atlanta, GA, 14, 249 (1985).
20. R.F. Zoft, "Radiation Curing V," AFP/SME Symposium on Radiation Curing, Boston, MA, Sept. 23-25 (1980).
21. W. C. Pekins, J. Rad. Cur., 8, 16 (1981).
22. K. B. Wischmann, Proceedings: 31st International SAMPE Symposium, April 7-10, Las Vegas, NV (1986).
23. C. Decker and A. D. Jenkins, Macromolecules, 18, 1241 (1985).
24. R. Phillips, Sources and Applications of Ultraviolet Radiation, Academic Press, New York (1983).
25. J. E. Moore, in UV Curing: Science and Technology Vol. 1, Ed. S. P. Pappas, pp. 134-84, Technology Marketinig Corporation, Stamford, CN (1978).
26. J. D. Wisnosky and R. M. Fantazier, J. Rad. Cur., 8, 16 (1981).

RECEIVED October 5, 1989

Chapter 2

Electron-Beam Curing of Polymeric Materials

James F. Kinstle

James River Corporation, 1915 Marathon Avenue, Neenah, WI 54956

    Electron beam induced reactions continue to grow in importance. This paper provides an introductory treatment of the equipment and materials options, including a mechanistic and kinetic view of the pertinent chemistry. Specific coverage includes several applications in polymer science, especially in curing of coatings.

Electrons are very reactive, and--properly harnessed--are valuable in inducing chemical reactions. Several types of electron induced reactions are commercially important. The reactions of principal interest here are polymerization and crosslinking. This article is an introduction to the electron beam induced reactions of monomers, oligomers, and polymers, emphasizing the curing of coatings. Coverage includes the basic equipment (for process engineers), and reaction mechanisms and materials (for chemists and formulators). Overall, the intent is to provide a perspective "framework" for those interested in electron beam curing of polymers and coatings. [Please note: this is not meant to be a complete review; references are given to a few arbitrarily selected works, and to those that are presented in the following section of this book.]

    Electrons are very energetic species. They are from the short wave length, high energy end of the electromagnetic radiation spectrum, as can be seen from Table I.

    Electrons can be obtained as one component of the radiation emitted from gamma and pile sources. The most common of these for polymerization and polymer crosslinking is the cobalt-60 source. Because of the high energy of the emitted particles and waves, these sources require extensive shielding. They also cannot be turned on and off at will. Alternatively, electrons can be isolated by electrostatic methods; e.g., by using a Van de Graaff generator.

0097–6156/90/0417–0017$06.00/0
© 1990 American Chemical Society

Table I. The Electromagnetic Spectrum

Wave Length λ, nm	Frequency ν, sec^{-1}	Energy eV	Source
1.25×10^{-5}	2.4×10^{22}	1.0×10^{8}	Hard X-rays
1.0×10^{-4}	2.9×10^{21}	1.2×10^{7}	High end γ range
6.0×10^{-4}	5.3×10^{20}	2.2×10^{6}	Y^{90} β electrons
4.0×10^{-3}	7.5×10^{19}	3.1×10^{5}	Co^{60} β electrons
1.25×10^{-2}	2.4×10^{19}	1.0×10^{5}	High end soft X-ray
2.5×10^{-2}	1.2×10^{19}	5.0×10^{4}	Low end γ range
6.8×10^{-2}	4.4×10^{18}	1.8×10^{4}	H^{3} β electrons
1.25×10^{-1}	2.4×10^{18}	1.0×10^{4}	Low end soft X-ray
3.1×10^{-1}	9.7×10^{17}	4.0×10^{3}	
1.0×10^{1}	2.9×10^{16}	1.2×10^{2}	
1.25×10^{2}	2.4×10^{15}	10	Photon UV
2.5×10^{2}	1.2×10^{15}	4.9	Hg lamp
1.25×10^{4}	2.4×10^{13}	1.0×10^{-1}	
1.25×10^{7}	2.4×10^{10}	1.0×10^{-4}	

Their energies can be determined by the technique used to isolate and focus them; the units can be turned off and on. Electron beams are conventionally generated now by so-called electron accelerators. These may be linear accelerators that provide pulsed electrons emitted as a scanning beam. The development of these accelerators coincided with increased understanding of free radical polymerization mechanisms, and the availability of an increasing number of monomers. Together, these allowed rapid development of an overall science and technology base in electron induced curing and crosslinking. Further development in sources has led to the continuous "blanket" approach, whereby electrons are generated from a filament that is self-shielded, and the electrons flood the surface of the irradiated substrate rather than being scanned.

Whatever the source of the electrons, there are important considerations in their utilization. Shielding is still required, both for the electrons and for the x-rays that are emitted when the electrons strike a metal. Energy of the emitted electrons can be determined by the design and operation of the generation equipment. Applications requiring high penetration use high energy electrons, but most coating curing applications use electrons in the 100-350 keV range. Note that these electrons have a huge excess of energy relative to the photons provided by a UV source (more about this later). Within this energy range, the electron flux is also controllable by design and operation, as is the speed that a substrate moves past the electron source; therefore the energy, dose rate, and total dose are readily specified and controlled.

When electrons strike an assembly of organic molecules, many events can be induced. They include:

AB
- dissociation ———> A^+B^-e, A^-B^+e, etc.
- e capture ———> AB^-, A^-B^-, etc.
- e ejection ———> AB^+ 2e, etc.
- excitation ———> AB^*

That this myriad of possibilities can take place upon irradiation dramatically illustrates the energy difference between these electrons and the photons generated in a UV system. Importantly, these electrons provide sufficient energy so that a specific initiator or sensitizer is not necessary. This has several ramifications. One is economic, in that photoinitiators and sensitizers tend to be expensive. Another is that one need not worry about photoinitiator fragments that smell, turn colors, and/or are extracted, or about residual sensitizer that can assist in starting unwanted degradation reactions in the product. However, the high energy of the electron (perhaps 100X that needed for bond breakage) can also cause fragments to form within the irradiated system. Even though the electron beam is a kind of "chemical sledgehammer", the chemist/formulator still has a fair number of options. In certain ultra-pure systems, electron-induced free cation type polymerizations may be conducted. Under less rigorous conditions, addition of a cationic initiator allows cationic (or mixed cationic-free radical) polymerization [see Chapter 32]. Under more usual conditions, the coatings formulations are not sufficiently "pure" to allow ionic reactions; ions get consumed by reaction with traces of water or other additives/impurities, and free radical events predominate. Further discussion will center on these free radical systems.

The polymerization that ensues after an electron initiated event is not very reflective of the process by which it was started. But some deviations do occur; e.g., at the high dose rates used in commercial practice, the rate of initiation is very high, and the overall rate of conversion can be dose rate dependent. The increasingly viscous, or even vitrifying, environment in which the polymerizations occur can also influence both rates and mechanisms of propagation and termination(1). [Also see Chapter 33 on vitrifying/crystallizing systems.] Reactions with oxygen can perturb the conversion. Still, the following rate expressions can be used to describe the "usual" coating curing process involving oligomer, multifunctional monomer, and monomer.

$$\text{Initiation } R_i = \phi I_a$$

where ϕ is a quantum efficiency and I_a is the amount of absorbed energy, alternatively the amount of incident energy times an absorption efficiency times the concentration of absorber.

$$\text{Propagation } R_p = k_p [\sim M°][M]$$

where k_p is the specific rate constant for propagation, $[\sim M°]$ is the concentration of growing radical, and [M] is the concentration of unreacted polymerizable group.

$$\text{Termination } R_t = k_t \, [{\sim}M^\circ]^2$$

The viscous effects mentioned above could have significant effects here. At steady state $R_i = R_t$, or $\phi I_a = k_t \, [{\sim}M^\circ]^2$. Solving for $[{\sim}M^\circ]^2$ (since this is very difficult to measure),

$$[{\sim}M^\circ] = \left[\frac{\phi I_a}{k_t}\right]^{1/2}$$

substituting back,

$$R_p = k_p[M] \left(\frac{\phi I_a}{k_t}\right)^{1/2}$$

Also remember that the polymerizable species [M] can be an absorbing species, and therefore it contributes to the numerator of the square root term. In spite of the complications, the equation can be used to assess system options. One can speed up the reaction by increasing R_i. One way to do this is by increasing the intensity of the radiation, but formation of too many radicals allows them to participate in termination. One can increase the concentration of monomer, but--while increasing rate--this means there is more monomer to cure. Molecules that efficiently form useful reactive species upon irradiation can be incorporated into the formulation (organic halides often are useful here). Note that some of these modifications enter the equation in a square root term; i.e., the pertinent concentration or effect must be quadrupled to cause a factor of 2 change in R_p. There are also formulation variables available, like the specific functional group being polymerized. In general, initiation occurs rapidly with all readily free radical polymerizable monomers and functional groups, though aromatic monomers are sluggish due to the energy dissipation mechanisms available through the π system. Polymerizabilities/propagation of groups is like that mentioned for the UV systems. In terms of predictable, fast, and clean polymerization, the order tends to be: acrylate>methacrylate>vinyl>vinylene> vinylidene>allyl. The monomers and oligomers used in electron beam curable coatings formulations are like those discussed earlier for UV curable systems. However, the formulator has much more latitude with additives in the electron beam systems since penetration of electrons is much greater than that of photons. So highly pigmented, filled, dyed, etc. systems of many mils thick can be cured with electrons in the "regular" 150-400 keV range.

This ability to penetrate can be used to advantage in other ways, too. A "buried" adhesive can be cured. Multiple chemical processes can be conducted at once, like curing a coating, grafting to a substrate, and crosslinking the substrate. The penetration also could be a negative efficiency factor, since those electrons that are transmitted through the substrate are usually wasted. One can, however, design the system so that a solid metal is immediately behind the irradiated article/web, so that the reflected/backscattered energy can also be utilized.

Electrons are useful for more than conversion of monomeric species to polymers. In fact, the following events all tend to take place upon irradiation of most curing systems:

[Reaction schemes showing:
- Monomer with X group → polymerization to give polymer with X side groups
- Polymer with Y side group → crosslinking
- Polymer with Y side group + additive A → crosslinking (two products shown)
- Polymer with Y side group + monomer with X → grafting
- Polymer with Z side group → + other fragments]

Again, opportunities are illustrated. Specifically, electron irradiation can be used to crosslink polymeric systems. This can be accomplished by irradiating the polymer alone, usually at a temperature above Tg (and above Tm if pertinent)(2). Or it can be accomplished by irradiating a polymer in the presence of a polymerizable monomer, as has been done for many years in the wire and cable industry; e.g., polyvinyl chloride plus a multifunctional monomer like trimethylolpropane-trimethacrylate(3). Electron irradiation can also be used to form graft copolymers. In fact, this can be accomplished in several ways(4); (a) by irradiating the polymer and then adding the monomer (the pre-irradiation grafting technique, in which case the newly grafted chain is formed from a reactive site on the preformed polymer) or (b) by irradiating the substrate polymer and the monomer together (the simultaneous irradiation grafting technique, in which the growing chain is grafted by the above mechanism plus by transfer and/or termination reactions). [Also see Chapter 36 on grafting.] Electron irradiation can also cause degradation reactions, including formation of monomer and/or other fragments. Selective crosslinking or degradation reactions can allow the use of these electron beam induced reactions in the resist field, too, where the shorter wave length (thus higher resolution) of electron, relative to UV is used to advantage(5). [Also see Chapters 34 and 35 on resists and effects on organic materials.]

Electron irradiation is useful in other areas, too. Foods can be irradiated, thereby allowing sterilization, and stabilization(6). For example, irradiation of potatoes kills any bugs, bacteria, etc., and eliminates germination/sprouting, which allows a great enhancement in storage-ability. Electron irradiation is increasingly used in sterilization of medical devices and equipment(7), and may even have promise in treatment of various waste streams(8).

Summary

Electrons can be generated in reliable devices and used in chemical processes. Important applications include polymer formation and crosslinking reactions. Mechanisms and kinetics of the reactions have been studied; at least a rudimentary understanding exists that can be used to guide the user. Monomeric, oligomeric, and polymeric reactants are available for exercise of the science and art. Applications of electron beam curing, especially in curing of protective and decorative coatings, is expected to continue to grow.

Literature Cited

1. (a) Thompson, D.; Song, J.H.; Wilkes, G.L.; J. Appl. Polymer Sci. (34), 1063 (1987). (b) Decker, C.; Moussa, K.; J. Appl. Polym. Sci. (34), 1603 (1987).

2. van Aerle, N.A.J.M.; Crevecoeur, G.; Lemstra, P.J.; Polymer Commun. (29), 128 (1988).

3. Bowmer, T.N.; Vroom, W.I.; J. Appl. Polym. Sci. (28), 3527 (1983), and prior 3 papers in their series.

4. (a) Kaji, K.; Hatada, M.; Yoshizawa, I.; Kohara, C.; Komai, K.; J. Appl. Polym. Sci., 37 2153 (1989). (b) Taher, N.H.; Hegazy, E-S.A.; Dessouki, A.M.; El-Arnaouty, M.B.; Radiat. Phys. Chem., (33,#2), 129 (1989).

5. (a) Thompson, L.F.; Willson, C.G.; Bowden, M.J.; "Introduction to Microlithography," ACS, Washington, D.C., 1983. (b) Eranian, A.; Bernard, F.; Dubois, J.C.; Makromol. Chem., Macromol. Symp. (24), 41 (1989).

6. (a) Josephson, E.S.; J. Food Safety (5), 161 (1983). (b) Swientek, R.J.; Food Processing, June 1985, pp. 82-90.

7. Bly, J.H.; Radiat. Phys. Chem. (33, #2), 179 (1989).

8. Trump, J.G.; Radiat. Phys. Chem. (24, #1), 55 (1984).

Also see Chapiro, A.; "Radiation Chemistry of Polymeric Systems," Interscience, NY, 1962, and Charlesby, A.; "Atomic Radiation and Polymers," Pergamon Press, London, 1960. Dole, M. (Ed.); "Radiation Chemistry of Macromolecules, Vol. I & II," Academic Press, NY, 1972, for historical perspective.

RECEIVED September 27, 1989

PHOTOINITIATORS

Photoinitiators

It is befitting to have the introductory section of research papers presented in this book deal with photoinitiators since they provide the vital link between the radiation source and the curable resin. The papers in this section are representative of the attention which is still given to the understanding and development of faster and more efficient initiator systems.

The first two papers in this section (Chapter 2 and 3) describe calculations related to the depth of penetration of light in photocurable coatings. The first paper (Chapter 2) describes estimates of light absorbance as a function of coating depth using a two flux model and taking into account component scattering and substrate reflection. Predictions resulting from the calculations were correlated to experimental data for photocuring of both a printing ink formulation and a clear coat. Data were collected using a unique stylus based analysis system. In the second paper (Chapter 3), a Beer-Lambert approach was employed to predict penetration of several different wavelengths of light from a single lamp source. The need to understand the effect of photoinitiator absorbance on the through cure and subsequent depth profile of physical properties of cured coatings is vividly illustrated.

The next three papers (Chapters 4-6) deal directly with the photochemistry and photophysics of three classes of photoinitiators: cleavage, abstraction (both oil and water soluble), and acid generating. These papers describe application of standard photophysical and photochemical techniques to characterize the basic excited state processes leading to the production of the reactive species responsible for initiating polymerization. It is by such detailed and careful application of standard photochemical analysis to study the reactive components in photocurable systems which allows progress in the development and proper application of new and more effective photoinitiators. As is pointed out by the authors of these three papers, it is possible to correlate the excited state photochemistry of photoinitiators to their effectiveness in actual cure systems.

The final contributions in this section (Chapters 7 and 8) describe the synthesis, characterization and application of new types of photoinitiators which show particular promise in pigmented coatings (Chapter 7), photoimaging systems (Chapter 7), and polymerization on functionalized surfaces (Chapter 8). In keeping with the recent trend of developing photoinitiators for specific applications, these papers describe highly reactive photoinitiators, in one case with enhanced red-shifted absorbance (Chapter 7) and in the other case with reactive groups designed for attachment of the photoinitiator chromophore directly to surfaces targeted for grafting (Chapter 8).

Chapter 3

Depth of Cure Profiling of UV-Cured Coatings

Leslie R. Gatechair and Ann M. Tiefenthaler

Ciba—Geigy Corporation, Ardsley, NY 10502

In this study Beer's law was utilized to determine the fraction of incident radiation absorbed (or transmitted) at various points within a UV cured coating. The calculations are carried out at each wavelength emitted by commercial curing lamps. Absorbance profiles help explain the difference in surface and through-cure effects and how cure varies with concentration, lamp selection, and coating thickness. Absorbance calculations were compared to experimental results, obtained through curing studies, with formulations containing different photoinitiator concentrations. We have found such calculations to be an effective educational tool in explaining the response of photoactive systems.

This contribution describes our most recent efforts to quantify the complex interactions of photoinitiators used in UV cured coatings with commercial radiation sources having multiple emission lines. The results of these calculations provide potential explanations of such effects as how the concentration of a photoinitiator affects the ratio of surface cure to body cure, how changing lamps can affect the choice of a photoinitiator, why optimum concentrations exist for some initiators, why some initiators consistently perform better under a given lamp, or in a given thickness than others, and why screening, the internal filter effect, can result in loss of adhesion or poor body cure. While the results of these calculations are currently not sufficient to provide a quantitative guide for formulating, percent incident radiation absorbed (PIA) calculations do provide an excellent educational tool for understanding the principles affected by changing many formulation and process variables.

We have recently described a computer program which was developed to model the absorption of various free radical photoinitiators ($\underline{1}$). Beer's law was utilized in these calculations to determine the fraction of incident radiation absorbed (or transmitted) at various points within a coating by a photoinitiator.

The calculations were carried out at each wavelength emitted by a selected lamp. Knowing the percent of incident radiation absorbed (PIA) at various points within the coating allowed the estimation of the relative concentration of initiating radicals. It should be noted that although these calculations apply to both free radical and cationic UV cured coatings, many of the discussions in this manuscript may apply only to coatings cured free radically in the presence of oxygen.

A variety of free radical photoinitiators are available for converting the energy of the radiation source into chemical energy for initiating the curing process (2-7). An understanding of the photochemistry of these initiators is necessary to aid in proper selection and formulating for a given application. In this study we have utilized a developmental initiator that is described later. The mechanism of photolysis, subsequent generation of radicals and application results are described in the contribution by Desobry and co-workers in this same symposium series (8).

THEORY

A common misconception is that the rate of photopolymerization is proportional to the concentration of the initiator. More correctly, the rate of photopolymerization will be a function of the intensity of absorbed radiation (9, 10). Unfortunately, this value is not easily available for measurement by common instruments. In ideal systems, theory suggests that the rate of a photopolymerization should be proportional to the square root of the photoinitiator concentration (11). However, formulated coatings behave far from ideal in this respect.

A variety of other factors also affect the polymerization rate: the average functionality of the formulation, the viscosity, the quantum yield of the photoinitiator, the fraction of free radicals produced by the initiator which react with monomer, the presence of inhibiting agents such as oxygen, and the presence of coinitiators such as tertiary amines. In the context of this paper these factors are neglected. For many comparisons, the bulk of these factors may be held constant if only the photoinitiator concentration is varied.

Such simplifications are less valid for the comparison of different photoinitiators. However, experience has suggested that calculations of PIA provide a reasonable guide to understanding photoinitiator performance. The power of curing lamps, reflector geometry, sample position and belt speed are also important. In this study we have maintained these at fixed levels.

A number of publications have proposed that an optimum concentration of a photoinitiator exists, such that the optical density (absorbance) of the resulting coating is 0.434 (9, 12-14). This is the optical density at which the maximum radiation will be absorbed in the lowest region of the coating. In principle, such a guideline only has meaning for ideal systems cured using monochromatic radiation. If the radiation source has two or more lines separated by at least 10-20 nm, the optical density will be different, the line intensities will likely be different, and the system is clearly nonideal. In commercial UV curing applications, industrial lamps may have 20 to 40 emission lines. More 'ideal' applications would be in imaging systems such as printing plates or photoresists for printed wiring boards. In these applications the radiation from a mercury bulb is often filtered by layers of polyester film and glass. For modeling the interaction of radiation with UV cured coatings, one must consider the intensity and wavelength of each line in the radiation source, as well as the optical density of the coating (due to the photoinitiator) at each wavelength. Since many of the photoinitiators used in these applications do not absorb beyond 380 nm, only a few lines between 330 and 366 nm are utilized.

As will be described below, using the optical density of 0.434 (or below) results in a relatively uniform absorption of radiation throughout the film. Assuming that each photon absorbed (or a constant fraction thereof) results in the formation of initiating radicals, one may anticipate uniform crosslinking throughout the volume of the coating. In practice, however, coatings cured by free radical polymerization are strongly inhibited by oxygen. Under such conditions, a high concentration of radicals at the surface offset the diffusion of oxygen into the coating during the curing process and result in improved film properties. Alternate solutions to oxygen inhibition are well known to the UV curing industry and include the use of inert gases, oxygen barrier films, and amine synergists (2).

Calculations. The Beer-Lambert law describes the dependance of the absorption of a compound (A) on the extinction coefficient (E) the concentration (C) and the path length of the sample (D) and provides the key to calculating the distribution of absorbed radiation.

$$A = ECD = \log I_o/I_t \qquad (1)$$

In these calculations I_o represents the intensity of radiation incident on the surface. If we wish to calculate the intensity of absorbed radiation at various points in the coating, we divide the coating into a series of n uniform segments having a thickness t = D/n. The intensity of radiation absorbed (I_a) or transmitted (I_t) by a segment is described by the equations below:

$$I_t = I'_t \, 10^{-ECt} \qquad (2)$$
$$I_a = I'_t \, (1-10^{-ECt}) \qquad (3)$$

Here, I'_t is the intensity transmitted by the preceding segment. As the extinction coefficient is dependent on wavelength, these calculations must be performed for each wavelength emitted by the radiation source and for each segment in the coating. Rubin has published a more detailed description of similar calculations (15).

To obtain a parameter relative to the intensity incident on the surface of the coating, we define PIA as the percent of incident radiation absorbed:

$$PIA = Ia/Io \times 100 \qquad (4)$$

Our Additive Absorbance analysis program performs PIA calculations for each wavelength and each segment of the coating. To simplify the analysis, we have also made the following assumption: if a photon of any wavelength is absorbed we assume that photochemical events occur with the same efficiency. This is equivalent to assuming that the quantum yield of initiating radicals is independent of wavelength. Although photochemical reactions are known having a quantum yield dependent on wavelength, information on the dependence of quantum yield of initiating radicals on wavelength is not available for most photoinitiators. We also assume that the same reactive species is formed from photons absorbed at any wavelength.

It should be noted that effects of reflectance, scattering and diffraction have been neglected. Calculations also assume (incorrectly) that the polymer is transparent to all wavelengths, and that there are no other absorbing species competing with the additive. Calculations of PIA which include such competitive absorption are not trivial (10, 16-18). We have also assumed that the absorbance of the additive does not change significantly during the irradiation. This assumption is reasonable considering the low degree of photolysis of the photoinitiator during most cure conditions.

EXPERIMENTAL

Developmental photoinitiator CGI-369 (Structure I) was obtained from the CIBA-GEIGY Additives Division as a yellow crystalline powder and used as received. Absorbance values were taken from absorption spectra run in ethanol at four concentrations ranging from .01 to 10.0 g/l. Extinction coefficients were calculated from Equation 1, at approximately 10 nm increments in units of liter/(cm x mole). Where extinction coefficients could be calculated at more than one concentration, the values were averaged. Extinction coefficients needed for specific wavelengths in the program were interpolated from nearest neighboring experimental values.

Curing Conditions. 1 mil (25.4 um) coatings were cured at 200 f/m under 1 focused Fusion H lamp operating at 300 W/in. Coatings were applied using a wire wound rod to Form N2A opacity cards (The Lenetta Company). Surface cure was measured as the minimum number of passes to achieve light scratch resistance over the black substrate region. The formulation below was used for the data in Table III.

Model Urethane Acrylate Formulation

Ebecryl 8800-20R	62.5 parts
TMPTA	22.9
TRPGDA	4.2
N-vinyl pyrrolidone	10.4

Pendulum hardness (Koenig) was measured on 1 mil (25.4 um) coatings applied to gray primed aluminum panels. All panels were exposed to 5 passes at 200 f/m under the Fusion H bulb. Dark substrates have been utilized to minimize reflection.

DISCUSSION

Figure 1 shows the extinction coefficient spectrum of developmental photoinitiator CGI-369. Such spectra are equivalent to absorption spectra but are independent of concentration and thickness.

Lamp Emission Spectra. Figure 2 shows the emission lines from 200-600 nm of a Fusion H medium pressure mercury arc lamp operating at 300 W/inch. Although emission spectra are normally plotted as Energy vs wavelength, we have utilized the Stark Einstein relation to convert energy to intensity, which is proportional to the number of photons emitted at each wavelength.

$$m = E\, Y/hc, \qquad (5)$$

where E = energy, Y = wavelength, h = Planck's constant, c = the speed of light, and m = the number of photons. To facilitate relative comparisons, the total emission intensity has been normalized to 100. Thus, each line represents the percent of emitted photons at a given wavelength. Since photochemical events are proportional to the number of photons absorbed (not energy absorbed) this correction makes the emission spectra more useful for PIA calculations.

Figure 3 shows the dependence of PIA throughout a coating, having 10 theoretical segments, on the total optical density of the film for the ideal case of monochromatic radiation. Below OD = 0.25, PIA is less than 6% and uniformly absorbed throughout the coating. As the OD approaches 0.434 the PIA increases at all points, but is almost twice as high near the surface as at the substrate. Increasing the OD beyond 0.434 results in significantly increased PIA near the surface, at the expense of reduced PIA deeper in the coating due to the internal filter effect in the upper segments. This extreme would probably favor improved surface cure at the possible expense of adhesion. The lower OD model would probably result in slower, but more uniform polymer formation. This model would be most useful under inert atmosphere curing conditions. The optimum PIA profile for a given property probably depends on what portion of the coating (ie. surface, body or substrate interface) is most directly associated with the property. We have recently demonstrated how different emission wavelengths and variation of initiator selection can cause similar effects ($\underline{1}$).

The first law of photochemistry states that "in order for photochemistry to occur, radiation must be absorbed." A useful corollary to this law for UV curing is that the only wavelengths which need be considered are those emitted by the radiation source. For the data in Table I we have used this corollary to select the 18 wavelengths of a Fusion H bulb mercury lamp. The molar extinction coefficients (molar epsilon) for CGI-369 were calculated at each wavelength. Optical Density (OD) was calculated using Equation 1. Calculations of % transmitted and % absorbed in Table I do not include the lamp emission intensity at each wavelength. At a level of 3% (typical for many commercial applications at 1 mil thickness) this photoinitiator absorbs radiation efficiently. The absorptivity in the blue region allows the use of the emission lines at 405 and 434 nm. CGI-369 also

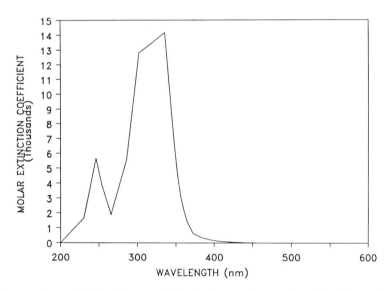

Structure I

Figure 1 Extinction coefficient spectrum for developmental photoinitiator CGI-369.

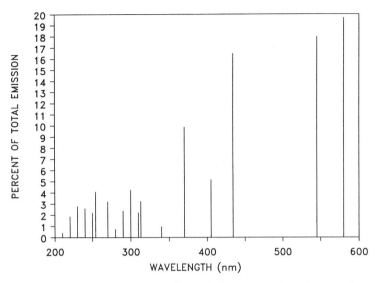

Figure 2 - Emission spectra of a Fusion H curing lamp adjusted to relative photon flux.

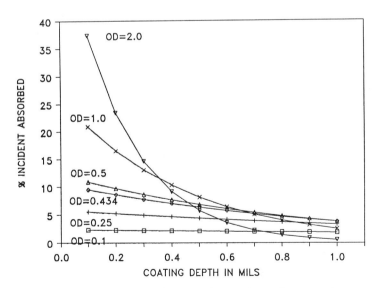

Figure 3 - The influence of optical density of a coating on the percent of incident radiation absorbed at various depths (monochromatic radiation).

utilizes the 365 nm line much more efficiently than the photoinitiators described in our earlier studies (1).

The amount of photochemistry occurring at a given point within a coating will be a function of the intensity of the radiation reaching that point, and the efficiency at which it is absorbed. Using equations 2 and 3, the percent of incident radiation transmitted and absorbed was calculated for each wavelength, in each segment. Table II shows the results of these calculations for a 1 mil coating divided into 10 segments cured with 3% CGI-369. These calculations include the lamp emission intensity.

The absorbed radiation in a given segment was summed over all wavelengths to determine the total percent of incident radiation absorbed in each segment. Plots of PIA vs. depth will be utilized in the discussions below. By convention, PIA calculated at depth = 0.2 mils represents the average PIA throughout the second segment, i.e. between 0.1 and 0.2 mils.

EFFECTS OF CONCENTRATION. Analyses of cured films have often indicated that most of the photoinitiator remains unreacted (19, 20). Surprisingly, the use of lower concentrations often result in insufficient cure. The concentration of a photoinitiator is not simply or directly related to curing efficiency. If the selection of initiator and lamp are held constant, the net effect of varying the concentration of the photoinitiator will be to determine the distribution of the initiating radicals throughout the coating.

Figure 4 shows the PIA calculations for 1 mil films cured under a Fusion H lamp having concentrations of CGI-369 ranging from 0.5 to 7.5%. These curves indicate that in the higher concentrations 5-10 fold higher PIA is obtained in the first 0.1 mil as compared to the typical PIA found in the lower half of the coating. This high PIA, resulting in a high concentration of initiating radicals, is needed to offset the reduced curing efficiency near the surface caused by oxygen inhibition. Lower photoinitiator concentrations which would create more uniform PIA throughout the coating are known to result in poor surface cure in air.

In a previous study of absorbance profiles for a high extinction coefficient photoinitiator, aminoacetophenone (AAP), we observed significant crossover of the PIA plots similar to those observed in Figure 3 (1). AAP is known to need careful formulating to avoid surface skin cure or wrinkling in thick or pigmented applications (1). The broader absorbance of CGI-369, coupled with the emission of the Fusion H bulb, reduces the internal filter effect. As a result, one can formulate fairly high concentrations of this photoinitiator with no surface wrinkling. This benefit is especially useful in pigmented applications (8).

CORRELATION OF PIA CALCULATIONS WITH CURING RESULTS. Table III shows the results of testing five concentrations of CGI-369 for scratch resistance and pendulum hardness in coatings cured under the Fusion H bulb. These test results are plotted in Figure 5. Both surface cure (scratch resistance) and through cure (pendulum hardness) were found to plateau at the higher concentrations of CGI-369 tested.

Table I. Absorbance and Transmittance Calculations
For Each Wavelength of a Fusion H Bulb
Medium Pressure Mercury Arc Lamp

Photoinitiator:	3% CGI-369	1 mil Thickness		
Wave-Length (nm)	Molar Epsilon	OD	% Trans.	% Absorbed
210	1637	0.3	45.6	54.3
220	1637	0.3	45.6	54.3
230	1642	0.3	45.5	54.4
240	4169	0.8	13.5	86.4
250	4766	0.9	10.1	89.8
254	3847	0.8	15.8	84.1
270	2627	0.5	28.4	71.5
280	4478	0.9	11.7	88.2
290	7393	1.5	2.8	97.1
300	11905	2.4	0.3	99.6
310	13131	2.7	0.1	99.8
313	13252	2.7	0.1	99.8
340	11580	2.4	0.3	99.6
365	858	0.1	66.2	33.7
405	120	0.024	94.3	5.6
434	25	0.0053	98.7	1.2
545	0	0.0000	100.0	0.0
580	0	0.0000	100.0	0.0

Table II. Calculations of Absorbance and
Transmittance by Thickness

Photoinitiator: 3% CGI-369   1 mil Thickness
Segment Thickness = 0.1 mils
Lamp = Fusion H

Segment No.	Coating Depth (mils)	% Trans. by Seg.	PIA % Incident Absorbed	
0	0.0	100	0.0	(Surface)
1	0.1	91	8.4	.
2	0.2	85	5.6	.
3	0.3	81	3.9	.
4	0.4	78	2.9	.
5	0.5	76	2.2	.
6	0.6	74	1.8	.
7	0.7	73	1.5	.
8	0.8	72	1.2	.
9	0.9	70	1.1	.
10	1.0	69	0.9	.
				(Substrate)

Note: Segment 0 represents the surface, thus no radiation is absorbed.

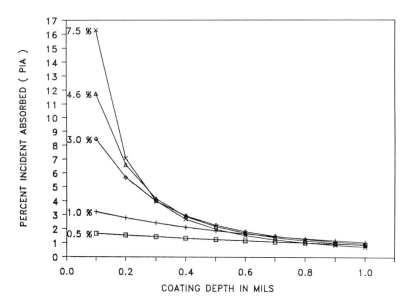

Figure 4 - Percent of incident radiation absorbed at various depths in a coating containing concentrations of CGI-369 ranging from 0.5 to 7.5% (Fusion H bulb radiation).

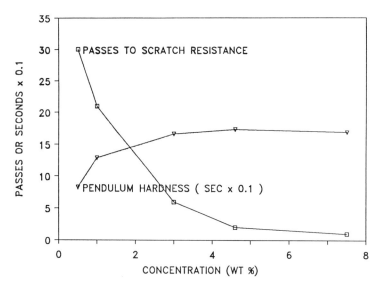

Figure 5 - The dependence of Surface Cure (scratch resistance) and Through Cure (pendulum hardness) on concentration of developmental photoinitiator CGI-369.

TABLE III. Photoinitiator Concentration Effect
on Cure Response

% CGI-369	Scratch Resistance (passes)	Pendulum Hardness (seconds)
0.5	25.	83.
1	21.	129.
3	6.	167.
4.6	2.	174.
7.5	1.	170.

Pendulum hardness results were found to correlate best with PIA values in the middle region (0.3-0.5 mils depth) of the coatings (see Figure 6). This corresponds to the common interpretation of pendulum hardness to be a measure of 'body cure'. Scratch resistance was found to correlate best with PIA in the second segment (0.2 mils), closer to the surface (see Figure 7). We interpret these results to indicate that the majority of radicals generated in the first segment are used to offset the diffusion of oxygen. This surface cure region is about 0.1 mils (2.5 um) thick, below which oxygen diffusion is significantly reduced and better cure is achieved. This could also indicate that our scratch resistance test is more an indication of cure just below the surface than right at the surface.

In a second study we have compared the performance of three commercial photoinitiators for surface cure in a commercial urethane acrylate formulation. The following photoinitiators were utilized in this study: HCPK = alpha-hydroxy-alpha-cyclohexylphenylketone (Structure II), BDMK = benzil-dimethyl-ketal (Structure III), and AAP (Structure IV). The extinction spectra of these initiators are in Figure 8.

Table IV shows curing results for a commercial urethane-acrylate formulation cured under the conditions of Figure 9 (0.5 mil thickness coating containing 3% photoinitiator).

Table IV. Cure Comparison Using a Mercury Lamp

Structure	Photoinitiator	Scratch Resistance (passes)
IV	AAP	8
III	BDMK	13
II	HCPK	14

Figure 9 shows a comparison of PIA calculations for the three photoinitiators integrating over the 18 lines of a medium pressure mercury arc lamp. In this case, all three photoinitiators have a higher PIA near the surface than close to the substrate. AAP generates the highest PIA at all points in the coating. BDMK results in intermediate levels of PIA, while HCPK is predicted to have the lowest concentration of radicals at any depth within the coating. Figure 9 suggests that under these conditions AAP will cure significantly faster than BDMK, which will cure slightly faster

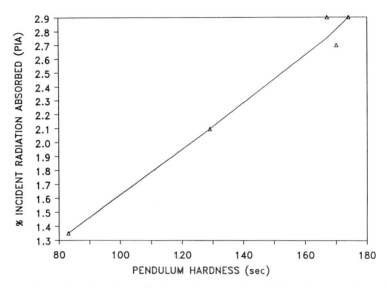

Figure 6 - Correlation of through cure (pendulum hardness) with PIA calculated at depth 0.4 mils. PIA calculated for concentrations of CGI-369 ranging from 0.5 to 7.5%.

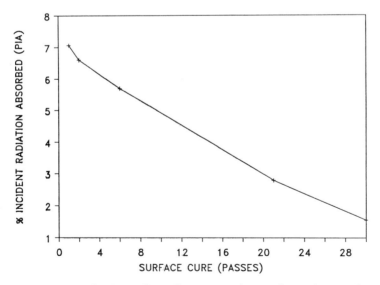

Figure 7 - Correlation of Surface Cure (scratch resistance) with PIA calculated at depth = 0.2 mils concentrations of CGI-369 range from 0.5 to 7.5%.

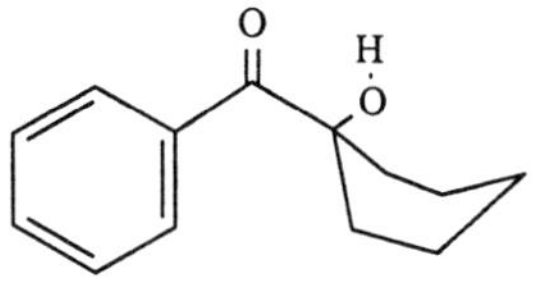

Structure II

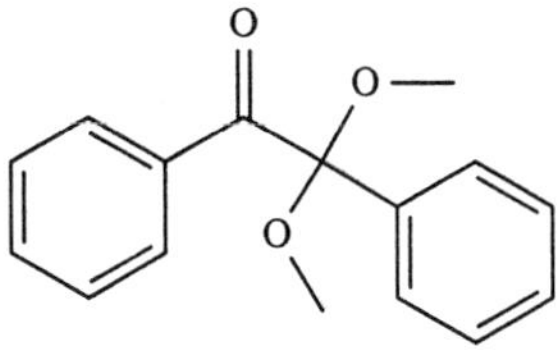

Structure III

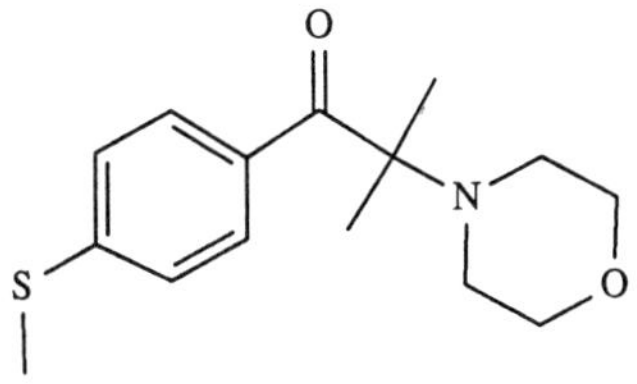

Structure IV

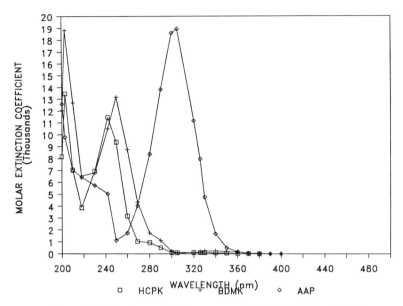

Figure 8 - Extinction coefficient spectra of HCPK, BDMK and AAP.

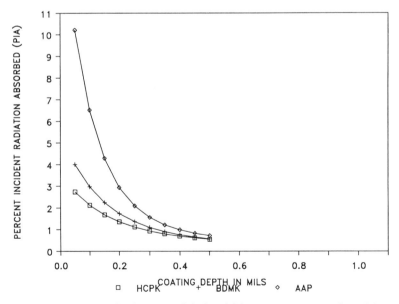

Figure 9 - PIA plotted for 3 commercial photoinitiators at a concentration of 3% in 0.5 mil coatings.

than HCPK. The scratch resistance test results in Table IV agree with this prediction.

CONCLUSIONS

Calculations of the percent of incident radiation absorbed aid in understanding the interaction between photoinitator concentration and the selection of a curing lamp. Varying the concentration of a photoinitiator was demonstrated to have a dramatic effect on where the radiation is absorbed within a coating, and hence where the initiating radicals are generated. The traditional belief of an optimum optical density of 0.434 does not apply to photosensitive systems cured in air or under a radiation source having multiple wavelengths. Although dependent on the test method, fair correlation exists between how model and commercial formulations cure, and the calculated PIA values below the surface of the coating. Formulators should be aware that changes in any of the variables which affect the PIA will result in the need for reoptimization of the photoinitiator level.

ACKNOWLEDGEMENTS

The authors thank Dr. M. Rembold, D. Wostratzky, and Dr. H. Angerer for discussions and critique during the development of the absorbance algorithms, and to J. Parchment and H. Evers for assistance with the experimental curing data. M. Synder is gratefully acknowledged for his efforts in refining the programming code. We thank the CIBA-GEIGY Corporation for permission to present this paper.

LITERATURE CITED

1. Gatechair, L. R. Proceedings of Radtech '88, April 26, 1988.
2. Gatechair, L. R. and Wostratzky, D., a) J. Radiation Curing, 10(3) 4-19,(1983). b) Proc. Radiation Curing VI, Sept. 20-23, 1-24,(1982).
3. Berner, G. and Sitek, F. Proceedings of the Conference of Dutch Paint Association, Woerden, Netherlands, December 17,(1985).
4. Pappas, S. P., "UV Curing Science and Technology", (S. P. Pappas, Ed.) Technology Marketing Corp. Norwalk, CT. USA, pp 1-25 (1985).
5. Vesley, G. F., J. Radiation Curing 13(1) 4-10 (1986).
6. Phan, X. T., J. Radiation Curing, 13(1) 11-17,1986.
7. Berner, G., Puglisi, J., Kirchmayr, R., Rist, G., J. Radiation Curing, April, 2-9(1979).
8. Desobry, V., Dietliker, K., Husler, R., Misev, L. Rembold, M., Rutsch, W. in this symposium.
9. Bush, R. W., Ketley, A. D., Morgan, C. R. and Whitt, D. G., J. Radiation Curing 7 (2): 20 (1980).
10. Shultz, A. R. and Joshi, M. G., J. Polym. Sci.: Polym. Phys, Ed. 22 (1984) 1753-1771.
11. Reference 4, p. 62.
12. Reiser, A. and Pitts, Photographic Sci. Eng. 20 (1976) 229.
13. Gutierrez, A. R., and Cox, R. J., Polymer Photochemistry 7 (1986) 517-521.

14. Thommes, G. A. and Webers, V. J., J. Imaging Sci., 29 (1985) 112-116.
15. Rubin, H., a) TAGA Proceedings, 279-301 (1976) b) pp 187-201 in reference 3.
16. Heller, H. J., European Polymer Journal-Supplement, 5(5) 105-132 (1969).
17. Shultz, A. R. and Andrady, A. L., J. Appl. Polym. Sci., 33 (1987) 2249-2252.
18. Guillory, J. P. and Cook, C. F., J. Polym. Sci.: Port A-1, 9 (1971) 1529-1536.
19. Hult, A., Yuan, Y.Y., Ranby, B. Polymer Degradation and Stability, 8(1984) 241-258.
20. Hult, A., Ranby, B., Polym. Degradation and Stability 8(1984) 89-105.

RECEIVED September 13, 1989

Chapter 4

Modified Kubelka–Munk Optical Treatment of Photon Flux in a Coating and Its Relationship to UV Curing as Measured with a UV N101 Cure Tester

James H. Nobbs[1] and Peter K. T. Oldring[2]

[1]Colour Chemistry Department, University of Leeds, Leeds LS2 9JT, England
[2]Thomas Swan & Company Ltd., Crookhall, Consett, Durham DH8 7ND, England

The Kubelka-Munk two flux model predicted significantly different magnitudes of photon flux within the layers of a coating than a model based on the Lambert-Beer law. Equations and calculation methods are described and results are given that illustrate the effect of substrate reflectance, layer thickness and the absorption and scattering of the layer components on the photon flux and the light absorbed at various levels within the coating. The predicted absorbance of a coating that gave the maximum light absorbtion in the layer adjacent to the substrate, increased as the substrate reflectance increased.
Reasonable agreement was found between the predicted curing rate behaviour of pigmented printing ink layers and transparent coatings and that measured by a UV N101 cure tester (Swan Instruments, Consett, England).

The first task when attempting a mathematical description of a chemical reaction is to establish a set of rate equations for the change in concentrations of the reactants and products. In the case of photo-chemical reactions the equations must also describe the intensity of the light at each point within the reactive system. The light intensity calculation is usually reduced to simple equations by careful design of the practical conditions under which the reaction is studied. This is not possible when attempting a description of a real system such as a semi-transparent UV curable layer coated on a substrate. To obtain a realistic estimate of the amount of light absorbed at each level in the coating it is necessary to use an optical model that takes into account:

1. Partial reflection of the external light incident on the air to coating interface.

2. Partial reflection of the internal light incident on the coating to air interface.

3. Absorption and scattering of the light from components present within the layer.

4. Absorption and scattering of the light by the substrate.

5. The angular distribution of the light intensity at each point within the coating.

The scattering and absorption of the light means that the angular distribution changes as it passes through the layer. The scattering process increases the angular distribution whereas the absorption process tends to reduce it. This makes a general solution to the problem quite complex. There are methods in radiation transfer theory that are capable of providing an exact calculation, however the equations that they provide are not presented in an explicit form and numerical methods of solution are often used.

If simplifying assumptions are made then it is possible to obtain solutions. A very simple case is to assume that the angular distribution of the light travelling away from the irradiated surface of the layer is the same as that travelling towards it. This is a two flux model and it has been found of value when calculating the reflectance of a layer containing a mixture of components. Computer programs that use this approach have found practical application in predicting the amounts of pigment required in a layer to match the colour of a standard material.

How close the model represents the actual situation depends on the optical properties of the layer and the way in which it is irradiated. The model would not apply very well to a narrow beam of light illuminating a highly scattering layer, since the light would become progressively diffuse as it travelled through the layer. In this case a three flux model would be more appropriate.

A two flux model would be reasonable for such a layer receiving light from a wide range of directions. The scattering would tend to maintain the diffuse nature of the radiation for both the fluxes.

In the following section the application of the Kubelka-Munk($\underline{1}$) two flux theory to UV curing calculations is described.

Optical Theory

Two Flux Model. The light within a layer is described in terms of energy fluxes rather than beams of light. The layer is assumed to be homogeneous and to have a large irradiated surface area compared to its thickness, so that the proportion of light lost from the edges may be neglected. Under these conditions the equations involve only a single dimension, the distance in the direction normal to the surface of the layer. The model is illustrated with the help of Figure 1 which refers to a thin section within a coating of overall thickness L. The section has thickness D and is at a distance x from the irradiated surface. At each section within the coating, light of a particular wavelength or narrow wavelength range is imagined to be split into two energy fluxes, a "down" flux I(x) travelling away from the irradiated surface and an "up" flux J(x) travelling towards the irradiated surface. The down flux includes all light with a positive component of velocity in the direction of increasing x. The up flux includes all light with a negative component of velocity in the direction of increasing x. The down flux just outside the irradiated surface of the layer is termed Io and the down flux just inside the layer is I(0).

The optical properties of the thin section are described by the fraction of an incident flux that is transmitted (T), reflected (Ro) and absorbed (A). Since energy is conserved it follows that

$$A = 1 - T - Ro \tag{1}$$

The assumption is made that the angular distribution of the light in the down flux is sufficiently similar to that of the up flux for the same values of T, Ro and A to apply to both fluxes. When a material of reflectance Rg is beneath the section then the reflectance R of the composite is given by

$$R = Ro + T^2 \cdot Rg/[1 - Ro \cdot Rg] \tag{2}$$

In order to determine the light absorbed by the thin section, the magnitude of the two fluxes I(x) and J(x+D) must be calculated.

$$\text{Light absorbed} = A \cdot [I(x) + J(x+D)] \tag{3}$$

To determine the values of I(x) and J(x+D) the coating is imagined split into three parts as illustrated in Figure 2. Section (1) is that above x, section (2) is at x and section (3) is that below x+D.

The transmittance of the upper section to down flux is termed T(1) and its reflectance to the up flux is termed Ro(1). Similarly T(2) and Ro(2) are defined for the centre section and T(3) and Ro(3) for the lower section.

The value of I(x), the down flux at the top of section 2, is the sum of the down flux transmitted by section 1 and the up flux reflected by section 1.

$$I(x) = I(0) \cdot T(1) + J(x) \cdot Ro(1) \tag{4}$$

The value of J(x), the up flux at the top of section 2 is the sum of the up flux transmitted by section 2 and the down flux reflected by section 2.

$$J(x) = I(x) \cdot Ro(2) + J(x+D) \cdot T(2) \tag{5}$$

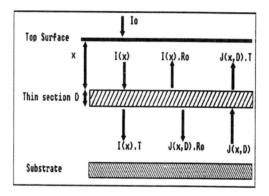

Figure 1. The energy flux at a thin section of thickness D and distance x from the irradiated surface of a coating layer.

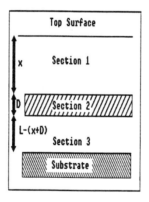

Figure 2. The three sections within the coating layer.

Similarly for I(x+D) and J(x+d).

$$I(x+D) = I(x).T(2) + J(x+D).Ro(2) \tag{6}$$

$$J(x+D) = I(x+D).R(3) \tag{7}$$

Where R(3) is the reflectance of section 3 including the contribution from the substrate (Rg). R(3) is calculated from Equation 2 which becomes,

$$R(3) = Ro(3) + T(3)^2.Rg/(1 - Ro(3).Rg) \tag{8}$$

The equations can be re-arranged to give I(x) and J(x+D) in terms of I(0) and the various reflectance and transmittance values.

$$I(x) = I(0).T(1)/[1 - Ro(1).R(2)] \tag{9}$$

Where $\quad R(2) = Ro(2) + T(2)^2.R(3)/[1 - Ro(3).R(3)] \tag{10}$

$$J(x+D) = I(x).T(2).R(3)/[1 - Ro(2).R(3)] \tag{11}$$

The amount of light E(x,D) absorbed by section 2 is given by:-

$$E(x,D) = [1 - T(2) - Ro(2)].[I(x) + J(x+D)] \tag{12}$$

The various values of T and Ro may be estimated from the Kubelka-Munk theory. The major assumptions of the theory are that the angular distribution of the light is the same for both fluxes, the coating is homogeneous and that the thickness of the layer is small compared to its length and width. The theory has been reviewed by the J. H. Nobbs(2) and the nomenclature used in that review will be used here.

The values of T and Ro for a layer of thickness X are expressed in terms of an absorption coefficient K and a scattering coefficient S. The values of K and S that should be used are the respective sums of the absorption and scattering coefficients of all the components in the coating, not just the photo-initiator.

$$T = [(1 - Q^2).\exp(-Z)]/[1 - Q^2.\exp(-2Z)] \tag{13}$$

$$Ro = [Q.(1 - \exp(-2Z))]/[1 - Q^2.\exp(-2Z)] \tag{14}$$

$$A = 1 - T - Ro \tag{15}$$

Where $\quad Z = X.[K.(K + 2S)]^{1/2} \tag{16}$

and $\quad Q = a - b, \quad a = (1 + K/S), \quad b = (a^2 - 1)^{1/2} \tag{17}$

<u>Boundary effects.</u> This is not quite the end of the calculation, the value of I(0), the down flux just inside the air to layer interface is not the same as the flux Io incident on the outside of the interface. The partial reflection of the light as it passes

through the boundary has to be taken into account. If Ts is the boundary transmittance for light travelling from air into the coating and Rs is the boundary reflectance for light travelling from the coating into air, then I(0) is related to Io by Equation 18.

$$I(0) = Io.Ts/[1 - Rs.R(1)] \qquad (18)$$

The Fresnel equations may be used to determine the boundary transmittance Ts for diffuse light incident from air into a layer and the boundary reflectance Rs for diffuse light incident from a layer into air. The values of Ts and Rs are 0.92 and 0.60 respectively for a layer of refractive index 1.50, a value typical for organic polymers. The high reflectance arises from the total internal reflection of light at high angles of incidence. R(1) is the reflectance of the underlying material and is given by

$$R(1) = Ro(1) + T(1)^2.R(2)/[1 - Ro(1).R(2)] \qquad (19)$$

Method of Calculation. The equations provide a method of estimating the magnitude of the fluxes at a position x within a layer and the amount of light absorbed by a section at that position. The effect of initiator concentration (via K), substrate reflectance (Rg), total layer thickness (L) and the scattering and absorption of light by pigments (K and S) may all be simulated by the model.

The values of K, S, Rg, x, D and L are selected and the transmittances(T) and reflectances(Ro) calculated from Equations 13 to 17 as follows.

Set X equal to	To determine the values of
X = x	T(1) and Ro(1)
X = D	T(2) and Ro(2)
X = L - (x - D)	T(3) and Ro(3)

The value of R(3), R(2) and finally R(1) are determined in sequence from Equations 8, 10 and 19 respectively. I(0), I(x) and J(x+D) may then be expressed in terms of Io using Equations 18, 9 and 11. Finally E(x,D) is obtained from Equation 12.

Single Flux Model. For comparison, calculations have also been made for a plane parallel beam of light incident normal to the surface of the layer. The surface transmission in this case will be Ts=0.9600 for a layer of refractive index 1.50. The calculation ignores the light scattering of the components of the layer and the reflection of the substrate hence I'(0) = Ts.Io. Setting "a" as the absorption extinction coefficient, the beam intensity at a distance x from the irradiated surface is given by the Lambert-Beer law as

$$I'(x) = I'(0).\exp(-a.x) \qquad (20)$$

and the amount of light absorbed within a section of thickness D at position x is given by

$$E'(x,D) = I'(x) \cdot (1 - \exp(-a.D)) \tag{21}$$

The relationship between the absorption coefficients K and a depends on the angular distribution of the flux assumed in the two flux model. The Kubelka-Munk theory assumes that the light has a diffuse angular distribution within the layer. In this case, when a flux travels a distance D along direction x, the average distance actually travelled by the light beams contained in the flux is 2.D. Since the rays travelling at an oblique angle to the layer will travel further, it follows that

$$K = 2.a \tag{22}$$

Results and Discussion

The results describing the distribution and absorption of light are expressed relative to the light flux Io incident on the top surface of the coating layer.

Distribution of Light Within a Layer. Table I shows values of $I(x)/Io$, $J(x)/Io$ and $E(x,D)/Io$ for a section of thickness 0.1 at various depths x within a layer of overall thickness 1. For the two flux model the values of the absorption and scattering coefficient for the layer are K=1 and S=0.2 respectively. The substrate reflectance was taken as Rg=0.50. For comparison, the amount absorbed $E'(x,D)$ according to the one flux model is also shown (a = 0.5).

Table I. Values of $I(x)/Io$, $J(x)/Io$ and $E(x,D)/Io$ for a section of thickness 0.1 at various depths x within a layer of thickness L = 1, K = 1, S = 0.2, and Rg = 0.50

x	$I(x)/Io$	$J(x)/Io$	$E(x,D)/Io$	$E'(x,D)/Io$
0.00	0.9941	0.1235	0.1059	0.0468
0.10	0.8840	0.1193	0.0952	0.0445
0.20	0.7862	0.1168	0.0858	0.0423
0.30	0.6995	0.1159	0.0776	0.0403
0.40	0.6226	0.1166	0.0705	0.0383
0.50	0.5544	0.1190	0.0644	0.0365
0.60	0.4940	0.1229	0.0592	0.0347
0.70	0.4405	0.1287	0.0549	0.0330
0.80	0.3932	0.1363	0.0512	0.0314
0.90	0.3514	0.1457	0.0484	0.0299

The distribution of light within the layer is illustrated in Figures 3 and 4. It can be seen from Figure 3 that the value of

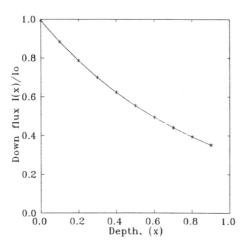

Figure 3. The calculated down flux I(x)/Io at various depths within a layer. Parameters L=1, K=1, S=0.2, Rg=0.5.

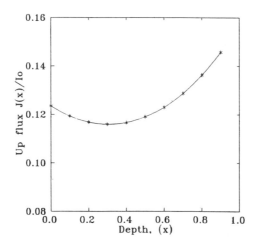

Figure 4. The calculated up flux J(x)/Io at various depths within a layer. Parameters L=1, K=1, S=0.2, Rg=0.5.

the relative down flux I(x)/Io decreases steadily as the depth within the layer increases. Figure 4 shows that the up flux J(x)/Io initially decreases with increasing depth, passes through a minimum and then increases steadily until the surface of the substrate is reached. The minimum only occurs when at least one of the components in the layer scatters light (S > 0). When there is no scattering the up flux is maximum at the substrate and decreases steadily towards the top surface. The ratio of the up flux to the down flux at the substrate is the substrate reflectance R_g.

Light Absorbed Within the Layer. The amount of light absorbed E(x,D)/Io by a section of thickness D=0.1 at various levels within the layer is shown in Figure 5. A steady decreases is calculated from 0.1059 just below the irradiated surface to 0.0484 adjacent the substrate. The values predicted by the one flux model are also shown in the plot. The plot shows that the two sets of values not only differ in absolute magnitude but also in the rate of change of light absorption with increasing depth. These differences suggest that the angular distribution of the irradiating light has a strong influence on the curing properties of a coating layer.

Influence of concentration. The effect of the concentration of photo-initiator on the amount of light absorbed at a particular depth in the layer can be simulated by varying the value of K while keeping the other parameters constant. Calculations were carried out for a section of thickness D=0.1 located at the top of the layer (x=0.0), half way through the layer (x=0.5) and also when adjacent to the substrate (x=0.9). The results are shown in Figure 6. The amount of light absorbed by the top layer is shown to increase steadily as the value of K increases. The predicted behaviour at the middle and lower part of the layer is very different. There is a rapid increase in light absorption to a maximum value at about K=1 and then a gradual decrease occurs as K is further increased.

The results clearly suggests that there is an optimum range of concentrations of initiator. For values of K up to about 0.5, a similar amount of light is absorbed in the top, middle and bottom sections of the coating and through cure can be expected. As K increases to 1.0 and higher, the topmost layer begins to absorb considerable more light than the middle and lower sections, which it is screening. The surface layer would cure at a significantly faster rate than the lower levels of the coating.

It is interesting that the middle and bottom level within the coating suggest roughly the same value of K=1 for maximum light absorption.

Effect of substrate reflectance. The time required for the through cure of a layer is determined by the slowest curing portion. If a coating is imagined split into two parts then the slowest curing portion would usually be the lower half section adjacent to the substrate, since it absorbs the least amount of

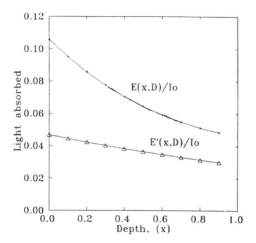

Figure 5. The calculated amount of flux absorbed by a section of thickness D=0.1 at various depths within a layer for the two flux E(x,D) and one flux model E'(x,D). Parameters L=1, K=1, S=1, Rg=0.5.

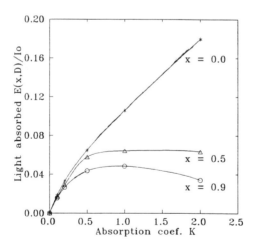

Figure 6. The calculated amount of light absorbed as a function of the coefficient K (or concentration of the absorbing species) at three depths within the layer. Parameters D=0.1, S=0.2, Rg=0.5.

light. It is the absorption of this section that is most
influenced by the reflectance of the substance.

In Figure 7 the light absorbed by the bottom half section is
shown for a calculation with three different levels of substrate
reflectance, low (Rg=0.0), medium (Rg=0.5) and high (Rg=0.9). In
each case the amount of light absorbed is plotted against the
absorption coefficient K. At a given value of K, the light
absorption can decrease by as much as 50% as the substrate
reflectance is reduced from 0.9 to 0.0. The plots are all similar
in shape to those in Figure 6 and show a maximum at a K value of
about 1.

Effect of film thickness. The influence of increasing the
thickness of the coating was simulated by considering the light
absorbed in a section of thickness D=0.5 adjacent to the substrate
and increasing the total coating thickness L. The value of x was
set as x=L-D. The results shown in Figure 8 show a very rapid
decrease in the light absorbed in the lower layer with increasing
film thickness.

Comparison with N101 Cure Tester Results

The N101 Cure Tester([3]) provides an instrumental method of
assessing cure properties of a film of coating material applied by
standard methods at the usual film thickness and on the substrate
which is normally used. The tester measures and records the force
required to move a stylus at a constant rate though or across a
film of coating material while it is being irradiated by UV light.
The radiation is from two 38 cm long low pressure mercury discharge
lamps mounted side by side, above the sample and parallel to its
direction of motion. The intensity of the radiation from each lamp
is adjustable and is continuously monitored. The type and
positioning of the lamps is such that the radiation is incident on
the surface from a wide range of angles, and the two flux model
would seem a reasonable first approximation.

The stylus is pressed with a constant vertical force against
the test surface. When performing a test, the stylus force
initially recorded is constant and reflects that of the stylus
moving against the surface wetted with an uncured liquid or paste
like coating material. As cure proceeds the force changes in a way
characteristic for that coating until on completion the force is
again constant. The final value is characteristic of the stylus
moving either over or through the surface of the cured coating
material. A typical trace of force against irradiation time for a
UV curable lithographic printing ink is shown in Figure 9.
A measure of the speed with which a material cures under the
selected test conditions is the time that elapses between switching
on the UV lamps and the test force reaching a value half way
between that of the uncured and cured coating. This time is termed
T50 and has been found to be a reliable way of characterising the
curing properties of a coating ([4,5]). A comparison between the
characteristic cure time (T50) measured by the N101 under various

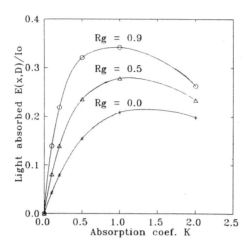

Figure 7. Light absorbed in a section of thickness D=0.5 adjacent to the substrate for three values of substrate reflectance Rg. Parameters L=1, x=0.5 and S=0.2.

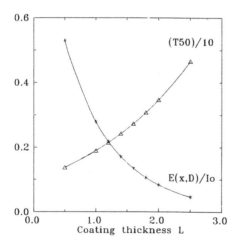

Figure 8. Light absorbed in the lower section (D=0.5) of a coating for various total coating thicknesses. Parameters K=1, S=0.2 and Rg=0.5.

conditions can be taken as a comparison between the time required to achieve a particular level of cure of the coating.

In order to compare the predictions of the optical model with N101 results, the calculated amount of light absorbed by a coating has to be converted into an equivalent T50 time. This requires a detailed knowledge of the reaction kinetics of the coating/substrate combination. Unfortunately this is not available, however the following simple model has proved useful in estimating T50 times from the optical calculations.

Simple Kinetic Model. It is thought that the N101 Cure Tester provides a measure of the state of cure of the section of coating adjacent to the substrate. If the absorption coefficient of the layer at the activating wavelength of the photo-initiating species is mainly due to the initiator, then the amount of light absorbed by this section can be related to the cure time of the coating. An estimate may be made by assuming that the time taken for this section to achieve a particular level of cure is inversely proportional to the polymerisation rate.

The model is approximate and takes no account of inhibition, depletion of initiator and other effects. The polymerisation rate constant is taken to be linearly related to the square root of the amount of light absorbed by this layer. On this basis plots of $1/\sqrt{E(x,D)}$ should show a similar trend to the T50 times recorded by the N101 tester for equivalent simulated and real experiments.

$$T50 \propto 1 / \sqrt{E(x,D)} \qquad (23)$$

As an example, the results shown in Figure 8 for the light absorbed at various thicknesses of layer have been re-expressed in terms of $1/[E(x,D)]$, the simulated T50 times. These values are also shown in Figure 8 plotted against coating thickness. The simulated T50 values increase progressively more rapidly as the film weight increases.

Direct comparison of the predicted T50 values and those measured on the N101 can be made by re-scaling the predicted values. For example, the L value of the calculation ought to be linearly related to an actual film weight of a test coating and the simulated and measured T50 values should also be linearly related. Without the advantage of a full kinetic model, the two constants of proportionality cannot be calculated. However the trend in T50 over a range of conditions can be compared by making a "best fit" calculation of the two proportionality constants. Such a comparison is made in Figures 10 and 11.

Film Weight. A series of prints were made of a commercial UV curable lithographic printing ink on a coated board. Two prints were made at six different film weights using a Duncan Lynch print proofer. The printed samples were immediately tested on the N101 tester under 254nm irradiation. The results are shown in Figure 10 together with a line predicted by the two flux model with K=1.0,

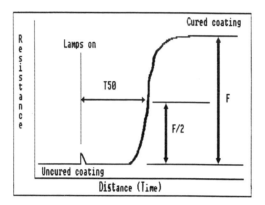

Figure 9. A typical trace obtained from the cure tester for a UV curable lithographic printing ink. Conditions 254/254nm irradiation, film weight 0.4 cm , coated board substrate.

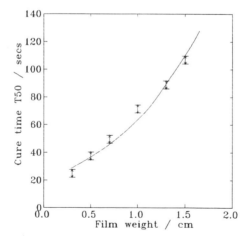

Figure 10. Comparison between predicted and measured T50 values for a UV curable lithographic printing ink. Parameters of predicted line K=1.0, S=0.2, D=0.5 and Rg=0.5. The proportionality constants used when plotting the predicted line were 0.55 and 20.5 for L and T50 respectively.

S=0.2, D=0.5 and Rg=0.5. The proportionality constants used when plotting the predicted line were 0.55 and 20.5 for L and T50 respectively. The agreement between the points and the line is remarkably good.

Initiator Concentration. The K value of the calculation ought to be linearly related to photo-initiator concentration and the simulated and measured T50 values should again be linearly related. A series of UV curable transparent varnishes were made containing range of levels of the photo-initiator Irgacure 184. The coatings were applied to aluminium foil board by an R-K automatic coater at a film thickness of 6um.

The samples were immediately tested on the N101 tester under 254nm and 355nm irradiation. The results are shown in Figure 11 together with the line predicted by the two flux model. The calculations were made with L=1, D=0.5, S=0.2 and Rg=0. The proportionality constants used when plotting the predicted line were 3.50 and 202.9 for K and T50 respectively. The agreement between the points and the line is again remarkably good.

Conclusions

The Kubelka-Munk(1) two flux model appears to be superior to a single flux model based on the Lambert-Beer law as it is capable of taking into account the scattering by components in the layer and the reflectance of the substrate. The model can provide estimates of the amount of light absorbed at any point within the coating and predict how this varies with film thickness, concentration of materials and reflectance of the substrate.

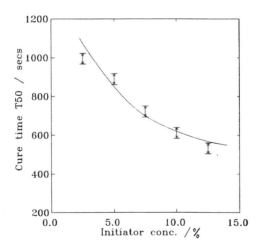

Figure 11.Comparison between predicted and measured T50 values for a UV curable transparent varnish. Parameters of predicted line L=1, D=0.5, S=0.2 and Rg=0. The proportionality constants used when plotting the predicted line were 3.50 and 202.9 for K and T50 respectively.

The model clearly shows that there is an optimum level of absorption coefficient (K = 1) that produces the maximum amount of light absorbed in the lower layers in the coating. The amount of light absorbed in the lower layers was also shown to be strongly dependent on the reflectance of the substrate.

Interpretation of the optical results by means of the simple kinetic model that relates light absorbed to cure time as measured by the N101 cure tester has proved useful. The agreement between prediction and measured results was remarkably good considering the large number of simplifying assumptions made. Although the test materials were chosen so as to satisfy most of the assumptions.

A more thorough test of the theory involving determination of the absorption and scattering coefficients in the layer and measurement of the flux transmitted and reflected needs to be made. In additon , a more realistic kinetic model should be applied to the interpretation of the results.

Literature Cited

1. Kubelka, P.; Munk, F. Z. Tech. Phys., 1931, 12, 593.
2. Nobbs, J. H. Rev. Prog. Coloration, 1985, 15, 66-75.
3. Thomas Swan and Co. Ltd., Consett, Co. Durham, DH8 7ND, England.
4. Nobbs, J. H.;Duerden, D.; Oldring, P. K. T. Radcure Europe, Munich, 1987.
5. Nobbs, J. H.; Guthrie, J. T.; Oldring, P.K.T.; Duerden, D. Proc. CCRA Int. Conf. Tokyo, 1988.

RECEIVED September 13, 1989

Chapter 5

Time-Resolved Laser Spectroscopy of Synergistic Processes in Photoinitiators of Polymerization

Jean-Pierre Fouassier and Daniel-Joseph Lougnot

Laboratoire de Photochimie Générale, Unité Associée au Centre National de Recherche Scientifique numéro 431, Ecole Nationale Supérieure de Chimie, 3 rue Alfred Werner, 68093 Mulhouse Cedex, France

> The synergic effects which are generally invoked to account for the specific features of a system of two ketones used as polymerization photoinitiators are reconsidered. The increase of reactivity observed when mixing these two initiators is reinterpreted in terms of a simultaneous energy and electron transfer in the pair. The relative efficiencies of these processes depends on the energy gap between the triplet states involved, which is known to be influenced by the polarity of the medium. A general discussion on the efficiency of various couples photoinitiator/photosensitizers is presented.

The wide use of photoinduced radical or cationic polymerization of unsaturated monomers in UV curing technologies [1-3] have stimulated many detailed investigations of the processes involved [4-10]. Time-resolved laser-spectroscopy has been extensively employed [7 and references therein][8][11-20] to study in real time, the dynamics of the excited states of molecules used as photoinitiators. Reaction models accounting for the photoinitiation steps of the polymerization have been proposed which allow relationships between the excited state reactivity and the practical efficiency as photopolymerization initiators to be discussed thoroughly.

Most of the photoinitiator families have been investigated during the past decade. In a general way, the main photoprocesses taking place in the excited states after absorption of light are, for the most part, well understood. However, the same does not hold true for the different reactions leading to excitation transfer in combinations of photoinitiators where the enhancement of the reaction efficiency is commonly referred to as "synergism" in the field of UV curing. The major problem encountered in the study of these reactions which take place in mixtures of several molecules exhibiting closely related absorption spectra is the selectivity of the excitation. This problem has been overcome very recently by using the emission of a Nd-YAG laser pumped dye laser [21] as the excitation light. In the present paper, a general discussion on synergistic effects will be presented, using a selection of systems recently studied in our laboratory.

HOW SYNERGISM CAN BE DESCRIBED ?

Under UV-light exposure, a radical photoinitiator is promoted to its first excited singlet state, S_1 and then, through a fast intersystem crossing, is converted to its lowest triplet state, T_1. This latter triplet state can yield radicals through cleavage, electron and proton transfer or H-abstraction. Extension of the spectral sensitivity of a photoinitiator, I, can be achieved by adding a photosensitizer, S. Such an energy transfer process must be exothermic with the energy level of the excited donor exceeding that of the acceptor by a few Kcal M^{-1} (with a difference of 3 Kcal M^{-1}, the energy transfer is almost diffusion controlled) :

Photosensitizer (T_1) + Photoinitiator (S_0) → Photoinitiator (T_1) + Photosensitizer (S_0)

However, the two terms *photoinitiator* and *photosensitizer* are often used in a more general sense to define any process involving either energy transfer (to generate from I the same radicals as those obtained through direct excitation) or electron transfer followed by proton transfer (to form new initiating radicals) (Scheme 1).

This mechanism is exemplified [22] by the mixture benzophenone **1** and Michler's Ketone **2** in which synergism is accounted for by the formation of a triplet exciplex which dissociates into a benzophenone ketyl radical and an aminobenzophenone radical as :

$$\underline{2}[S_0] \longrightarrow \underline{2}[S_1] \longrightarrow \underline{2}[T_1] \xrightarrow{1} \text{exciplex}$$

Chemical reactions with the radicals formed during the primary processes are also found to be responsible for synergistic effects, *e.g.* in a combination of **1** and 1-benzoyl cyclohexanol **3** in aerated medium. A possible explanation [23] is based on the decomposition by **1** of the hydroperoxides arising from the oxidation of the radicals generated through cleavage of **3**. The weak reactivity of these peroxy radicals compared to alkoxy and hydroxy radicals (Scheme 2) and the concommittant reduction of air inhibition due to the consumption of dissolved oxygen by excited benzophenone can be also invoked to account for the experimental observations.

3

Very recently, it was shown that mixing substituted thioxanthones **4** or **5** with the morpholino ketone **6** extends the photosensitivity towards the near visible part of the spectrum and accelerates the curing of pigmented coatings [24].

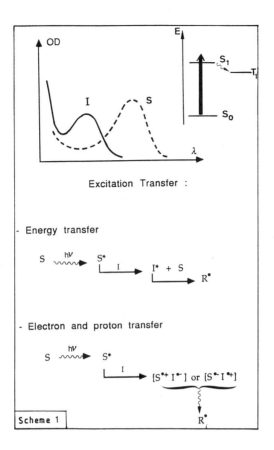

Scheme 1

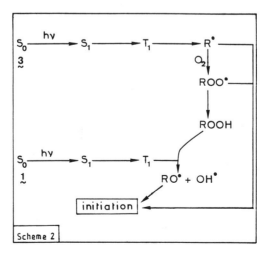

Scheme 2

$R_1 = R_3 = R_4 = H$
$R_2 = CH(CH_3)_2$ } **4**

$R_1 = CH_3$
$R_2 = R_4 = H$
$R_3 = COOEt$ } **5**

6: $CH_3S-C_6H_4-C(O)-C(CH_3)_2-N$(morpholine)

Photo DSC experiments conducted in a clear coating at $\lambda \geq 400$ nm - where only the substituted thioxanthone **7** ($R_1 = R_2 = R_3 = H$; $R_4 = COO(CH_2CH_2O)_8 H$) absorbs - demonstrates the increase of the polymerization efficiency [25]. Under exposure, the mixture of **7** + **6** leads to a considerable polymerization enthalpy whereas, in the presence of **7** or **6** alone, the exothermic signal remains very small. The same is true when using a laser light at $\lambda = 440$ nm [26] for the excitation of a mixture **5** + **6**. No polymerization occurs in the presence of **5** or **6**. The relative reactivity of **6** at 363 nm and **5** + **6** at 440 nm shows a 35 : 1 ratio, thus defining a low quantum efficiency of excitation transfer.

Evidence for the α-cleavage of **6** either in the absence or in the presence of thioxanthone derivatives has been shown in NMR-CIDNP spectra [24][25] and through GCMS [26], which supports at least, the view of an energy transfer process between **6** and **7** (or **4**) : excitation with a filtered light at $\lambda \geq 400$ nm leads to the formation of methyl thiobenzaldehyde which proceeds, in non hydrogen donating solvents, from a cage process in the radical pair generated after cleavage.

In photopolymerization experiments of methylmethacrylate (MMA) in degassed toluene solution, the increase of the initiation quantum yield ϕ_i ($\phi_i \propto R_p^2$) does not exceed 10-15% [26].

TIME-RESOLVED LASER-SPECTROSCOPY OF EXCITATION TRANSFER BETWEEN THIOXANTHONES AND 6

Investigating directly the processes involved between **4** or **5** and **6** during the exchange of excitation energy requires a pumping pulsed source of light which can excite **4** or **5** without exciting **6** (Figure 1) : this is readily achieved by using a light beam delivered by a dye laser pumped with a Nd-YAG laser [21].

A - EXCITED STATE PROCESSES IN 5

The triplet state of **5** is long-lived in deaerated solution [26] ; the absorption maximum of the T-T transition is blue shifted upon going from a non-polar to a polar solvent (670 nm in toluene and 625 nm in methanol). The fluorescence intensity increases and is red shifted with increasing solvent polarity : this can be readily understood in terms of a change of the spectroscopic character of the lowest lying excited state (an inversion of the relative position of the $n\pi^*$ and $\pi\pi^*$ states) or of a modification of the dipole moment (in the excited singlet and triplet state) upon photoexcitation.

Fluorescence and triplet absorption are quenched when methyldiethanolamine (MDEA), tert- butylmorpholine (TBM) and thioanisole (TA) are added to a solution of **5** and ketyl species are formed. The data are listed in Table 1. Scheme 3 summarizes the main photoprocesses occuring in the excited states of **5**.

B - PRIMARY PROCESSES IN 6

In the case of **6**, a fast cleavage occurs from a very short lived triplet state ($\leq$ 10 ns) generating a methylthiobenzoyl-morpholino isopropyl radical pair that undergoes disproportionation or separation (Scheme 4). The absorption of the methylthio benzoyl radical [26] is observed as well as the generation of a ketyl structure through a quenching of the triplet state by TBM. The usual bimolecular quenching of triplet states by monomer is also observed. Relevant data are presented in Table 1.

C - INTERACTION BETWEEN 5 AND 6

The interaction between **5** and **6** was investigated through laser pumped dye laser spectroscopy at λ = 430 nm and steady state fluorescence measurements. Figure 2 clearly shows that the thioether moiety is mostly responsible for the deactivation process occuring in the S_1 state and which is admittedly related to a back electron transfer.

On the contrary, the triplet state of **5** is deactivated by the morpholino moiety of **6** as indicated by the quenching efficiency shown in Figure 3. This interaction is drastically affected by the solvent (Table 1) : when going from toluene to methanol, the quenching rate constant increases by almost an order of magnitude whereas the ketyl radical absorption is only detected when the experiment is carried out in toluene. These results reveal that excitation transfer occurs mostly through energy transfer in methanol (and in polar solvents) and electron transfer in toluene (and in non polar solvents). The relative efficiencies of these two competitive processes (Scheme 5) depend on the stabilization of the lowest excited state T_1 by solvatation (that governs the spectroscopic character of T_1) and on the relative positions of the triplet energy levels of the donor **5** and the acceptor **6** (which determines the efficiency of an energy transfer process).

Changing **5** for **4** shows that either in toluene or in methanol, energy transfer would predominate. The conclusions drawn for **4** and **5** in both solvents are in good agreement with the findings of the GCMS study [26] : in these experiments, methylthio benzaldehyde was detected in **4** + **6** and **5** + **6** in methanol and in **4** + **6** in toluene but not in **5** + **6** in toluene.

This difference in behavior of **4** and **5** (and probably of other substituted thioxanthones) likely reflects the small changes in the position of the triplet state energy levels of **4** or **5** and **6** and the change in reactivity due to a more or less mixing of the $n\pi^*$ and $\pi\pi^*$ levels of **4** or **5**, the former influencing the energy transfer and the latter the electron transfer. The close values determined in EPA glasses at 77°K [24][25] for **4**, **5** and **6** (61.4, 58.4 and 61 Kcal/mol respectively) suggest that energy transfer is feasible and leads to speculation that slight solvent effects can modify the balance between energy transfer vs. electron transfer.

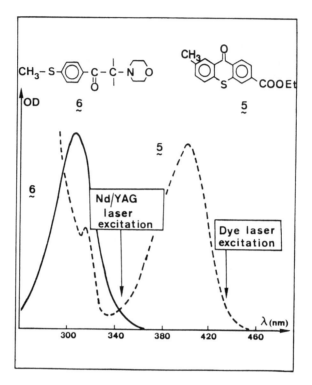

Figure 1. Ground state absorption spectra of <u>5</u> and <u>6</u> in methanol.

Table 1 : Rate constant of cleavage k_α, electron transfer k_e in toluene, excitation transfer k_T of <u>4</u> and <u>5</u> with <u>6</u>

	k_α (s^{-1})	$10^{-9} k_e$ (M^{-1}s^{-1})			$10^{-9} k_q$ (M^{-1}s^{-1})	$10^{-9} k_T$ (M^{-1}s^{-1})	
		MDEA	TBM	TA		in toluene	in methanol
<u>4</u>			0.1			0.06	0.055
<u>5</u>		3	0.15	<10^{-5}	2×10^{-4}	0.012	0.11
<u>6</u>	≤ 10^8		0.4		2×10^{-3}		

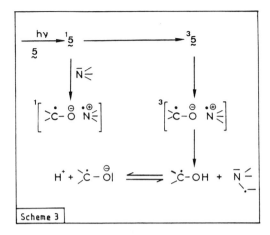

Scheme 3

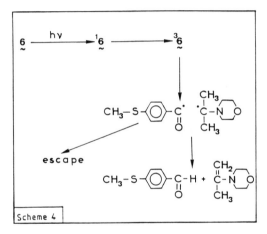

Scheme 4

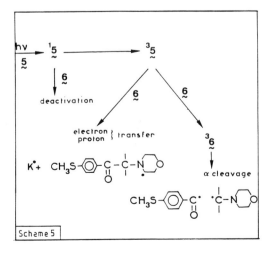

Scheme 5

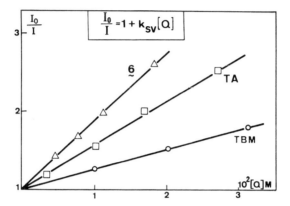

Figure 2. Fluorescence quenching of $\underline{5}$ in aerated methanol according to the usual Stern Volmer plot. λ_{exc} = 375 nm; λ_{an} = 470 nm; [Q] is in mol^{-1}.

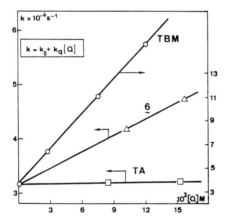

Figure 3. Triplet quenching of $\underline{5}$ in methanol as λ = 625 nm; [Q] is in mol^{-1}; k is the triplet state lifetime.

The reality of the excitation exchange between **5** and **6** is demonstrated by the fact that a polymerization can be initiated under laser irradiation at 440 nm (a wavelength which is only absorbed by **5**) in the presence of **5** + **6** but not in presence of **5** alone. The relative quantum yield of polymerization obtained is similar to that of the system **5** + morpholine (at λ = 363 nm under CW Ar+ laser) but considerably less than that of **6** at the same irradiation wavelength [26]. This observation is in good agreement with the assumption of an electron transfer process predominating in non-polar media.

SYNERGISTIC EFFECTS IN OTHER SYSTEMS

Synergistic effects have also been demonstrated in other oil-soluble systems such as **4** + **8** [25]. Photopolymerization of acrylamide conducted in aqueous solution [28] or in solution [29] in the presence of **4** + **9** or **4** + **10** [27] or **11** + **12** can be initiated at a wavelength where the photosensitizer absorbs (Table 2).

For all the energy acceptors **6**, **7**, **8**, **10**, **12**, the usual ground state absorption of the corresponding unsubstituted ketone is replaced by that of a transition exhibiting a chage transfer character ; the same holds true for **9**. The lowest lying triplet state is presumably affected, as revealed by the completely modified triplet-triplet absorption spectra of these compounds and the drastic changes in the rate constants of cleavage, electron transfer and monomer quenching [30][31]. That explains the considerable lowering of the T_1 energy level by about 10 K cal M^{-1} and the occurence of the energy transfer process. However, careful consideration of Table 2 indicates that **13** can act as an excitation acceptor while **14** can not.

13 (structure: Ph-C(=O)-C(CH₃)(OCH₃)-N(morpholine))

14 (structure: Ph-C(=O)-C(CH₃)₂-OH)

Other ketones, **15**, **16** and **17**, cannot undergo cleavage upon photosensitization because of their high triplet state energies. Photoreduction of the thioxanthone derivative through electron transfer in the singlet and the triplet state is only observed for **15** and **16** [26]. Sensitization of the polymerization is very inefficient for **5** + **14** ; **5** + **15** ; and **5** + **16**.

15 (structure: Ph-C(=O)-C(OCH₃)₂-Ph)

16 (structure: Ph-C(=O)-C(OH)(CH₂OSO₂CH₃)-Ph)

17 (structure: Ph-C(=O)-C(CH₃)=N-O-C(=O)-C₂H₅)

The efficiency of the sensitization is only slightly affected by the nature of the thioxanthone as shown in experiments carried out for **6** + **4** ; **6** + **7** ; and **6** + **18** (in **18** $R_1 = R_3 = H$; $R_4 = COOMe$). The photoinitiator plays also a very significant role.

Table 2 shows that sensitization can substantially improve the cure speed in pigmented films (8 : 1 ratio for **6** + **7** [25]) to a higher extent than in solution polymerization [26]. This apparent discrepancy is not so important since the experiments are carried out under very different conditions of light absorption : due to the absorption of the TiO_2 pigment, the photoinitiator **6** alone receives few photons whereas, in the presence of **7**, the light absorption of the film becomes very important. In that case, even with a low quantum yield of excitation transfer between **6** and **7** (as calculated in model solution photopolymerization studies), the apparent effect of the sensitization is intense due to the high dose of energy absorbed around 400 nm by thioxanthone.

Values of triplet energies E_T are listed in Table 3. At a first approximation, a parallel can be drawn between the lowering of the E_Ts of the photoinitiators and the increase of reactivity shown in Table 2 : this explains why sensitization of **14** and **15** is not efficient. Introduction of a methyl thioether group on the benzoyl moiety leads to a strong change of the energy levels of this chromophore : as already explained the characteristic and weak $n\pi^*$ transition of lowest energy which is typically observed in usual ketones is replaced by an intense $\pi\pi^*$ transition which displays a certain amount of charge transfer character involving the sulfur atom. Such a change is also expected to occur in the triplet state levels and may explain the energy stabilization of the lowest lying triplet state (going from a $n\pi^*$ to a $\pi\pi^*$ CT character). This explanation seems to hold when comparing **14** to **8** and should be also correct for **15** and **10**. However, even if this effect is present in **6**, one is forced to say that its contribution is weak since the E_T of the parent unsubstituted compound **13** is

Table 2: Role of the addition of a photosensitizer of the thioxanthone series. Reactivity obtained for the curing of a white pigmented lacquer (in m/mn). Rate of polymerization (in a.u.) in solution: MMA in toluene (a) and acrylamide in water (b) at $\lambda \geq 400$ nm [29]

Photosensitizer			Reactivity	R_p a	R_p b	ref.
13	no		3.3			[25]
13	yes	7	30			
6	no		5			
6	yes	7	40			
8	no		< 1			
8	yes	7	5			
14	no		3.3			
14	yes	7	5			
6	no			52		[26]
6	yes	(4)		59		
6	yes	(7)		65		
6	yes	(18)		58		
9	no			0.7		[28]
9	yes	(4)		2		
10	no			11.5		[28]
10	yes	(4)		14.5		
12	no				12	[29]
12	yes	(11)			15	

Table 3: Triplet energy levels of different compounds determined in EPA matrix at 77 K [25]

Thioxanthones	Photoinitiators	E_T (Kcal M^{-1})
	6	61
	8	63
	1 3	62
	1 4	71
	1 5	66.2
4		61.4
5		58.4
7		61
1 8		63
	1 9	65
	2 0	63
	2 1	63

almost the same (Table 3) : this is the reason why the substitution of **13** with an ether group (**19**) or an amino group (**20**) does not affect the E_Ts and leads to suitable photoinitiators whose decomposition can be sensitized [25]. On the contrary, the introduction of an amino **21** or a thioether group **8** (and not an ether group) on **14** lowers E_T [25].

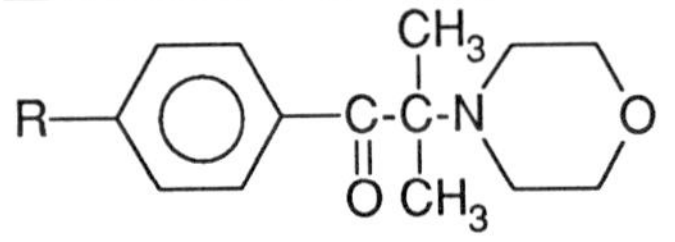

$$R = OCH_3 \quad \mathbf{19}$$

$$R = N(CH_3)_2 \quad \mathbf{20}$$

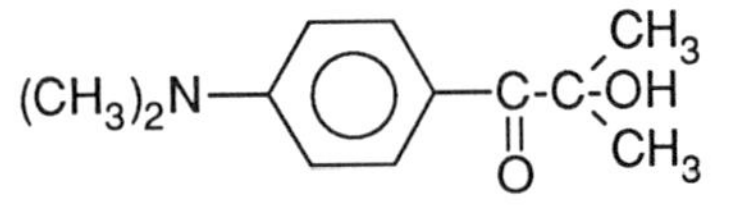

21

Moreover, the fact that **8** and **21** (although having the same E_T) exhibit a different reactivity in photopolymerization for a clear coating with filtered light (1 : 7 ratio) suggests once again that energy transfer is not the only factor governing the excitation transfer.

In summary, it is apparent that *i)* the presence of a thioether or an amine group at the *para* position of a benzoyl chromophore (as in **8** or **21**) or *ii)* the introduction of a morpholino moiety at the α carbon of the carbonyl group (as in **13**) is advantageous to lower E_T and favors the sensitization process (although a concomitant electron transfer in **6** and **13** – dependent on the experimental conditions as shown in model systems – cannot be completely ruled out). The same behavior is observed in **22** and **23** (by changing the morpholino group for a sulfonyl group).

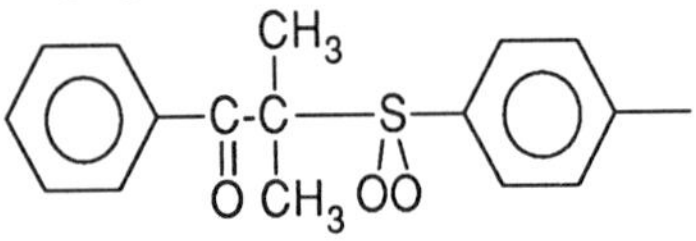

22 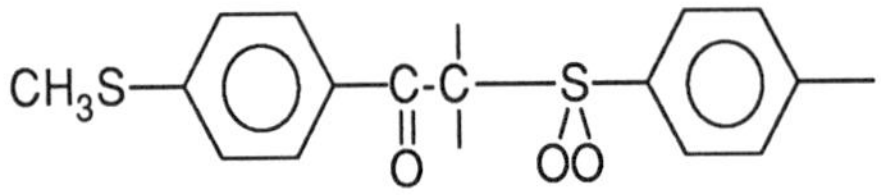   **23**

As expected, photosensitization of the decomposition of **23** in the presence of 2-methyl thioxanthone was recently demonstrated [32]. However, in phenacyl phenyl sulfide **24**, the triplet energy level remains high (73,5 Kcal M [33]) :

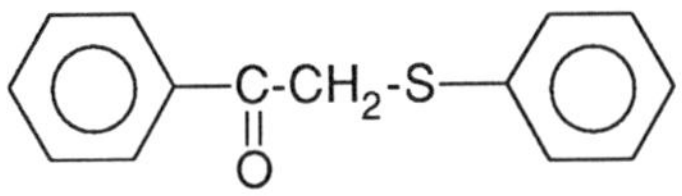

24

whereas in the aryl-aryl sulfide **9**, E_T is presumably low since sensitization occurs (Table 2).

CONCLUSION

A thorough investigation of the interaction between thioxanthones and a large variety of photoinitiators with or without a thioether group as well as a study of the influence of the solvent on the excitation transfer will be the subject of a forthcoming paper. Here and now, it can be stated that synergistic effects which are generally invoked to account for the specific features exhibited by mixtures of photoinitiators should receive a scientific description in terms of a simultaneous energy and electron transfer, the balance of which depends on the system used.

Acknowledgments
AKZO, Ciba Geigy, Fratelli Lamberti, Merck and Ward Blenkinsoap are greatly acknowledged for the gift of samples : 16 ; 5, 6, 7, 13, 18,TBM ; 10, 22, 23 ; 8, 12 ; 9, 11 respectively.

REFERENCES
[1] C.G. Roffey, Photopolymerization of Surface Coatings, J. Wiley Sons, New York, 1982.
[2] S.P. Pappas, UV Curing : Science and Technology, Techn.Mark.Corp., Stamford, 1978.
[3] J.F. Rabek, Mechanisms of Photophysical and Photochemical Reactions in Polymer : Theory and Practical Applications, Wiley, New York, 1987.
[4] J.V. Crivello, J.L. Lee, Polymer J., 17, 73 (1985).
[5] H.J. Hageman, Progress in Org. Coatings, 13, 123 (1985).
[6] S.P. Pappas, Progress in Org. Coatings; 13, 35 (1985).
[7] W. Schnabel, in Applications of Lasers in Polymer Science and Technology, J.P. Fouassier and J.F. Rabek Eds., CRC Press, Bota Raton, USA, in press.
[8] J.P. Fouassier, in Photopolymerization Science and Technology, N.S. Allen Ed., Elsevier Publ., London, in press.
[9] H.J. Timpe, Sitzungsberichte der Akademie des Wissenschaften der DDR, Akademie-Verlag Berlin, 13/N, 1986.
[10] W. Rutsch, G. Berner, R. Kirchmayer, R. Husler, G. Rist, in Organic Coatings : Science and Technology, G.D. Porfitt, A.V. Patsis Eds., 8, M. Dekker, New York, 1986.
[11] D.J. Lougnot, C. Turck, J.P. Fouassier, Macromolecules, 22, 108 (1989).
[12] D.J. Lougnot, J.P. Fouassier, in Applications of Lasers in Polymer Science and Technology, J.P. Fouassier and J.F. Rabek Eds., CRC Press, Boca Raton, USA, in press.
[13] J.P. Fouassier, D.J. Lougnot, J. Polym. Sci. : Part A : Polym.Chem., 26, 1021 (1988).
[14] J.P. Fouassier, Makromol. Chem., Macromol.Symp., 18, 157 (1988).
[15] J.P. Fouassier, P. Jacques, M.V. Encinas, Chem.Phys.Lett., 148, 309 (1988).
[16] J.P. Fouassier, D.J. Lougnot, J.Chem.Soc., Faraday Trans.1, 83, 2935 (1987).
[17] J.P. Fouassier, D.J. Lougnot, J. Appl. Polym. Sci., 34, 477 (1987).
[18] J.P. Fouassier, D.J. Lougnot, I. Zuchowicz, P.N. Green, H.J. Timpe, K.P. Kronfeld, U. Muller, J. Photochem., 36, 347 (1987).
[19] P. Jacques, D.J. Lougnot, J.P. Fouassier, J.C. Scaiano, Chem. Phys. Lett., 129, 205 (1986).
[20] J.P. Fouassier, P. Jacques, D.J. Lougnot, T. Pilot, Polym.Photochem., 5, 57 (1984).
[21] J.P. Fouassier, D.J. Lougnot, A. Payerne, F. Wieder, Chem.Phys.Lett., 135, 30 (1987).
[22] V.D. Mc Ginniss, T. Provder, C. Kuo, A. Gallopo, Macromolecules, 11, 405 (1978).
[23] S.P. Pappas, Rad. Phys. Chem., 25, 633 (1985).
[24] W. Rutsch, G. Berner, R. Kirchmayer, R. Husler, G. Rist, N. Buhler, Proc. Radcure Basel, 1985.
[25] K. Dietliker, M.W. Rembold, G. Rist, W. Rutsch, F. Sltek, Proc. Radcure Munich, 1987.
[26] J.P. Fouassier, D. Burr, to be published.
[27] J.P. Fouassier, unpublished data.
[28] J.P. Fouassier, to be published.
[29] M. Koehler, J. Ohnggemach, Proc. Radcure Munich, 1987.
[30] J.P. Fouassier, D.J. Lougnot, J.C. Scaiano, Chem. Phys. Lett., to be published.
[31] J.P. Fouassier, in Focus on Photophysics and Photochemistry, J.F. Rabek Ed., CRC Press, in press.
[32] G. Li Bassi, L. Cadona, F. Broggi, Proc. Radcure Baltimore, 1986.
[33] P.J. Wagner, M.J. Lindstrom, J.Amer.Chem.Soc. 109, 3062 (1987).

RECEIVED October 1, 1989

Chapter 6

Photochemistry and Photopolymerization Activity of Water-Soluble Benzophenone Initiators

Norman S. Allen[1], F. Catalina[2], J. L. Mateo[2], R. Sastre[2], W. Chen[3], P. N. Green[4], and W. A. Green[4]

[1]Department of Chemistry, Faculty of Science and Engineering, John Dalton Building, Manchester Polytechnic, Chester Street, Manchester M1 5GD, United Kingdom
[2]Instituto Plasticos y Caucho, CSIC, Juan de la Cierva 3, 28006 Madrid, Spain
[3]Institute of Photographic Sciences, Academia Sinica, Beijing, China
[4]Ward-Blenkinsop and Company Ltd., Halebank Industrial Estate, Lower Road, Widnes, Cheshire WA8 8NS, United Kingdom

> Quantitative photophysical, photochemical and photo-
> polymerisation data are presented on five novel water
> soluble benzophenone photoinitiators. Phosphorescence
> quantum yields, triplet lifetimes and transient formation
> on conventional flash photolysis correlate with the
> ability of the initiators to photoinduce the polymer-
> isation of 2-hydroxyethylmethacrylate and a commercial
> monoacrylate resin in aqueous media. The results
> indicate that the lowest excited triplet state of
> the initiator is abstracting an electron from the
> tertiary amine cosynergist probably via a triplet
> exciplex followed by hydrogen atom abstraction. This
> is confirmed by a detailed analysis on the effect of
> oxygen, pH and the ionisation potential of the amine
> on transient formation and photopolymerisation. Using
> photocalorimetry a linear correlation is found between
> the photopolymerisation quantum yields of the initiators
> and their photoreduction quantum yields in aqueous media.
> One of the photoinitiators, 4-methylsulphoniumbenzo-
> phenone sodium salt, exhibited a photoreduction quantum
> yield of 1.5 which is associated with secondary processes
> due to the reaction of the alkylamino radicals with
> a ground-state ketone molecule.

In previous papers we reported on the photochemistry and photopolymerisation activity of a series of water- and oil-soluble thioxanthone
derivatives (1-6). In that work, all the thioxanthone structures
were found to operate essentially by a primary photochemical process
of electron abstraction via an exciplex with an amine co-synergist.
However, whereas the oil soluble structures were found to operate
solely through their lowest excited triplet state (1,2,6) the water-

soluble structures appeared to operate via both their lowest excited singlet and triplet states (3-5). Electron transfer between a synergistic mixture of a thioxanthone photoinitiator and an amine has recently been investigated using laser flash spectroscopy and been found to be highly dependent on the polarity of the environment (7). In this paper we have examined the photopolymerisation and photopolymerisation activity of five water soluble para-substituted benzophenone derivatives of the structures 1-5 shown in Table I. Luminescence, photopolymerisation (using photocalorimetry) and photoreduction quantum yields have been measured for these initiators as well as transient formation on conventional flash photolysis in an attempt to understand the nature of the excited state involved in the photoinitiation process and the subsequent mechanistic features of the polymerisation. The benzophenone chromophore is a strong triplet sensitiser and the activity of these derivatives in water makes an interesting comparison, under the same experimental conditions, with our earlier work on the thioxanthones where the nature of the substituents were similar. Previous work on a water-soluble benzophenone sulphonium salt in inducing the photopolymerisation of acrylamide has shown that photoreduction occurs primarily through the lowest excited triplet state of the initiator (8-10). Our work here with 2-hydroxyethylmethacrylate and a commercial water-soluble monoacrylate resin not only supports this conclusion but also shows that in aqueous media the presence of an amine co-synergist is important for effective photoreduction and photopolymerisation. In fact one of the photoinitiators examined here is highly photoactive with a measured photoreduction quantum yield greater than unity.

Experimental

Materials. Samples of the benzophenone salts of the structures (1-5) (Table I) were supplied by Ward-Blenkinsop and Company Limited, Widnes, Cheshire, UK and patent cover is pending. The chemical names of the salts are as follows : (1) 4-(3-N,N,N-trimethylammoniumpropoxy)benzophenone methylsulphonium salt; (2) 4-(2-hydroxy-3-N,N,N-trimethylammoniumpropoxy)benzophenone chloro salt; (3) sodium (4-benzoyl)phenylmethylsulphonate; (4) 4-(3-sulphoniumpropoxy)benzophenone sodium salt and (5) 4-(2-hydroxy-3-sulphoniumpropoxy)benzophenone sodium salt.

The water used in this study was double distilled and the amines (N-methyldiethanolamine, tri-n-butylamine, triethylamine, diethylamine, N-diethylmethylamine, dicyclohexylamine and N,N'-dimethylaniline) were obtained from the Aldrich Chemical Company Limited, UK as was the monomer 2-hydroxyethylmethacrylate. The monoacrylate water-soluble resin RCP-1784 was supplied by Lankro Chemicals Limited, Eccles, manchester, UK and used as received. The 2-propanol (Analar grade) was obtained from Fisons Limited, UK.

Spectroscopic Measurements. Absorption spectra were obtained using a Perkin-Elmer Model 554 Spectrophotometer and phosphorescence spectra and mean lifetimes were obtained at 77 K using a Perkin-Elmer LS-5 Luminescence Spectrometer coupled to a 3600 data station. Phosphorescence quantum yields were obtained by the relative method using benzophenone (ϕ_p = 0.74 in ethanol glass at 77 K) as a standard (11).

Flash Photolysis. Transient absorption spectra were recorded using a microsecond kinetic flash photolysis apparatus equipped with two 16 kV xenon-filled flash lamps (operated at 10 kV) and a 150 W tungsten-halogen monitoring light source. Transient decay profiles were stored using a Tetronix DM6 storage oscilloscope. Solutions were deoxygenated using white-spot nitrogen gas (below 5 ppm oxygen).

Photoreduction Quantum Yields. Absolute quantum yields of photoreduction ($\emptyset r$) for the initiators were determined in water at 10^{-5} M in the presence of 2-N,N-diethanolamine at a concentration of 10^{-2} M using an irradiation wavelength of 365 nm selected from a Philips high pressure mercury lamp (HB-CS 500 W) and a Kratos GM252 monochromator. Sample cells were thermostatted at 30°C and the solutions were deoxygenated using nitrogen gas (5 ppm oxygen). The absorbed light intensity was measured using an International Light Model 700 radiometer previously calibrated by Aberchrome 540 actinometry (12,13) Photolysis of the benzophenones was monitored by measuring the change in UV absorption of the main π-π^* maximum at 290 nm.

Photopolymerisation. The kinetics of photoinitiated polymerisation of the monoacrylate resin were monitored using a differential scanning photocalorimeter developed at the CSIC on the basis of a modified Perkin-Elmer DSC-4 calorimeter (14). The monoacrylate resin was mixed with 20% v/w of water followed by the addition of 10% v/w of N,N-dimethylaniline. The latter amine was chosen for study here owing to practical limitations with this technique for the resin used. The water-soluble benzophenones were dissolved in the resin to give an absorbance of 0.3 at 365 nm using a cell path length of 1.5 mm.

The light intensity was measured by monitoring the temperature increase due to light absorption by a thin graphite coating in a standard aluminium pan placed in the sample holder of the equipment. Prior to irradiation samples were allowed to equilibrate to the operating temperature of 40°C for 10 minutes. The DSC traces were output to a Perkin-Elmer Model 3600 data station and recorded on a Hewlett-Packard 7470A graphics plotter. The kinetics were cross-checked gravimetrically after irradiation and good agreement was found.

Photopolymerisation efficiencies were also determined gravimetrically by irradiating a quartz cell containing 10 cm³ of a 50:50 v/w mixture of 2-hydroxyethylmethacrylate-water, 0.3% w/v of initiator (2) and 1.5% w/v of amine for 20 minutes. The light source used was a Thorn high pressure Hg-W fluorescent lamp (150 W) set at a distance of 10 cm and the reaction mixture was deoxygenated by continually bubbling with white spot nitrogen gas. After irradiation, the mixture was poured into ethanol for precipitation, and this was followed by centrifuging, filtration and drying to constant weight in a hot air over at 60°C.

Results and Discussion

The photoreduction quantum yields for the five benzophenone photoinitiators are compared in Table I together with their respective mean triplet lifetimes.

Table I. Extinction Coefficients (E), Triplet Lifetimes(T) and Photoreduction Quantum Yields (Ør) of the Water Soluble Benzophenones

Reference	4-Substituent	E365 (M^{-1} cm^{-1})	T (ms)	Ør
1	$-OCH_2CH_2CH_2N^+(CH_3)_3SO_3-CH_3$	38.46	17.4	0.60
2	$-OCH_2CH(OH)CH_2N^+(CH_3)_3CI^-$	38.78	12.4	0.40
3	$-CH_2SO_3^-Na^+$	25.65	6.4	1.50
4	$-OCH_2CH(OH)CH_2SO_3^-Na^+$	48.68	8.8	0.93
5	$-OCH_2CH_2CH_2SO_3^-Na^+$	55.78	11.6	0.73

These data show two interesting features. Firstly, all the photoreduction quantum yields are high, the most significant being a value of 1.5 for structure 3. This interesting behaviour has been associated with the following secondary reaction paths between a ground-state ketone molecule and an alkylamino radical which follows the primary process of hydrogen atom abstraction from the amine by the photoexcited triplet state of the carbonyl groups (15,16).

$$p\text{-}R\text{-}Ar\text{-}CO\text{-}Ar + {}^\cdot C(CH_3)HNR_2 \begin{array}{l} \xrightarrow{|1|} p\text{-}R\text{-}Ar\text{-}CO^-\text{-}Ar + CH(CH_3)=N^+R_2 \\ \\ \xrightarrow{|2|} p\text{-}R\text{-}Ar\overset{\cdot}{\underset{|}{C}}\text{-}Ar \\ O\text{-}CH(CH_3)NR_2 \end{array} \qquad (1)$$

The first step (1) involves electron abstraction from the alkylamino radical by the ground state ketone to give a carbonyl radical anion and an immonium ion while the second step (2) involves the formation of an addition product.

The second interesting feature of the results is the clear distinction between the cationic and anionic type photoinitiators. The latter structures exhibit greater activity and one possible explanation, is that the cationic structures exhibit a much higher degree of intermolecular and intramolecular self-quenching owing to the presence of the quaternary amine salt. This effect, although reasonable, does not appear to be reflected by their longer triplet lifetimes.

Photopolymerisation Activity. From the DSC curves the quantitative data in Table II were obtained for each of the initiators. The polymerisation rates Rp were calculated from the initial slopes of the DSC curves using a least-squares approach. These data were then converted into actual photopolymerisation quantum yields (Øm) by dividing by the real absorbed light intensity of each sample (Ia), the values of Øm are expressed as the number of grammes of monomer polymerised by 1 mole of absorbed photons. In agreement with the photoreduction quantum yields, it is seen that the cationic structures are much less active than the anionic types, although this disagrees with recent work on acrylamide photopolymerisation (17). The latter effect may be due to the nature of the monomer. From the data in Table II it is seen that there is a correlation between the photoreduction and photopolymerisation quantum yields. This confirms our earlier conclusions with regard to the importance of a phtoreduction process in photopolymerisation involving an amine co-synergist. The structure of the benzophenone appears to be very important in controlling this relationship.

Table II. Quantitative Photocalorimetric Data for the Benzophenone Initiators

Reference	Time at Peak	R_p (g L^{-1} s^{-1})	I_a (10^5 einsteins L^{-1} s^{-1})	$\emptyset_m$
1	1.42	0.64	3.23	19700
2	1.52	0.74	3.78	19520
3	1.49	0.87	3.55	24463
4	1.67	0.77	3.52	21886
5	1.65	0.77	3.78	20372

The results in Table III show the effect of various amines of different ionisation potential on the photopolymerisation of 2-hydroxyethylmethacrylate in nitrogen saturated water initiated by the benzophenone with the structure 2. As found earlier for the water soluble thioxanthones (6), the percentage photoconversion decreases with increasing ionisation potential of the amine. In all of these experiments, oxygen had a strong quenching effect on the photopolymerisation. It would appear, therefore, that in aqueous media the photoinduced polymerisation of the acrylate monomer occurs solely via the lowest excited triplet state of the benzophenone molecule to form an exciplex with the amine co-synergist (Scheme I). This contrasts significantly with our earlier photopolymerisation results on the water-soluble thioxanthones (6) where photoconversion of the same monomer was unaffected by the presence of oxygen, indicating the possibility of photoactivity via the lowest excited singlet state.

Table III. Photopolymerisation of 2-Hydroxyethylmethacrylate in Nitrogen Saturated Water by Initiator 2 in the Presence of Various Amines

Amine	Ionisation Potential (eV)	% Conversion (20 mins)
N-Methyldiethanolamine	7.20	2.31
Tri-n-butylamine	7.40	2.06
Triethylamine	7.85	0.10
Diethylamine	8.40	0.03
Dicyclohexylamine	9.20	0.01

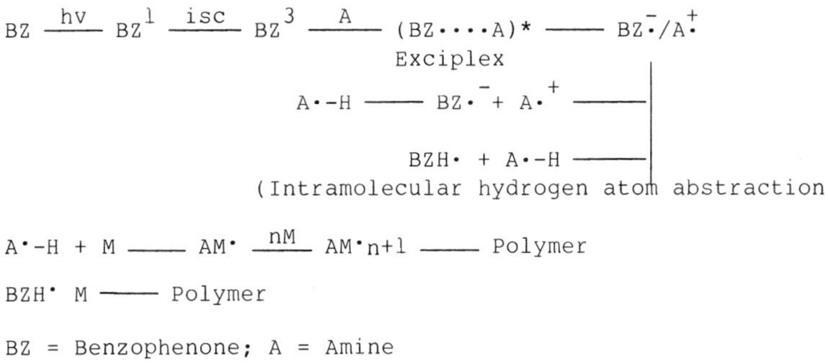

$$BZ \xrightarrow{h\nu} BZ^1 \xrightarrow{isc} BZ^3 \xrightarrow{A} (BZ\cdots A)^* \longrightarrow BZ\bar{\cdot}/A\bar{\cdot}^+$$
$$\text{Exciplex}$$
$$A\cdot\text{-H} \longrightarrow BZ\bar{\cdot} + A\cdot^+ \longrightarrow$$
$$BZH\cdot + A\cdot\text{-H} \longrightarrow$$
(Intramolecular hydrogen atom abstraction

$$A\cdot\text{-H} + M \longrightarrow AM\cdot \xrightarrow{nM} AM\cdot_{n+1} \longrightarrow \text{Polymer}$$

$$BZH\cdot\ M \longrightarrow \text{Polymer}$$

BZ = Benzophenone; A = Amine

Scheme I

Spectroscopic Properties. The phosphorescence quantum yields of the initiators at 77 K in 2-propanol are compared in Table IV. These were obtained upon excitation at the absorption maximum of 290 nm for all the compounds. Both the absorption and phosphorescence spectra of the initiators were similar to those of benzophenone. The quantum yields on the otherhand were variable. The phosphorescence emission spectra of all the compounds had maxima at 414, 443 and 474 nm respectively. No fluorescence was observed which is typical of aromatic ketones indicating a high rate of intersystem crossing. The high rate is further supported by the variable but high phosphorescence quantum yields.

Table IV. Phosphorescence Properties of the Benzophenone Initiators in 2-Propanol

Initiator	Phosphorescence Emission Spectra nm	$\emptyset p$ (secs)
1	414,443,474	0.98
2	"	0.41
3	"	0.79
4	"	0.67
5	"	0.76

In fact, structure 1 exhibits the highest recorded quantum yield
of 0.98 whereas structure 2 exhibits the lowest at 0.41. Comparison
of the phosphorescence quantum yields with the photopolymerisation
quantum yields shows a good correlation. Thus, for the cationic
types structure 1 has a higher phosphorescence quantum yield than
structure 2 and this is consistent with its higher photopolymerisation
activity. The same correlation applies to the three anionic structures. From these results it would appear that a high triplet yield
is consistent with high photoreactivity, although it should be
pointed out that the yields of other competing processes are important
in any overall assessment. However, as discussed above, the presence
of an amine was important for photopolymerisation to occur, which
suggests that the reactivity is more related to the yield of triplet
exciplex. The triplet exciplex will then dissociate as indicated
in Scheme I to give the reactive alkylamino radical.

Flash Photolysis. End-of-pulse transient absorption spectra of
the benzophenone initiators in water and 2-propanol are summarised
in Tables V and VI. the latter shows the effect of the presence
of triethylamine on transient formation. In both solvents, weak
transient formation was observed with a broad maximum between 500
and 600 nm and a much stronger band below 350 nm. Apart from structure 1, the transient is stronger in 2-propanol than it is in water
and is consistent with earlier work on benzophenone where the same
transient was associated with the ketyl radical formed by the lowest
excited triplet state of the benzophenone abstracting a hydrogen
atom from the solvent (18,19). Transient formation due to hydrogen
abstraction from water is unusual and may well be due to intermolecular and/or intramolecular hydrogen atom abstraction from the
molecules themselves. The presence of triethylamine enhances transient formation in both solvents and induces a red shift in the
spectrum owing to the radical anion formed by electron abstraction
via the triplet exciplex shown in Scheme I. The radical anion will
then undergo intermolecular hydrogen atom abstraction from the
amine and enhance ketyl radical formation. The red shift in the
transient spectrum is analogous to that reported earlier for the
benzophenone radical anion (18,19).

Table V. Transient Absorption Maxima of Benzophenone Initiators
(5×10^{-5} M) in the absence and presence of
Triethylamine (10^{-4} M) in Nitrogen Saturated Water

Initiator	No Amine		Amine	
	Wavelength (nm)	Absorbance	Wavelength (nm)	Absorbance
1	520	0.012	640	0.030
2	540	0.010	620	0.015
3	540	0.010	620	0.013
4	560	0.012	540	0.023
5	560	0.006	560	0.016

Table VI. Transient Absorption Maxima of Benzophenone Initiators (5×10^{-5} M) in the absence and presence of Triethylamine (10^{-4} M) in Nitrogen Saturated 2-Propanol

Initiator	No Amine		Amine	
	Wavelength (nm)	Absorbance	Wavelength (nm)	Absorbance
1	535	0.008	560	0.022
2	500	0.015	530	0.028
	560	0.017	590	0.022
3	585	0.018	560	0.050
4	500	0.017	500	0.024
	580	0.028	580	0.027
5	540	0.014	550	0.027

The most interesting feature of the results is the greater enhancement in the transient spectrum in the presence of the amine for the more photoactive initiator (1) compared with that of the less effective initiator (2). Because of variations in absorption maxima, such correlations are difficult but nevertheless provide useful information with regard to the involvement of the triplet exciplex. Further confirmation of the formation of the radical anion is shown by a number of experimental observations. The results in Table VII show the effect of different amines on transient absorption for initiator (1) in 2-propanol. Here transient absorption at 540 nm increased with decreasing ionisation potential of the amine. Finally, the results in Table VIII show the effect of pH on transient absorption in water for initiator (1). Here the pH was varied using dilute HCl and NaOH solutions. In agreement with earlier work on benzophenones (18) and thioxanthones (6), the transient absorption increases and undergoes a significant red shift with increasing pH from 2 to 12, confirming the involvement of the radical anion.

The role of excited state complexes in photopolymerisation has been the subject of some controversy. Although many workers have supported the involvement of a charge-transfer complex between aromatic carbonyl compounds and tertiary amines through which electron transfer and subsequent hydrogen abstraction may take place (20-22), others believe a mechanism involving direct hydrogen abstraction from the amine by the photoexcited triplet state of the carbonyl compound (23). Although the results presented here do not directly implicate a triplet exciplex, they nevertheless appear to indicate involvement of an electron transfer process which strongly supports this hypothesis.

Table VII. Transient Absorption Maxima at 540 nm for Inititiator 1 (10^{-5}M) Compared with the Ionisation Potential of Various Amines (1.48×10^{-5}M) in Nitrogen Saturated 2-Propanol

Amine	Ionisation Potential (eV)	Transient Absorption 540 nm
Tri-n-butylamine	7.40	0.039
Triethylamine	7.85	0.036
N-Diethylmethylamine	8.10	0.035
Diethylamine	8.40	0.034
Dicyclohexylamine	9.20	0.031

Table VIII. Transient Absorption Maximum of Initiator 1 (5×10^{-5}M) at Various pH Values in Nitrogen Saturated Water

pH	Wavelength (nm)	Absorbance
2	540	0.0019
4	560	0.0057
6	560	0.0072
8	560	0.0073
10	600	0.0197
12	620	0.0368

Conclusions

The results clearly show that the photoinitiating activity of the water soluble benzophenone in the presence of an amine is mainly associated with the ability of the lowest excited triplet - * state to abstract an electron from the amine co-synergist via an intermediate exciplex shown in Scheme I. The radical anion will then induce hydrogen atom abstraction to give a ketyl radical and an alkylamino radical. The latter is mainly responsible for inducing polymerisation of the acrylic monomer and supports earlier work on the benzophenone-triethylamine tetramine-induced photopolymerisation of methylmethacrylate during which terminal amine groups were detected (24).

Acknowledgments

The authors thank NATO Scientific Affaires, Brussels, for a travel grant.

Literature Cited

1. Allen, N. S., Catalina, F., Green, P. N., Green, W. A., Eur. Polym. J., 1986, 22, 347.
2. Allen, N. S., Catalina, F., Green, P. N., Green, W. A., Eur. Polym. J., 1985, 21, 841.

3. Allen, N. S., Catalina, F., Moghaddam, B. M., Green, P. N., Green, W. A., Eur. Polym. J., 1986, 22, 691.
4. Allen, N. S., Catalina, F., Green, P. N., and Green, W. A., Eur. Polym. J., 1986, 22, 793.
5. Allen, N. S., Catalina, F., Green, P. N., Green, W. A., Eur. Polym. J., 1986, 22, 871.
6. Allen, N. S., Catalina, F., Green, P. N., Green, W. A., J. Photochem., 1987, 36, 99.
7. Fouassier, J. P., Lougnot, D. J., Paverne, A., Wieder, F., Chem. Phys. Letts., 1987, 135, 30.
8. Fouassier, J. P., Lougnot, D. J., Zuchowiez, I., Eur. Polym. J., 1986, 22, 933.
9. Fouassier, J. P., Lougnot, D. J., Polym. Photochem., 1983, 3, 79.
10. Fouassier, J. P., Lougnot, D. J., Jacques, P., J. Photochem., 1982, 19, 59.
11. Demas, J. N., Crosby, G. A., J. Phys. Chem., 1971, 75, 91.
12. Heller, H. G., Lanagan, J. R., J. Chem. Soc., Perkin Trans. 1., 1981, 341.
13. Heller, H. G., British Patent, 7/1464603.
14. Sastre, R., Catalina, F., Conde, M., Mateo, J. L., Revs. Plast. Mod., in the press.
15. Cohen, S. G., Parola, A., Parson, G. H., Chem. Revs., 1973, 73, 141.
16. Merlin, A., Lougnot, D. L., Fouassier, J. P., Polym. Bull., 1980, 3, 1.
17. Fouassier, J. P., Lougnot, D. L., Zuchowicz, I., Green, P. N., Timpe, H. J., Kronfield, K. P., Muller, U., J. Photochem., 1987, 36, 347.
18. Beckett, A., Porter, G., Trans. Faraday Soc., 1983, 59, 2038.
19. Porter, G., Wilkinson, F., Trans. Faraday Soc., 1961, 57, 1686.
20. Granchak, S. M., Kondratenko, P. A., Dilung, I. I., Toer. Eskp. Khim., 1984, 20, 43.
21. Granchak, S. M., Chermerskaya, Z. F., Dilung, I. I., Vysokomol, Soedin. Ser. A., 1985, 27, 276.
22. Guttenplan, J. B., Cohen, S. G., Tetrahedron Letts., 1972, 2163.
23. Merlin, A., Lougnot, D. J., Fouassier, J. P., Polym. Bull., 1980, 3, 1.
24. Ghosh, P., Bandyopadhyay, A. R., Eur. Polym. J., 1984, 11, 1117.

RECEIVED September 13, 1989

Chapter 7

Cationic Photoinitiators

Rearrangement Reactions from Direct Irradiation of Diarylhalonium Salts

Nigel P. Hacker and John L. Dektar

IBM Almaden Research Center, 650 Harry Road, San Jose, CA 95120–6099

> The solution photochemistry of diaryliodonium, diarylbromonium and diarylchloronium salts has been studied. Direct irradiation of the diphenylhalonium salts yields 2-, 3- and 4-halobiphenyls, halobenzene, benzene, acetanilide, biphenyl and acid. Similarly, irradiation of di-(4-tolyl)-halonium salts gives the respective halobitolyls, 4-halotoluene, 4-methylacetanilide, bitolyls and acid. The halobiphenyls are formed by in-cage fragmentation-recombination reactions, whereas halobenzene and benzene are cage-escape products. Cleavage of the carbon-halogen by both homolytic and heterolytic pathways is implicated by the identification of benzene or toluene and the respective anilides as escape products.

Onium salt photoinitiators are increasingly being used for radiation curing of polymers in electronic applications.[1-3] Irradiation of onium salts in polymeric media generates acid which on further processing can crosslink acid sensitive monomers (e. g. epoxy functional resins),[4] or cleave acid sensitive groups (e. g. poly(p-t-butoxycarbonyloxystyrene)).[5] Previous mechanistic studies on onium salts proposed both homolytic and heterolytic cleavage pathways to account for product formation.[6] For example, irradiation of triphenylsulfonium salts gives diphenylsulfide, biphenyl and substituted benzenes, and homolytic,[7] and heterolytic,[8] cleavage of the carbon-sulfur bond have been proposed from photoproduct analysis. We have recently reported that in addition to diphenylsulfide, rearrangement products result from photolysis of triphenylsulfonium salts.[9] This new rearrangement reaction generates acid and rationalizes the observation, by others, that acid formation exceeds diphenylsulfide formation.[10] We report here the direct photolysis of diarylhalonium salts and formation of 2-, 3- and 4-halobiaryls by in-cage fragmentation-recombination reactions, in addition to the escape products, haloarenes.

Experimental

2- and 3-Bromobiphenyl were obtained from Columbia Organics. 4-iodobiphenyl was obtained from Eastman Organic Chemicals. 2-, 3- and 4-Chlorobiphenyl, and

2-iodobiphenyl were obtained from Lancaster Synthesis. 4-Iodotoluene was obtained from Alfa. Acetone and acetonitrile were American Burdick and Jackson UV grade, and were used as received. All other chemicals were obtained from Aldrich Chemical Co. Diaryliodonium salts and the rearrangement products were prepared as previously reported.[11] The other halonium salts were prepared by the procedure of Olah et. al.[12]

Capillary GLC analysis was performed on a Hewlett-Packard 5890 chromatograph equipped with a Hewlett-Packard 7673A autosampler and a Hewlett-Packard 3396A integrator. The column used in all analyses was a J & W Scientific DB-1 (cross-linked methyl silicone) 0.4 μ by 0.18 mm by 20 m.

0.01 M Acetonitrile solutions were irradiated in a Rayonet reactor (λ = 254 nm), for exploratory studies, or with with a 500 W mercury-xenon lamp focussed through a monochromator (λ = 248 +/- 4 nm), for relative quantum yield studies. Two aliquots were analyzed by GLC before irradiation to account for any partial decomposition (usually found to be less than 0.5%). Three 3.00 mL aliquots were placed in Suprasil cuvettes, sealed with a rubber septum, and purged with argon for 8 min immediately prior to irradiation. After irradiation, the samples were transferred to tubes containing 1.00 mL of hexanes containing a small amount of n-tetradecane as internal standard, and 10.00 mL of 0.5 M NaH_2PO_4. The tubes were stoppered and thoroughly mixed. After standing for 4 hr, the hexane layer was removed and analyzed by capillary GLC. The integrator was calibrated against similar concentrations of authentic samples of the photoproducts, which were treated to a similar work-up as the photolysis solutions.

Results and Discussion

Exploratory photolysis of 0.01M acetonitrile solutions of diphenyliodonium triflate gave 2-, 3- and 4-iodobiphenyls, benzene, acetanilide and biphenyl. The iodine-containing photoproducts have strong absorbances compared with the onium salt (Figure 1), and low conversions are necessary to prevent secondary photolysis. The iodonium salt consumption is small and could not be accurately determined by HPLC analysis. However, it is known that onium salts form a complex with cobalt thiocyanate which absorbs at 624 nm.[13] Iodonium salt consumption could be accurately determined by mixing the reaction mixtures with an aqueous solution of $CoCl_2$ and NH_4CN, and monitoring the disappearance of the cobalt thiocyanate complex (Figure 2).[11] Acid formation was measured by a non-aqueous photometric method using 4-nitrophenoxide indicator.[9] Under these conditions, it was found that the total iodine containing photoproducts were 4.93×10^{-4} M, the acid formed was 5.5×10^{-4} M, and that 5.33×10^{-4} M of the iodonium salt was consumed. This excellent agreement for volatile product formation, acid formation, and onium salt consumption, has also been observed upon irradiation of sulfonium salts.[14] Similarly, irradiation of diphenylbromonium or diphenylchloronium hexafluorophosphates also yields 2-, 3- and 4-bromobiphenyls or 2-, 3- and 4-chlorobiphenyls, benzene, acetanilide, biphenyl and acid (Figure 3). The relative quantum yields for formation of halogen-containing photoproducts from irradiation of halonium salts in acetonitrile solutions are shown in Table 1. The relative quantum yield (Rel ϕ) for product formation from iodonium salts is much lower than from chloronium or bromonium salts. Chloronium and bromonium salts are

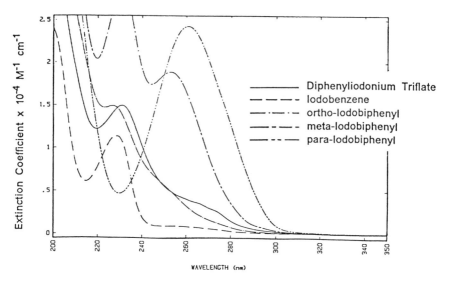

Figure 1: UV Absorption Spectra of Diphenyliodonium Triflate and Photoproducts

$$CoCl_2 + 2\ NH_4SCN \longrightarrow Co(SCN)_2 + 2\ NH_4Cl$$
$$Co(SCN)_2 + Ph_2I^+\ X^- \longrightarrow \{Ph_2I^+\}_2Co(SCN)_2X_2$$

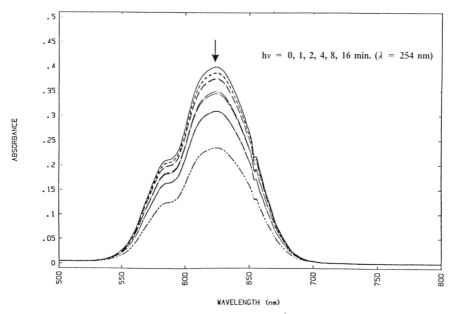

Figure 2: UV Absorption Spectra Monitoring $Ph_2I^+CF_3SO_3^-$ Consumption from Photolysis Mixtures

known to be less thermally stable than iodonium salts in solution.[16] To determine if thermal decomposition was responsible for these differences in Rel ϕ, solutions of these salts subjected to the same treatment as the irradiated samples but kept in the dark. These solutions showed less than 0.5 % decomposition. Also, substantially more rearrangement products are obtained from irradiation of iodonium salts than chloronium or bromonium salts, and the iodonium salts have an enhanced selectivity for ortho rearrangement products.

Product versus time studies reveal that the rate of biphenyl formation increases faster than the rate of formation of halogen containing products which suggests that biphenyl is a secondary photoproduct. However, while biphenyl is only a trace product from irradiation of the bromonium and chloronium salts, significantly more biphenyl is produced from iodonium salt photolysis. This result suggests that some biphenyl may be formed as a primary photoproduct from iodonium salts. The major reaction pathway for the cage-escape phenyl moieties is reaction with solvent. We have previously shown that triplet sensitization of triphenylsulfonium salts gives 100 % escape products and that the cage-escape phenyl radical reacts with solvent to form benzene.[20] The relative yield of benzene (to halobenzene) is 3.5 % for the diphenylchloronium salt, 9 % for the bromonium salt and 20 % for the diphenyliodonium salt. In contrast, cage-escape phenyl cation reacts with the acetonitrile solvent to give acetanilide. Unfortunately, quantification of acetanilide is not precise, leading to errors up to 20%. The relative yield of acetanilide (to halobenzene) is 110 % for the diphenylchloronium salt, 91 % for diphenylbromonium salt, and 78 % for diphenyliodonium salt. However within experimental error, benzene and acetanilide account for all the escape moiety from direct photolysis of diphenylhalonium salts.

The 4,4'-ditolylhalonium salts were studied to give a better understanding of the escape aryl moieties. Photolysis of solutions of 4,4'-ditolyliodonium, 4,4'-ditolylbromonium, or 4,4'-ditolylchloronium hexafluorophosphate gives 2- and 3-iodobitolyl, 2- and 3-bromobitolyl, or 2- and 3-chlorobitolyl, toluene, 4-methylacetanilide, isomeric bitolyls and acid (Figure 4). Two of the isomeric bitolyls, 3,4'-bitolyl and 2,4'-bitolyl, are secondary photoproducts. While the 3,4'-bitolyl is a trace product from halonium salt photolysis, the 2,4'-bitolyl is not detected at low conversions. However prolonged photolysis of the halonium salt does produce detectable amounts of the 2,4'-bitolyl, as does irradiation of 3-halobitolyl. 4,4'-Bitolyl is a primary photoproduct and can be formed from dimerization of 4-tolyl radical, or by ipso attack of 4-tolyl radical with halotoluene radical cation. As with the diphenylhalonium series, significantly more bitolyls are obtained from photolysis of the iodonium salts than the bromonium or chloronium salts. The cage-escape tolyl moieties, 4-tolyl radical and 4-tolyl cation, react with solvent to yield toluene and 4-methylacetanilide respectively. The relative yield of toluene (to 4-halotoluene) is 6.4 % for the 4,4'-ditolylchloronium salt, 11 % for the 4,4'-ditolylbromonium salt and 38 % for the 4,4'ditolyliodonium salt. Similarly, the relative yield of 4-methylacetanilide (to 4-halotoluene) decreases from 4,4'-ditolylchloronium salt to 4,4'-ditolylbromonium salt and to the 4,4'ditolyliodonium salt.

Diphenylbromonium, diphenylchloronium and 4-chlorodiphenylchloronium hexafluorophosphates have previously been evaluated as photoinitiators, but the

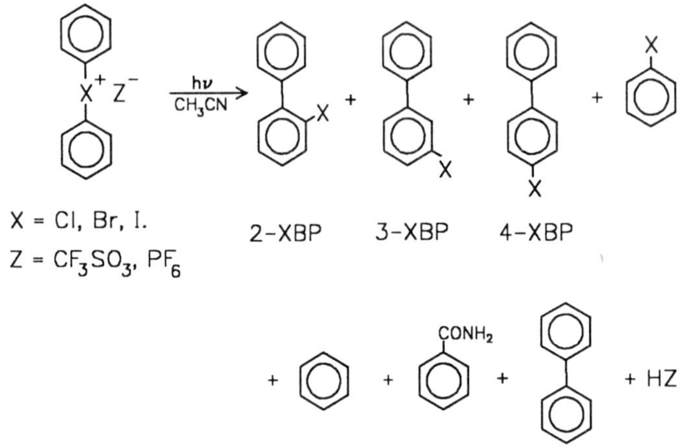
hν
CH_3CN
X = Cl, Br, I.
Z = CF_3SO_3, PF_6
2-XBP
3-XBP
4-XBP
$CONH_2$
+ HZ

Figure 3: Products from Photolysis of Diphenylhalonium Salts

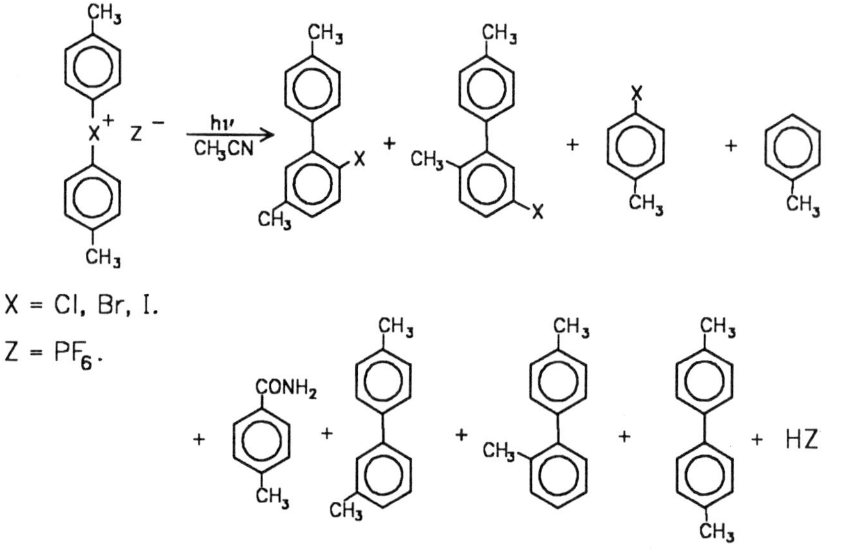
hν
CH_3CN
X = Cl, Br, I.
Z = PF_6.
$CONH_2$
+ HZ

Figure 4: Products from Photolysis of 4,4′Ditolylhalonium Salts

Table 1. Product Quantum Yields upon Direct Irradiation of 0.01 M Salt Solutions in Acetonitrile, $\lambda = 248$ nm (Ph = Phenyl, 4-Tol = 4-Tolyl)

Compound	Rel ϕ	ArX	2-XBP	3-XBP	4-XBP
Ph_2Cl^+ PF_6^-	0.96	84	8	5	3
Ph_2Br^+ PF_6^-	1.00	81	9	6	4
Ph_2I^+ $CF_3SO_3^-$	0.66	74	19	3	4
$4\text{-}Tol_2Cl^+$ PF_6^-	0.89	85	5	10	-
$4\text{-}Tol_2Br^+$ PF_6^-	0.90	82	11	6	-
$4\text{-}Tol_2I^+$ PF_6^-	0.55	71	23	6	-

rearrangement reaction was not observed.[16] Earlier studies on diaryliodonium salt photolyses have reported recombination products. Irradiation of 4,4'-di-*tert*-butyldiphenyliodonium tetrafluoroborate [17] and and diphenyliodonium hexafluoroarsenate [18] gave only traces of the respective iodobiaryls. However, a more recent study on diphenyliodonium hexafluorophosphate, reported an 18 % yield of 4-iodobiphenyl and 2-iodobiphenyl in a 80:20 ratio.[19] Our results here indicate that direct photolysis of diaryliodonium salts gives iodobiaryls in 20 - 30 % yield. In the case of the diphenyliodonium salt we observe all three iodobiphenyl isomers in a ratio of 75:25 for the 2-isomer to the 3- and 4-isomers. This is similar to the results from our studies on triphenylsulfonium, triphenylselenium and triphenyltelluronium salts, where the 2-substituted biphenyl is always the major isomer.[9, 14] We also see the rearrangement products, the isomeric halobiaryls, in 15 - 20 % yield from irradiation of the diarylbromonium and diarylchloronium salts. Our results here indicate that direct photolysis of diarylhalonium salts is similar to the photochemstry of triarylsulfonium salts. Acid is produced from halobiaryl formation by rearrangement of the halonium salt to give the cyclohexadienyl cation. Aromatization of the cyclohexadienyl cation gives halobiaryl and acid. The cyclohexadienyl cation can be formed by in-cage recombination from either of the aryl cation - haloarene or aryl radical - haloarene radical cation pairs (Figure 5). These pairs of intermediates can also form acid by the previously reported cage-escape reactions.[1-3, 6, 17, 18]

The escape arene fragments represent evidence for both homolytic and heterolytic cleavage of the carbon-halogen bond. The homolytic pathway gives phenyl radical and halobenzene radical cation from irradiation of the halonium salt. Phenyl radical reacts with solvent to give benzene, whereas dimerization to biphenyl is a minor process. Similarly, toluene is formed from reaction of 4-tolyl radical, the escape arene fragment from homolysis of the 4,4'-ditolylhalonium salts, with solvent. The heterolytic pathway gives aryl cation and haloarene from irradiation of the halonium salt. The aryl cation reacts with solvent to give the respective anilide (acetanilide or 4-methylacetanilide). An anilide were previously reported as a trace photoproduct from irradiation of 4,4'-di-*tert*-butyldiphenyliodonium tetrafluoroborate and a minor photosolvolysis pathway was proposed for it's formation.[17] Our results indicate that anilides are major primary photoproducts from direct irradiation of diarylhalonium salts and that solvolysis occurs via the phenyl cation. The homolytic cleavage to give aryl radical accounts for a maximum of 40 % of the observed escape products from photolysis of any of the six diarylhalonium salts studied.

Table 2. Product Distribution from Direct Irradiation of 0.01 M Sulfonium Salt Solutions, λ = 248 nm (Ph = Phenyl, 4-Tol = 4-Tolyl)

Compound	Solvent	% Ar$_2$S	% Rearr.	% ArH	% ArR
Ph$_3$S$^+$ CF$_3$SO$_3^-$	CH$_3$CN	40	60	--	70
Ph$_3$S$^+$ CF$_3$SO$_3^-$	CH$_3$OH	38	62	--	77
4-Tol$_3$S$^+$ CF$_3$SO$_3^-$	CH$_3$CN	32	67	52	72
4-Tol$_3$S$^+$ CF$_3$SO$_3^-$	CH$_3$OH	32	68	40	45

Triplet sensitization of sulfonium salts proceeds exclusively by the homolytic pathway, and that the only arene escape product is benzene, not biphenyl or acetanilide.[20] However, it is difficult to differentiate between the homolytic or heterolytic pathways for the cage reaction, formation of the isomeric halobiaryls. Our recent studies on photoinduced electron transfer reactions between naphthalene and sulfonium salts, have shown that no *meta-* rearrangement product product is obtained from the reaction of phenyl radical with diphenylsulfinyl radical cation.[21] Similarly, it is expected that the 2- and 4-halobiaryl should be the preferred products from the homolytic fragments, the arene radical-haloarene radical cation pair. The heterolytic pathway generates the arene cation-haloarene pair, which should react less selectively and form the 3-halobiaryl, in addition to the other two isomers. The increased selectivity of 2-halobiaryl over 3-halobiaryl formation from photolysis of the diaryliodonium salts versus the bromonium or chloronium salts, suggests that homolytic cleavage is more favored for iodonium salts than bromonium or chloronium salts. This is also consistent with the observation that more of the escape aryl fragment is radical derived for diaryliodonium salts than for the other diarylhalonium salts.

There are some major differences between the photochemistry of diarylhalonium salts and triarylsulfonium salts (Table 2). Considerably more rearrangement products are detected from irradiation of sulfonium salts. The rearrangement products account for 60 - 70 % of the reaction of sulfonium salts whereas rearrangement is only 10 - 25 % of the halonium salt reaction. The intermediates produced from heterolysis of the onium salt are halobenzene or diphenylsulfide and phenyl cation. The reaction between phenyl cation and the respective arene to give the rearrangement product can be considered an electrophilic substitution. It is well known that diphenylsulfide is much more reactive towards electrophiles than halobenzene, and so should be more likely give recombination products. The escape phenyl moiety gives products mainly derived from phenyl cation from direct irradiation of the halonium salts salts, whereas phenyl cation accounts for only 50 - 70 % of the escape phenyl moiety from sulfonium salts. This suggests that the phenyl radical pair of intermediates is more stable than the phenyl cation pair for sulfonium salts than for halonium salts. The phenyl cation - halobenzene pair can convert to the phenyl radical pair by an electron transfer (Figure 6). The observation that products derived from both pairs of intermediates are detected, suggests that there may be interconversion between the phenyl cation pair and phenyl radical pair. From the oxidation potentials of the intermediates, phenyl radical, diphenylsulfide, the relative stability of the phenyl cation - diphenylsulfide pair to the phenyl radical - diphenylsulfinyl radical cation pair is estimated to be ΔE = -16 kcal / mole with

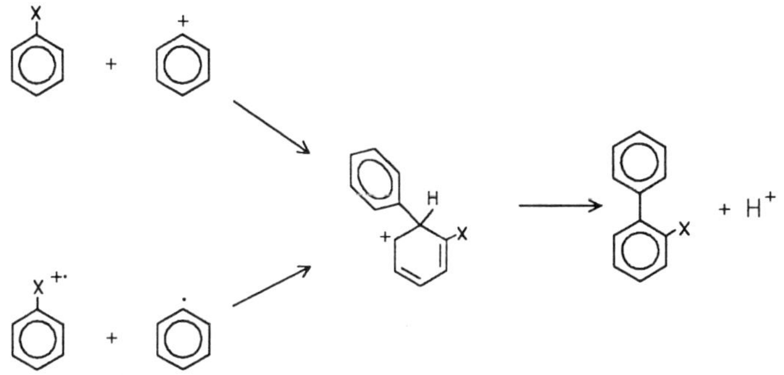

Figure 5: Acid Generation from Formation of Rearrangement Products

$$Ph_2X^+ Z^- \xrightarrow{h\nu} \left[Ph_2X^+ Z^-\right]^* \longrightarrow \overline{PhX \quad Ph^+ \quad Z^-}$$

$$\overline{PhX \quad Ph^+ \quad Z^-} \xrightleftharpoons{e^- \text{ transfer}} \overline{PhX^{+\cdot} \quad Ph^\cdot \quad Z^-}$$

$$\overline{PhX \quad Ph^+ \quad Z^-} \xrightarrow{RH} Ph-PhX + PhX + PhR + HZ$$

$$\overline{PhX^{+\cdot} \quad Ph^\cdot \quad Z^-} \xrightarrow{RH} Ph-PhX + PhX + PhH + HZ$$

Figure 6: Mechanism for Photodecomposition of Diarylhalonium Salts

the phenyl radical - diphenylsulfinyl radical cation pair more stable. From similar estimates for the halogen series, the phenyl radical - halobenzene radical pair is only 0 - 6 kcal / mole more stable than the phenyl cation - halobenzene pair. Thus it is more likely that the phenyl cation pair of intermediates will participate in the photochemistry of diarylhalonium salts than triarylsulfonium salts.

Conclusions

Direct irradiation of diarylhalonium salts results in formation of 2-, 3- and 4-halobiaryls by an in-cage fragmentation-recombination reaction, in addition to the escape product, haloarene. The cage and escape products are formed from both homolytic and heterolytic fragmentation of the carbon-halogen bond, with the heterolytic process being more dominant. Iodonium salts give more in-cage products and more homolytic cleavage products than bromonium or chloronium salts. The Rel ϕ is lower for iodonium salt photolysis than bromonium or chloronium photodecomposition.

References

1. (a) J. V. Crivello, "UV Curing: Science and Technology", Ed. S. P. Pappas, Technology Marketing Corporation, Stamford, 1978, p. 23.
   (b) J. V. Crivello, CHEMTECH, 1980, **10**, 624.
   (c) J. V. Crivello, Polym. Eng. Sci., 1983, **23**, 953.
   (d) J. V. Crivello, Adv. Polym. Sci., 1984, **62**, 1.
   (e) J. V. Crivello, Makromol. Chem., Macromol. Symp., 1988, **13/14**, 145.
2. Y. Yagci and W. Schabel, Makromol. Chem., Macromol. Symp., 1988, **13/14**, 161.
3. (a) S. P. Pappas, Prog. Org. Coat., 1985, **13**, 35.
   (b) S. P. Pappas, J. Imag. Tech., 1985, **11**, 146.
4. J. V. Crivello, J. H. W. Lam and N. C. Volante, J. Radiat. Curing, 1977, **4**, 2.
5. H. Ito and C. G. Willson, ACS Symposium Series No. 242, "Polymers in Electronics", Ed. T. Davidson, American Chemical Society, Washington D. C., 1984, p. 11.
6. (a) J. W. Knapzyck and W. E. McEwen, J. Org Chem., 1970, **35**, 2539.
   (b) J. W. Knapzyck, J. J. Lubinkowski and W. E. McEwen, Tetrahedron Lett., 1971, 3739.
7. J. V. Crivello and J. H. W. Lam, J. Polym. Sci., Polym. Chem. Ed., 1979, **17**, 977.
8. R. S. Davidson and J. W. Goodin, Eur. Polym. J., 1982, **18**, 589.
9. (a) J. L. Dektar and N. P. Hacker, J. C. S. Chem. Comm., 1987, 1591.
   (b) N. P. Hacker and J. L. Dektar, Polym. Prepr., 1988 **29**, 1591.
10. S. P. Pappas, B. C. Pappas, L. R. Gatechair, J. H. Jilek and W. Schnabel, Polym. Photochem., 1984, **5**, 1.
11. J. L. Dektar and N. P. Hacker, J. Org. Chem., submitted.
12. G. A. Olah, T. Sakakibara and G. Asenio, J. Org. Chem., 1978, **43**, 463.
13. H. A. Potratz and J. M. Rosen, Anal. Chem., 1949, **21**, 1276.
14. J. L. Dektar and N. P. Hacker, unpublished results.
15. H. Irving and R. W. Reid, J. Chem. Soc., 1960, 2078.
16. J. V. Crivello and J. H. W. Lam, J. Polym. Sci., Polym. Lett. Ed., 1978, **16**, 563.
17. J. V. Crivello and J. H. W. Lam, Macromolecules, 1977, **10**, 1307.

18. S. P. Pappas, B. C. Pappas, L. R. Gatechair, J. H. Jilek and W. Schnabel, *J. Polym. Sci., Polym. Chem. Ed.*, 1984, **22**, 69.
19. (a) R. J. DeVoe, M. R. V. Sahyun, N. Serpone and D. K. Sharma, *Can. J. Chem.*, 1987, **65**, 2342.
    (b) R. J. DeVoe, M. R. V. Sahyun, E. Schmidt, N. Serpone and D. K. Sharma, *Can. J. Chem.*, 1988, **66**, 319
20. J. L. Dektar and N. P. Hacker, *J. Org. Chem.*, 1988, **53**, 1833.
21. J. L. Dektar and N. P. Hacker, *J. Photochem. Photobiol., A. Chem.*, 1989, **46**, 233.

RECEIVED September 13, 1989

Chapter 8

Novel Photoinitiator for Modern Technology

V. Desobry, K. Dietliker, R. Hüsler, L. Misev, M. Rembold, G. Rist, and W. Rutsch

Additive Research, Ciba—Geigy Ltd., CH—1701 Fribourg, Switzerland

>   2-Benzyl-2-dimethylamino-1-(4-morpholinophenyl)-butanone-1 (BDMB) has been synthesized and shown to be an efficient photoinitiator for UV curing applications. Photochemical and CIDNP investigations suggest that photochemical decomposition occurs mainly via α-cleavage. In comparison with other photoinitiators, BDMB provided superior results in pigmented systems and imaging applications.

Advances in the field of UV-curing necessarily engender the development of specialized photoinitiators that meet the specific needs of new applications. Whereas the prepolymer determines many physical characteristics of the cured film such as gloss, hardness, solvent and scratch resistance, etc., the photoinitiator must ensure the proper curing of the film. The selection of the photoinitiator is especially important when absorbing species such as pigments or stabilizers are added to the prepolymer.

Our research in this area has focused upon the tailoring of the photoinitiator not only to the formulation but also to the currently available light sources. In a preceeding publication [1], we discussed the specific applications of three structurally distinct photoinitiators. We now present a new alpha-cleavage photoinitiator 2-benzyl-2-dimethylamino-1-(4-morpholinophenyl)-butanone-1 (BDMB) which promises great utility in various branches of the graphic arts and printing technology.

Synthesis and Properties of BDMB

Synthesis. The synthesis of BDMB and its analogs requires an efficient strategy for the construction of the amino-substituted quaternary center. α,α-Disubstituted aminoketones cannot be prepared by the simple reaction of amines with the corresponding α-bromoketones. While α,α-dimethyl derivatives are easily obtained by the addition of morpholine to intermediate epoxyether [2] [3], this transformation fails when the epoxyether is substituted with sterically demanding alkyl groups. To circumvent this problem, we employed the intramolecular Stevens rearrangement [4] [5] to create the quaternary center, thereby gaining access to a great variety of highly substituted α-aminoketones (see Figure 1) [6].

Absorption Characteristics. Clearly, only compounds having strong absorptions in the emission range of the light source can serve as efficient photoinitiators. This allows the direct excitation of the photoinitiator to an excited state where its efficient conversion to reactive species (radicals in the case of alpha-cleavage type photoinitiators) is essential. The absorption spectra of two commercial photoinitiators BDK (I) and MMMP (II) [7] as well as BDMB are reproduced in Figure 2. It is apparent that BDMB, which exhibits a strong absorption at 322 nm., 16 nanometers higher than the structurally related II, is the photoinitiator which best matches the emission lines of the medium pressure mercury lamp. These absorption characteristics also allow its application in the curing of pigmented systems as well as in resist formulations and flexographic printing plates.

Photochemistry

The photochemistry of α-amino acetophenone derivatives has been shown by various groups to be strongly dependent upon the α-carbon and nitrogen substituents. Unsubstituted α-(dialkylamino)-acetophenones ($-CO-CH_2-N(alkyl)_2$) undergo an efficient elimination reaction upon irradiation (Figure 3a) to afford acetophenone and imines as the sole products [8] [9] [10].
However, when the lone pair on the amine is a part of a π-system (N-Acyl [8] [11] [12], N-Tosyl [12] or N-Phenyl [13] [14]) 3-azetidinols are obtained, via cyclization of an intermediate 1,4-diradical, unless steric factors prevent the formation of the four-membered ring (Figure 3b).
These results, and the observation that the photoelimination of α-(dialkylamino)acetophenones is not suppressed by the standard triplet quenchers can be explained by a mechanism involving electron transfer from the amine to the carbonyl group [8] (Figure 3a). Lowering the ionization potential of the amine, i.e. by acetylation, sulfonylation, etc., disfavors the electron transfer pathway and results in γ-hydrogen abstraction.
Irradiation of N-phenylacetophenones (Figure 3c) furnishes products resulting from direct β-cleavage [13] [14]. Geminally disubstituted derivatives (R' = CH_3) undergo photolytic decomposition via both α- and β-cleavage pathways [13]. Predominant α-cleavage is observed upon photolysis of II [3] [15]. This trend in reactivity can be explained by the interplay of two factors:

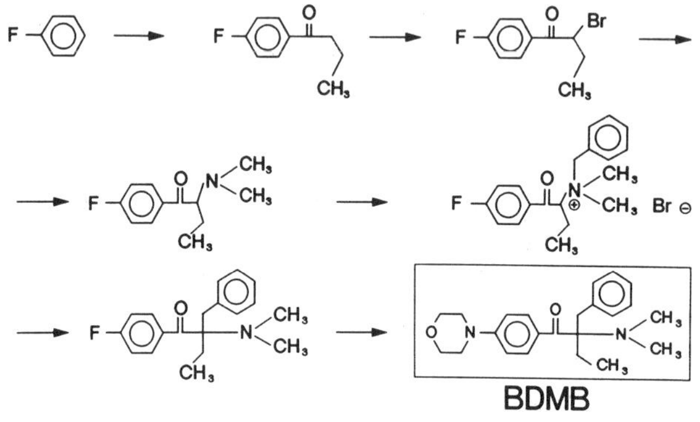

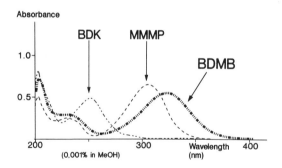

Figure 1. Synthesis of BDMB.

Figure 2. Absorption spectra of benzildimethylketal (BDK), 4-methyl-thiophenyl-2-morpholino-2-methyl-propanone-1 (MMMP) and BDMB (concentration: 0,001 % in methanol).

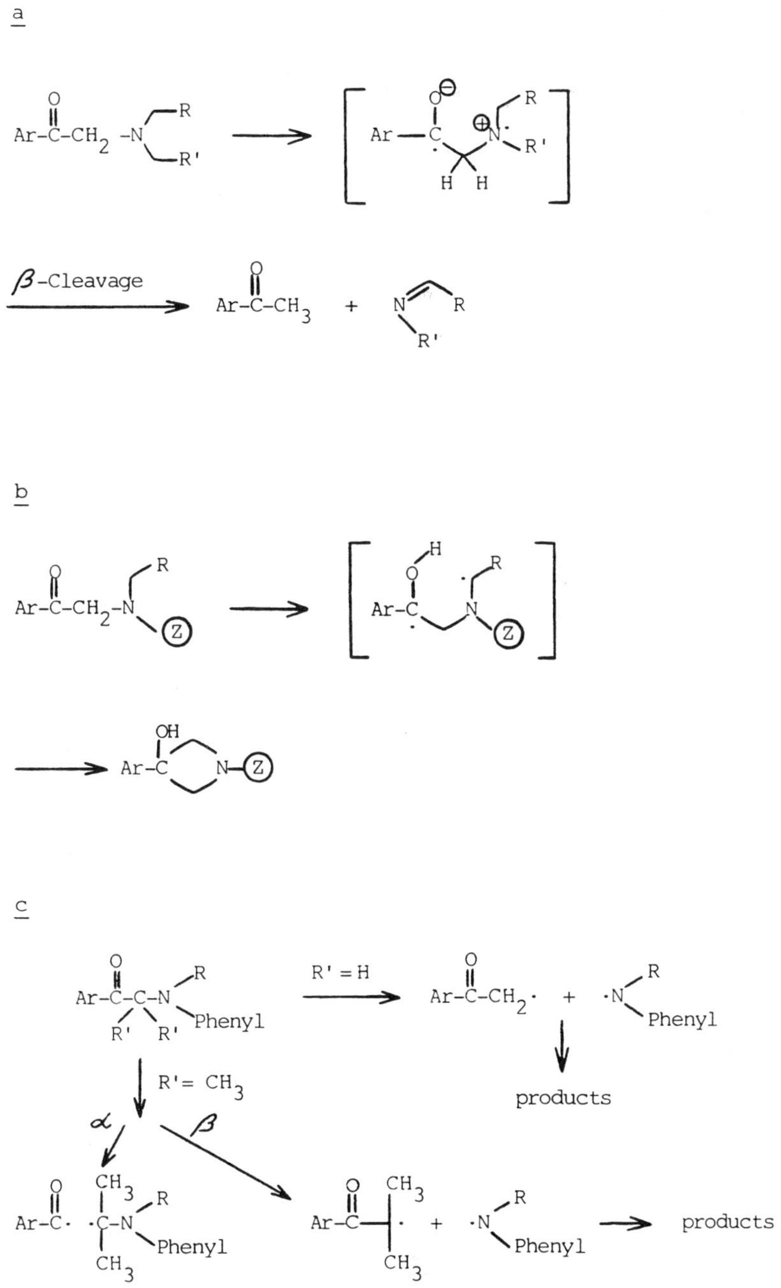

Figure 3. Photochemical reaction pathways of α-amino acetophenones.

1. increasing substitution at the α-carbon favors α-cleavage due to the weakening of the acyl-α-carbon bond [16]; and
2. electron donating groups (alkoxy, amino, etc.) in the α-position can stabilize the transition state of α-cleavage by interaction of a lone pair of electrons with the breaking carbon-carbon bond [17].

In view of the continuum of reactivity exhibited by this class of compounds, the photochemistry of BDMB was investigated.

CIDNP Spectrum of BDMB. The CIDNP technique can provide much useful information on processes in which radical species are formed and has been used to determine the species responsible for the initiator activity of compounds such as I and II [2] [3] [18] [19]. The NMR spectrum taken during photolysis of BDMB was obtained under similar conditions as employed for the study of I and II [20] [21]. The enhanced singlet absorption at 9,63 ppm (Figure 4) is assigned to 4-morpholinobenzaldehyde III which must result from an initial α-cleavage of BDMB to engender a benzoyl and aminoalkyl radical pair (Figure 5). The aldehyde polarization, according to Kaptein's rules [22], and the ESR parameters for these two radicals are in full agreement with a triplet state precursor. By comparing the spectra obtained in different solvents, it was determined that the radical pair is formed essentially via an unimolecular process, thus ruling out the photolytic decomposition of BDMB via intermolecular electron transfer or other bimolecular reactions. The two quartets at 4.46 and 4.62 ppm which exhibit enhanced absorptions are assigned to the olefinic protons of IVa and b. The singlets at 5.28 ppm (emission) and 5.33 ppm (enhanced absorption) are attributed to the olefinic protons in Va and b. The pattern of these signals is due to the combination of cage reactions leading to products exhibiting absorption polarization and escape reactions furnishing the same products but exhibiting emission polarization. The spectrum in deuterated cyclohexane (Figure 4) and the CIDNP experiments in other solvents lead to the conclusion that the formation of Va and b occurs preferentially via escape reactions whereas for the polarizations of IVa and b the contribution of the cage reaction is slightly larger.

Preparative Photochemistry. Irradiation of BDMB on a preparative scale [23] leads to a product mixture that would be expected based upon the results of the CIDNP experiments (Figure 6). In benzene, the main photoproducts are 4-morpholinobenzaldehyde III (21 %) and 1-phenyl-butan-2-one VII (32 %). The latter compound is believed to arise from hydrolysis of the initially formed enamines during work-up. On the basis of our experiments, however, other pathways for its formation cannot be completely ruled out. The deaminated derivative VIII was also isolated in 11 % yield indicating that competing reactions - i.e. Norrish type II or direct β-cleavage - also take place. Surprisingly N,N-dimethyl-4-morpholinobenzamid VI was isolated in 19 %. The mechanism of the formation of VI will be the subject of subsequent investigations. Irradiation of BDMB in isopropanol affords the same products albeit in slightly different ratios.

Dodecanthiol IX has been often used as a scavenger for non-cage benzoyl radicals [24] [25]. Irradiation of BDMB in benzene with a

8. DESOBRY ET AL. *Novel Photoinitiator for Modern Technology* 97

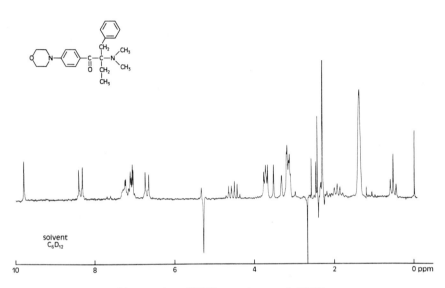

Figure 4. CIDNP spectrum of BDMB.

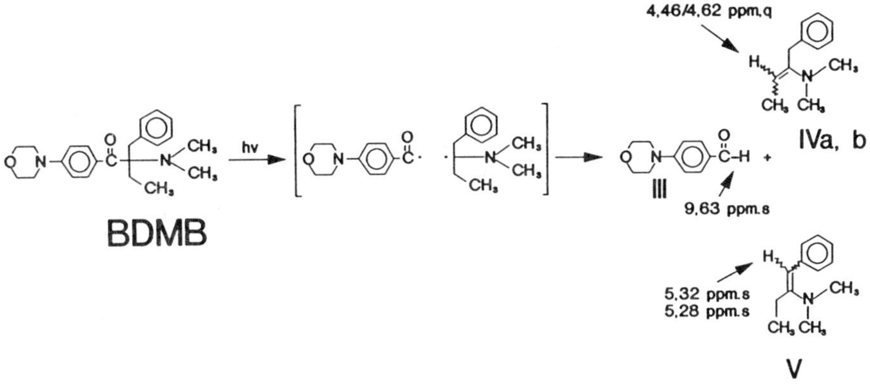

Figure 5. Decomposition products of BDMB detected by CIDNP.

large excess of IX afforded III in 88 % yield (Figure 7) as well as products resulting from the reaction of the benzoyl radical with 1-dodecanthiyl radicals (X, 6 %). The trapping product of the α-aminoalkyl radical was also obtained (XI, 50 %).

The decomposition of various photoinitiators into carbon centered radicals has also been illustrated by Hageman [25] who used TMPO (2,2,6,6-tetramethylpiperidinoxyl) as a trapping agent. Addition of a threefold excess of TMPO to a solution of BDMB results in the formation of XII in high yield (91 %) (Figure 8). VII could also be isolated. This product may arise from an instable primary addition product (XIII) of TMPO and the α-amino-alkyl radical which is hydrolysed during work-up. These results also confirm that the dominant pathway of decomposition is that of α-cleavage.

Trapping Reactions with 2-t-Butylacrylic Acid Methylester XIV. This trapping reaction, which mimics the initiation step of the polymerisation process, has been used to obtain information on the reactivity of the primary radicals formed upon irradiation [1] [26]. Photolysis of BDMB in the presence of a threefold excess of XIV affords the benzoyl derivative XV in 87 % yield (Figure 9). Again VII was also isolated (64 %), whereas no stable addition product of α-aminoalkyl radical could be identified. This result suggests that the benzoyl radical is mainly responsible for the polymerisation of vinylic monomers and is in agreement with previous studies on benzil ketals and benzoin ethers [27]. But, as α-aminoalkyl radicals have also been shown to initiate acrylate polymerization [28], further investigations will be devoted to the elucidation of the role of this primary photoproduct in the overall polymerisation process.

Application Studies with BDMB

White Pigmented Lacquer. BDMB was incorporated into the formulation at concentration levels up to 5 %, and the resulting low viscosity lacquer was applied to an aluminium foil to form a 30 g/m^2 film. Curing was achieved by passing the sample through a conveyor system equipped with two 80 watt/cm medium pressure mercury lamps. The reactivity of the UV-curable coating was quantified in terms of the maximum conveyor belt speed yielding sufficient cure. The latter is determined by the resistance of the coating against rubbing with tissue paper.

As demonstrated in Figure 10, BDMB markedly exceeds its structural analog II in photopolymerization reactivity. Even though addition of an efficient triplet sensitizer, isopropyl-thioxanthone (ITX) boosts the performance of the latter by a factor of six, its reactivity is still inferior to that of BDMB. This observation can be rationalized by comparing the absorption spectra of BDMB, II, ITX and titanium dioxide. While II is efficiently screened from incident UV radiation by titanium dioxide, BDMB still absorbs above the cut-off wavelength of titanium dioxide. The marked increase in reactivity of II in the presence of ITX results from triplet sensitization by ITX [26] which also absorbs at longer wavelength than titanium dioxide.

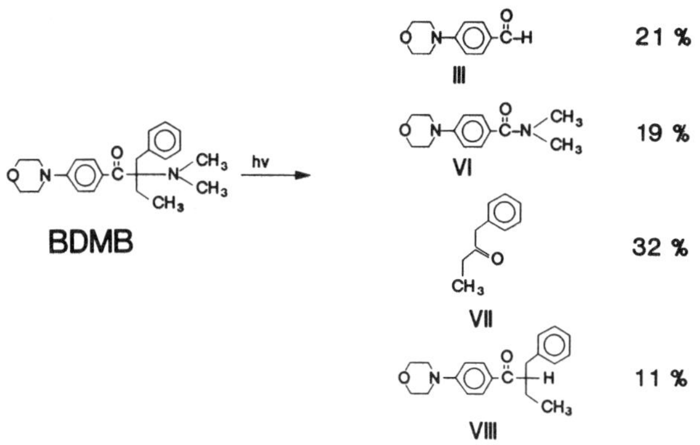

Figure 6. Preparative photolysis of BDMB.

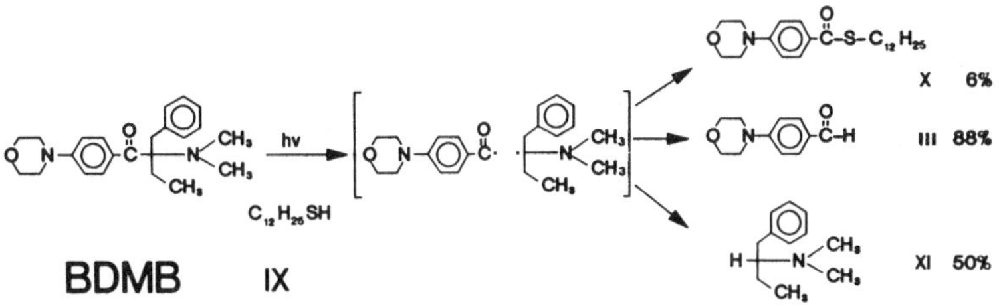

Figure 7. Photolysis of BDMB: Trapping with H-donor.

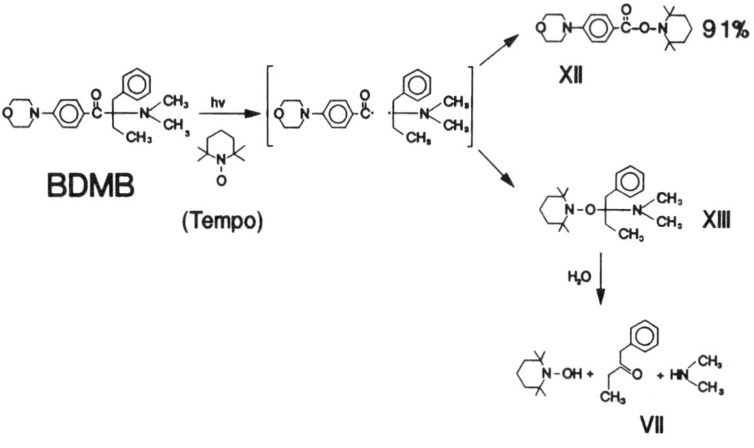

Figure 8. Photolysis of BDMB: Trapping with TMPO.

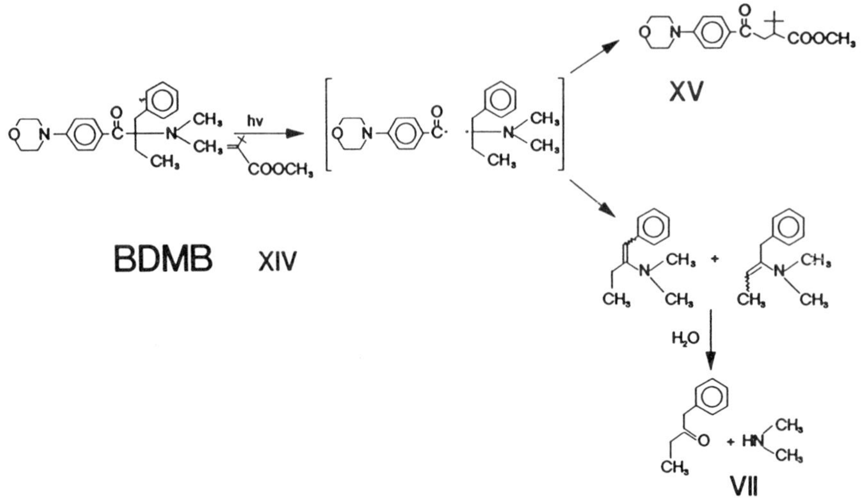

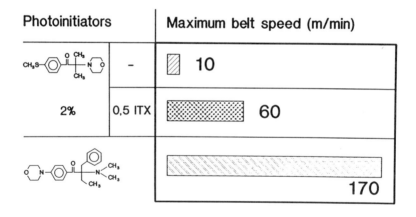

Figure 9. Photolysis of BDMB: Trapping with a non-polymerisable acrylate.

Photoinitiators		Maximum belt speed (m/min)
CH₃S-⟨⟩-CO-C(CH₃)₂-N(morpholino)	–	10
2%	0,5 ITX	60
O(morpholino)-⟨⟩-CO-C(CH₂Ph)(CH₃)-N(CH₃)₂		170

Figure 10. Comparison of BDMB and II in a white pigmented lacquer. Dry film thickness: 15 microns
Radiation source: 2 x 80 W/cm medium pressure Hg lamps.

Offset Printing Inks. BDMB also proved to be slightly more efficient than II in the UV-curing of urethane based blue pigmented systems (Figure 11).

The high performance of these α-amino substituted acetophenone photoinitiators is best explained by the fact that both possess sufficient absorption in the region where the incident light is not filtered by the pigment.

Resist Formulations. In this study we used two aqueous resist systems, an etch and a solder mask. Both are based on polymeric acrylate resins with carboxylic groups and multifunctional acrylate monomers as crosslinking agents. The acid numbers of the etch and the solder formulation are 90 and 45 mg KOH/g respectively.

The sensitivity to photopolymerisation was determined by the highest step on an optical density wedge at which complete curing of the formulation was observed. Results are graphically represented in Figure 12. Here BDMB is compared to I, a benzil ketal widely used for this application. While the performance of the two photoinitiators is comparable in the etch resist, use of BDMB in the solder resist results in efficient curing 7 density steps higher than I. This corresponds to an approximate 11-fold increase in reactivity in this less acidic formulation. This observation and further investigations indicate that the reactivity of BDMB is impaired in an acidic environment due to protonation of the tertiary amine substituent α to the keto group.

Flexographic Printing Plate. This photoimaging system is based on a styrene butadiene copolymer and a multifunctional acrylate monomer as a crosslinking agent. Prior to exposure of the front side to near UV radiation through a negative photomask, the reverse side is entirely exposed to form a solid base. Again, as with the aforementioned resist formulation we chose to compare BDMB with I since the latter is widely used in flexography printing plates. As shown in Figure 13 a cured product of better quality can be obtained utilizing lower concentrations of BDMB.

These results can be easily explained by examining Figure 14 and 15. The first figure shows the overlap between the emission band of the BASF Nyloprint bulb (a light source commonly employed in printing plate technology) and the absorption bands of BDMB, I and II. The latter two exhibit a small overlap, whereas BDMB can absorb a significant part of the emitted light. In other words BDMB (Figure 14) having a much larger extinction coefficient at 366 nm can be employed at lower concentrations affording a formulation of high optical density.

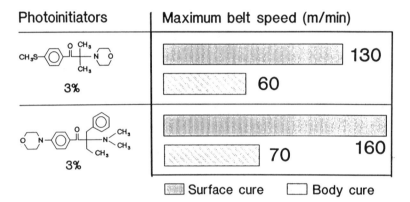

Figure 11. Comparison of BDMB and II in a blue offset printing ink. Dry film thickness: 1,5 microns
Radiation source: 80 W/cm medium pressure Hg lamp.

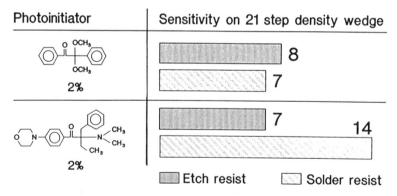

Figure 12. Comparison of BDMB and I in two resist formulations (exposure distance: 30 cm).

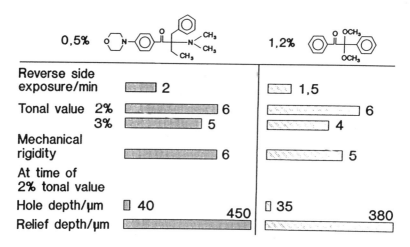

Figure 13. Comparison of BDMB and I in a flexographic printing plate (radiation source: BASF Nyloprint bulb, 350-400 nm).

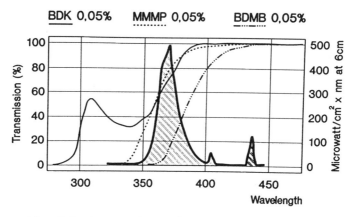

Figure 14. Emission spectrum of a BASF Nyloprint bulb vs. absorption of BDK (I), MMMP (II) and BDMB.

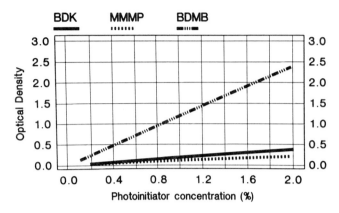

Figure 15. Influence of the initiator concentration on the optical density at 370 nm for a relief height of 600 microns.

Literature Cited

1. Desobry, V.; Dietliker, K.; Hüsler, R.; Rutsch, W.; Loeliger, H. Polym. Paint J. 1988, (Suppl.), 125.
2. Kirchmayr, R.; Berner, G.; Hüsler, R.; Rist, G. Farbe und Lack 1982, 88, 910.
3. Rutsch, W.; Berner, G.; Kirchmayr, R.; Hüsler, R.; Rist, G.; Bühler, N. In Organic Coatings - Science and Technology Vol 8; Parfitt, G; Patsis, A., Eds.; M. Dekker: New York, 1986; p. 175.
4. Stevens, T.S.; Creighton, E.M.; Gordon, A.B.; MacNicol, M. J. Chem. Soc. 1928, 3193; Stevens, T.S. ibid. 1930, 2107; Stevens, T.S.; Snedden, W.W.; Stiller, E.T.; Thomson, T. ibid. 1930, 2119; Thomson, T.; Stevens, T.S. ibid. 1932, 55; Dunn, J.L.; Stevens, T.S. ibid. 1932, 1926; Thomson, T.; Stevens, T.S. ibid. 1932, 1932.
5. Chantrapromma, K.; Ollis, W.D.; Sutherland, I.O. J. Chem. Soc. Chem. Commun. 1978, 670, and references cited therein.
6. Desobry, V.; Dietliker, K.; Hüsler, R.; Rembold, M.; Sitek, F. Eur. Patent Appl. 284561, 1987.
7. I and II are products of CIBA-GEIGY Ltd., commercialized under the names IRGACURE 651 and 907 respectively.
8. Padwa, A.; Eisenhardt, W.; Gruber, R.; Pashayan, D. J. Am. Chem. Soc. 1971, 93, 6998.
9. Wagner, P.J.; Kemppainen, A.E.; Jellinek, T. J. Am. Chem. Soc. 1972, 94, 7512.
10. Hyatt, J.A. J. Org. Chem. 1972, 37, 1254.
11. Gramain, J.-C.; Ouazzani-Chahdi, L.; Troin, Y. Tetrahedron Lett. 1981, 22, 3185.
12. Gold, E.H. J. Am. Chem. Soc. 1971, 93, 2793.
13. Allworth, K.L.; El-Hamamy, A.A.; Hesabi, M.M.; Hill, J. J. Chem. Soc. Perkin Trans. I 1980, 1671.
14. Hill, J.; Zakaria, M.M.; Mumford, D. J. Chem. Soc. Perkin Trans. I 1983, 2455.
15. Meier, K.; Rembold, M.; Rutsch, W.; Sitek, F. In Radiation

Curing of Polymers; Randell, D.R., Ed.; Special Publication No 64. Royal Society of Chemistry: London 1987; p 196.
16. Turro, N.J. Modern Molecular Photochemistry: Benjamin/Cummings: Menlo Park, CA 1978; p. 528.
17. Heine, H.-G.; Traenker, H.-J. Progr. Org. Coat. 1975, 3, 115.
18. Borer, A.; Kirchmayr, R.; Rist, G. Helv. Chim. Acta 1978, 61, 305; Kirchmayr, R.; Berner, G.; Rist, G. Farbe und Lack 1980, 86, 224.
19. Yankelevich, A.Z.; Potapov, V.K.; Hageman, H.-J.; Kuznets, V.M.; Pershin, A.D.; Buchachenko, A.L. Izv. Akad. Nauk. SSSR, Ser. Khim. 1982, 513; Chem. Abstr. 1982, 96, 217061u.
20. BDMB was dissolved in perdeuterated solvents (C_6D_6, C_6D_{12}, CD_3CN, $(CD_3)_2CDOD$).
21. CIDNP spectra were recorded using a varian XL 100 spectrometer. In situ irradiation was effected with UV light (1 kw - high pressure Hg lamp - Philips SP-1000). To avoid IR or visible components, an aqueous filter solution of $NiSO_4$ and $CoSO_4$ was employed.
22. Kaptein, R. J. Chem. Soc. Chem. Commun. 1971, 732; Kaptein, R. J. Am. Chem. Soc. 94, 6251 (1972).
23. 10^{-2} M solutions of BDMB in benzene or isopropanol were irradiated with a Philips HPK 125 mercury lamp located centrally in water cooled Pyrex Finger. Irradiation was continued until disappearance of the starting material. After evaporation of reaction solvent, the products were separated using column chromatography on silicagel.
24. Lewis, F.D.; Magyar, J.G. J. Am. Soc. 1973, 95, 5973.
25. Hageman, H.J.; Overeem, T. Makromol. Chem., Rapid Commun. 1981 2, 719.
26. Dietliker, K.; Rembold, M.; Rist, G.; Rutsch, W.; Sitek, F. Radcure Europe 87, Conf. Proc. 3th; 3/37; Assoc. Finish. Processes SME: Dearborn, MI, 1987.
27. Groeneborn, C.J.; Hageman, H.-J.; Overeem, T.; Weber, A.J.M. Makromol. Chem. 1982, 183, 281.
28. Hageman, H.-J. Progr. Org. Coat. 1985, 13, 123.

RECEIVED September 13, 1989

Chapter 9

Coreactive Photoinitiators for Surface Polymerization

M. Koehler and J. Ohngemach

E. Merck, P.O. Box 4119, D-6100 Darmstadt 1, Federal Republic of Germany

Coreactive photoinitiators are prepared by the introduction of a reactive functionality to the (2-hydroxy-2-propyl)phenone initiator unit. Triethoxysilyl, epoxy and azido groups cover a broad range of coreactivity toward different substrates and surfaces. Triethoxysilyl substituted initiators give a covalent coupling reaction on silica gel as a model. By this method the photoinitiator activity is transferred to the solid support and a surface polymerization on such modified silicas becomes possible. Undiluted monomers (DMAEMA, NVP) and their ethanolic solutions are investigated in photopolymerizations under unstirred conditions. The amount of polymer which segregates on the silica is regulated by the monomer concentration and the initiator content of the silica is also of importance in the case of monomer solutions. The extent of the initiation and polymerization on silica (measured as weight percent polymer of the solid material) is explained by a radical transfer which depends on the intermolecular distance of the reactants.

Most of the photoinitiators for UV curing of acrylate based formulations belong to the chemical class of aromatic ketones which are able to generate reactive radicals by an UV radiation induced process.

2-Hydroxy-2-methylpropiophenone Ia is known for about ten years as an effective photoinitiator (1). A benzoyl and a (2-hydroxy-2-propyl) radical are formed by a photochemical α-cleavage (2-4).

The (2-hydroxyethoxy) substituted phenone Ib and its acrylate Ic are examples of more advanced (2-hydroxy-2-propyl)phenone initiators (5, 6).

R—⟨C₆H₄⟩—C(=O)—C(CH₃)₂—OH

Ib, R = O-CH$_2$-CH$_2$OH

Ic, R = O-CH$_2$-CH$_2$-O-C(=O)-CH=CH$_2$

New compounds are prepared by the introduction of a triethoxysilyl, epoxy or azido group linked over a short spacing unit to the phenyl ring (7). A combination of the basic photoreactivity with such a second functionality within one molecule allows interesting properties. The covalent (chemical) coupling of the photoinitiator unit to different molecules or materials seems to be possible with the socalled "coreactive photoinitiators". Trialkoxysilyl compounds are known to give condensations with surface silanol groups, epoxides can add to hydroxyl, amino or carboxylic funtionalities and azides are able to eliminate nitrogen and produce a highly reactive nitrene for CH-insertions.

Surface coupling reactions on porous materials with an inner surface will be investigated first. Silica gel was chosen as a model substrate with surface silanol groups which react under condensation with triethoxysilylsubstituted initiators. The resulting initiator modified silicas are used to study the polymerization of monofunctional monomers which should occur on the surface of the solid material.

Experimental Part

Materials

Coreactive photoinitiators Id-g are experimental products (designated ZLI) of E. Merck, Darmstadt, West-Germany:
4-[3-(Triethoxysilyl)propoxy]phenyl (2-hydroxy-2-propyl) ketone
  Id (ZLI 4404),
4-[2-(Triethoxysilylpropoxy)ethoxy]phenyl (2-hydroxy-2-propyl) ketone
  Ie (ZLI 4434),
4-Oxiranylmethoxyphenyl (2-hydroxy-2-propyl) ketone
  If (ZLI 4402),
4-(2-Azidoethoxy)phenyl (2-hydroxy-2-propyl) ketone
  Ig (ZLI 4405).
Silica gel (Si 60, E. Merck) was used as received. Dimethylaminoethylmethacrylate (DMAEMA, Röhm) and N-Vinylpyrrolidone (NVP, E. Merck) were used as monomers for photopolymerizations without further purification.

Methods

Preparation of photoinitiator modified silicas: A certain amount (1 or 5 g) of the triethoxysilane photoinitiator (Id or Ie) was dissolved in 20 ml ethanol and 10 g silica gel was suspended in the solution. After evaporation of the solvent the solid material was heated for 15 minutes at 150 °C under stirring. The photoinitiator modified silicas IIa-d were washed five times with 100 ml ethanol to remove adsorbed materials and dried overnight in vacuo.

Photopolymerization: 0,5 g silica II was mixed with 2 ml of the liquid monomer or its solution in ethanol in a flat shaped weighing bottle (35 x 30, 15 ml). The photopolymerization was performed in the presence of air by irradiating the mixture for a certain time (3 minutes in a typical experiment) with a high pressure mercury lamp (TQ 105, Heraeus) at a distance of 5 cm without agitation at room temperature. The solid material III (with silica gel shape) was collected quantitatively by filtration, washed with 100 ml ethanol on the filter crucible and dried overnight in vacuo. The weight increase of the sample corresponds well with the polymer formation on the materials III calculated from the results of the elemental analysis. (All experiments were performed twice to control the experimental error.)

Elemental analysis (e.a.): Elemental microanalysis of the silica-based materials II, III (carbon, nitrogen) were obtained with a Perkin Elmer 240 B elemental analyzer. The percentage photoinitiator on silica was calculated from the percentage carbon (found) on the material II divided by the percentage carbon of the corresponding silica-coupled initiator species, assumed as Id,e $-3(C_2H_5O)$ (multiplied by 100). The percentages poly-DMAEMA on material IIIa and poly-NVP on IIIb were calculated from the percentage nitrogen (found) on the material III divided by the percentage nitrogen of the corresponding monomer (multiplied by 100).

Infrared spectra (i.r.): Infrared spectra of the materials II and III were obtained with a BRUKER IFS48 FT-IR spectrometer purged with nitrogen gas and using a mid band mercury cadmium telluride (MCT) detector. A HARRICK 'Praying Mantis' diffuse reflectance attachment with two ellipsoidal mirrors collected the diffuse reflectance spectra. To avoid residual radiation bands all samples were diluted with KBr powder (Uvasol quality, E. Merck) so that the sample concentration was about 10 %.

Thermal gravimetric analysis (t.g.a): The weight loss of photoinitiator modified silicas II as a function of the temperature was measured with a Du Pont Instruments 9900 Computer / Thermal Analysis System. The measurement was carried out at a heating rate of $2°\ min^{-1}$ and an air flow rate of $0.15\ l\ min^{-1}$.

Synthesis of coreactive photoinitiators

The starting materials for the synthesis of the coreactive initiators Id-g already contain the photolabile (2-hydroxy-2-propyl) carbonyl group which shows an excellent stability during the following preparations.

Triethoxysilane derivatives are prepared by the platinum catalyzed addition of triethoxysilane (8) to the corresponding allylic ethers. The silanes Id,e are isolated and purified by column flash chromatography, reduced yields (at 50 %) might be due to irreversible coupling reactions with the column material or polycondensation. The pure compounds are colourless oils which are stable on storge in a closed bottle.

The epoxide If was obtained by etherification of 4-hydroxyphenyl-2-hydroxy-2-propyl ketone with epichlorohydrin in the presence of potassium carbonate and a phase transfer catalyst. Crystallization from diethylether yields the pure compound If as a white solid with a melting point of 53-54 °C.

Synthesis of triethoxysilyl substituted (2-hydroxy-2-propyl) phenones

$$R(O-CH_2-CH_2)_nO-CH_2-CH=CH_2 + HSi(OC_2H_5)_3$$

$$\downarrow CODPt^{II}Cl_2, TCE$$

$$R(O-CH_2-CH_2)_nO(CH_2)_3Si(OC_2H_5)_3 \qquad Id,e$$

I	n
d	0
e	1

$R:$ ⟨phenyl⟩-C(=O)-C(CH$_3$)$_2$-OH

$CODPt^{II}Cl_2$: (1,5-Cyclooctadiene)platinum(II)chloride
TCE : Tetrachloroethylene

Synthesis of 4-Oxiranylmethoxyphenyl- and 4-(2-Azidoethoxy)phenyl-
(2-hydroxy-2-propyl)ketone If,g

$$R-OH + Cl-CH_2-CH\overset{O}{-}CH_2 \xrightarrow{K_2CO_3, TDA} R-O-CH_2-CH\overset{O}{-}CH_2 \quad If$$

TDA: Tris (3,6-dioxaheptyl)amine

$$R-O-CH_2-CH_2-O-\overset{O}{\underset{O}{S}}-CH_3 + NaN_3 \xrightarrow{(DMSO)} R-O-CH_2-CH_2-N_3 \quad Ig$$

DMSO: Dimethylsulfoxide

$R:$ C(=O)-C(CH$_3$)$_2$-OH

The azide Ig was prepared from the methanesulfonic acid ester of the initiator Ib. Nucleophilic displacement of the mesylate group by the azido anion in dimethylsulfoxide leads to the formation of the compound Ig obtained as a yellow oil after purification by column flash chromatography. The structures of the phenones Id-g are in correspondence with the results of the elemental analysis and the spectral data.

Photoreactivity Tests

Due to the definition coreactive photoinitiators combine two different reactivities in one molecule and it has to be investigated first whether the photoinitiator reactivity is not remarkably decreased by an intra- or intermolecular influence of the coreactive group.
Therefore the phenones Id-g were compared in their curing behaviour with the unsubstituted parent compound Ia in an usual acrylate based UV curable system (experimental conditions see ref. 9). From the results (Table I) it is concluded that the phenones Id-f have a similar activity as the parent compound Ia, although differences became obvious at short irradiation time. The initiator activity of the azide Ig was extremely low compared to the phenones Ia and Id-f. This unexpected result might be explained by the effect of the azido group. To get more insight to the photoreactivity of the azide Ig an ethanolic solution was irradiated for one hour. A new crystalline compound was isolated and characterized as 4-(2-azidoethoxy)benzoic acid.

$$\text{Ig} \xrightarrow{h \cdot \nu \text{ (ethanol)}} N_3\text{-}CH_2\text{-}CH_2\text{-}O\text{-}\langle\text{\_}\rangle\text{-}COOH$$

The acid should be formed by an α-cleavage with subsequent oxidation and hydrogen abstraction of the resulting radical. The azido group shows a remarkable stability under these conditions but the situation might be different in a UV-curable formulation.

Coupling of initiators to silica

Coreactive photoinitiators seem to be promising tools to investigate UV induced processes of ketone structures with a permanent bond to solid substrates. Under the conditions of the thermal substrate coupling (coreaction) the photoinitiator activity should be conserved and a subsequent photochemical α-cleavage in the presence of a suitable monomer should initiate a polymerization process.

Several alkoxysilane reagents have been coupled recently to silica and related materials (10-13) but the introduction of a photoinitiator group to a surface has not been reported so far. Silica gel in a quality which is commonly used for column chromatography (particle distribution 40-63 μm) was selected as a starting material and the triethoxysilane initiators Id,e were applied at two different concentrations to influence the initiator loading of the silica. A certain amount of the silanes Id,e will react on heating and the excess is removed by repeated solvent washing.

The reaction might be due to condensation with silanol groups and polysiloxane formation.

$$SiO_2\Big\}\text{--}OH + (C_2H_5O)Si(CH_2)_3O(CH_2\text{-}CH_2\text{-}O)_n R$$
$$\text{Id,e}$$
$$\downarrow \text{heating}$$
$$SiO_2\Big\}\text{--}O\text{-}Si(CH_2)_3\text{-}O(CH_2\text{-}CH_2\text{-}O)_n R$$
$$\text{IIa-d}$$

$$R : \langle\text{\_}\rangle\text{-}\overset{O}{\overset{\|}{C}}\text{-}\overset{CH_3}{\underset{CH_3}{C}}\text{-}OH$$

It is assumed that the ethoxy groups are lost by partial hydrolysis, condensation with surface silanols and by self-condensation. Initial amounts of 0.5 and 0.1 weight equivalent silane Id,e per silica were used in the preparation of the four silicas IIa-d.

The structure of coupled agents was investigated by diffuse reflectance Fourier Transform Infrared spectroscopy and it has been outlined that a weak sharp band around 3740 cm^{-1} is due to the free Si-OH surface group (10). Such an absorption was not detectable in the i.r. spectrum of silica IIa but it is obvious in the case of IIc (at 3738 cm^{-1}) due to the presence of unreacted silanol groups (Figure 1).

Elemental analysis (e.a.) and thermal gravimetric analysis (t.g.a.) were used as alternative methods to calculate the percentage photoinitiator species per silica (Table II). An inaccuracy in the results of the elemental analysis is possible by the influence of adsorbed ethanol (from the washing process) on the carbon values. In the thermal gravimetric analysis absorbed species should be removed below 200 °C.

Table I. Pendulum Hardness of UV Cured Layers (measured according to DIN 53 157) Depending on Initiator Compound and Irradiation Time

Irradiation Time (s)	Initiator Compound					Remark
	Ia	Id	Ie	If	Ig	
1.0	205	111	127	178	-	(a)
2.5	214	155	213	216	42	(a)
5.0	215	214	216	220	88	(a)
10.0	210	209	208	213	193	(a)

(a) Pendulum Hardness, in seconds.

Table II. Photoinitiator Loading of the Modified Silicas II a-d (The weight equivalent silane I d,e affects the final initiator content.)

	Weight Percent Initiator on Silica			n	Weight Equivalents (a) Id,e per Silica
	by e.a.	by t. g. a.			
IIa	12.9	11.9	IIa	0	0.5 Id
IIb	12.2	11.9	IIb	1	0.5 Ie
IIc	6.3	7.4	IIc	0	0.1 Id
IId	8.5	8.8	IId	1	0.1 Ie

(a) g Initiator / g Silica

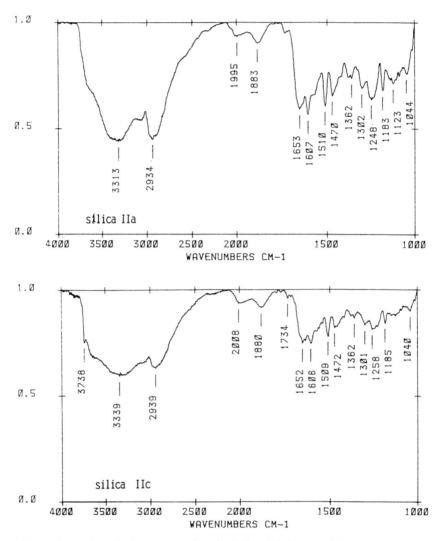

Figure 1. Transmission spectra (i.r.) of photoinitiator modified silica IIa and silica IIc. The absorbance at 1653 cm^{-1} (IIa) and at 1652 cm^{-1} (IIc) are characteristic for the photoinitiator (co stretching).

The corresponding e.a. and t.g.a. values are found in a comparable range and it is evident how the initial ratio of coupling agent I toward silica affects the final loading. A remarkable influence of the spacer lenght between the triethoxysilyl group and the initiator phenyl ring (Id, n = 0 and Ie, n = 1) on the coupling properties was not expected.

In Figure 2 typical examples for the t.g.a. behaviour of the photoinitiator modified silicas IIa,c are given. The weight loss below 200 °C is ascribed to the removal of absorbed species and the weight loss between 200 and 800 °C is characteristic for the thermal degradation of the photoinitiator.

For a comparison the interaction of the initiators 2,2-dimethoxy-2-phenylacetophenone (DMPA), (2-hydroxy-2-propyl)phenone Ia (both noncoreactive) and the epoxide If with silica was also investigated. Silica was treated with 0.5 weight equivalent of these compounds in the same manner as previously described for the silanes Id,e. The resulting materials were analyzed by e.a. (Table III). The results of e.a. are primarily given as percent carbon because the structure of the organic species on silica is not absolutely clear (a trihydroxy compound is assumed as derivative of If).

In the case of DMPA no significant amount of the organic compound was detectable on the silica. A surprisingly high carbon content was found on the silica derived from the epoxide If (SiO_2 + If) and for the parent compound Ia (SiO_2 + Ia) the carbon amount of 2.5 % seems to be substantial. The calculated percentage of the initiator (Ia and water adduct of If) is given in brackets on the previous table.

As a proof for permanent coupling and extractable initiator content 2 g samples of IIa-d, (SiO_2 + Ia) and (SiO_2 + If) were refluxed in 50 ml ethanol overnight. The solids were collected by filtration, dried and analyzed (e.a.) again. The corresponding filtrates were concentrated and the residues investigated by thin layer chromatography (t.l.c.).

The triethoxysilyl compounds Id,e with a silica specific coreactivity gave covalent coupling and the materials IIa-d were not extractable. The phenone Ia was found as a residue in the (SiO_2 + Ia) extract but the carbon value of the extracted silica was not decreased (probably by the adsorption of ethanol at refluxing). The i.r. spectrum of this silica was almost identical with a reference of untreated material Si 60.

A different behaviour was observed with the epoxide derived material (SiO_2 + If). The carbon value is decreased after extraction but the characteristic initiator absorptions are still present in the i.r. spectrum. A new compound which was identified as the water adduct of the epoxide If is found in the ethanolic extract by t.l.c. The trihydroxy compound was also formed by refluxing the epoxide If with water under acidic conditions in the absence of silica.

$$If + H_2O \xrightarrow{(H^{\oplus})} HO-CH_2-\underset{OH}{CH}-CH_2-O-\underset{}{\bigcirc}-\underset{CH_3}{\overset{O\ CH_3}{\underset{|}{C}-\underset{|}{C}-OH}}$$

It is assumed that the nonextractable part is due to such highly polar initiator species derived from the epoxide by ring opening. The interaction to the silica should be based mainly on physical forces (hydrogen bonding).

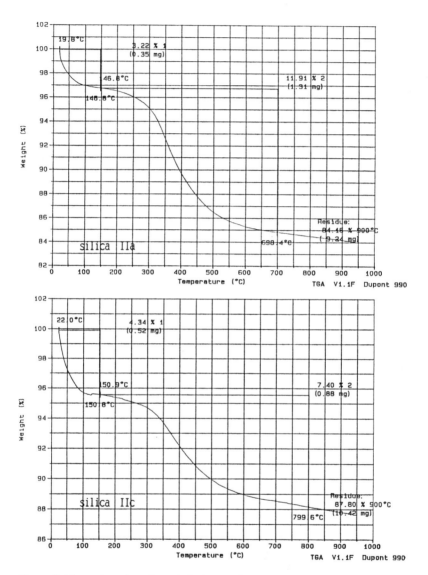

Figure 2. Thermal gravimetric analysis (t.g.a.) of the modified silicas IIa and IIc. The percent photoinitiator per silica is directly measured by the weight loss of a sample during controlled heating in an air stream.

Surface polymerization on photoinitiator modified silica

The silicas IIa-d were used for photopolymerization of monofunctional monomers which were applied as undiluted liquids or as ethanolic solutions. Several monomers were tested first without a solvent on their ability to polymerize.

$CH_2=CH-COO-CH_2-CH_2-OH$ (2-Hydroxyethyl)acrylate HEA

$CH_2=\overset{CH_3}{\underset{}{C}}-COO-CH_2-CH_2-OH$ (2-Hydroxyethyl)methacrylate HEMA

$CH_2=CH-C_6H_5$ Styrene

$CH_2=\overset{CH_3}{\underset{}{C}}-COO-CH_2-CH_2-N(CH_3)_2$ (2-Dimethylaminoethyl)methacrylate DMAEMA

(N-Vinyl)pyrrolidone NVP

HEA underwent a vigorous polymerization reaction on irradiation in the presence of silica II which led to a sticking and lumping on particles. Such a process can hardly be controlled and the resulting material was not very useful for a quantitative characterization (e.a., weight increase). Nevertheless the reaction is due to the photoinitiator because in a control experiment with untreated silica no polymerization took place. The corresponding methacrylate HEMA did not react and the silica II was recovered unchanged (but still active toward other monomers). Styrene also did not polymerize under the applied conditions.

Satisfactory results were obtained with the nitrogen containing monomers DMAEMA and NVP. The poly-DMAEMA (IIIa) and poly-NVP (IIIb) silicas with more than 30 weight percent polymer conserve the silica morphology and were easy to separate from the monomer excess.

As an additional result it was confirmed that photoinitiator silicas II which are derived from silanes Id,e with different spacers show an almost identical behaviour. To simplify the further discussion

the former four silicas IIa-d can be designated as only two different materials: IIa,b = IIh (higher initiator content)
IIc,d = IIl (lower initiator content)

Undiluted DMAEMA and NVP

From the weight percent poly-DMAEMA on silica formed under condensed phase conditions (undiluted DMAEMA) a minor influence of the initiator silica type and the irradiation time became apparent (Table IV).

The polymer content of the materials IIIa was found in close range between 30 and 37 wt % and it seemed to be impossible to increase the polymer loading by extended irradiation. As a proof on the validity of the experimental conditions several polymerizations were run with undiluted DMAEMA on the silicas IIh,l and an average value of 34.1 wt % with a standard deviation of 1.71 was calculated as a maximum capacity of the silica for poly-DMAEMA.

On SEM images (obtained with a Hitachi Model S-570 Scanning Electron Microscope) no signifacant difference was observed between the morphologies of a typical material IIIa and normal silica gel at different magnifications in a range from 150 to 1 µm. It is concluded that the polymer coating on the silica is homogeneous and a considerable part of the polymeric material is located at the inner surface (specific surface of Si 60 is given (14) at 550 $m^2 g^{-1}$).

In this case the surface polymerization on photoinitiator modified particles occured without agglomeration and the polymerization seems to be restricted mainly on the particle surface (immobilized polymer) but the formation of a soluble polymer (dissolved in the monomer excess) cannot be excluded. This behaviour depends also on the chemical structure of the monomer (compare agglomeration with HEA) and it is an objective of further work to investigate various monomers.

The activities of the silicas (SiO_2 + Ia) and (SiO_2 + If) toward DMAEMA were examined for comparison. Surprisingly they provide similar amounts of polymer as obtained with the silane modified materials IIh,l and the polymer was not extractable from the silica. After the intensive extraction (refluxing with ethanol overnight) the silica (SiO_2 + Ia) lost most of its initiator activity when it was used again in a photopolymerization experiment.

The amount of poly-NVP on the materials IIIb is found in a range comparable to the IIIa series above. The macroscopic shape of the materials IIIa and b is also very similar.

The i.r. spectra of the components IIIa and IIIb are represented in Figure 3. The spectrum of IIIb still shows a less intensive absorption of the initiator carbonyl group at 1643 cm^{-1}. In the spectrum of IIIa the absorption of the wavelength 1721 cm^{-1} is attributed to poly-DMAEMA and in the spectrum of IIIb the absorption at 1706 cm^{-1} is attributed to poly-NVP.

In general from these experiments with undiluted DMAEMA and NVP it became obvious that the polymerization process which tends to a certain limit of polymer formed on the silica does not depend much on the amount of photoinitiator. The initiation (radical transfer from the initiator radical pair to the neighbour monomer) and subsequent polymerization (radical center transfer along the growing polymer chain) are promoted by the close contact of the molecules.

Table III. Interaction of the Phenoles Ia, If and DMPA with Silica

	Materials	% Carbon (Initiator)	% Carbon (a) (Initiator)
SiO$_2$ ↗ ↖ Ia If ↓ ↓ (SiO$_2$ + Ia) (SiO$_2$ + If)	(SiO$_2$ + Ia)	2.5 (3.4)	3.1 (-)
	(SiO$_2$ + If)	7.8 (11.4)	1.4 (7.0)
	(SiO$_2$ + DMPA)	-	-

(a) after repeated extraction (refluxing with ethanol overnight)

Table IV. Weight Percent Poly-DMAEMA on Silica Depending on Initiator Silica Type and Irradiation Time

Photoinitiator Modified Silica/Monomer	Irradiation Time (min)	Weight Percent Polymer on Silica
IIh / DMAEMA	3	32.0
IIh / DMAEMA	1	30.9
IIh / DMAEMA	10	35.9
IIl / DMAEMA	3	32.6
(SiO$_2$ + Ia) / DMAEMA	3	35.4
(SiO$_2$ + If) / DMAEMA	3	37.1
IIh / NVP	3	33.3
IIl / NVP	3	36.9
(SiO$_2$ + If) / NVP	3	40.1

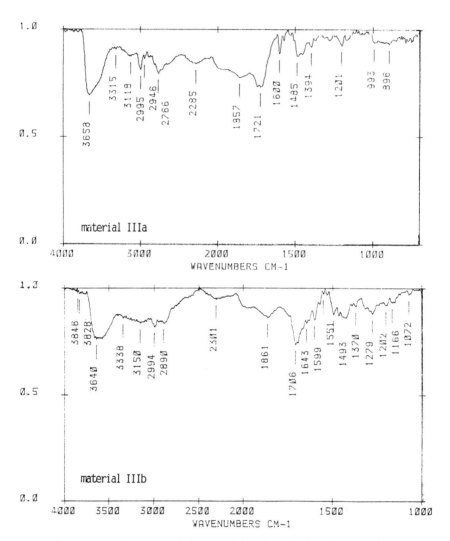

Figure 3. Infrared spectra of the materials IIIa and IIIb. The absorbance at 1721 cm^{-1} (IIIa) is attributed to poly-DMAEMA and at 1706 cm^{-1} (IIIb) to poly-NVP.

Initiation by radical transfer: For the process it should not be very important which radical ($R^1\cdot$ or $\cdot R^2$) will primarily react with a monomer molecule. A covalently surface-bound polymer only results through initiation by the radical $R^1\cdot$.

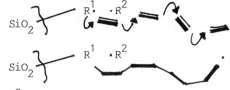

$R^1\cdot$ $\cdot R^2$ initiator radical pair formed by photochemical α-cleavage of the molecule $R^1 - R^2$
monomer,   oligomer radical

For such a process it is not necessary to assume a considerable motion or diffusion of the reacting species which is less probable under the applied conditions.

Recently the photoinitiated polymerization of multifunctional monomers in condensed phase (usually called UV curing) was characterized as an extremely fast reaction with a kinetic chain length in the order of 100 000 acrylate functions per initiating radical(15).

The results of the photoinitiated surface polymerization indicate short chain lengths which might be due to a high degree of radical combination on the porous substrate (cage conditions).

Monomer dilution

In the presence of a solvent (ethanol) a different situation was expected because the average distance between the reacting molecules (initiator-monomer as well as monomer-monomer) is enlarged. Ethanol was considered as a diluent only, other influences on the reaction are excluded.

The resutls, listed in Table V, were obtained from the initiator silica IIh with the higher loading and DMAEMA concentrations in the range between 50 and 10 volume percent monomer. The final polymer content of the resulting materials IIIa' is remarkably decreased in comparison to IIIa. A similar dilution effect is noted with NVP solutions. For a comparison see the results with 100 % (undiluted) monomers: IIh,1/DMAEMA and IIh/NVP.

In the i.r. spectrum of a material IIIa' (12.9 wt % poly-DMAEMA on silica) the typical photoinitiator carbonyl absorption is present at 1642 cm^{-1} (Figure 4). Such a material should be still active toward undiluted DMAEMA and NVP if the photoinitiator group is accessible to the applied monomer. The corresponding photopolymerization experiments provided the materials IIIa and IIIa'/b'. With DMAEMA the polymer content was increased to 32.6 wt % poly-DMAEMA (IIIa) and NVP led to a composite IIIa'/b' with 27.7 wt % poly-NVP.

Table V. Weight Percent Poly-DMAEMA (III a',a'') and Poly-NVP (III b') on Silica Obtained with Monomer Solutions

IIh + DMAEMA $\xrightarrow{(C_2H_5OH)}$ IIIa'	
vol. % DMAEMA (in ethanol)	wt % poly-DMAEMA (on silica)
50	28.0
30	21.9
20	17.4
10	12.9
IIh + NVP $\xrightarrow{(C_2H_5OH)}$ IIIb'	
vol % NVP (in ethanol)	wt % poly-NVP (on silica)
50	30.2
30	26.9
20	19.8
10	15.1
IIl + DMAEMA $\xrightarrow{(C_2H_5OH)}$ IIIa''	
vol. % DMAEMA (in ethanol)	wt % poly-DMAEMA (on silica)
50	6.7
30	3.9
20	2.8
10	2.2

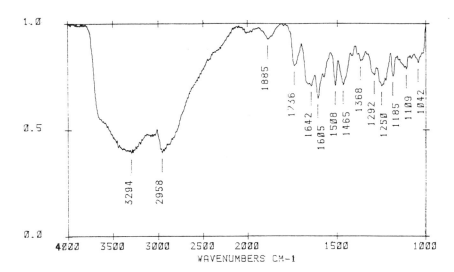

Figure 4. Infrared spectrum of the material IIIa'.

Second polymerization on material IIIa' with undiluted DMAEMA and NVP

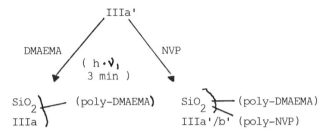

The influence of reduced initiator loading became obvious in experiments with the initiator silica III and DMAEMA solutions. The poly-DMAEMA contents of the resulting materials IIIa'' are much lower compared to IIIa'. The decreased number of initiation centers on the silica surface together with the monomer dilution result in a reduced probability for a successful radical transfer (initiation).

In general the experiments with monomer solutions provide deeper insight to the mechanism of the photoinitiation at the silica/monomer interface. A diffusional mixing of the reacting molecules can be excluded if the initiator is covalently anchored to a solid material. The situation is different from a homogeneous reaction which is usually performed by stirring the starting materials. In the absence of stirring and under the dimensional constraint of a surface reaction the initial conditions and the nearest-neighbour distance of the reactant species become of importance (16).

The next reaction scheme describes the different behaviour of two surface coupled initiator radical pairs and randomly distributed monomer molecules at the interface.

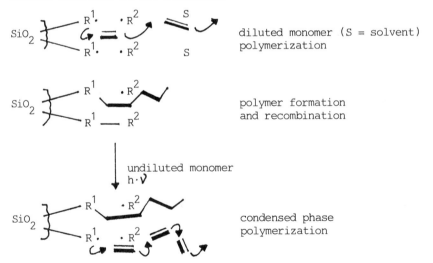

If an initial radical pair gets no opportunity for a radical transfer it will recombine after a certain lifetime. By monomer dilution the intermolecular distance between the immobilized initiator and

the monomers is enhanced. The degree of polymerization is limited and a considerable part of recombined initiator is still available. A second initiation on such a material (IIIa') was possible by changing to condensed phase conditions. The radical transfer (initiation) and the subsequent polymerization are forced now by the reduced intermolecular distance of the species.

Conclusions

From the first experiments with surface immobilized initiators (derived from coreactive photoinitiators) some general aspects became apparent.

A radical chain process which is started on the solid surface results in a segregation of organic material (usually a polymer) with a strong interaction to the substrate. The surface initiation may be a very complex process from a mechanistic point of view but the practical application already becomes obvious.

Surface modifications of particles which are commonly used as materials for chromatography, as pigments or fillers are of considerable technical interest. By the selection of appropriate monomers and their surface polymerization different chemical functionalities or physical properties are introduced. Silica gel with an extended surface area (Si 60, E. Merck) is a useful substrate for the study of photoinitiated surface polymerizations but less porous particles, microcrystalline materials or fibers are the objects of investigations with a more practical background.

In principle macroscopic surfaces of different chemical nature should be accessible to coreactive initiator coupling and subsequent photopolymerization. The capacity of a macroscopic surface like glass, metal, ceramics or plastics for initiator coupling might be very low compared to porous silica gel. Nevertheless few initiation centers could lead to a high degree of polymerization especially if the monomer is applied as condensed phase.

Acknowledgment

The Analytical Central Laboratory of E. Merck is gratefully acknowledged for the performance of thermal gravimetric analysis, diffuse reflectance infrared spectra and elemental microanalysis of the silica based materials.

Literature Cited

1. J. Gehlhaus and M. Kieser, German Pat. 2722 264 (1978)
2. J. Eichler, C.P. Herz, J. Naito and W. Schnabel, J. Photochem. 12, 225 (1980)
3. A. Salmassi, J. Eichler, C.P. Herz and W. Schnabel, Polym. Photochem. 2, 209 (1982)
4. J.P. Fouassier, P. Jaques, D.J. Lougnot and T. Pilot, "Lasers, Photoinitiators and Monomers: a Fashionable Formulation", Photochemistry and Photophysics in Polymers, Elsevier Appl. Sci. Publ. London, 1984, p. 57
5. M. Koehler, J. Ohngemach and G. Wehner, German Pat. Appl. 3512179 (1985)

6. W. Baeumer, M. Koehler and J. Ohngemach, Proc. Radcure Conference 1986, 4-43
7. M. Koehler, J. Ohngemach, E. Poetsch, R. Eidenschink, G. Greber, D. Dorsch, J. Gehlhaus, K. Dorfner and H.-L. Hirsch, Europ. Pat. Appl. 0 281 941 (1988)
8. L.N. Lewis and N. Lewis, J. Amer. Chem. Soc. 108, 7228 (1986)
9. M. Koehler and J. Ohngemach, Polymers Paint Colour J. 178, 203 (1988)
10. S.R. Culler, H. Ishida and J.L. Koenig, J. Colloid. Interface Sci. 106, 334 (1985)
11. J.D. Miller and H. Ishida, Langmuir 2, 127 (1986)
12. Y. Okahata, K. Ariga, H. Nakahara and K. Fukuda, J. Chem. Soc., Chem. Commun. 1986, 1069
13. C.S. Goldstein, K.D. Weiss and R.S. Drago, J. Amer. Chem. Soc. 109, 758 (1987)
14. E. Wellner, D. Rojanski, M. Ottolenghi, D. Huppert and D. Avnir, J. Amer. Chem. Soc. 109, 575 (1987)
15. C. Decker and K. Moussa, J. Appl. Polymer Sci. 34, 1603 (1987)
16. R. Kopelman, Science 241, 1620 (1988)

RECEIVED October 27, 1989

NOVEL RADIATION PHOTOCURABLE SYSTEMS

Novel Radiation Photocurable Systems

The papers in this section are devoted to describing the development and characterization of photocurable systems which, while not in widespread use at the present time, have the potential of leading to new products and applications with enhanced properties. It is the continual process of innovative research, exemplified by the reports herein, which places the radiation curing industry in a rapid growth position within the industrial coatings and plastics sector.

The first two papers (Chapters 9 and 10) deal with the use of photocurable systems to modify the properties of conventional polymers such as polyethylene, polypropylene, and ethylene-propylene-diene (EPDM) elastomers. In one case (Chapter 9), the efficiency of graft polymerization of styrene on polyethylene and polypropylene was found to increase significantly in the presence of very small amounts of trifunctional acrylates and methacrylates. In the second (Chapter 10), a high degree of crosslinking of such important commercial polymers as high density polyethylene, low density polyethylene, and EPDM was attained by exposing melt-prepared samples containing low weight percent trifunctional crosslinking agents and photosensitizers to a high pressure mercury lamp source. By careful analysis, the authors of Chapters 9 and 10 have developed mechanistic models for the unique grafting and crosslinking processes. Certainly, the use of photocurable systems to modify the properties of existing engineering polymers provides a new avenue of opportunities for the radiation curing industry and the relevance of Chapters 9 and 10 should be noted with particular care.

Chapters 11-14 give results for four types of non-conventional photocurable resins which to date have not enjoyed widespread industrial use. The systems in Chapters 11-13 deal with non-acrylated functionalities which are readily cured by exposure to lamp intensities commonly used for curing acrylates. The diarylketone/oligomeric amine/triallylcyanurate systems in Chapter 11 and the multifunctional thiol/norbornene resin compositions in Chapter 12 are characterized by rapid cure times and formation of highly crosslinked networks. In the case of the thiol/norbornene composition (Chapter 12), excellent physical properties accompanied by low shrinkage are obtained. In addition, when norbornene functionalized silicones were employed, elastomers with physical properties comparable to silicone acrylates were formed. The acrylated melamines in Chapter 13 yield photocured coatings with adhesion properties superior to acrylated epoxy and urethane systems formed under comparable conditions. They are also characterized by very rapid curing and low shrinkage. Chapter 14 describes the preparation and characterization of photocurable acetylenic containing polymers which yield water resistant hydrophobic coatings with excellent adhesion to steel. Interestingly, films containing no acrylated diluent could be cured rapidly in both the absence and presence of a benzophenone photosensitizer. Additionally, photocuring of formulations containing multifunctional acrylate and methacrylate diluents provides latitude in tailoring coatings for a variety of uses.

Chapter 15 provides evidence for the origin of the accelerating effect of the chlorendic dianhydride group on the rate of photocuring polyester coatings. The chlorine and carbon based radicals generated in the presence of a typical photoinitiator can ultimately lead to increases in polymerization rate and crosslinking and the production of hard, durable coatings.

The photoinduced generation of a series of ordered polymers containing heteroaromatic groups in the main chain is described in Chapter 16. The results suggest that it is indeed feasible to produce photochemically high temperature resistant polymers via cydization reactions. Such highly ordered polymers have potential impact as nonlinear-optical materials.

In summarizing the results presented in Chapters 9-15, one is left with the impression that there is still room for introduction of new chemistries into the radiation curing industry.

Chapter 10

Effect of Monomer Structure on Concurrent Grafting During Radiation Curing

Stephen J. Bett[1], Paul A. Dworjanyn[2], Barry A. Fields[2], John L. Garnett[2], Stan V. Jankiewicz[2], and David F. Sangster[3]

[1]Polycure Pty. Ltd., Brookvale, NSW 2100, Australia
[2]School of Chemistry, University of New South Wales, Kensington, NSW 2033, Australia
[3]Lucas Heights Research Laboratories, Division of Polymers and Chemicals, Commonwealth Scientific and Industrial Research Organisation, Menai, NSW 2234, Australia

The role of multifunctional acrylates as additives in both grafting and curing reactions initiated by UV and ionising radiation is discussed. The model grafting system studied was styrene in methanol to the polyolefins. The role of multifunctional acrylates, particularly TMPTA and TPGDA, in accelerating these grafting processes is reported. Synergistic effects in grafting enhancement between these multifunctional acrylates and electrolytes is discussed The significance of monomer structure in these enhancement reactions is considered and a novel mechanism proposed for these processes. The relevance of this grafting work to the related fields of curing and cross-linking especially under electron beam conditions is discussed, particularly the role of multifunctional acrylates in the presence of commercial additives such as silanes. Using TMPTA and TPGDA as representative monomers, it is shown that concurrent grafting with curing can be important in UV and EB rapid polymersation work.

Radiation curable formulations are essentially solvent free. In order to overcome problems associated with viscosity reduction of the formulation needed for application pruposes, monomers, especially the multifunctional acrylates (MFAs) are useful. The presence of such monomers also influences the degree of cross-linking and the speed of cure in the finishing process (1 - 4). One of the major difficulties encountered in these radiation rapid cure reactions is to achieve good adhesion to the substrate in the very short time associated with the polymerisation. The possibility that concurrent grafting can occur with curing especially on plastics and cellulosic substrates assists the adhesion process. Typical of the MFAs used in curing reactions is trimethylolpropane triacrylate (TMPTA). When TMPTA is utilised as an additive in analogous radiation grafting reactions, the grafting yield is enhanced for a typical system such as the copolymerisation of styrene to cellulose and polyolefins initiated by sensitised UV and ionising radiation

(5). Under these conditions there appears to be a relationship between radiation grafting and curing, common mechanistic pathways occurring in both systems (4).

It is the purpose of this paper to examine the role of functionality of MFA and MFMA in the enhancement process in reactions using the copolymerisation of styrene in methanol to the polyolefins under the influence of UV as model system. The results will be extrapolated to ionising radiation work. In addition, synergistic effects involving these MFAs and other additives in both UV and radiation grafting processes will be reported. The significance of the data in UV and electron beam curing reactions will be discussed.

Experimental

The grafting procedures have previously been outlined (6). Monomers used were trimethylolpropane trimethacrylate (TMPTMA), propyleneglycol triacrylate (PGTA), tripropyleneglycol diacrylate (TPGDA), allyl methacrylate (AMA), triethyleneglycol dimethacrylate (TEGDMA), tetrahydrofurfuryl methacrylate (TFMA) and TMPTA. Benzoin ethyl ether (BEE) was used as sensitiser. Polyethlene was low density film (0.12mm) while polypropylene was isotactic film (0.10mm).
The model monomer grafting system used was the copolymerisation of styrene in methanol to the polyolefins as representative trunk polymers. In a typical experiment isotactic polypropylene film (5 x 3 cm, thickness 0.10mm) which had previously been Soxhelt extracted in benzene for 72h was immersed in solution (25mL) of monomer in solvent containing the necessary additives including photosensitizer (UV only). For the UV irradiations lightly stoppered pyrex tubes were utilized. These were positioned in a rotating rack surrounding a 90W medium pressure mercury vapor lamp. At the completion of the irradiation, the polypropylene strips were quickly removed from the tubes to minimize post-irradiation effects, washed in cold methanol, extracted in benzene in a Soxhlet for 72h to remove homopolymer, then dried at 40°C for several hours to constant weight, the percentage graft being calculated as the percentage increase in weight of the grafted strip. A similar procedure was used for the ionising radiation grafting work, experiments being performed in the spent fuel element and cobalt-60 facilities of the Australian Atomic Energy Commission, dose rates being determined by ferrous sulfate dosimetry (G(Fe)=15.6).

For the swelling studies of styrene in polyethylene, film samples of polyethylene of uniform size and weight were swollen in methanolic solutions of tritiated styrene which had been prepared by a unique tritium catalytic exchange method (Garnett, J.L. and Long, M.A., Univeresity of NSW, Sydney - unpublished data). The swelling reaction was performed in a water bath at 25°C. At completion of the swelling, substrates were rapidly blotted dry and immediately immersed in an extraction solution of 1,4-dioxane. After one week the dioxane solutions were sampled and styrene content determined by liquid scintillation counting. Repeated extractions did not give further yield of monomer. This method was found to be convenient and extremely sensitive. Results are expressed as mg styrene absorbed per gram of film with a detection limit of one mg styrene per gram of film and a reproducibility, based on ten samples per data point, of ± 1%. Counting of tritiated styrene solutions was achieved using a low background, double headed coincident liquid scintillation counter. Bray's solution was used as liquid scintillator.

For the radiation rapid cure experiments, appropriate resin mixtures containing oligomers, monomers, flow additives and sensitisers (UV) were applied to the substrate as a thin coating, the material placed on a conveyor belt and then exposed to the UV and EB sources. The time taken to observe cure for each of the samples was then measured on a relative basis. The UV system used was a Primarc Minicure unit with lamps of 200W per inch. Two EB facilities were utilised namely a 500KeV Nissin machine and a 175KeV ESI unit.

Results

MFA Additive Effects in Grafting Initiated by UV.

Inclusion of trifunctional acrylate monomers in grafting solutions in additive amounts (1% v/v) generally resulted in a very large increase in percent graft, consistent with previous preliminary work with TMPTA (5,7). TMPTA was marginally superior to TMPTMA with PGTA less effective (Table I). The magnitude of the enhancement was particularly significant at the styrene concentration corresponding to the Trommsdorff peak (8). When lithium nitrate is added to the monomer solution, a synergistic effect in grafting enhancement with the acrylate monomers is observed at certain styrene concentrations (Table II) consistent again with earlier TMPTA data (5).

Table I. Photografting of Styrene in Methanol to Polypropylene in presence of Trifunctional Monomers [a]

| Styrene | Graft (%) | | | |
(% v/v)	B	B+TM	B+T	B+P
20	5	95	100	50
30	24	282	297	113
40	45	513	496	289
50	31	285	283	131
60	22	180	281	98
70	15	–	–	–

[a] Irradiated 8 h at 24cm from 90W lamp at $20°C$; B= BEE (1% w/v); TM= TMPTMA (1% v/v); T = TMPTA (1% v/v); P = PGTA (1% v/v).

With the lower functionality MFAs and MFMAs, namely TPDGA, TEGDMA, AMA and TFMA, the grafting enhancement is decreased when compared with the trifunctional series in Tables I and II, the difunctional acrylates being more effective than their monofunctional analogues (Table III). When lithium nitrate is used as a co-additive with the difunctional acrylates, grafting is enhanced further at certain styrene concentrations. In this group of difunctional acrylates, the TPGDA result is particularly important since this is one of the most widely used monomers in radiation rapid curing work.

Table II. Synergistic Effect of Trifunctional Monomers with Lithium Nitrate in Photografting Styrene to Polypropylene [a]

Styrene	Graft (%)				
(% v/v)	L	B+L	B+L+TM	B+L+T	B+L+P
20	1.9	10	111	126	55
30	1.4	38	376	529	147
40	1.7	20	274	337	146
50	2.0	14	144	221	73
60	4.5	14	93	81	46

[a] Irradiation conditions and symbols as in Table I; relevant control data also in Table I.; L = lithium nitrate (0.25M)

Table III. Synergistic Effect of Difunctional and Monofunctional Monomers with Lithium Nitrate in Photografting Styrene to Polypropylene [a]

Styrene	Graft (%)								
(% v/v)	N.A.	A	A+L	TE	TE+L	TP	N.A.[b]	TF[b]	TF+L[b]
20	5	36	59	25	105	17	10	11	80
30	24	136	305	61	296	99	36	38	280
40	45	331	88	217	141	339	116	181	118
50	31	98	40	132	101	167	53	112	65
60	22	–	–	104	64	118	40	76	27

[a] Irradiation conditions and symbols as in Table I; N.A/= no additive ; A = AMA ; TE = TEGTMA ; TP = TPGDA

[b] Irradiated 16h; TF = TFMA.

Other co-additives in addition to lithium nitrate have also been discovered (9) to accelerate these grafting processes. Typical of these is mineral acid and representative synergistic UV grafting data for TMPTA and sulfuric acid is shown in Table IV. Sulfuric acid is a polar substrate and behaves in a manner similar to lithium nitrate in these grafting reactions.

Table IV. Synergistic Effect of TMPTA and Acid as Additives in UV Grafting Styrene to Polyethlene [a]

Styrene	Graft (%)			
(% v/v)	N.A.	TMPTA	H^+	TMPTA+H^+
20	28	28	14	41
30	101	52	126	78
40	189	321	193	266
50	124	412	107	525
60	37	133	31	188

[a] Irradiated 24 h at 24cm from 90W lamp at 20°C
BEE (1% w/v) sensitiser ; TMPTA (1% w/v) ; methanol solvent; N.A. = no additive ; H^+ = H_2SO_4 (0.2M).

Additives are also useful in radiation rapid cure work. Thus in formulations used in UV and electron beam (EB) curing, additives are utilised to control slip, gloss, flow etc. Typical of these are the fluorinated ester and the silane used in the present study (Table V). When these commercial additives are included in the monomer grafting solution, the results show that the surface active fluorinated compound enhances graft whereas the silane is a retarder, presumable due to the repulsion effect of the silicon atom. Inclusion of TMPTA in the grafting solution containing all of the above commercial additives leads to large enhancement in graft, the effect of the TMPTA overcoming the retarding properties of the silane.

Table V. Effect of TMPTA in Presence of Commercial Organic Additives on Photografting Styrene to Polypropylene[a]

Styrene (% v/v)	Graft (%)			
	B+U	B+Si	B+FE	B+Additives+T
20	5	5	5	260
30	30	18	23	588
40	46	31	53	711
50	19	13	16	368
60	13	9	19	283

[a] Irradiation conditions and symbols as in Table I; U = urea (1% w/v) ; Si = silane (1% v/v); Z-6020 ex-Dow; FE = fluorinated alkyl ester (1% v/v), FC-430 ex-3M; T = TMPTA (1% v/v)

Additive Effects with MFAs in Grafting Initiated by Ionising Radiation.

When ionising radiation is used as source of initiation for grafting instead of UV (10-14), analogous additive effects to those previously discussed, have been found. Thus inclusion of sulfuric acid in methanolic solutions of styrene leads to an enhancement in copolymerisation to a polyolefin, such as polyethylene, when irradiated by cobalt - 60 gamma rays (Table VI).

Lithium salts when used as additives also increase the grafting yields for the same reaction, lithium perchlorate being more efficient that lithium nitrate (Table VII). A comparison of the effectiveness of acid versus lithium perchlorate in the same grafting reaction is shown in Table VIII, the salt being more effective than acid only at the lower monomer concentrations i.e. 25%.

When MFAs such as TMPTA are included in the monomer solution enhancement in grafting to polyethylene is observed at certain styrene concentrations (Table IX). In the presence of both acid and TMPTA as additives, a synergistic effect in the same grafting reaction is observed (Table IX), consistent with the UV data, thus these additive effects appear to be a general phenomenon in UV and radiation grafting processes.

Table VI. Acid Enhancement in Radiation Grafting Styrene to Polyethylene in Methanol [a]

Styrene (% v/v)	Graft (%)	
	N.A.	H+
10	30	40
20	84	118
30	244	267
40	150	156
50	114	120
60	92	98
80	71	73

[a] N.A. = no additive; H+ = H_2SO_4 (0.1M); Dose rate = 3.30×10^4 rad/h; Dose = 2.0×10^5 rad; Temperature 29°C.

Table VII. Effect of Lithium Salts on Radiation Grafting of Styrene in Methanol to Polyethylene [a]

Styrene (% v/v)	Graft (%)		
	N.A.	Li ClO_4	Li NO_3
15	34	42	28
20	60	77	68
25	78	140	122
30	142	160	102
35	139	116	91
40	117	92	80

[a] N.A. = no additive; Lithium salts (0.2M); Dose rate = 3.30×10^4 rad/hr; Dose = 1.8×10^5 rad; Temperature 20°C

Table VIII. Comparison of Acid with Lithium Salt on Radiation Grafting of Styrene in Methanol to Polyethylene [a]

Styrene (% v/v)	Graft (%)		
	N.A.	H_2SO_4	Li ClO_4
15	31	32	44
20	64	70	81
25	103	148	192
30	187	240	196
35	193	212	140
40	150	157	114

[a] N.A. = no additive; H_2SO_4 (0.1M); $LiClO_4$ (0.2M); Radiation conditions as in Table VII

Table IX. Synergistic Effect of Acid and TMPTA as Additives in Grafting Styrene to Polyethylene Film Initiated by Ionising Radiation[a]

Styrene (% v/v)	Graft (%)			
	N.A.	H+	TMPTA	H$^+$+TMPTA
30	37	51	39	54
40	76	81	73	106
50	109	134	137	181
60	89	73	105	101
70	68	62	59	95

[a] N.A. = no additive; H$^+$ = H$_2$SO$_4$ (0.2M); TMPTA (1% v/v); Dose rate = 4.1 x 10^4 rad/hr; Dose = 2.4 x 10^5 ras

Discussion

Mechanism of MFA Additive Effect in UV and Radiation Grafting

The present results suggest that the enhancement effect due to MFAs in both UV and ionising radiation work is a general phenomenon. Previously the enhancement had only been observed with TMPTA and divinylbenzene (5) in very preliminary studies with both UV and ionising radiation initiating systems. Consistent with the mechanism previously proposed for this enhancement effect (5), the present results may be attributed to branching and cross-linking of the polystyrene chains which result when compounds with more than one polymerisable group are present. According to this theory, branching and cross-linking occur when one polymerisable group of the polyfunctional monomer has been bonded to a growing polystyrene chain. This growing chain has then acquired unsaturated functionality and reactivity because one or more polymerisable groups are still present on the polyfunctional monomer. These unsaturated groups may then participate in extra polymerisation or scavenging reactions with adjacent growing polystyrene chains, leading to a highly cross-linked network with many of the polystyrene chains having numerous branching points. Such a model can explain why the graft copolymer yield increases so dramatically in the presence of polyfunctional monomers.

Mechanism of Acid and Salt Effect in UV and Radiation Grafting.

An important observation recently made concerning acid and salt effects in UV and radiation grafting to polyethylene is that in the swelling of polyethylene in the presence of methanolic solutions of styrene, partitioning of styrene into polyethylene is significantly improved by the inclusion of mineral acid or lithium salt in the grafting solution. Styrene labelled with tritium was used for these sophisticated experiments which indicate that most swelling occurs within the first few minutes of exposure of backbone polymer to solution (Table X). At 25° C, at least 80% of the swelling is achieved during the first hour, the system asymptotically approaching equilibrium after 10 hours (Table XI). In each series of data, the lithium salt is more efficient than the sulfuric acid in enhancing the partitioning of monomer into the polyethylene. As representative

Table X. Variation in Styrene Absorption by Polyethylene with time: Initial Swelling Behaviour[a]

Time of Swelling (Minutes)	Styrene Absorption (mg styrene / g polyethylene)		
	N.A.	H_2SO_4	$LiClO_4$
0	0	0	0
0.08	5.1	5.3	5.6
0.5	11.9	12.4	12.9
2.0	24.9	27.2	31.3
4.0	38.0	41.2	43.1
equilibrium[b]	45.4	54.0	54.4

[a] Styrene in methanol (30% v/v) was used as model monomer solution ; N.A. = no additive ; H_2SO_4(0.1M); $LiClO_4$(0.2M)
[b] Value determined after 13 hours swelling; Temperature 25°C.

monomer solutions for these swelling experiments, 30% styrene in methanol was chosen since for most radiation conditions, peak grafting and enhancement occurs around this region.

Table XI. Variation in Styrene Absorption by Polyethylene with Time: Asymptotic Approach to Equilibrium[a]

Time of Swelling (Hours)	Styrene Absorption (mg styrene / g polyethylene)		
	N.A	H_2SO_4	$LiClO_4$
0	0	0	0
0.5	38.8	42.8	43.8
1	40.5	44.3	47.0
2	42.5	45.8	48.0
12	45.6	53.3	54.8
20	45.4	54.0	54.4

[a] Conditions as in Table X.

The swelling behaviour exhibited in Tables X, XI indicate that a major portion of the equilibrium styrene absorption occurs in the first five minutes of exposure. It is this early, rapid absorption phase that is representative of the diffusion situation operative during the similtaneous grafting process. Using this time scale of five minutes, the variation in styrene absorption for monomer concentrations in the region of the Trommsdorff peak in the presence of the two electrolytes has been examined and the results shown in Table XII. The data indicate that at styrene concentrations of 30% v/v, H_2SO_4

Table XII. Effect of Electrolytes on Styrene Absorption into Polyethylene from Solutions near the Trommsdorff Region[a]

Styrene conc. (% v/v)	Styrene Absorption (mg styrene / g polyethylene)		
	N.A.	H_2SO_4	$LiClO_4$
15	20.0	20.8	22.1
20	27.4	28.6	30.3
25	33.3	35.0	36.8
30	37.9	42.4	44.2
35	45.6	48.0	52.6

[a] Conditions as in Table X. Exposure time 5 minutes.

increases styrene absorption by 12% over additive free solutions Under the same conditions, 0.2M $LiClO_4$ increases styrene absorption by 17%. Clearly these results support the hypothesis that the presence of electrolytes significantly affects the concentration of monomer within the graft region.

Based on the above results and more extensive evidence recently reported (15 - 17) a new model has been proposed to explain the effect of electrolytes in enhancing radiation grafting. In any grafting system at any one time, there is an equilibrium concentration of monomer absorbed within the grafting region of the backbone polymer. This grafting region may be continually changing as grafting proceeds. Thus in grafting styrene to polyethylene during the initial part of the reaction, the grafting region will be essentially olefinic in nature, however, as reaction proceeds, the grafting region will become more styrenated. The degree to which monomer will be absorbed by this grafting region will therefore depend on the chemical structure of the region at the specific time of grafting. The swelling data in Tables X - XII indicate that increased partitioning of monomer occurs in the graft region when appropriate electrolytes are dissolved in the bulk grafting solution. Thus higher concentrations of monomer are available for grafting at a particular backbone polymer site in the presence of these additives. The extent of this improved monomer partitioning depends on the polarities of monomer, substrate and solvent and also on the concentration of acid. It is thus the effect of these ionic species on partitioning which is essentially responsible for the observed increase in radiation grafting yields in the presence of such additives. In the specific instances where ionising radiation is the initiating source, radiolytically produced free radicals are formed and, in the presence of electrolytes such as acid particularly, could lead to enhancement in the grafting processes, however such processes do not appear to be the predominant pathway for increasing grafting yields i.e. the partitioning phenomenon would still be expected to prevail under ionising radiation conditions.

Combined Effect of Acids and Salts with MFAs.

The results show that the pathways wereby the electrolytes (acids and salts) and MFAs enhance UV and radiation grafting are different, thus the two classes of additives can act in concert to give appreciably increased copolymersation yields. This observation is important in a preparative context since, in the presence of these additives, lower total radiation doses are required to achieve a particular percentage graft, thus radiation degradation effects on the backbone polymers are minimised during grafting. In addition, the enhancement due to these sysergistic effects reaches a maximum at the Trommsdorff peak which is the region where the length of the grafted chains is also a maximum. Thus, these additives can influence the nature and structure of the copolymer former and may lead to the synthesis of new polymeric materials.

With respect to absolute grafting yields, TMPTA and TMPTMA demonstrate the highest enhancement in UV grafting of all the MFAs and MFMAs studied. Although their grafting profiles are almost coincident and both monomers are present at the same concentration, TMPTMA has a higher molecular weight, thus it is more effective on an equivalents basis than TMPTA in increasing grafting yield. This difference may reflect the higher hydrocarbon functionality of the TMPTMA, leading to higher compatibility with the olefinic polypropylene, resulting in increased swelling of this backbone polymer, higher absorption of monomer and enhanced grafting reactivity.

Significance of MFA Grafting Work in Curing Applications

Radiation rapid cure (RRC) process involve curing oligomer/monomer mixtures in a fraction of a second either under (i) a high electron field from a low energy EB machine or (ii) sensitised UV from high pressure sources of up to 400 watts/inch intensity. In these RRC systems the ability to achieve concurrent grafting with cure is important in order to achieve good adhesion to many substrates e.g. cellulosics and plastics such as the polyolefins. In the present grafting studies monomers such as TMPTA, TMPTMA and TPGDA have been deliberately used since these materials are common components in RRC formulations. Such monomers exhibit a dual function in curing processes, namely fast rates of polymerisation and cross-linking. TMPTA and TPGDA are used predomonantly for fast cure whereas TMPTMA is slower to cure but is utilised as a highly efficient cross-linking agent especially with EB.

The present data show that TMPTA is a valuable additive for acceleratinggrafting reactions involving monomers such as styrene to the polyolefins with both UV and ionising radiation. Inclusion of monomers like TMPTA in RRC formulations offers the potential of enhancing concurrent grafting with cure. This concept is particularly relevant to those RRC formulations where additives are needed to achieve specific properties such as gloss, slip and flow. Typical of such commercial additives are the fluorinated alkyl ester and silane used in the present study (Table V). When these commercial additives are included in a monomer grafting solution, the results show that the surface active fluorinated compound enhances graft whereas the silane is a retarder, presumably due to the repulsion effect of the silicon atom. However the data show that if TMPTA is included in such grafting solutions, dramatic increases in graft are observed even in the presence of the silane retarder (Table V). The mechanistic role of TMPTA in RRC formulations is thus more complicated than hitherto considered. TMPTA not only speeds up cure and cross-linking, it can also markedly affect the occurrence of concurrent grafting during cure. Hence careful choice of MFA in RRC mixtures could lead to graft enhancement during cure with improved properties in the finished product.

The difficulty with RRC mixtures incorporating a range of such additives is that the final material is a compromise in properties of the components i.e. each additive will not necessarily perform at maximum efficiency because of possible detrimental interactions with the other components in the system. For example, TMPTA, although one of the fastest curing MFAs, is generally not used commercially because it is too brittle in polymerised form for many flexible applications. It also possesses a questionable Draize value which reflects the degree to which TMPTA influences skin irritancy factors. As a consequence, TPGDA is frequently preferred to TMPTA in RRC formulations as a reactive diluent. TPGDA possesses none of the above disadvantages of TMPTA, however in accelerating grafting reactions it is less efficient than TMPTA. Even with lower reactivity in grafting, the overall properties of TPGDA are a suitable compromise for RRC formulations. TPGDA can thus be a valuable additive in RRC work leading to enhancement in concurrent grafting during cure and yielding improved properties in the final product.

Acknowledgments

The authors thank the Australian Institute of Nuclear Science and Energy, the Australian Research Grants Committee, the Australian Nuclear Science and Training Organisation and Polycure Pty Ltd. for financial assistace.

Literature Cited

1. Garnett, J.L. J. Oil.Col. Chem.Assoc. 1982, 65, 383
2. Pappas, S.P. UV Curing: Science and Technology; Technology Marketing Corpn. : Norwalk, Conn., 1978
3. Senich, G.A.; Florin, R.E. Rev. Macromol. Chem. Phys. 1984, C24(2), 234
4. Bett, S.J.; Garnett, J.L. Proc.Radcure Europe '87, Assoc. Finishing Processes, Munich, West Germany, 1987
5. Ang, C.H.; Garnett, J.L.; Levot, R.; Long, M.A. J.Polym.Sci. Polym. Lett.Ed. 1983, 21, 257
6. Garnett, J.L. Radiat. Phys.Chem. 1979, 14, 79

7. Micko, M.M.; Paszner, J. J.Rad. Curing. 1974, 1(4), 2
8. Odian, G.; Sobel, M.; Rossi, A.; Klein, R. J. Polym.Sci. 1961, 55,663
9. Dworjanyn, P.A.; Garnett, J.L. J. Polym. Sci Polym. Lett.Ed. 1988,26,135
10. Geacintov, N.; Stannett, V.T. ; Abrahamson, E.W. ; Hermans, J.J. J.Appl.Polym.Sci. 1960, 3, 54
11. Arthur, J.C. Jr. Polym. Preprints 1975,16,419
12 Kubota, H.; Murata, Y.; Ogiwara, Y. J.Polym.Sci. 1973,11,485
13. Tazuke, S.; Kimura, H; Polym.Lett. 1978,16, 497
14. Hebeish, A.; Guthrie, H.T. The Chemistry and Technology of Cellulosic. Copolymers; Springer-Verlag: Berlin Heidelberg, 1981.
15. Chappas, W.J.; Silverman, J. Radiat.Phys.Chem. 1979,14,847.
16. Garnett, J.L.; Jankiewicz, S.V.; Long, M.A.; Sangster, D.F. J.Polym.Sci. Polym. Lett.Ed. 1985,23,563
17. Garnett, J.L.; Jankiewicz, S.V.; Lavot, R.; Sangster,D.F. Radiat.Phys.Chem. 1985,25,509

RECEIVED October 27, 1989

Chapter 11

Photoinitiated Cross-Linking of Polyethylenes and Diene Copolymers

Bengt Rånby

Department of Polymer Technology, The Royal Institute of Technology, S–100 44 Stockholm, Sweden

During recent years we have studied the photoinitiated crosslinking of various synthetic polymers, e.g. polyethylenes (HD, LLD and LD), ethylene-propylene-diene elastomers (EPDM) and unsaturated polyesters, using a UV-absorbing initiator (mainly aromatic ketones) and a multifunctional crosslinker (e.g. an allyl ether) as additives. The crosslinking is largely related to hydrogen abstraction from polymer and crosslinker, giving free radicals which form crosslinks by combination. High degrees of crosslinking have been obtained (measured as gel content by extraction) after short irradiation times (10 to 60 sec). Kinetic studies of the crosslinking reaction indicate a second order reaction at the early stages and a lower order reaction with increasing gel content of the polymer. The mechanism of photocrosslinking of EPDM elastomers has been studied in more detail using model compounds for the dienes: ethylidenenorbornane (1), dihydrodicyclopentadiene (2) and 2-heptene (3). ESR studies of the spin trapped radicals formed has established that EPDM elastomers containing (1) and (2) dienes are crosslinked mainly by hydrogen abstraction to allyl radicals via combination. Only EPDM elastomers containing (3) crosslink by abstraction like (1) and (2) as well as free radical addition to the double bonds. The reported crosslinking reactions are rapid and convenient and require only inexpensive equipment. Therefore, photoinitiated crosslinking is promising for industrial applications both for polyethylenes and EPDM elastomers.

During the last few years we have developed methods for rapid crosslinking of polyethylenes[1], ethylene-propylene-diene elastomers (EPDM)[2] and linear unsaturated polyesters[3] using UV light for initiation. Efficient crosslinking to high gel content has been obtained for these polymer systems, which contain small amounts of UV-absorbing initiator and a multifunctional monomer (a crosslinker), with short UV irradiation times. These photochemical methods for crosslinking of important commercial polymers seem to offer an attractive alternative to crosslinking by chemical initiation and high-energy irradiation which now are used. A first full report on photocrosslinking of polyethylenes and polyesters was presented in 1987 (printed in 1988)[4]. The present paper reports data on the photo-crosslinking reactions of polyethylenes and EPDM elastomers.

Experimental

The polymer samples are mixed as melts with a UV-absorbing photo-initiator (about 1 wt% of benzophenone or a photofragmenting aromatic ketone) and a crosslinker (about 1 wt% of a multifunctional allyl ether or acrylate) in a Brabender plasticorder to a homogeneous compound. Samples for crosslinking are pressed to sheets of 0.3, 2, 5 and 10 mm thickness at temperatures above the melting point of the compound.

The sheets are irradiated with UV light from a high pressure mercury lamp (1 or 2 kW HPM from Philips) in nitrogen atmosphere at a thermostated temperature (from room temperature for the elastomers to the melting range for the polyethylene) in a UV-CURE irradiator (Fig. 1). By moving the lamp L to different distances from the sample S, the intensity of the UV light can be varied about 50 times. The degree of crosslinking is measured as gel content by extraction with boiling xylene (for polyethylenes) or as swelling in cyclohexane (for diene copolymers). Crosslinking is also related to density for polyethylenes, measured in a density gradient column.

Photocrosslinking of Polyethylenes

Sheets of polyethylenes (LDPE, LLDPE and HDPE) compounded with benzophenone (BP), 4-chlorobenzophenone (4-CBP) or 4,4'-dichloro-benzophenone (4,4'-DCBP) and triallylcyanurate (TAC) as described were UV-irradiated. With 1 wt% 4-CBP and 1 wt% TAC, HDPE sheets are crosslinked most efficiently to the highest gel content (about 90%) while LLDPE and LDPE reach gel contents of 70 to 80%. The cross-linking reaction is a function of the irradiation temperature as shown in Fig. 2. One can see that accessibility and mobility of the polyethylene chains increase sharply at the melting point of HDPE (130 to 140°C) and more gradually for LLDPE and LDPE (up to about 110°C).

The mechanism of crosslinking in this system is well established. The UV quanta (340-360 nm) are absorbed by the BP molecules and excited to a singlet state (S) which is short-lived and rapidly reverts to the triplet state (T) by intersystem crossing. The T state is a rather long-lived biradical which abstracts hydrogen from surrounding molecules, e.g. polyethylene chains, which gives macroradicals on the chains (formula 1).

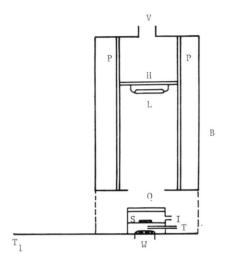

Fig. 1 Schematic diagram of the UV-irradiation equipment UV CURE:
B -- shield, H -- lamp holder, I -- N_2 inlet, L -- UV lamp, P -- stand, Q -- quartz plate, S -- sample, T -- temperature detector, T_1 -- track, V -- ventilation and W -- heating wire.

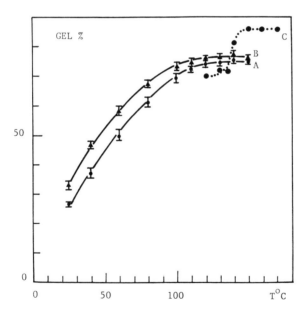

Fig. 2 Effect of irradiation temperature on crosslinking of polyethylene with 1.5 % TAC, 1 % photoinitiator and 15 sec. UV irradiation: LDPE (A with 4-CBP and B with 4,4'-DCBP) and HDPE (C with 4-CBP).

$$\underset{Ph\quad Ph}{\overset{O}{\underset{\|}{C}}} \xrightarrow[UV]{h\nu} \underset{Ph\quad Ph}{\overset{|O|S}{\underset{\|}{C}}} \xrightarrow{ISC} \underset{Ph\quad Ph}{\overset{|\overset{\cdot}{O}|T}{\underset{\cdot}{C}}} + \underset{HCH}{\overset{HCH}{\underset{|}{CH}}} \longrightarrow \underset{Ph\quad Ph}{\overset{OH}{\underset{|}{C}}} + \underset{HCH}{\overset{HCH}{\underset{|}{\cdot CH}}} \quad (1)$$

Ketyl radical

Two PE chain radicals may combine and form a crosslink. When TAC is present the BP T state may abstract hydrogen from the very reactive allyl groups. Allyl radicals are formed (formula 2) and take part in the crosslinking reaction by combining with chain radicals or other allyl radicals. Due to the high reactivity of the allyl groups, TAC has a pronounced effect on the rate of crosslinking. The ketyl radicals are rather unreactive, some combine to the dimer pinacol and some add to chain radicals and prevent crosslinks from forming. Allyl groups may also add a radical which initiates free radical polymerization.

$$\underset{Ph\quad Ph}{\overset{|\overset{\cdot}{O}|}{\underset{\|}{C}}} + R-O-CH_2-CH=CH_2 \longrightarrow R-O-\overset{\cdot}{C}H-CH=CH_2 + \underset{Ph\quad Ph}{\overset{OH}{\underset{|}{C}}} \quad (2)$$

Allyl radical   Ketyl radical

The rate of crosslinking is related to the chain length of the PE sample. This is shown in Fig. 3 where the rate of crosslinking is shown for a HDPE sample of low DP before and after adding 10 and 20 wt% respectively of a high DP sample of HDPE[5]. The added crosslinker has a strong effect on the crosslinking reaction, giving high gel content at short irradiation times (Fig. 4). Two HDPE sheets (5 mm thick) are compared, one containing 1 wt% 4-CBP (A) and one containing both 1 wt% 4-CBP abd 1 wt% TAC (Fig. 5). The sheets are irradiated for 20 sec. and then sliced by microtome to 0.25 mm thick sections for which the gel content is analysed by extraction. The added TAC is most effective in deeper layers. In deeper layers, the UV intensity is low due to absorption by BP in the top layers. Because of the high reactivity of the allyl hydrogens, the presence of TAC is increasing the rate of crosslinking also in deeper layers where only few triplet state BP molecules are formed. TAC as only additive gives insignificant crosslinking due to very weak UV absorption.

Considering the reactions in formulas (1) and (2) and the foregoing discussion of the possible reaction mechanisms, it seems likely that the photoinitiated crosslinking is a second order reaction, i.e. the rate should be a linear function of the UV intensity in square. Assuming that the rate of crosslinking (which is difficult to measure) is proportional to the rate of gel formation (which is easy to measure), the data in Fig. 6 support this conclusion at low UV intensities, i.e. at

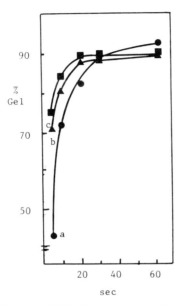

Fig. 3 Effect of different HDPE blends on crosslinking. 4-CBP: 1 wt%, TAC: 1 wt%, T: 155°C.
a: Pure Lupolen ($M_n = 2.7 \times 10^4$), b: Lupolen/Hostalen = 9/1 (Hostalen, $M_n = 9.8 \times 10^4$), c: Lupolen/Hostalen 8:2.

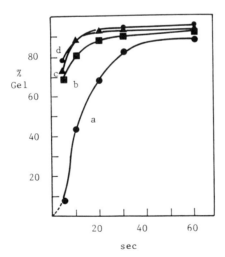

Fig. 4 Effect of TAC on crosslinking rate. 4-CBP: 1 wt%, T: 155°C, TAC: a__0 wt%, b__0.5 wt%, c__1 wt%, d__2 wt%.

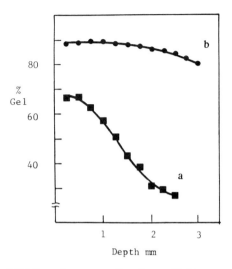

Fig. 5 Effect of TAC on homogeneity of crosslinking measured as gel content. 4-CBP: 1 %, T: 155°C, TAC: a--0%, b--1%.

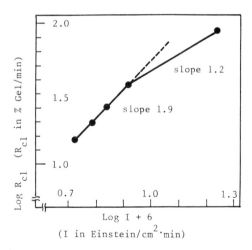

Fig. 6 A log-log plot of the rate of crosslinking measured as gel content versus light intensity for HD polyethylene.

the early stages of crosslinking. At high intensities, the amounts of gel are larger and many crosslinks are formed inside the gel phase without increasing the amount of gel. This would explain why the slope of the curve decreases from 1.9 (~2) to 1.2 at increasing UV intensity. These results are further discussed and supported in a series of forthcoming papers[5].

Photocrosslinking of Diene Copolymers

The photoinitiated crosslinking of three ethylene-propylene-diene copolymers (EPDM elastomers) has been investigated in a series of studies[6,7,8,9]. Commercial EPDM samples were used containing the following dienes; ethylidene-norbornene (ENB), dicyclopentadiene (DCPB) and hexadiene (HD). The main photoinitiators used were benzoyl-1-cyclohexanol (PI) from Ciba-Geigy and 2,4,6-trimethyl-benzoyl diphenylphosphineoxide (APO) from BASF, which both are photofragmenting (formulas 3 and 4):

$$\text{(3)}$$

$$\text{(4)}$$

The PI initiator absorbs at 320-340 nm and the acylphosphineoxide with R = phenyl group (APO) absorbs at 350-400 nm. The four radicals formed initiate polymerization by addition to acrylic monomers or abstraction of hydrogen from allyl groups. To increase the rate of crosslinking, various crosslinking agents were added to the elastomer, e.g. trimethylolpropane triacetate (TMPTA), penta-erythritole tetraallylether (PETAE), triallylcyanurate (TAC), dilimonene-dimercaptane (DSH) and dodecyl-bismaleimide (MI).

After mixing the EPDM elastomer with initiator and crosslinking agent in the Brabender plasticorder at room temperature, 2 mm thick sheets were pressed at 100°C and embedded in an epoxy resin matrix with one side uncovered. The sample sheets were irradiated for 10 min. at 15 cm distance from the UV lamp in N_2 atmosphere at room temperature in the UV-CURE (Fig. 1). After crosslinking, the rubber sheets were cut into 20 µm thin films by a Leitz microtome for IR analysis. Sample sheets were also pressed to 0.5 mm thickness, UV-irradiated for 10 min., extracted and placed in cyclohexane for swelling for one week. The crosslinking density of the gel was calculated according to Flory-Rhener's theory.

Because saturated ethylene-propylene copolymers are not crosslinked in this process, it is concluded that the crosslinking reaction is definitely related to the double bonds in the diene monomer units. However, the double bonds of the EPDM copolymers are not consumed in the crosslinking process, only decreased to some extent. The IR transmission spectra of the thin sections (20 μm) of EPDM containing ENB units has absorption peaks at 1688 and 808 cm^{-1} (assigned to C=C double bonds) both before and after crosslinking (Fig. 7). Solid state MAS NMR spectra do not give spectra sufficiently resolved to interpret the reaction mechanism.

The use of model compounds, spin trapping and ESR spectroscopy has solved the problem. Three model compounds corresponding to the three dienes in the EPDM samples were used (formula 5):

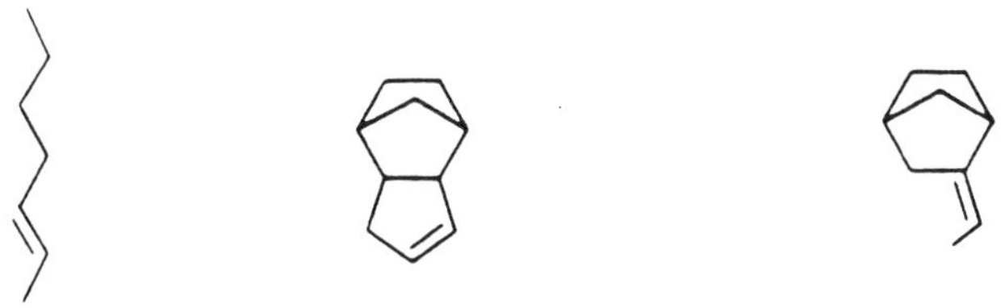

(5)

2-heptene   dihydrodicyclopentadiene   ethylidenenorbornane

As spin trap in these experiments, pentamethylnitrosobenzene (PMNB) was used. The first problem was to find if the benzoyl radical as an initiator could add to the double bonds of the model compounds. Benzoyl radicals produced from benzil by a laser beam output of 308 nm added to 2-heptene (formula 6) but not to the other two model compounds. In addition allyl radicals are formed by hydrogen abstraction (formula 7).

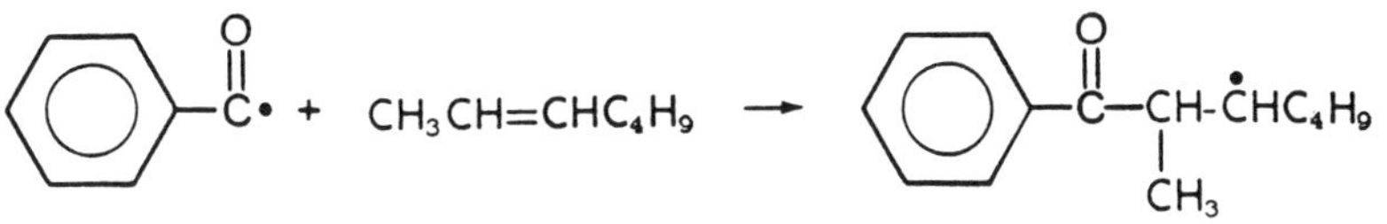

Ph-C(=O)• + CH$_3$CH=CHC$_4$H$_9$ → Ph-C(=O)-CH(CH$_3$)-ĊHC$_4$H$_9$ (6)

$$CH_3CH=CHC_4H_9 \xrightarrow{Ph-\overset{\bullet}{C}=O} \bullet CH_2CH=CHC_4H_9$$

$$CH_3CH=CHCH_2C_3H_7 \xrightarrow{Ph-\overset{\bullet}{C}=O} CH_3CH=CH\overset{\bullet}{C}HC_3H_7$$

(7)

Spin trapping with PMNB was applied to the radicals derived from initiator decomposition (formula 3) and their subsequent reactions with the model compounds (formula 5). Both initiator radicals could be trapped and identified. When model compounds were present during UV-irradiation, new radicals were identified from the ESR spectra. For dihydrocyclopentadiene (DHCPD) only one trapped radical was found and for ethylidene norbornane (ENB) two radicals. By comparison with computer simulated ESR spectra, it is concluded that the radicals of these model compounds are all allyl radicals (formula 8 and 9) formed by hydrogen abstraction from the models. Radical (8 a) has two stereoisomers but they have closely the same ESR spectra when trapped and cannot be separated. Radical (8 b) has two resonance structures (shift of double bond in the ethylidene group) but only one radical (8 b) is trapped, probably due to steric hinderance for trapping the methin radical. The DHCPD radical (formula 9) has two steric forms because the two allylic hydrogens are not identical. Once they are formed, the spin trap can only approach from one side and only one of the steric forms is trapped as shown in the ESR spectrum.

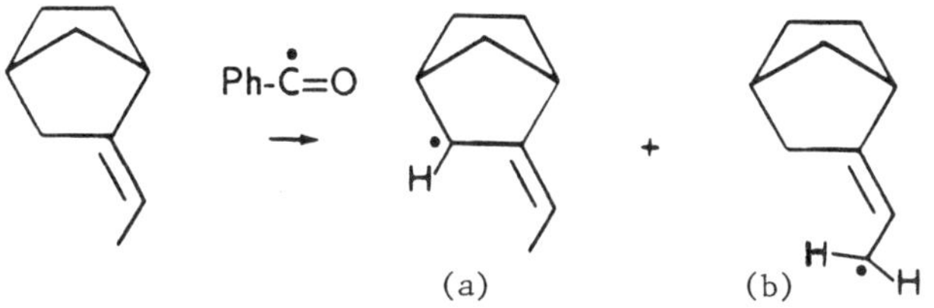

(8)

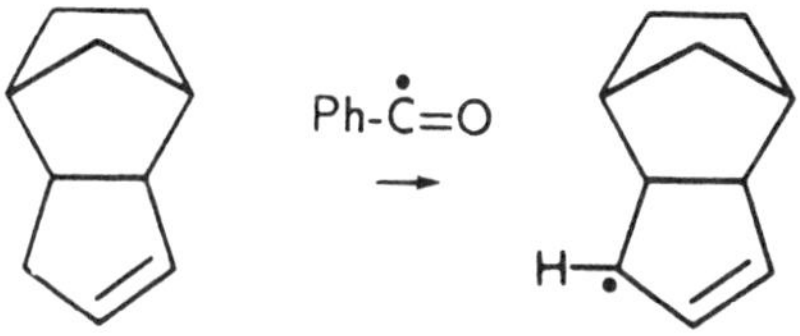

(9)

From the experiments with the model compounds it is concluded that the dominant crosslinking reaction for EPDM with ENB and CPD monomer units is formation of allyl radicals and combination of allyl radical pairs. For EPDM with hexadiene (HD) units, the model experiments indicate two crosslinking reactions: addition of initiator radicals and combination (a slow reaction due to steric hinderance), or hydrogen abstraction to allyl radicals and combination of allyl radical pairs, respectively.

The resulting crosslinking measured as swelling ratio in cyclohexane (25°C) show clearly that ENB is the most effective diene and HD the least effective (Fig. 8). The low rate of reaction for the HD copolymer is interpreted as due to addition of initiator radicals which in this case is competing with allyl radical formation.

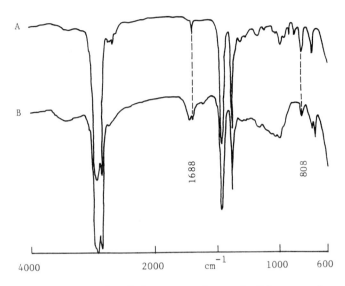

Fig. 7 IR spectra of ethylene-propylene-ethylidene norbornene copolymer (EPDM): (A) before and (B) after photocrosslinking.

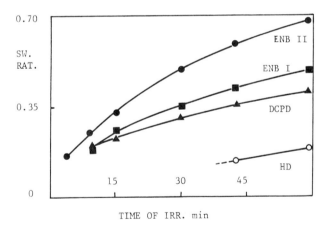

Fig. 8 Four different EPDM elastomers containing benzoyl-1-cyclohexanol have been photocrosslinked by UV-irradiation for different periods of time to products of different swelling ratios in cyclohexane (25°). ENB = ethylidenenorbornene, DCPD = dicyclopentadiene, HD = hexadiene are the dienes in the elastomer samples studied. ENB (I) contains 2.6 mol% diene, ENB (II) 10.0 mol%, DCPD 1.5 mol% and HD 1.6 mol% diene.

Acknowledgement

The research projects on photoinitiated crosslinking of polyolefins and diene copolymers have been supported by grants from the National Swedish Board for Technical Development (STU), Tour & Andersson AB (a subsidiary of Incentive AB) and a fellowship from the Wenner-Gren Foundation (to Chen Yong Lie) which is gratefully acknowledged.

References

1. Rånby, B. Photoinitiated Reactions of Organic Polymers, in Polymer Science in the Next Decade, Intern. Symposium Honoring Herman F. Mark on his 90th Birthday, May 1985 (Eds. O. Vogl and E.H. Immergut), p. 121-133, J. Wiley, New York, N.Y. 1987.
2. Rånby, B.; Hilborn, J. Photoinitiated Vulcanization of Rubber, in Proceedings Intern. Rubber Conf. IRC 86, Suppl. Vol. p. 16-27 (1986), PGI Service AB, Värnamo, Sweden.
3. Rånby, B., Shi, W.F. Polymer Preprints (ACS, Div. Polymer Chem.) 28:1, 297-298 (1987).
4. Rånby, B.; Chen, Y.L.; Qu, B.J.; Shi, W.F. Photoinitiated Crosslinking of Polyethylenes and Polyesters, in IUPAC Intern. Symposium on Polymers for Advanced Technologies, Jerusalem, Israel, August 1987, VCH Publ., New York, N.Y. 1988 (Ed. M. Lewin), p. 162-181.
5. Chen, Y.L. Photocrosslinking of Polyethylene, Inaug. Dissertation, Sept. 1988, The Royal Institute of Technology, Stockholm, Sweden, submitted to J. Polymer Sci., Polymer Chem. Ed. 1989.
6. Hilborn, J. Photocrosslinking of EPDM Elastomers, Inaug. Dissertation, May 1987, The Royal Institute of Technology, Stockholm, Sweden.
7. Hilborn, J.; Rånby, B. Photocrosslinking of EPDM Elastomers, in IUPAC Intern. Symp. on Polymers for Advanced Technologies, Jerusalem, Israel, August 1987 (Ed. M. Lewin), p. 144-161. VCH Publ., New York, N.Y. 1988.
8. Hilborn, J.; Rånby, B. Rubber Chem. Technol. 81, 568 (1988).
9. Hilborn, J.; Rånby, B. Macromolecules, 22, 1154 (1989).

RECEIVED October 27, 1989

Chapter 12

Nonacrylate Curing Mechanisms

Photoinitiated Cross-Linking Using the Amine—Ene Reaction

G. K. Noren and E. J. Murphy

DeSoto Inc., P.O. Box 5030, Des Plaines, IL 60017

The use of polymer bound amine synergists for UV-cure of non-acrylate unsaturated resins initiated by intermolecular hydrogen abstraction photoinitiators has been investigated and has been utilized to effect crosslinking of organic coatings. Benzophenone was found to be the most efficient photoinitiator for this system. Tertiary amine concentrations of about 10^{-4} equivalents per gram of resin were found to give the best cure at the lowest dosages. Good cure was obtained with methyl or ethyl substituted amines and high functionality unsaturated resins. Thermal mechanical analysis of one film showed a distance of 8.3×10^{-8} cm between crosslinks.

The area of ultra-violet curable resins has undergone considerable growth during the last 15 to 20 years. Useful resin systems which can be cured by ultra-violet light initiated free radical mechanisms can be classified as follows:
  1. Unsaturated polyester resin/styrene
  2. Thiol-ene
  3. Multifunctional (meth)acrylates

The polyester/styrene systems are low in price but have slow cure speed and present environmental and health problems due to the high volatility of the styrene monomer. They are used primarily in wood finishing. Thiol-ene systems are high in price and cure much faster but also present health problems due to the thiol component in the system. They are used in PVC flooring and in gaskets. Acrylate systems are moderately priced and are also fast curing but can cause skin irritation and have recently come under scrutiny for long term toxicity. Methacrylates are not considered to create health problems but are very slow curing. (Meth)acrylate materials have been used in many application areas.

The photoinitiation of polymerization by the use of intermolecular hydrogen abstraction is well known.([1]) Diaryl ketones, such as benzophenone, are used as the photoinitiator and a coinitiator consisting of a low molecular weight tertiary amine having at least one abstractable hydrogen atom on a carbon atom that is alpha to the nitrogen is required. These amine synergists or photoactivators form an exiplex with the excited state of the diaryl ketone and subsequently transfer a hydrogen atom to the diaryl ketone producing a ketyl radical and an alpha-amino radical. (Scheme I) While the ketyl radical has proved to be quite inefficient as an initiator, the alpha amino radical has been shown to be an effective initiator for acrylate polymerization.(2) We wish to report the application of this initiation mechanism to a system consisting of polymeric or oligomeric tertiary amines and unsaturated oligomers which do not contain (meth)acrylate functionality.

Results and Discussion

Photoinitiator Studies. A tertiary amine pendent acrylic copolymer (ACRYLIC 1) was synthesized by free radical polymerization and used for the evaluation of various initiators. Thus, 414 g of a solution containing 43% methyl methacrylate, 32% 2-ethylhexyl acrylate, 5% hydroxypropyl acrylate and 20% dimethylaminoethyl methacrylate and 1.9% (based on monomers) of AIBN was added over a 2.5 hour period to 172 g of n-butyl acetate at reflux. The resulting copolymer solution was diluted to 55% NVM with n-butyl acetate which resulted in a final viscosity of about 4000 cps. The copolymer had a theoretical equivalent weight per nitrogen of 785 corresponding to a functionality of 110 based on a theoretical molecular weight of about 87,000. Triallylcyanurate (equivalent weight per double bond = 83) dissolved in MEK (80% NVM) was used as the unsaturation source for these experiments. The ratio of unsaturation equivalents to amine equivalents (UE/AE) was 4 to 1 and a photoinitiator level of 4% based on resin solids was used. The degree of cure was determined by the solvent resistance as measured by the number of methyl ethyl ketone double rubs (MEKDR) of 1 mil films cured on an IST unit (pulsed xenon lamp UV source) at 3 J/sq-cm dosage. These results are shown in Table I.

Table I. Effect of Photoinitiator on The Cure of ACRYLIC 1/Triallylcyanurate System

Photoinitiator(4% by wt)	MEKDR @ 3 J/sq-cm
Benzophenone	30
Xanthone	23
Isopropylthioxanthone	17
Methylthioxanthone	16
4,4'-dichlorobenzophenone	15
10-thioxanthone	9
Benzopinacol	8

Based on these results benzophenone was chosen as the photoinitiator to be used in further studies.

Using the same oligomer system as above the effects of benzophenone level and dosage were studied. The results are shown below in Table II.

Table II. The Effect of Benzophenone Level and UV Dosage on ACRYLIC 1/Triallylcyanurate System

Dosage, J/sq-cm	MEKDR @ Benzophenone Level (PHR)		
	3	6	9
1	5	6	8
2	12	16	22
3	24	32	52

From this data it can be seen that both increased amounts of benzophenone and higher dosages result in increased degrees of cure. A photoinitiator level of 4% was considered practical and representative and used in further experiments.

The Effect of the Structure of the Unsaturation Source. The effect of the structure of the unsaturated component was studied using two types of amine sources. The first was the copolymer ACRYLIC 1 described above. The second was a linear polyurethane (LPUR) prepared by reacting bisphenol A (1.0 mol) with trimethylhexane diisocyanate (1.5 mol) in urethane grade MEK at 70 C for 4 hours in the presence of dibutyl tin dilaurate. Dimethylethanol amine (1.1 mol) was then added at 50 C and reacted at 70 C for 18 hours. The resultant material was 55% NVM in urethane grade MEK, had an equivalent weight per nitrogen of about 630 and had a theoretical functionality of two. The studies were done using a UE/AE ratio of 4/1 and 4% of benzophenone photoinitiator based on solid resin. The results are shown in Table III.

Table III. Effect of Unsaturation and Amine Sources on the Degree of Cure at 3 J/sq-cm

Unsaturation Source	Eq. Wt.	MEKDR Amine Type	
		LPUR($f = 2$)	ACRYLIC 1($f = 110$)
Diallyl phthalate($f = 2$)	123	4	80(a,b)
Triallyl trimellitate($f = 3$)	110	6	25
Poly(diallyl phthalate) ($f = 10$)	453	81	--
Poly(diallyl isophthalate) ($f = 17$)	317	150	--
Unsaturated polyester ($f = 4$)	213	8	12
Allyl terminated unsaturated polyester (f = High)	594	>200	>150(c)

a. Dosage was 3.5 J/sq-cm.
b. 6% Benzophenone based on solids.
c. 8% Benzophenone based on solids.

Based on this data it was indicated that the higher the functionality the better the degree of cure for allyl compounds. The use of an unsaturated polyester in place of an allyl compound appears to be detrimental to the cure but a synergistic effect occurs when both terminal allyl unsaturation and internal maleic type unsaturation exist together. In fact, the system containing the allyl terminated unsaturated polyester (ATUPE) and the linear polyurethane had >200 MEKDR at 2 J/sq-cm and 30 at 1 J/sq-cm. When the ACRYLIC 1 copolymer was substituted for the LPUR, the films had >150 and 31 MEKDR at 2 and 1 J/sq-cm respectively. The allyl terminated unsaturated polyester was chosen for further study. No advantage was observed for either amine.

The Effect of Amine Structure. In order to evaluate the effects of amine functionality and substitution on the crosslinking reaction, several amine terminated urethane oligomers were synthesized by the same procedure as used in the synthesis of the LPUR. The oligomers were prepared in Urethane Grade MEK. These amines were evaluated with the allyl terminated unsaturated polyester as the unsaturation source. The composition and the properties of the oligomers are shown in Table IV.

Table IV. Oligomers Used for the Study of Amine Structure

No.	Components	Mole Ratio	Eq. Wt.	%NVM	Viscosity(cps)
1	TMP/IPDI/DMEA	1/2.8/2.8	335	73	10,200
2	TMP/IPDI/DEEA	1/2.8/2.8	361	73	4,080
3	TMP/IPDI/DIPEA	1/2.8/2.8	388	80	5,680
4	Hexane Diol/ TMDI/ DEEA	2/3/2	551	80	6,500
5	TMP/IPDI/DEEA	1/3/3	384	74	12,000

TMDI = Trimethylhexamethylene diisocyanate
TMP = Trimethylol propane
IPDI = Isophorone diisocyanate
DMEA = Dimethylethanolamine
DEEA = Diethylethanolamine
DIPEA = Diisopropylethanolamine

Formulated coatings were made at a UE/AE ratio of 4/1 using 6% benzophenone as the photoinitiator. The degree of cure of the films after exposure to dosages of 1 and 2 J/sq-cm was again measured as the number of MEK double rubs. The results of the testing of these systems are shown in Table V.

Table V. Effect of Amine Structure on The Cure of Oligomers

Resin No.	Amine Type	Variable	MEKDR @ Dosage, J/sq-cm	
			1	2
1	DMEA	Methyl	100	>150
2	DEEA	Ethyl	92	>150
3	DIPEA	Isopropyl	45	>150
4	DEEA	f = 2	55	>150
5	DEEA	f = 3	40	>150
ACRYLIC 1	Dimethylamino	f = 110	31	>150

Very little difference was observed in the degree of cure between methyl and ethyl substitution on the nitrogen. However, in the case of the isopropyl group somewhat slower curing was observed. This was probably due to both statistical and steric effects. There also seems to be a slight advantages to lower degrees of functionality and having the functionality located on a terminal position rather than pendant to a chain.

The Effect of Unsaturation Equivalents to Amine Equivalents Ratio. The ratio of unsaturation equivalents to amine equivalents should be an important factor in controlling the cure speed of this system and would be an important factor in the determination of the contribution that the addition of the hydrogen on the alpha-carbon atom of the amine makes to the crosslinking reaction. This effect was studied by keeping the initiator concentration constant while varying the UE/AE ratio. This set of results was again obtained with the ACRYLIC 1 copolymer as the amine source and triallyl cyanurate as the unsaturation source. The benzophenone photoinitiator concentration was held constant at 4% over the whole range of UE/AE ratios studied. Curing was accomplished at 3 J/sq-cm and 6 J/sq-cm. The results are shown in Figure 1. A maximum was observed at an UE/AE ratio of 10/1 which is equivalent to 8 hydrogen atoms per double bond. However, if only 1 hydrogen is active, which might be expected due to steric hindrance, a degree of polymerization (DP) per allyl of 10 would result. This is a reasonable DP for allyl polymerization.($\underline{3}$) This UE/AE ratio also corresponds to an amine concentration of 6.19×10^{-4} equivalents of nitrogen per gram. This concentration of amine is almost equal to that used in the benzophenone initiated photopolymerization of methyl acrylate (2.3×10^{-4}) ($\underline{2}$) and also corresponds well to the level of tertiary amine synergist (4.2×10^{-5} to 4.2×10^{-4}) used with benzophenone in epoxy acrylate crosslinking systems.($\underline{4}$)

At this point we decided to separate the terminal allyl unsaturation from the internal unsaturation being contributed by the maleic type double bonds in the unsaturated polyester. A simplex mixture design was used to study a system consisting of the ACRYLIC 1 copolymer, an allyl terminated polyester (ATPE) and an unsaturated polyester (UPE). The allyl terminated polyester was prepared by reacting diethylene glycol (1 mol), azelaic acid (1 mol) and trimethylolpropane diallyl ether (0.3 mol) at 160 to 200°C while removing water as the xylene azeotrope. The ATPE had an equivalent weight per double bond of 560 and a functionality of about four. The unsaturated polyester was prepared by reacting diethylene glycol (1.5 mol), fumaric acid (1.5 mol) and benzoic acid (0.44 mol) at 160 to 200°C while removing water as the xylene azeotrope. The UPE had an equivalent weight per double bond of 332 and a functionality of about 6-7. The three component mixtures were prepared containing 4% benzophenone and were cured at 3 J/sq-cm. The results are shown in Figure 2 which indicates a maximum at a mixture consisting of 66.6% UPE, 16.7% ATPE and 16.7% of ACRYLIC 1. This mixture is at an UE/AE ratio of about 10/1 which corresponds to 2.1×10^{-4} equivalents of nitrogen per gram. This also compares favorably with the concentration used for low molecular weight amine synergists.

Thermal Behavior of Cured Films. A film from triallylcyanurate and ACRYLIC 1 copolymer was analyzed by DSC and TGA. The sample was cured at 3 J/sq-cm. The glass transition temperature was $-42°$ C (under Nitrogen) and the isothermal weight loss after 1 hr at 200°C in air was 31%.

Finally, Thermal Mechanical Analysis was performed on the

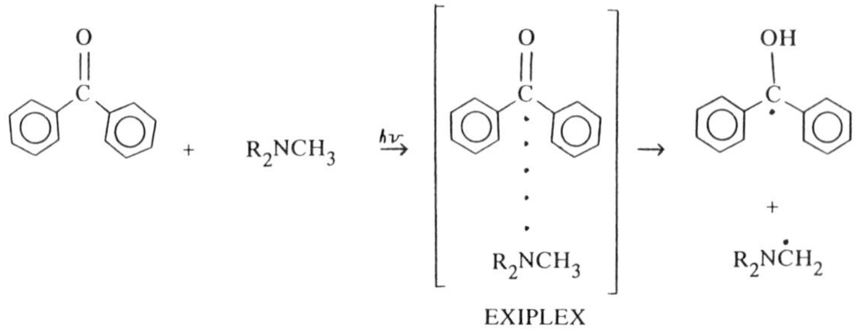

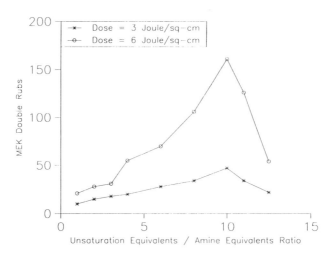

Scheme I

Figure 1. Effect of UE/AE ratio on the degree of cure of ACRYLIC 1/triallylcyanurate system.

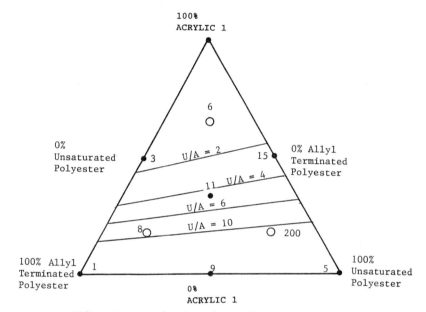

Figure 2. Simplex experimental design study of the degree of cure of ACRYLIC 1/ ATPE/UPE blends as measured by MEK double rubs.

sample that had the best performance in the simplex design. The sample was cured at 3 J/sq-cm. The results are shown in Figure 3. A calculation of the crosslink density using the modulus data at 120°C gives a value of 1.75×10^{21} per cc or 8.3×10^{-8} cm between crosslinks. Theoretical calculations using statistical methods yield crosslink densities of 10^{20} to 10^{22} for trimethylol propane triacrylate, pentaerythritol triacrylate and urethane oligomers at 60% conversion.($\underline{5}$) Thus, amine-ene systems have an equivalent crosslink density to acrylates.

Conclusions

Formulations containing tertiary amine substituted oligomers and non-acrylate unsaturated resins cured well at moderate dosages (1-2 J/sq-cm). Based on the effective concentration of amine (10^{-4} equivalents of nitrogen per gram) it appears that the amine substituted oligomers function as polymer bound amine synergists. Benzophenone was found to be the best photoinitiator for these systems. The amine nitrogen can contain either methyl or ethyl groups and still function equally well as a coinitiator. Increasing the functionality of the unsaturated resin also increases the degree of cure.

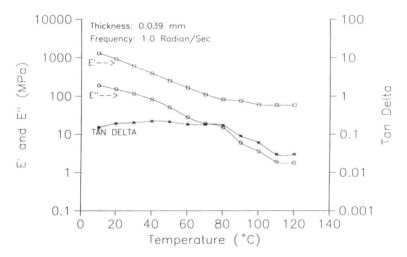

Figure 3. Thermal Mechanical Analysis of a film containing 16.7% ACRYLIC 1, 16.7% ATPE and 66.6% UPE cured at 3 J/sq-cm.

Acknowledgments

The authors would like to thank DeSoto, Inc. for the opportunity to present the findings of this research. We would also like to extend our appreciation to R. S. Conti and J. J. Krajewski for helpful discussions and advice and to R. W. Johnson for conducting the TMA work.

Literature Cited

1. Pappas, S. P. J. Radiation Curing, July 1987, page 6.
2. Sander, M. R.; Osborn, C. L.; Trecker, T. J. J. Polym. Sci., Polym. Chem. Ed., 1972, 10, 3173.
3. Lenz, R. L. Organic Chemistry of Synthetic High Polymers, Interscience, New York, 1967, p 294.
4. Christensen, J. E.; Wooten, W. L.; Whitman, P. J. J. Radiation Curing, July 1987, page 35.
5. Ishikawa, H.; Motomura, M. paper presented at the 66th Annual Meeting of the Federation of Societies for Coatings Technology, Oct. 19-21, 1988, Chicago, IL.

RECEIVED September 13, 1989

Chapter 13

Photoinitiated Cross-Linking of Norbornene Resins with Multifunctional Thiols

Anthony F. Jacobine, David M. Glaser, and Steven T. Nakos

New Technologies Group, Loctite Corporation, 705 North Mountain Road, Newington, CT 06111

A series of [2.2.1]bicycloheptenyl (norbornene) functional prepolymers have been prepared via the cycloaddition reaction of cyclopentadiene monomer with corresponding acrylics. When these materials are formulated with an appropriate multifuntional thiol crosslinker and photoinitiator and irradiated, a rapid, exothermic, crosslinking reaction takes place. When the acrylic precursors are organic resins, the derived polymers behave like toughened plastics. The choice of a norbornene functional polydimethylsiloxane precursors gives elastomeric products.

Over the years, interest in thiol-olefin (thiolene) polymerizations has waxed and waned. Early on, pioneering studies by Marvel (1-3) lay dormant until the late nineteen fifties when researchers at Esso realized the tremendous possibilities of this type of crosslinking reaction (4-6). However this polymer technology never really developed until researchers at W. R. Grace and Co. recognized the potential of linking this earlier work with the emerging field of photoinitiated polymerizations (7-10). The polymer technology developed by these researchers resulted in the granting of over one hundred and fifty United States patents and numerous technical articles and reports (11). For the most part, the research was focused on the photoinitiated addition of various multifunctional thiols such as pentaerythritol tetramercaptopropionate to allylic functionalized resins such as allyl ethers and allylic urethanes and ureas (Figure 1).

0097–6156/90/0417–0160$06.00/0
© 1990 American Chemical Society

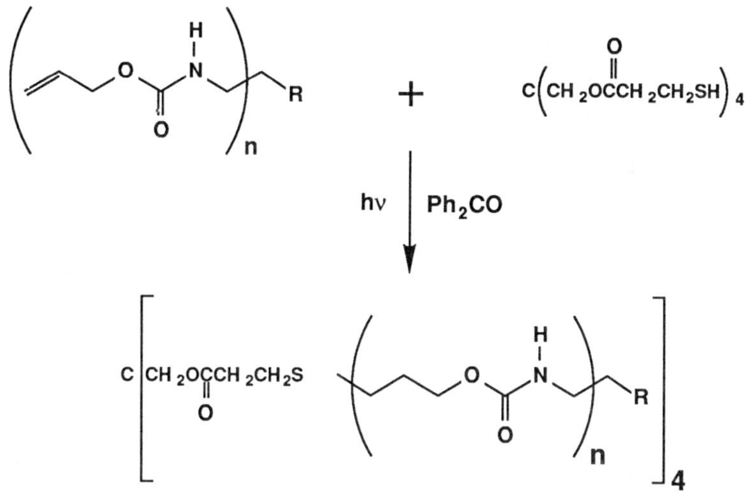

FIGURE 1: THIOLENE ADDITION POLYMERIZATIONS

Later on, attempts to balance the advantages of the thiolene polymerization mechanism with esthetic considerations showed that lower levels of thiols used as "initiators" with conventional photopolymer systems such as acrylics gave many of the advantages of the technology and attenuated the undesirable aspects (8) (Figure 2).

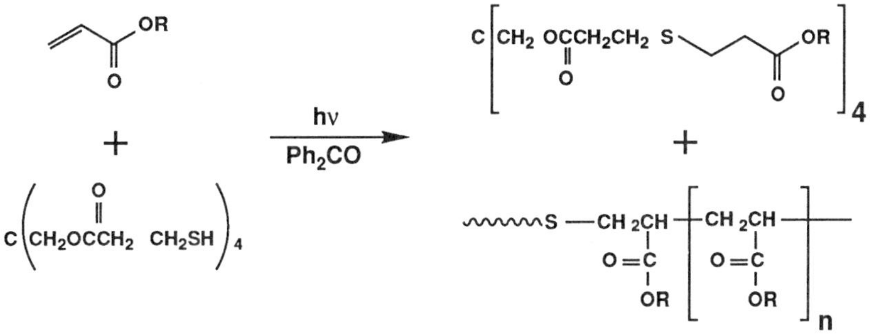

FIGURE 2: THIOL INITIATED ACRYLIC POLYMERIZATION

Probably the most important observation of this research was that in practical terms, the polymerization was not inhibited by ambient oxygen (12) and thus was differentiated from photoinitiated acrylic or methacrylic polymerizations. Structural limitations on the reactivity of the olefinic component were noted (13) and the observation that internal olefins exhibited lower reactivity, in general, agreed with earlier studies that showed that cyclic olefins such as cyclohexene or cyclopentene had low relative reaction rates (14) with thiols when compared with terminal olefins. The present study has determined that certain cyclic olefins, such as [2.2.1]bicycloheptenyl (norbornenyl) derivatives, exhibit high reactivity toward the thermal or photocatalyzed addition of thiols. In addition, this study has shown that norbornenyl functional resins may be readily prepared from either the reaction of polyols or polyamines with norborn-2-ene-5-carboxylic acid chloride or via the reaction of acrylic derivatives with cyclopentadiene monomer.

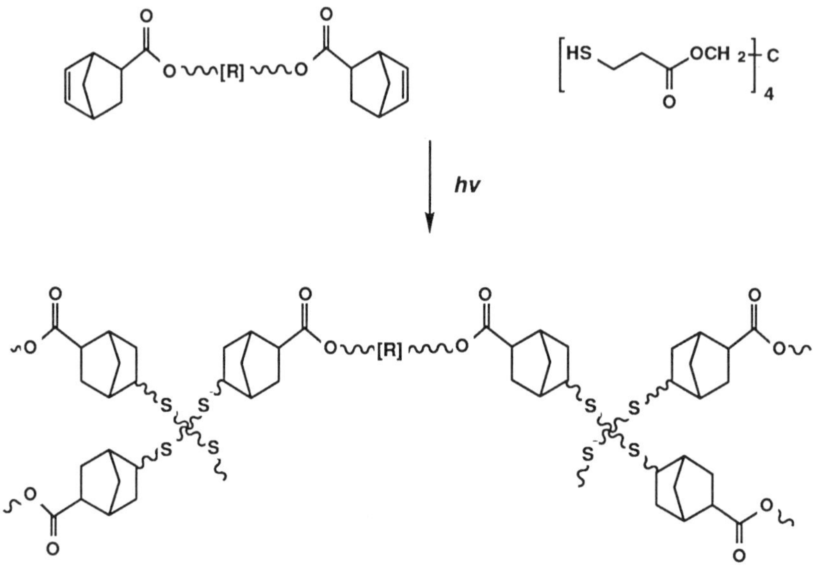

FIGURE 3: PHOTOCROSSLINKING OF NORBORNENE RESINS WITH THIOLS

Photocrosslinking of the derived norbornenyl functional resins with conventional multifunctional thiols in the presence of a photoinitiator gives crosslinked polymers that behave like toughened plastics (Figure 3). Elastomeric material can be obtained from crosslinking norbornenyl siloxane fluids with thiol siloxanes.

EXPERIMENTAL

General: Most reagents were used as received without further purification. Cycloaddition reactions were carried out under inert atmosphere and great care was taken to exclude oxygen from the cyclopentadiene cracking still. Norbornene functional resins were mixed with equivalent amounts of crosslinking thiols such as pentaerythritol tetra-(3-mercaptopropionate) (Argus Chemical Co., Evans Chemetics) and photoinitiator to give UV curable formulations. Physical testing was performed on an Instron™ Tester. NMR spectra were recorded on a JEOL

FX-90Q instrument and infrared spectra were recorded on a Nicolet MX-1 or Digilab FTS 2 FT-IR instrument. All norbornene resins and intermediates were completely characterized via HPLC, NMR, FT-IR, and GC analytical techniques and gave satisfactory results. Silicone prepolymers were compounded with 35% of a reinforcing fumed silica filler and photoinitiator. All formulations were cured under a medium pressure mercury vapor lamp until constant durometer values were obtained. Differential photocalorimetric studies were carried out on a DuPont Model 930 photocalorimeter in air at several temperatures. Dynamic mechanical analyses were performed on a Polymer Laboratories DMTA instrument. Organic resins were analyzed as single cantilevered beams at 1 Hz frequency with a temperature scan of 5°C per minute. Silicone elastomers were analyzed under the same conditions.

Norborn-2-ene-5-carbonyl Chloride: Acryloyl Chloride (90. 5 g, 1.0 mol) is stirred under nitrogen in toluene (400 mL) in an externally cooled 1-L, four necked, round bottomed flask equipped with an efficient condenser, a constant pressure addition funnel, a thermometer, and mechanical stirring. Freshly cracked cyclopentadiene monomer (15) (72.6 g, 1.1 mol) is added dropwise at such a rate that a reaction temperature of 55-60° C is maintained. Upon completion of the addition, the reaction mixture is slowly warmed to 90° C and stirred for one hour. The cooled reaction mixture is then concentrated under reduced pressure and distilled *in vacuo* to give 143 g (92% Th.) of the product as a clear liquid bp 50-55° C (0.9 mm Hg).

2-Ethyl-2-(hydroxymethyl)-1,3-propanediol Tri-(norborn-2-ene-5-carboxylate) (TMP TN): Trimethylolpropane Triacrylate (338 g, 1.0 mol) is stirred under nitrogen in a 1-L, four-necked, round-bottomed flask equipped with a Freidrichs condenser, a thermometer, a constant pressure addition funnel, and mechanical stirring at 40° C. Freshly cracked cyclopentadiene monomer (217 g, 3.3 mol) is added at such a rate that the temperature of the reaction slowly climbs to about 90° C by the end of the addition. The reaction mixture is stirred at this temperature for two hours. The extent of reaction is monitored by the decrease in the infrared absorption band at 1636 cm^{-1}. When the reaction is judged to be complete (no change in the infrared absorption), excess cyclopentadiene is removed by vacuum concentration of the resin. The yield of resin is 534 g (quantitative conversion).

Bis-2,2-[4-(2-[Norborn-2-ene-5-carboxy]ethoxy)phenyl]propane, (Ethoxylated Bis Phenol A Di-(norborn-2-ene-5-carboxylate, EBPA DN): Ethoxylated Bis Phenol A Diacrylate (*Sartomer 349*, 700 g, 1.44 mol, 2.88 eq., determined by NMR) is stirred in a 2-L, four-necked flask equipped as described above. Cyclopentadiene monomer (198 g, 3.0 mol) is added at such a rate as to keep the reaction temperature at about 90° C at the end of the addition. The extent of reaction is monitored by HPLC (acetonitrile-water, UV detector 254 nm). When the reaction is complete the reaction mixture is warmed to 120° C and excess cyclopentadiene monomer and dimer is removed by vacuum concentration. The norbornene functionalized resin is recovered in quantitative yield, nmr spectroscopy is used to determine the absence of acrylic unsaturation and the equivalent weight of the resin (integration of bicyclic unsaturation *versus* ether methylene groups).

2-(Chlorodimethylsilyl)-propenyl-norborn-2-ene-5-Carboxylate:
Acryloxypropenylchlorodimethylsilane (24) (97.48 g, 0.476 mol) was stirred under nitrogen at 50°C in a round-bottomed flask equipped with a thermometer, an efficient condenser, and a constant-pressure addition funnel. Freshly distilled cyclopentadiene monomer (33.0 g, 0.50 mol) was added dropwise at such a rate that the temperature slowly climbed to 90°C. When the addition was complete, the reaction was aged at 70°C for two hours. The solution was then distilled *in vacuo* to yield the product as a mixture of isomers (colorless liquid, bp 98-100°C 0.1mm Hg). Yield: 106.2 g (82% Th., four isomers detected by GLPC).

Norbornene Terminated Poly(dimethylsiloxane) Prepolymers: Silanol fluid (RP 48V 3500 Rhone Poulenc Chimie, 500 g, 0.038 eq SiOH) was stirred under nitrogen at 70°C with triethylamine (5.28 g, 0.052 mol). Norborn-2-ene-5-carboxypropenyl-2-chlorodimethylsilane (12.86 g, 0.048 mol) was added. The reaction mixture was stirred at 75°C for three hours. The reaction mixture was then diluted with hexane and filtered through diatomaceous earth. Removal of solvent on a rotary evaporator gave the product as a clear, colorless fluid.

General Preparation of Methyl(mercaptopropylsiloxy) dimethylsiloxy Silicone Copolymers (Silicone thiols): Mixtures of tetrakis(mercaptopropyl)tetramethylcyclotetrasiloxane, octamethylcyclotetrasiloxane, decamethyltetrasiloxane, and trifluoromethanesulfonic acid were stirred under nitrogen at 100°C for twenty hours. The reaction mixtures were then cooled and diluted with toluene and washed with aqueous sodium bicarbonate. Concentration on a rotary evaporator followed by filtration through diatomaceous earth yields the products as clear oils.

Determination of Relative Reaction Ratios of Olefins: Comparison of relative reaction rates was done by gas chromatographic analysis of reaction mixtures with addition of a suitable internal standard such as biphenyl. A mixture of the olefin to be compared and norbornene (both olefins in excess) were reacted with a deficiency of ethyl mercaptoacetate at various temperatures. Relative retention times were compared with those of independently prepared and characterized mercaptoacetate-olefin adducts. Comparison of the relative amounts of each product in the reaction mixture gave the relative, competitive reaction rate. The independently synthesized adducts were also examined to determine that the mode of addition (Markownikov *vs.* anti-Markownikov) of thiol or thiyl radical was not dependent on reaction conditions.

RESULTS AND DISCUSSION

<u>MODEL STUDIES:</u> Early in this study it appeared that [2.2.1]bicyclic olefin resins added conventional crosslinking thiols in a rapid, exothermic, manner. These results appear to contradict earlier reports that internal olefins and cyclic olefins such as cyclohexene and cyclopentene react only slowly with thiols. In reality, [2.2.1]bicyclic olefins represent a separate class of reactive olefins. These results are also consistent with reports (<u>16-19</u>) that bicyclic olefins such as norbornadiene are quite reactive to the addition of monofunctional thiols and thiyl radicals. In order to quantify the relative reactivity of norbornene resins with other "standard" ene components, a model study of the addition reaction was undertaken. A "typical" thiol (ethyl mercaptoacetate) was examined in a series of competitive reactions in which there was a deficiency of olefin (Figure 4). Olefin substrates that were compared were norbornene, styrene, butyl vinyl ether, [2.2.2]bicyclooctene and phenyl allyl ether. The results of that study are listed below in Table I.

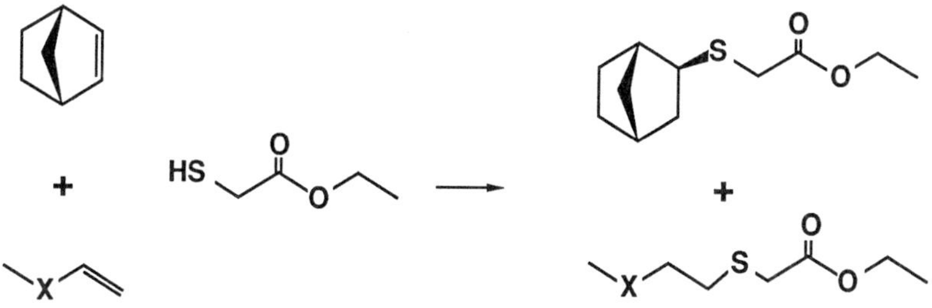

FIGURE 4: COMPETITIVE THIOL ADDITION

TABLE I
RELATIVE OLEFIN ADDITION RATES

OLEFIN	RELATIVE RATE
Norbornene	10.0
Styrene	6.6
[2.2.2]Bicyclooctene	1.4
Butyl vinyl ether	1.0
Phenyl allyl ether	0.3

These results are probably due to a combination of several effects. The enhanced reactivity of the [2.2.1]bicyclic system is most likely due to a steric ring strain factor (16-18) which is complicated by a stereoelectronic effect (19). In addition to these factors, it is probable that the kinetics of the initial addition of thiol or thiyl radical to the olefin is "less reversible" in the case of [2.2.1]bicyclic olefins This was confirmed by our model studies described above. The product distribution in the competitive reaction studies was essentially invariant with reaction temperature in the range of 10°C - 90°C. There appeared to be no equilibration of the discreet olefin thiol adducts.

The effect of ring strain on the reactivity of the norbornene ring system with respect to thiols was elegantly demonstrated by Cristol and his co-workers in their studies of the addition of thiophenols and other radical precursors to norbornadiene and norbornene (16-18). This effect is also observed in other reactions of [2.2.1]bicyclics. Turner and co-workers (25) observed that the heats of hydrogenation within a series of olefins (both cyclic and internal) reached a maximum for norbornadiene and norbornene and is about 5 Kcal/mol higher than "unstrained" olefins (TABLE II). It is interesting to note that the heat of hydrogenation for [2.2.2]bicyclooctene is in this lower range indicating it is relatively strain free.

TABLE II
OLEFIN HEATS OF HYDROGENATION (25)

Olefin	Heat of Hydrogenation (Kcal/mol)
trans-butene	27.6
cis-butene	28.5
cyclooctene	22.9
bicyclo[2.2.2]octene	28.0
norbornene	33
norbornadiene	68

Since "normal" thiolene polymerizations are well known not to be inhibited by the presence of oxygen, we also focused our model study on the nature of the interaction of norbornene, thiol, and oxygen. When the addition-cooxidation reaction of thiophenol and norbornene is catalyzed by slowly bubbling air through the reaction mixture, two products are obtained. The major product (ca. 60%) is the expected addition product *exo*--2-thiophenylnorbornane (from *cis-exo* addition of PhS and H). The other product is the cooxidation product, *exo*--2-phenylsulfinyl-*exo*--3-hydroxynorbornane (Figure 5).

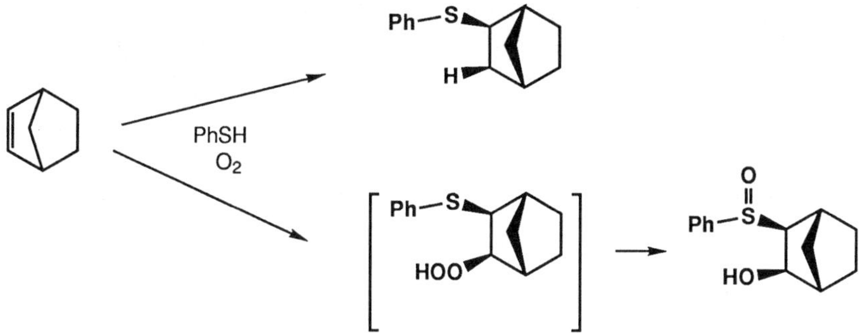

FIGURE 5: THIOL-OLEFIN CO-OXIDATION

The accepted mechanism of thiolene reaction with oxygen is generally illustrated as the stepwise, reversible, addition of thiyl radical followed by reaction of the incipient β-carbon radical center with oxygen to give a β-hydroperoxysulfide. This type of hydroperoxysulfide has been isolated by Oswald (4-6) and rearranges to give the hydroxysulfoxide. It is interesting to note in the case of norbornene that the addition of oxygen is *exo* specific (determined by high field NMR). This stereochemical result could indicate either the presence of a configurationally stable carbon radical to give this product or perhaps a pre-equilibrium between oxygen and thiyl radical to give Ph-S-O-O-H which adds in a concerted, *exo* specific manner. *Exo* specific additions of sulfur and selenium to norbornene are currently under extensive investigation by Bartlett (26-27) and others (28). Convincing arguments are made for a bimolecular concerted addition of an S_3 unit. This type of concerted addition of -S-H or -S-O-O-H is not inconsistent with the recently published mechanistic studies of Szmant and co-workers which presents NMR evidence for the "complexation" of the thiol S-H group with the electron rich pi system of olefins (20-23).

CROSSLINKED NETWORK MODEL CALCULATIONS: Our research has shown that tetrathiol crosslinked norbornene resins form a densely crosslinked, three dimensional network. Recently there has been considerable interest in crosslinked networks from a theoretical and practical point of view (29-31). As part of our study we attempted to analyze the polymer network using the Miller-Macosko formalism as applied by Bauer (32). For the purposes of this analysis we assumed that the curable formulation was an A_2 (ene) B_4 (thiol) type system. We also assumed, based on HPLC analysis of EBPA DN and acrylate precursor batches that the norbornene resins was a mixture of oligomers consisting of difunctional olefin (85%) and monofunctional olefin (15%). The thiol crosslinker was assumed to be essentially tetrafunctional. Furthermore, we made the not unreasonable assumption that there would be no thiol-thiol or norbornene-norbornene reactions. In one case, a chain extending dithiol (glycol dimercaptopropionate, GDMP) was used as part of the thiol component to probe the effect of chain extension on gel point. The results of the network calculations are illustrated graphically in Figures 6, 7, and 8 which plot extent of reaction versus molecular weight of the mixture.

The effect of the monofunctional oligomer is clearly seen as the gel point is shifted from the ideal value of about 58% to 63% (32). A similar but more pronounced shift in gel point is also noted when a chain extending dithiol (GDMP) is used to replace half of the crosslinker. In this case, illustrated in Figure 7, the gel point is shifted to approximately 78% conversion of ene.

In the case of norbornene functional siloxane fluids a much more complex case arises due to the oligomeric nature of both the thiol siloxane and the ene siloxane components which are reacting in the mixture. Gel permeation chromatography of both reactants confirms the expectation that each is a broad distribution of products (M_w/M_n for the thiol siloxane is typically 2.2 and about 1.8 for the norbornene siloxane). For the purpose of the calculation we assumed an idealized distribution as noted in Figure 8. The relationship between percent conversion of the ene oligomer and molecular weight of the mixture is also depicted in Figure 8. Current work in our laboratories is aimed at correlating the predicted, calculated molecular weights and actual experimental results. We are particularly interested in factors effecting conversion beyond the gel point.

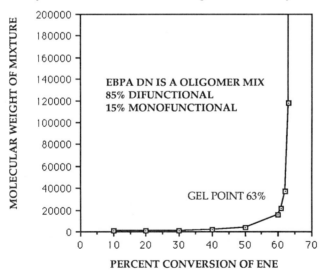

FIGURE 6: GEL POINT CALCULATION FOR A CROSSLINKED THIOL-NORBORNENE RESIN SYSTEM

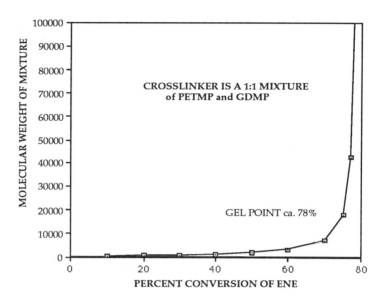

FIGURE 7: GEL POINT CALCULATION FOR A 50% CHAIN EXTENDED SYSTEM

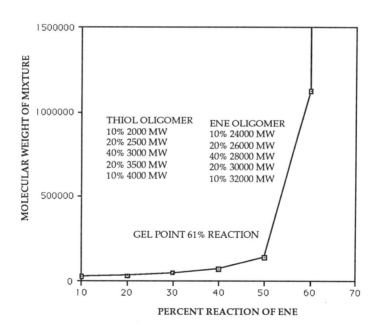

FIGURE 8: GEL POINT CALCULATION FOR NORBORNENE SILICONE OLIGOMER SYSTEM

THERMAL ANALYSIS STUDIES: The results of our model study indicates that [2.2.1]bicyclic olefins do exhibit a high relative reactivity to the addition of thiols. We have also initiated a study of the polymerization reaction with differential photocalorimetry (photo DSC). The results of this study indicate that the enthalpy of the photopolymerization increases to a maximum of ca. 224 J/g at 320°K and decreases thereafter as the temperature continues to increase (illustrated in Figure 9). The enthalpy per equivalent of ene and thiol is roughly 100,000 J/eq (23.9 Kcal/eq) at 320°K.

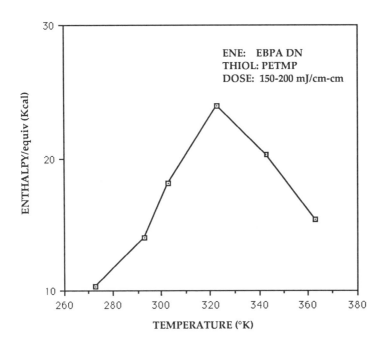

FIGURE 9: ENTHALPY OF CROSSLINKING REACTION VERSUS TEMPERATURE

We suspect the cause of this behavior is due to two distinct temperature effects. The lower enthalpy values at lower temperature is probably due to restricted molecular motion which limits curing below T_g. As the curing temperature gradually increases so does the enthalpy which reaches a maximum, as curing is carried out 10-15°C above T_g.

The rapid decrease in enthalpy above 320°K is more difficult to interpret. This phenomena is still under investigation in our laboratories and we cannot speculate on this phenomena at this time.

Dynamic mechanical thermal analysis of several of the norbornene functional organic resins and the silicone resins gave relatively unremarkable results. The maximum for tan δ peaks were in good agreement with T_g determined by DSC. The silicone elastomer (with 35% fumed silica as reinforcing filler) exhibited a T_g of ca. -90°C and a T_m at ca. -30°C which is typical for this type of polymer.

BULK PHYSICAL PROPERTIES TESTING: The results of physical testing on various norbornene functionalized resin backbones are summarized in Table III and Table IV. In general, the polymers derived from ethoxylated bisphenol A backbone show reasonable tensile strength and modulus. The shrinkage of these polymers on curing is generally 3-6% depending, of course, on the ene component and its functionality.

TABLE III
POLYMER PHYSICAL PROPERTIES[†]

RESIN	T_g °C	TENSILE (MPa)	ELONGATION (%)	MODULUS (MPa)
EBPA DN	38	30.35	3.83	1434
EBPA DN w 2.5% TMP TN	39	33.10	4.04	1500
EBPA DN w 5% TMP TN	41	33.58	3.85	1603
EBPA DN w 10% TMP TN	42	42.6	3.88	1839

[†]Resins crosslinked with pentaerythritol tetramercaptopropionate under medium pressure Mercury lamp. Testing according to ASTM D-639 Type 1.

TABLE IV
SELECTED POLYMER PHYSICAL PROPERTIES

RESIN	T_g	SHRINKAGE	DUROMETER	n_d
EBPA DN	38° C	3.17 %	83 Shore D	1.52
HDDN	22	3.16	73 Shore A	1.54
PEG[400] DN	-22	3.16	71 Shore A	1.52
ETMPTN	35	5.44	78 Shore D	1.54

NORBORNENE TERMINATED SILICONE FLUIDS: Our efforts to expand this technology have also focused on the preparation α,ω-norbornenyl functional silicone fluids as precursors to photocured elastomers. The results of this work indicate that indeed, thiol-norbornene photoinitiated crosslinking will produce high elongation elastomers with physical properties comparable to those obtained from the photopolymerization of silicone acrylates (24). These properties are listed below in Table V. The synthetic methodology that we

TABLE V
PHYSICAL PROPERTIES OF NORBORNENYL SILICONE CROSSLINKED WITH A SILICONE MULTITHIOL

RESIN	TENSILE (MPa)	ELONGATION (%)	TEAR (pli)	DUROMETER Shore A
28,000 MW N-ALKENYL TELECHELIC	7.51	546	150	37
28,000 MW N-ALKYL	5.40	350	126	33

Each Contains a 3000 MW Pentafunctional Thiol Crosslinker and 35% w/w Fumed Silica Filler

have developed allows the preparation of either norbornene-propenyl (alkenyl) or norbornene-ethyl (alkyl) terminated silicone fluids. Silicone fluids with pendant norbornenealkyl functionality are also available via standard hydrolytic or equilibrium polymerization techniques. Polymerization studies have indicated that an optimum in cured properties is achieved when the alkylthiol functional silicone crosslinker is ca. 3000 MW and pentafunctional.

CONCLUSIONS

The preceding study has demonstrated [2.2.1]bicycloheptenyl functionalized resins can be useful and interesting ene components in photoinitiated thiolene polymerizations. The addition of thiols to the unsaturation of this bicyclic system appears to be rapid and exothermic. The relative rates of this addition compared with allylic derivatives and vinyl ethers are quite favorable. The organic resins can be readily prepared from either polyols, polyamines, or acrylic precursors (Figure 10) and the yields are generally quite good. When acrylate esters are used as precursors, the cycloaddition reaction occurs spontaneously and no catalysis of the reaction is necessary.

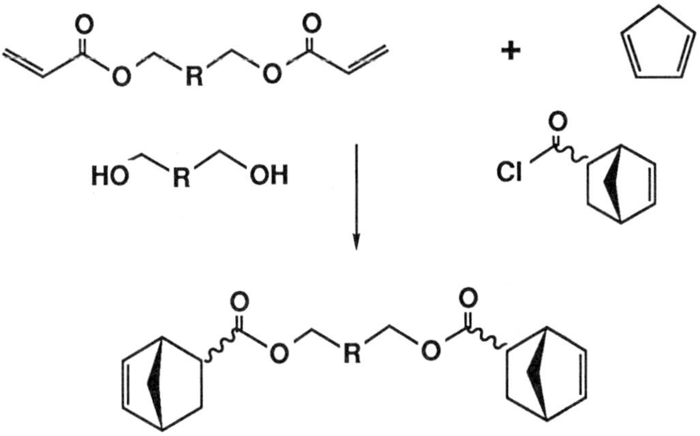

FIGURE 10: NORBORNENE RESIN PREPARATION

The residual levels of acrylic functionality after the cycloaddition reaction are usually quite low (< 0.5%).

Norbornene functional dimethylsiloxane resins are also accessible by this type of chemistry (Figure 11). Acryloxypropenyl terminated polydimethylsiloxane fluids also undergo a cycloaddition reaction with cyclopentadiene monomer to give a good yield of the desired silicone fluid. Endcapping of silanol fluids with the appropriate N-alkyl or N-alkenyl halosilane will also give the desired prepolymer. When these silicone prepolymers are photocrosslinked with a crosslinking thiol functional siloxane, elastomers with high tensile strength and elongation are obtained.

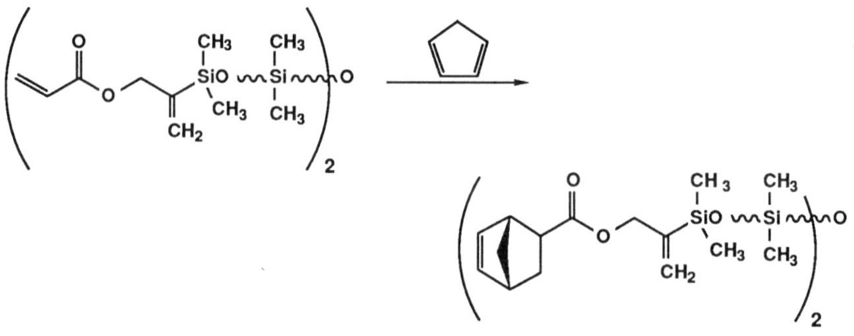

FIGURE 11: NORBORNENE FUNCTIONALIZED SILOXANES

Norbornene functional resins, unlike many other cyclic olefinic materials (e.g. cyclohexenyl or methylcyclohexenyl functional resins), will undergo a rapid photoinitiated crosslinking reaction with multifunctional thiols that may also be initiated with standard thermal initiators such as benzpinacol, AIBN, etc. in conjunction with photoinitiation. The polymers that are derived from the photocrosslinking process exhibit reasonable physical-mechanical properties and low shrinkage on cure. Since a wide variety of structural types are available as acrylate multifuntional norbornene precursors, materials ranging from toughened plastics to high elongation elastomers can be obtained by judicious choice of precursor resin structure precursor.

LITERATURE CITED

1. Marvel, C. S. and Chambers, R. R., *J. Am. Chem. Soc.* **1948,** 70, 993.
2. Marvel, C. S. and Aldrich, P. H., *J. Am. Chem. Soc.*, **1950,** 72, 1978.
3. Marvel, C. S. and Markhart, A. H., *J. Polym. Sci.*, **1951,** 6, 711.
4. Oswald, A. A., Griesbaum, K., and Hall, D. N., *Polymer Preprints,* **1967,** 8, 743.
5. Griesbaum, K., *Angew. Chem. Internat. Edit.,***1970,** 9, 273.
6. Oswald, A. A. and Noel, F., *J. Org. Chem.*, **1961,** 26, 3948.
7. Kehr, C. L. and Wszolek, W., Paper presented to the Division of Organic Coatings and Plastics Chemistry, American Chemical Society, Dallas, Texas, April 1973.
8. Gush, D. P. and Ketley, A. D., Chemical Coatings Conference II, Radiation Coatings, National Paint and Coatings Association, Washington, D. C., May 10, 1978.
9. Guthrie, J. L. and Rendulic, F. J., U. S. Pat. 3,908,039 (1975). Morgan, C. R., U. S. Pat. 4,020,233 (1977).
10. Ketley, A. D., and Morgan, C. R., U. S. Pat. 4,125,644 (1978).
11. Morgan, C. R., Magnotta, F., and Ketley, A. D., *J. Poly. Sci. Poly. Chem. Ed.*, **1977,** 627.

12. Hoyle, C. E., Hensel, R. D., and Grubb, M. B., *Polymer Photochemistry,* **1984,** *4,* 69.
13. Reference 6, p. 638.
14. Walling, C. and Helmreich, W., *J. Am. Chem. Soc.,* **1965,** *81,* 1144. Ohno, A. and Oae, S., *Organic Chemistry of Sulfur,* Plenum Press, New York, 1977.
15. Moffet, R. B., *Organic Synthesis, Coll. Vol. IV,* 238, J. Wiley and Sons, New York, 1963.
16. Cristol, S. J., Brindell, G. D., and Reeder, J. A., *J. Am. Chem. Soc.,* **1958,** *80,* 635.
17. Cristol, S. J., Russell, T. J., and Davies, D. I., *J. Org. Chem.,***1965,** *30,* 207.
18. Davies, D. I., Parfitt, L. T., Alden. C. K., and Claisse, J. A., *J. Chem. Soc. (C)* , **1969,** 1585.
19. Ito, O., and Matsuda, M., *J. Org. Chem.,* **1984,** *49,* 17. Ito, O., *Nippon Gomu Kyokaisha,* **1988,** *61(3),* 201.
20. D'Souza, V. T., Nanjundiah, R., Baeza, J., and Szmant, H. H., *J. Org. Chem.,* **1987,** *52,* 1720.
21. D'Souza, V. T., Iyer, V. K., Szmant, H. H., *J. Org. Chem.,* **1987,** 52, 1725.
22. D'Souza, V. T., Nunjundiah, R., Baeza, J., and Szmant, H. H.,*J. Org. Chem.,* **1987,** *52,* 1729.
23. Chung, M. I., D'Souza, V. T., and Szmant, H. H., *J. Org. Chem.,* **1987,** *52,* 1741.
24. Okamoto, Y., Crossan, D. I., Ferrigno, K., and Nakos, S.T., in *Adhesives, Sealants and Coatings For Space and Harsh Environments,* Plenum Press, New York, 1988.
25. Turner, R. B., Meador, W. R., and Winkler, R. E., *J. Am. Chem. Soc.,* **1957,** *79,* 4116.
26. Bartlett, P. D., and Ghosh, T., *J. Am. Chem. Soc.,* **1988,** 110, 7499.
27. Bartlett, P. D., and Ghosh, T., *J. Org. Chem.,* **1987,** *52,* 4937.
28. Dirlikov, S. K., Paper presented to the Division of Organic Chemistry of the American Chemical Society, Los Angeles, California, September 1988, Abstract ORGN 18.
29. Kloosterboer, J. G., *Adv. Poly. Sci.,* **1988,** *84,* 1.
30. Macosko, C. W., *Br. Polym. J.,* **1985,** *17,* 239.
31. Dusek, K. and MacKnight, W. J., in *Cross-Linked Polymers,* Dickie, R. A., Labana, S. S., and Bauer, R. S., Ed., American Chemical Society Symposium Series 367, Washington, D. C., 1988.
32. Bauer, D. R., *Journal of Coatings Technology,* **1988,** *60,* 53.

RECEIVED September 29, 1989

Chapter 14

Acrylated Melamines in UV-Curable Coatings

Joel J. Gummeson

Monsanto Chemical Company, 730 Worcester Street, Springfield, MA 01151

Acrylated melamines have both acrylic and alkoxy functionality. The acrylate functionality allows the melamine to be UV cured by a free radical mechanism. The alkoxy groups may be cured by a condensation mechanism.

The triazine ring has six reactive sites that can be used to prepare materials with a range of unsaturated functionality. The result is that a number of acrylated materials can be prepared to meet a variety of process requirements.

This paper demonstrates the utility of incorporating acrylated melamines into UV curable coating formulations. It reports on coating properties obtained by replacement of common acrylated oligomers by acrylated melamines.

BACKGROUND

Acrylated melamines can be made by reacting acrylamide with etherified melamine resins thereby incorporating acrylic functionality onto the triazine ring.[1] Table 1 describes all of the Acrylated Melamine (AM) resins used in the paper. Some work was done with the base resin without solvent or diluent (AM 1-3) and other work refers to the diluted resin (AM 4-10). Table 2 compares the properties of two AM resins. Figures 1-2 show typical structures of these base resins. The AM in figure 1 has been modified with oleamide in addition to acrylamide. These base resins are available in a variety of diluents and solvents. Figure 3 shows the viscosities of these resins in N-vinyl pyrrolidone (NVP).

The experimental work presented in this paper is intended to show the utility of AM oligomers in UV cure formulations and the effect combined acrylamide content has on the UV cure properties of acrylated melamines. This work will consider three cure processes: UV, thermal, and dual cure. In this case dual cure is UV followed by a thermal bump.

Table 1. Description of acrylated melamines used

Reference in Paper	Resin Structure	Moles Acrylamide	Percent AM	Diluent
AM 1	A	2.0	100	None
AM 2	B	2.5	100	None
AM 3	B	1.5	100	None
AM 4	B	0.5	75	TPGDA
AM 5	B	1.0	75	TPGDA
AM 6	B	1.5	75	TPGDA
AM 7	B	2.5	75	TPGDA
AM 8	A	2.0	75	NVP
AM 9	B	2.5	75	NVP
AM 10	B	2.5	75	Dowanol PM

Table 2. Typical physical properties of the AM resins

Base Resin	AM 1	AM 2
Molecular Weight (No. Avg.)	1130	970
Average Functionality	3.6	4.0
Solution Properties @ 75% Resin Solids in NVP:		
Density @25 C	1.099	1.093
Viscosity @25 C	10,000 CPS	16,200 CPS

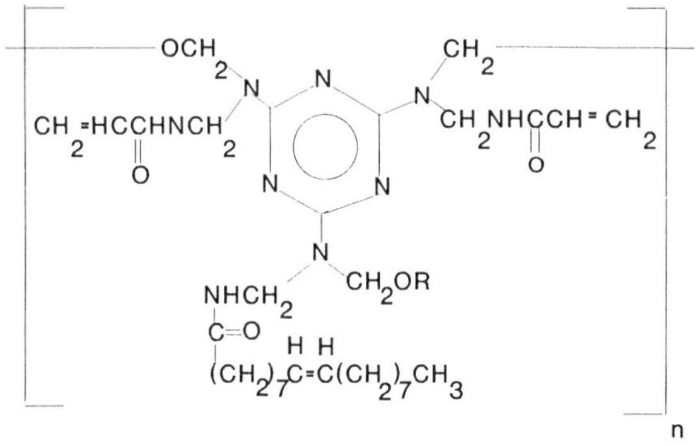

Figure 1. Type A acrylated melamine typical structure.

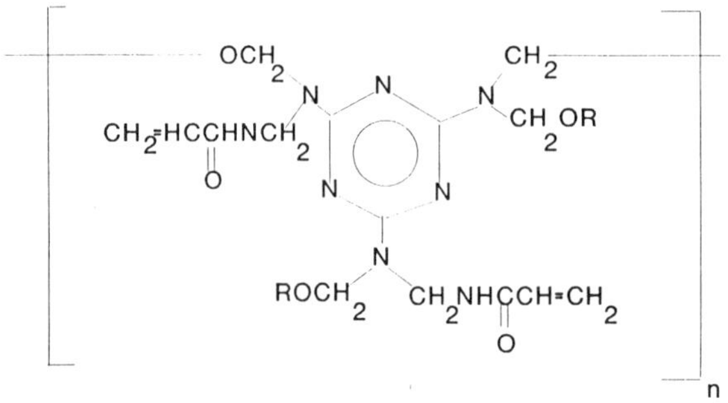

Figure 2. Type B acrylated melamine typical structure.

EXPERIMENTAL

Acrylated Melamine resins were prepared by reacting different levels of acrylamide with a fully alkylated coetherified melamine resin containing methyl and butyl groups at an approximate ratio of 1:1. They were prepared with 0.5, 1.0, 1.5, and 2.5 moles of acrylamide per triazine ring. Throughout the paper a bis-phenol A epoxy acrylate resin (Ebecryl 3700 from Radcure Specialties) and an aliphatic urethane acrylate resin (Ebecryl 8800 from Radcure Specialties) are used as controls.

UV Cure. Shrinkage upon UV Cure was determined for AM 8, AM 9, and the epoxy acrylate, all at 75% in NVP. Shrinkage was determined from the density difference before and after UV cure. The density measurements were made in pycnometers at 25° C.

AM 8, and AM 9 were evaluated in an overprint varnish formulation for high speed litho applications. The viscosities of the formulations were measured with a Brookfield RVT viscometer. Curing was done with an RPC UV processor with two 200 watt/inch medium pressure Hg vapor lamps or with a Fusion Systems processor with one 300 watt/inch "H" bulb.

Thermal Cure. Thermal cure of an AM 10 was compared with that of Resimene 755 a methyl butyl coetherified melamine resin[2]. Both melamines were formulated in a clear topcoat at 25, 40, and 50% of total resin solids (TRS) with Joncryl 500 (S.C. Johnson & Sons) as the polyol coreactant.

Dual Cure. Films were prepared for Dynamic Mechanical Analysis (DMA). All films were cast on release paper with a 4.5 mil draw down bar, and partially cured with two 200 watt/inch lamps at half power and a belt speed of 200 ft/min. The films were intentionally under cured to facilitate cutting with minimal flaws. After the films were cut into ½-inch test pieces, they were cured with two 200 watt/inch lamps at 100 ft/min, equal to 260 millijoules/cm^2 dose. The instrument used for the DMA work was a Rheometrics RSA II Solids Analyzer. All tests were made at a frequency of 11 hz with a nominal strain of 0.05%, under nitrogen. Both temperature scans, at 2°C/minute, and isothermal runs were made.

An FTIR was used to observe the changes in UV cured films. The same coating formulations were used for the epoxy acrylate and oleamide modified AM but also included 3% PTSA. An initial scan was made for these materials on salt plates. The plates were rescanned after a UV exposure of 260 mj/cm^2 and again as the films were cured at 150° C.

RESULTS

UV Cure

Viscosity. The viscosities of AM resins decrease with decreasing acrylamide content. Figure 4 shows the viscosity of the 1.5 mole base resin (AM 3) without solvent or diluent. Figure 5 shows the

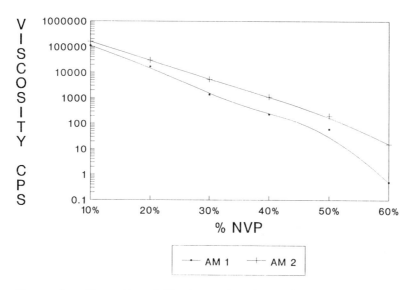

Figure 3. Viscosity of the AM resins in N-Vinyl Pyrrolidone.

Brookfield RVT, Thermosel, #27 Spindle

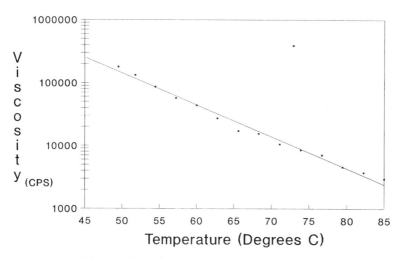

Figure 4. Viscosity of AM3 base resin.

viscosity of the base resin in TPGDA (AM 6) as a function of resin content. A wide range of viscosities is possible by adjusting the combined acrylamide content and the amount of monomer. However, solubility in TPGDA decreases with increasing acrylamide content and resins with more than 2 moles of acrylamide are not completely soluble.

Viscosities of overprint varnish formulations containing AM or epoxy acrylate are shown in Table 3. Viscosities of the AM formulations are in the same range as that of the epoxy acrylate.

Shrinkage. The results in Table 4 show that the shrinkage of AM resins is less than the epoxy acrylate control. It has been reported that for UV Curable coatings used in mirror construction, it is desirable for good adhesion to have coatings with less than 10 percent shrinkage.[3]

Film Properties. The overprint varnish formulations of Table 3 were applied to clay coated paper stock over black lithographic ink with a #3 rod and cured with the RPC processor. Properties evaluated one day after cure, are shown in Table 5. The data show the epoxy acrylate cures faster than the AM resins, and in this high speed lithographic formulation cured at a low energy level (46 mj/cm^2) a high degree of crosslinking is not expected. This is reflected in the low MEK double rubs seen for all formulations. Even though the AM resins cure slower they still exhibit improved crosshatch and scratch adhesion, and improved abrasion resistance. More through cure can be obtained with increased radiation levels or in the case of AM resins by addition of thermal cure.

In another experiment, formulations with oligomers in TPGDA were coated over Bonderite 40 steel as shown in Table 6. Panels were cured with exposures ranging from 200 to 2300 mj/cm^2 with a Fusion "H" lamp. Figure 6 shows the average Tukon hardness with AM 6 is higher than with the urethane acrylate but significantly lower than with the epoxy acrylate. The film hardness for all films is very consistent for cures over the entire exposure range. Table 6 shows the results of cross hatch tape adhesion and mandrel bend tests for the films cured with approximately 2000 mj/cm^2 of UV energy. Both the AM and epoxy films have good adhesion. The flexibility of the AM film is intermediate between that of the urethane and epoxy acrylate films.

Accelerated Weathering. The color and Q-U-V aging of AM 6 was compared with the epoxy acrylate and aliphatic urethane acrylate. For this evaluation all resins were used at 70% in TPGDA and formulated with 4% Darocur 1173 as the photoinitiator. Formulations were drawn down on white clay coated board with a #14 rod and cured with the Fusion processor. The color of the films was measured with a D25D2 Color/Difference meter from Hunterlab. Figure 7 shows initial and aged film color for the AM is significantly lower than that of the urethane acrylate, and far lower than that of the epoxy acrylate. Some improvement in epoxy acrylate color can be obtained by blending with AM 6.

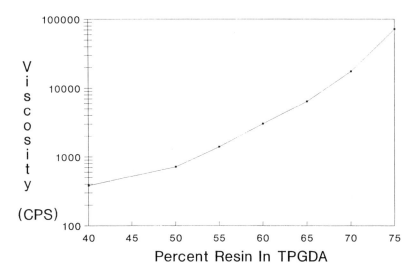

Figure 5. Viscosity of AM6 in TPGDA.

Table 3. Overprint varnish formulations containing AM or epoxy acrylate

Formulation	I	II	III
3700 @ 75% TRS in NVP (1)	32	--	--
AM 2 (at 75% in NVP)	--	32	--
AM 1 (at 75% in NVP)	--	--	32
TMPTA	36	36	36
PETA	11	11	11
Triethanolamine	8	8	8
Benzophenone	6	6	6
1173 (2)	5	5	5
L-7602 (3)	2	2	2
Viscosity (CPS @25 C)	205	242	188

(1) Ebecryl 3700 from Radcure Specialties
(2) Darcur 1173 from EM Industries
(3) SILWET L-7602 from Union Carbide

Table 4. Shrinkage of AM resins compared to epoxy acrylate

	AM 1	AM 2	Epoxy Acrylate
Sample 1	6.3%	5.8%	6.5%
Sample 2	5.4%	5.1%	8.3%
Sample 3	6.1%	5.8%	6.7%
Average	5.9%	5.6%	7.2%
Std. Dev.	0.330	0.386	0.806
Variance	0.7	0.7	0.7

The .7% variance was calculated statistically by pooling the standard deviation values for all four experiments and is based on a 95% confidence level.

Table 5. Cured film properties for the overprint varnish formulations

Formulation	I	II	III
Max. Belt Speed (Ft/Min) For Tack Free Surface With 1-200 Watt Lamp	450	250	250
(Cured With One 200 Watt/Inch Lamp @ 200 Ft/Min)			
MEK Double Rubs [1]	30	20	15
Cross Hatch Tape Adhesion Over Black Litho Ink	10% Loss	No Loss	No Loss
Scratch Adhesion Coin Rubs Over Black Litho Ink	12	50	25
Taber Abrasion [2] Mg Coating Loss	11.8	9.4	5.7

(1) Coated On White Clay Coated Paper With #3 Rod
(2) CS-10 Taber Abrasion Wheel With 500g Weight

Table 6. AM and epoxy acrylate based films cured over bonderite 40 steel

Formulation	VIII	IX	X
75% 8800 in TPGDA	75	--	--
75% 3700 in TPGDA	--	75	--
AM 6	--	--	75
TPGDA	5	5	5
NVP	10	10	10
Darocur 1173	5	5	5
Benzophenone	4	4	4
L-7602	1	1	1
Tape Adhesion (3M #610)	50% loss	No Loss	No Loss
Mandrel Bend	Passes 1/8 Inch	Fails 1/2 Inch	Passes 1/2 Inch

Films applied to Bonderite 40 panels (4 mil dry film) and UV Cured with one 300 watt/inch Fusion "H" bulb at a belt speed providing 2000 mj/sq cm of UV radiation.

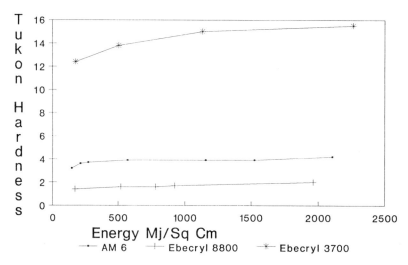

Figure 6. Tukon hardness of AM and epoxy acrylate based films.

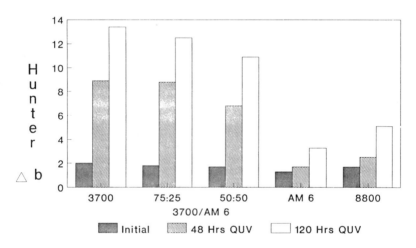

Figure 7. Color of epoxy acrylate, urethane acrylate, and AM films compared.

Thermal Cure. Since the alkoxy functionality of AM resins is significantly lower than that of unmodified melamine resins, a series of experiments were run to demonstrate that the AM resin retains the ability to cure by condensation reactions.

Figure 8 compares the Tukon hardness of the melamine and AM 10 films. Films of each material were drawn on primed Bonderite 40 with a #4 blade and cured for 30 minutes at 250° F. Both formulations were catalyzed with 0.8% PTSA on TRS. For the melamine cured films, hardness increases slightly as melamine content is increased from 25 to 40%, but then is reduced at 50% melamine content, probably due to plasticization by unreacted excess melamine functionality. For the acrylated melamine, hardness increases monotonically, probably because of the lower alkoxy functionality and higher Tg of the AM resin compared to the melamine resin. At 40% AM the Tukon hardness is similar to that achieved with a 25% melamine resin.

Table 7 shows a comparison of other properties between the 40% AM and 25% melamine containing films. With the exception of condensing humidity, the AM film has comparable properties. The lower humidity resistance of the AM film might be due to residual unsaturation. It is likely that under thermal cure conditions and with a lack of free radicals there was little reaction of the double bonds. The purpose of this experiment was to show that condensation cure was possible with AM. Under "dual cure" conditions an effort would be made to cure both functionalities of the AM.

Dual Cure

Demonstration of Dual Cure. To demonstrate thermal cure after UV cure three sets of films were prepared for Dynamic Mechanical Analysis (DMA). In the first set, four formulations were based on AM 1, AM 2, the epoxy acrylate, and the urethane acrylate. Formulations used the test resin at 65% in TPGDA, and included 3% Darocur 1173 as a source of free radicals. DMA plots for the epoxy acrylate and urethane acrylate are shown in Figure 9. The elastic or storage modulus (E') in the rubbery region is often used as a measure of crosslink density. The maximum of the tan delta plot, not shown, is taken as the Tg.[4] In this case the Tg's for the epoxy acrylate and urethane acrylate are 96 and 70° C respectively.

The DMA plot for the AM is shown in Figure 10. It shows a very broad transition, with Tg about 120° C, on the first scan. The broadness is most likely because the film is curing during the scan. This can be seen from the increase in E' between 140 and 180° C, and from the increase during the 10 minute isothermal hold at 180° C. The result is a Tg that is continuously increasing. When the film is rescanned, the E' curve has higher values indicating an even higher crosslink density. A broad tan delta peak, not shown, gives a Tg of about 160° C for the second scan. This demonstrates that temperature scanning techniques are not good tools for measuring the Tg of AM films.

Figure 11 shows that for the epoxy and urethane acrylates the elastic modulus remains constant during isothermal DMA runs at 150° C. However, the elastic moduli of AM films increase significantly for about 30-40 minutes indicating thermal cure resulting in an

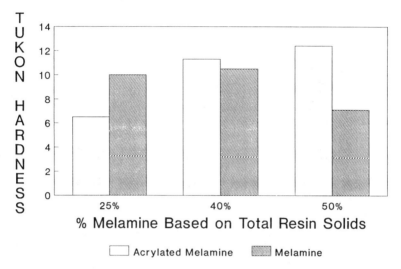

Figure 8. Tukon hardness of melamine and AM10 films compared.

Table 7. Properties of 40% AM and 25% melamine films compared

	Acrylated Melamine 40% on TRS (1)	Melamine 25% on TRS (1)
Dry Film Thickness	1.30 mil	1.30 mil
20 Degree Gloss	93	92
Tukon Hardness @ 24 Hours	11.5	10.0
24 Hr. Condensing Humidity	No Change	No Change
10 Days Condensing Humidity	Slight Haze (2)	No Change

(1) Coreactant is Joncryl 500 From S.C. Johnson & Sons
(2) Haze disappears on standing.

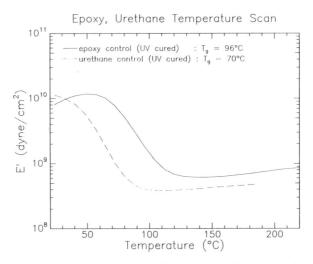

Figure 9. DMA temperature scan for UV cured epoxy and urethane acrylate films.

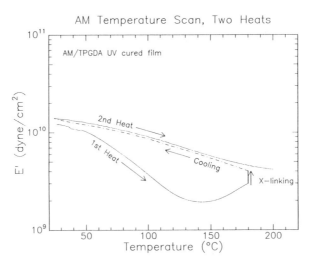

Figure 10. DMA temperature scan for a UV cured AM Film.

increase in crosslink density. Care must be taken in interpreting the magnitude of the increase in E' however. Where E' is proportional to crosslink density in the rubbery region, that is not true in the glassy region. Therefore the temperature chosen for the isothermal experiment will have a significant effect on the result. Whether the temperature is above or below Tg will determine if E' is being measured in the rubbery or glassy region. Order of magnitude increases in E' can result if the reaction drives the film into the glassy region. In this experiment one would expect the Tg of the AM film to be approaching 150° C, thus moving the measurement from the rubbery to the transition region. For the second set of films, blends of AM 1 and epoxy acrylate in ratios of 50:50, 20:80, 10:90, and 5:95 were used at 68% in TPGDA. Films were prepared as above and analyzed with isothermal DMA runs. Figure 12 shows that an increase in crosslink density can be achieved with as little as 10% AM in epoxy acrylate films.

For the third set of films, the 0.5, 1.0, and 1.5 mole acrylamide resins at 70% in TPGDA were formulated with 3% Darocur 1173. These formulations also included 2% CG21-746 UV deblockable PTSA from Ciba Geigy. Films were prepared as above and evaluated using isothermal DMA. Figure 13 shows that as the level of acrylamide functionality is increased from 0.5 to 1.5 moles the initial crosslink density is increased, with the largest increase between 0.5 and 1.0 moles of acrylamide. Thermal cure occurs with all resins, with the fastest increase in modulus seen with the 0.5 mole AM which has the most remaining alkoxy functionality.

FTIR experiments show a loss of ether functionality at 990 cm^{-1} at 150° C which is further evidence that condensation of alkoxy functionality is occurring. A loss of acrylic functionality at 1640 cm^{-1} is observed for the AM as well as for the epoxy acrylate during thermal cure. This may be the result of thermal generation of radicals or of radical trapping in the photoploymerized acrylic network which has been reported.[5]

Properties of Dual Cured Films. The properties of dual cured films containing the oleamide modified AM and epoxy acrylate were compared. The formulations used oligomers at 65% in TPGDA and included 3% Darocur 1173 and 3% PTSA. They were applied to clay coated board over black lithographic ink with a #3 rod and cured by a UV exposure of 250 mj/cm^2 followed by a thermal cure of 5 minutes at 150° C. The data in Table 8 show that the 5 minute cure at 150° C increases MEK resistance of the AM formulation, indicating additional cure. Even with increased cure the AM film maintains excellent cross hatch tape adhesion. The epoxy acrylate formulation, in contrast, loses all adhesion during the thermal bump. While it is recognized that a 5 minute at 150° C thermal cure may be impractical in many cases, useful improvements in properties may be attainable with lower cures and catalysis.

Thermal treatment of UV cured films was also used to improve stain resistance. In this work the urethane acrylate, AM6, and their blends were reduced to 70% in TPGDA and formulated with 4% Darocur 1173 as the photoinitiator. The coatings were applied over white clay coated board with a 2 mil bar and cured as in the accelerated weathering study. Initial color was determined by the Hunter

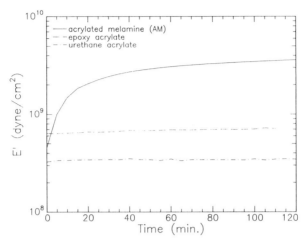

Figure 11. Isothermal scan at 150 degrees C for UV cured films.

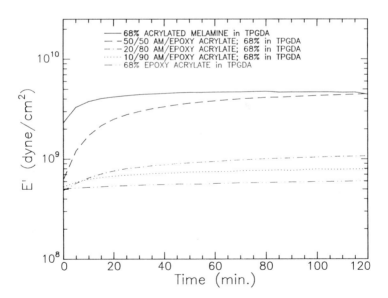

Figure 12. Isothermal scan at 150 degrees C for UV cured films of epoxy acrylate modified with AM.

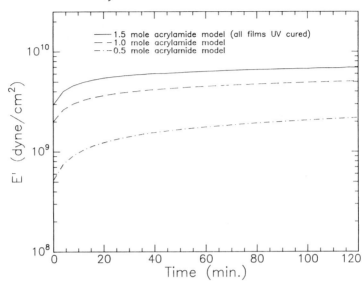

Figure 13. Isothermal scan at 150 degrees C for UV cured films of AM containing varied levels of combined acrylamide.

Table 8. The effect of a 5 min/150 degree C post UV thermal bump on the film properties of epoxy acrylate and AM

Formulation	XI	XII
AM 1 (1)	100	--
Ebecryl 3700 (1)	--	100
Darocure 1173	3	3
40% PTSA	5	5
UV Cured Only		
Tape Adhesion	0% loss	16% loss
MEK Double Rubs	196	> 300
UV + Thermal Bump		
Tape Adhesion	0% loss	100% loss
MEK Double Rubs	> 300	> 300

(1) Used at 68% TRS in TPGDA

UV Cured With Two 200 Watt/Inch lamps @ 100 Ft/Min

measurement. The coatings were then stained for 2 hours with yellow mustard. After wiping off the mustard, the residual stain was measured with the Hunter meter. Figure 14 shows that addition of AM 6 to the aliphatic urethane will significantly reduce mustard staining when a post UV thermal bump is used. The reduction in mustard stain increases with increasing levels of AM.

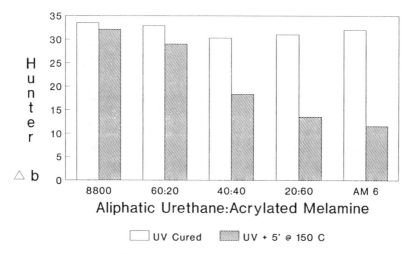

Figure 14. The effect of mustard staining on AM modified films of urethane acrylate.

CONCLUSIONS

Acrylated melamines have properties which make them suitable for use in UV curable coatings. Properties can be varied by adjusting the combined acrylamide content and the inclusion of other modifiers. Formulation viscosities are satisfactory for normal application methods, and films cure to a high degree of hardness at practical irradiation levels. Benefits demonstrated include improved tape and scratch adhesion, abrasion resistance, color, and Q-U-V durability.

Acrylated melamines can also cure by condensation reactions. The condensation reactions will proceed either before or after a UV cure. Use of acrylated melamines in a dual cure process allows a unique combination of melamine condensation and acrylate free radical addition chemistries. This can result in increased hardness, improved stain resistance, and durability.

ACKNOWLEDGMENTS

The author wishes to thank the Monsanto Physical and Analytical Science Center and Paul D. Garrett for the DMA analysis. Contributons and insights from Loren W. Hill, Robert T. Jones, David M. Lademan, Neil J. Morelli and George E. Sheldrick are also appreciated.

Literature Cited

(1) Strazik, W.F.; Leblanc, J.R.; Santer, J.O. U.S Patent 4 293 461, 1980.
(2) <u>Resimene Amino Crosslinker Resins for Surface Coatings</u>, Monsanto publication No. 6515D
(3) Sirkock, R.J.; Niederst, K.W.; Weissberg, A.B.; Greigger, P.P.; Henning, C.C. U.S. Patent 4 745 003, 1988.
(4) Hill, L.W.; Kozlowski, K. <u>J. Coatings Tech</u> 1987, <u>59</u>, 751.
(5) Decker, C.; Moussa, K. <u>J. Polym. Sci.</u> 1987, <u>A2</u>, 739.

RECEIVED July 17, 1989

Chapter 15

Coatings Curable with Low-Emission Radiation

Xiaosong Wu, Yan Feng, and Stoil K. Dirlikov

Coatings Research Institute, Eastern Michigan University, Ypsilanti, MI 48197

Radiation and thermally curable coatings based on linear polymers of propargyl ethers of biphenols have been developed and characterized for corrosion protection. Polymers and copolymers based on dipropargyl ether of Bisphenol A are probably the most attractive candidates. Radiation cure of their coatings proceeds for a couple of seconds at room temperature without photosensitizer and very fast in the presence of benzophenone. Thermal cure proceeds at $100^{\circ}C$ for about 15 mins.. The coatings have excellent adhesion to metals, hardness, flexibility, impact strength, solvent resistance, low water absorption, and corrosion resistance. They appear very attractive for corrosion protection and lithographic applications. The polymers are soluble in (meth)acrylate reactive diluents which allows the preparation of radiation curable coatings with low volatile organic compound.

Long term performance of a good coating for corrosion protection requires:
1. A primary chemical bonding between the coating and (metal/metal oxide) substrate, and
2. Resistance of the coating bonding to disturbances caused by water in the pH at which corrosion occurs.
   The hydrophilic thermosetting materials do not usually satisfy well these two requirements. Water concentrates at their coating/substrate interface, causes corrosion, delamination, and coatings failure. As a result of these problems in the hydrophilic thermosets, there is a strong need for a hydrophobic material.
   Hydrophobic coatings based on acetylenic functionalities are especially attractive. Experimental results have indicated that an acetylene bond forms a primary

bonding with different metal substrates. Acetylene terminated resins have very low water absorption (in the range of 0.5%) and excellent adhesion to metal surfaces.

The majority of the radiation cured coatings are based on polymerization of acrylic and methacrylic functionalities. Acetylenic functionality has not been practically explored (1,2).

The objectives of our research are directed towards the development of radiation and/or thermal curable materials based on linear polymers of propargyl derivatives of biphenols and their evaluation as hydrophobic coatings for corrosion protection. Polymers and copolymers based on dipropargyl ether of Bisphenol A are probably the most attractive candidates:

$$HC\equiv C.CH_2O-\phi-C(CH_3)_2-\phi-OCH_2.C\equiv CH$$

$$\Downarrow$$

$$+[C\equiv C.CH_2O-\phi-C(CH_3)_2-\phi-OCH_2.C\equiv C]_n+$$

These types of polymers have been reported by Hay, et al. (1,2) but they have never been characterized as coatings for corrosion protection. The two conjugated triple bonds of the type: $-CH_2.C\equiv C-C\equiv C.CH_2-$ are specially suitable for low emission radiation curable coatings because of the low energy required for their crosslinking.

The preparation of several propargyl monomers and diacetylene polymers and their initial characterization as radiation cured coatings for corrosion protection is discussed in the present report.

Monomer Preparation

The propargyl monomers are available in a pure state from a one-step, easy preparation which proceeds in quantitative yield from industrially produced inexpensive starting raw materials: Bisphenol A (62 cents per pound) and propargyl chloride ($5.00 per pound).
Several different methods for their preparation have been previously described in the literature (1-3):
1. Dipropargyl ether of Bisphenol A is produced in pure state with quantitative yield from Bisphenol A and pro-

pargyl bromide by refluxing their solution in acetone for 72 hours in the presence of potassium carbonate (1,2). Other monomers have been prepared in a similar manner.

2. Our initial results show that the same reaction proceeds much more rapidly in other polar solvents, as dimethylformamide, dimethylacetamide, dimethylsulfoxide, and even methanol. It is completed for two hours at room temperature in dimethylacetamide and dipropargyl ether of Bisphenol A is again obtained in pure state with quantitative yield.

3. Dipropargyl ether of Bisphenol A has also been produced with practically quantitative yield from Bisphenol A and propargyl bromide by refluxing their aqueous solution for two hours in the presence of sodium hydroxide (3). Other monomers have been obtained in a similar manner. This method has the advantage of shorter reaction times and utilization of an aqueous reaction medium. The main disadvantage of the method is Claisen rearrangement which occurs during the preparation. It results in the formation of a large amount of Claisen rearranged products: 3- or 3,3'-substituted phenols (about 50 percent), which decreases the thermostability of the final thermosets and increases their water absorption:

$$HO-\text{C}_6H_3(CH_2C\equiv CH)-C(CH_3)_2-\text{C}_6H_3(CH_2C\equiv CH)-OH$$

All of these methods are suitable for an industrial scale-up and propargyl terminated monomers might be available at a potentially low price.

The use of propargyl chloride for preparation of propargyl ethers of bisphenols has several advantages over propargyl bromide. It is cheaper, less toxic, and shock insensitive. It contains more propargyl residue per pound in comparison to propargyl bromide because chlorine is much lighter than bromine. Propargyl chloride, therefore, is especially attractive for an industrial scale-up of the process. Propargyl bromide is also industrially produced but at a much higher price ($18.00 per pound), and it is more toxic. There are also reports that it is shock sensitive and it is recommended to be used as 80% solution in toluene.

We have prepared several different propargyl monomers according to the procedure described in the experimental part for dipropargyl ether of Bisphenol A, with a general formula:

$$HC\equiv C.CH_2O-\phenyl-X-\phenyl-OCH_2.C\equiv CH$$

1). X = $C(CH_3)_2$, dipropargyl ether of bisphenol A,
2). X = $C(CF_3)_2$, dipropargyl ether of 4,4'-(hexafluoro-isopropylidene)diphenol
3). X = CO, dipropargyl ether of 4,4'-dihydroxybenzophenone,
4). X = SO_3, dipropargyl ether of 4,4'-sulfonyldiphenol,
5). X = S, dipropargyl ether of thiodiphenol,
6). X = -, dipropargyl ether of biphenol, and
7). dipropargyl ether of resorcinol.

All monomers prepared here are single compounds. Some of them are crystalline compounds (Table I) which are easily purified by known methods to satisfy the strict requirement of the electonics industry.

Table I. Melting Points of Crystalline Dipropargyl Bisphenol Monomers

X	M.P. (C^o)
$C(CH_3)_2$	84-85
CO	85-86
SO_2	189-191
-	175

The monomers have excellent thermal stability below 100^oC (for about a week) and infinite shelf-life at room temperature. They are expected to be less toxic than epoxies, isocyanates, etc. monomers which readily react with the amino groups of all biological molecules: proteins, etc.. Finally, they are neither hydroscopic nor reactive with moisture which is another advantage over water sensitive monomers such as epoxies, isocyanates, etc..

Polymerization

The polymerization of dipropargyl ether of bisphenol A as well as of the other monomers has been carried out according to the so-called oxidative polymerization or Glaser coupling by oxidation of the two terminal hydrogen atoms at the two triple bonds of the monomers and the formation of two conjugated carbon-carbon triple bonds in the polymers. The exact polymerization procedure is similar to that described by Hay (1,2) in polar solvents. For instance, we use a mixture of dimethylformamide and pyridine as a solvent and copper (I) chloride as a catalyst in the presence of N,N,N',N'-tetramethylethylenediamine. Oxygen gas has been bubbled through the reaction mixture. The polymerization is completed for 10-15 min. at room temperature. All polymers have been isolated in quantitative yield by precipitation in methanol or water.

The infrared spectra of the initial monomer, dipropargyl ether of bisphenol A and its linear polymer, are shown in Figure 1. The strong absorption band at 3271 cm^{-1} in the spectrum of the monomer corresponds to the C-H stretching vibrations of its triple bonds. This absorption band practically disappears in the spectrum of the polymer, since it does not have $C\equiv C$-H bonds. The degree of polymerization, based on the intensity of this absorption band, is more than 250 which corresponds to polymers with molecular weight in the range of 10,000 - 25,000.

It is well established that the $C\equiv C$ stretching vibrations of mono-substituted acetylenes, $C-C\equiv C$-H is observed in the infrared spectra as an absorption band in the range of 2100-2150 cm^{-1} with medium intensity. The corresponding vibrations of the disubstituted acetylenes $C-C\equiv C$-C proceed without change in the dipole moment of their triple bonds and, therefore, they are usually not active in the infrared spectra. If active, their absorption bands have very low intensity. In good agreement, the absorption band at 2117 cm^{-1} with medium intensity in the infrared spectrum of the monomer, which corresponds to the $C\equiv C$ stretching vibrations of its triple bonds, practically disappears in the spectrum of the polymer.

Lower molecular weight polymers with lower viscosity have been prepared by introducing mono-functional monomers (for instance, propargyl ether of p-cresol) into the polymerization feed which terminates the polymerization growth. The molecular weight of these polymers as well as their viscosity has been varied with the amount of the mono-functional monomer. These polymers or oligomers have lower viscosity and they may be more suitable for coatings applications.

All polymers appear quite stable at room temperature and we have not observed their premature cross-linking at storage for several months in solid state or solution.

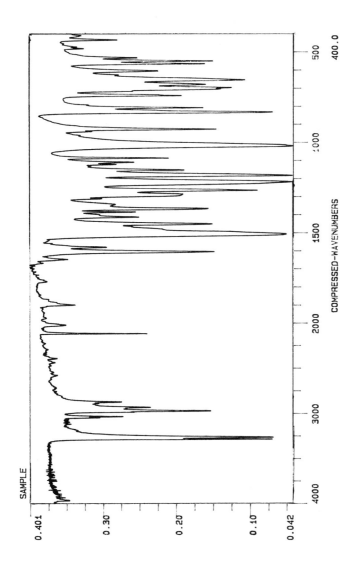

Figure 1A. Infrared Spectra of Dipropargyl Ether of Bisphenol A.

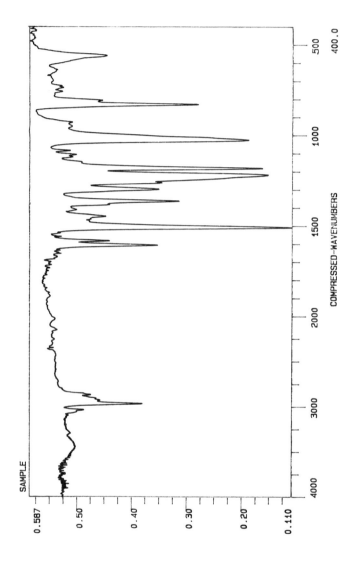

Figure 1B. Infrared Spectra of the Linear Polymer of Bisphenol A.

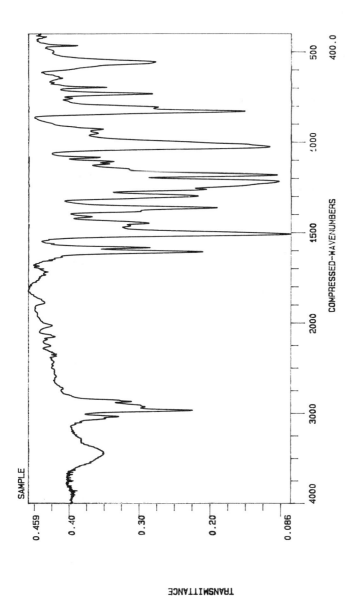

Figure 1C. Infrared Spectra of the Corresponding Cured Coating of Bisphenol A, Radiation Time of 5 Seconds in KBr.

Coatings Preparation

The coatings have been prepared on standard cold rolled steel panels (4" x 8" x 0.020") from 5 to 20% polymer solutions in different organic solvents: MEK, acetone, toluene or chlorinated solvents.

After evaporation of the solvent, the coatings are thermally or radiation cured. Radiation cure has been carried out with a medium pressure mercury lamp for 1, 3, 5, 8, 10, 13, 15, and 20 seconds in both the absence and the presence of a photosensitizer. The curing proceeds for several seconds in the absence of a photosensitizer, and very fast in the presence of benzophenone, a common photosensitizer. The two conjugated triple bonds of the linear polymers evidently require very low energy and easily cross-link with standard coatings equipment for radiation cure. The coatings, cured at radiation time longer than 10 seconds, have a pale yellow color. The thermal cure (without radiation) proceeds at $100^\circ C$ for about 15 minutes.

Coatings Characterization

All coatings used here for characterization are clear coatings based on freshly prepared polymer of dipropargyl ether of Bisphenol A. They have been prepared from solution in organic solvents in the absence of photosensitizer. The dependence of the properties of these coatings on radiation cure time is given in Table II.

If the polymer is stored in solid state and exposed to daylight for a long period of time (months) the properties of its coatings (with the exception of hardness) practically do not change with time of storage. It indicates that slow cross-linking might occur in solid state as observed by Hay (1). Such cross-linking process in storage, however, has not been observed so far if the freshly prepared polymer is stored in solution.

Adhesion. Adhesion has been studied according to the so-called cross-hatch or tape test (ASTM D3359-78) on coatings with thickness of about 1.5 mil.

The results show excellent adhesion of these coatings to standard steel surface. Adhesion of 5B corresponds to maximum (100%) adhesion. At radiation time longer than 15 seconds, however, the adhesion slightly decreases.

Hardness. Hardness has been studied by Pencil Hardness (ASTM D3363-74) and Rocker (Sward) Hardness Test (ASTM D2134-66). Coatings with thickness of about 1.5 mils have been used for Pencil Hardness. Slightly thicker coatings with thickness in the range of 1.5-1.8 mils are used for Rocker Hardness Test.

The results from both tests show that the coatings hardness is excellent. Pencil hardness is at least 3H. Our equipment, however, does not allow us to measure

Table II. Dependence of the Properties of the Coatings Based on Dipropargyl Ether of Bisphenol A Homopolymer on Cure Time

Cure time, seconds	0	1	3	5	8	10	13	15	20
Adhesion	5B	5B	5B	5B	5B	5B	5B	4B	3B
Hardness:									
Pencil	-	3H	3H	>3H	>3H	>3H	>3H	>3H	>3H
Rocker	117	91	97	91	87	90	-	70	72
Impact strength, lb/in.:									
direct	>160	>160	>160	>160	>160	>160	>160	>160	>160
reverse	>160	>160	>160	150	150	140	-	140	140
MEK rubs, number	-	11	13	19	-	160	-	-	-
salt spray, days	12	18	20	18	26	20	-	26	26

hardness higher than 3H because the lead slides back into the holder. The increase of hardness at radiation cure time longer than 3 seconds is probably associated with the formation of denser polymer cross-linked network. Rocker hardness shows about 95 $\pm$ 5. It slightly decreases at radiation time longer than 15 seconds.

Both pencil and rocker hardness depend on the time of storage of the solid initial polymer. "Older" polymers stored in solid state and exposed on daylight for more than six months produce softer coatings.

Flexibility. Flexibility has been studied as integrity of the coating on a standard metal panel bent on 180° around a standard cylinder with a diameter of 1/8" according to ASTM method D4145-83.

The coatings have excellent flexibility which does not depend on radiation time. When examined by lens or optical microscope at 15 x 10 magnification, the coatings do not show any crack formation or tape off after bending.

Impact Strength. Impact strength has been studied according to ASTM D2794-84 on coatings with thickness of about 0.5 mil on cold rolled steel panels with thickness of 1/32".

The results indicate that the coatings have excellent impact strength. The direct and reverse impact strengths are approximately >160 and 150 lb./in., respectively, in the whole range of radiation cure time of 0 to 20 seconds. Reverse impact strength only decreases slightly at cure time longer than 10 seconds.

Water Absorption. Water absorption is below 1% in water at room temperature to saturation.

Solvent Resistance. Solvent resistance has been tested by the so-called MEK rubbing test on very thin coatings with thickness in the range of 0.2-0.3 mils.

The number of MEK rubs represents the average value of four independent MEK resistance tests. MEK rubs increase with radiation time and coatings with radiation cure time longer than 10 seconds require more than 100 MEK rubs.

Corrosion Resistance. Corrosion resistance has been tested by Cleveland Humidity Chamber and Salt Spray Test on very thin coatings with thickness of 0.2 mils.

Cleveland Humidity Test (still in progress) shows no traces of corrosion for more than two months and all coatings keep their integrity in the whole range of radiation time of 0 to 20 seconds. It is an excellent result for a clear coating.

Salt Spray Test results show that the first traces of corrosion appear after more than 20 days for coatings with radiation time longer than 5 seconds. Longer radiation cure time improves the corrosion protection.

<u>Infrared Spectra</u>. The infrared spectra of the initial linear polymer based on dipropargyl ether of Bisphenol A and its radiation cured coatings (for 5 seconds) are shown on Figure 1. Both spectra are similar with the exception of a single absorption band in the spectrum of the cured coating. This weak band at 3445 cm^{-1} is assigned to the O-H stretching vibrations. We believe that in addition to the main curing reaction through crosslinking of the diacetylene bonds of the polymer, a small amount of free phenol group is formed by Claisen rearrangement under the conditions of radiation curing as it has been already observed for the propargyl groups of the propargyl terminated resins ($\underline{4}$). The proposed structure of the rearranged units of the polymer is schematically shown here:

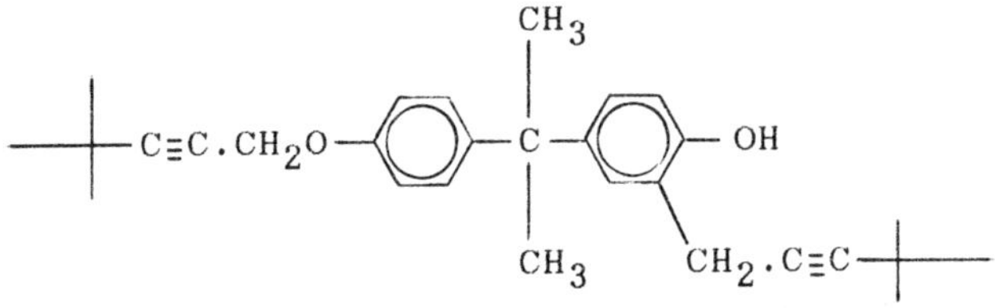

$$\underset{\underset{\text{CH}_3}{|}}{\overset{\overset{\text{CH}_3}{|}}{-}}-C\equiv C.CH_2O-\langle\bigcirc\rangle-\underset{\underset{\text{CH}_3}{|}}{\overset{\overset{\text{CH}_3}{|}}{C}}-\langle\bigcirc\rangle\underset{CH_2.C\equiv C-\underset{|}{\overset{|}{-}}}{\overset{-OH}{}}$$

The three very weak absorption bands at 463, 692, and 731 cm^{-1} correspond to traces of toluene which has been used for preparation of the coatings.

Careful comparison of both spectra does not show any other new absorption bands in the infrared spectra of the coating and only changes in the band intensities are observed. It is, therefore, difficult to establish the mechanism of curing of these polymers directly from the infrared spectra without model compounds.

This initial characterization shows that all basic properties of the coatings based on linear polymers of dipropargyl ether of Bisphenol A appear very attractive for many applications and especially as primers for metal surface corrosion protection. Radiation time of 3 to 5 seconds in the absence of a photosensitizer is enough for achieving optimum coating properties. Cure time longer than 15 seconds usually results in slight deterioration of the coatings properties, probably as a result of the formation of a denser cross-linked polymer network or a partial degradation.

In addition, our initial results have shown that the linear homopolymer of dipropargyl ether of Bisphenol A also forms good radiation curable coatings from solution in organic solvents on wood and paper as well as pigmented coatings with titanium dioxide on cold rolled steel.

Low VOC Coatings

(Meth)acrylate Reactive Diluents. It has been found that the initial linear propargyl polymers are also soluble in reactive diluents based on different acrylate or methacrylate monomers which are commonly used in radiation curable coatings. Reactive diluents are preferred over the organic solvents, because they do not evaporate. They replace the volatile organic solvent and are very attractive for low VOC (volatile organic compound) coatings for reducing the air-pollution.

The different reactive diluents, which we have used for coatings preparation, have been divided into four groups according to the solubility of the homopolymer based on dipropargyl ether of Bisphenol A and the viscosity of the resulting solutions (Table III). The first

Table III. Solubility of the Homopolymer Based on Dipropargyl Ether of Bisphenol A in Different Reactive Diluents

Reactive Diluents	Solubility
1,4-butanediol dimethacrylate	yes
N-vinyl-2-pyrrolidone	yes
neopentylglycol methacrylate	yes
tetraethyleneglycol diacrylate	yes
dicyclopentenyl acrylate	yes
iso-bornyl acrylate	yes
trimethylolpropane trimethacrylate	yes
beta-carboxyethyl acrylate	no
iso-octyl acrylate	no

two reactive diluents: 1,4-butanediol dimethacrylate and N-vinyl-2-pyrrolidone dissolve the polymer rapidly - in about half an hour. They produce solutions with lower viscosity and appear more suitable than the other reactive diluents. The second group, which consists of four reactive diluents: neopentylglycol dimethacrylate, tetraethyleneglycol diacrylate, dicyclopentenyl acrylate and iso-bornyl acrylate dissolve, the polymer slower and the viscosity of their solutions is slightly higher. Trimethylolpropane trimethacrylate requires much longer time (several days) for dissolving the polymer and its solutions have the highest viscosity. The two remaining reactive diluents (fourth group): beta-carboxyethyl acrylate and iso-octyl acrylate do not dissolve the polymer.

The solutions based on linear homopolymer of dipropargyl ether of Bisphenol A in reactive diluents (first group) rapidly cure for several seconds in the presence of a photosensitizer (benzophenone) and form excellent compatible, homogenous, transparent, amorphous coatings on cold rolled steel panels. The coatings prepared from the reactive diluents in the absence of photosensitizer, however, require longer radiation time than coatings prepared from solutions in organic solvents.

The properties of these coatings are under evaluation.

Diacetylene Reactive Diluents. An attempt for preparation of liquid reactive diluents (for low VOC coatings), based on compounds with reactive diacetylene bond, similar to that in the linear polymers, has been carried out as well. For this purpose, the propargyl ethers of a mixture of meta- and para-cresols in 1:1 ratio has been used for preparation of their dimer by oxidative coupling. This "oxidative dimerization" has been carried out under the same conditions at which the linear polymers have been prepared as described in the experimental part. The resulting diacetylene dimer consists from a mixture of three isomers: para/para-para/meta-, and meta/meta-cresols dimer:

$$\text{CH}_3\text{-C}_6\text{H}_3\text{-OCH}_2\text{.C}\equiv\text{C-C}\equiv\text{C.CH}_2\text{O-C}_6\text{H}_3\text{-CH}_3$$

We expected that this mixture would not crystallize and could be used as a reactive diluent of our polymers. The resulting isomer mixture, however, is solid at room temperature and the three dimer isomers presumably cocrystallize. Furthermore, it thermally cures at a temperature below melting. It, therefore, can not be used as a reactive diluent. We are exploring several other possibilities in this direction.

Experimental

Monomer Preparation. Preparation of dipropargyl ether of bisphenol A: a solution of 228.29 g. (1 mole) of bisphenol A in two liters of acetone was reacted with 285.53 g (2.4 moles) of propargyl bromide in the presence of

331.70 g. (2.4 moles) of potassium carbonate, by heating under reflux for 72 hours. After filtering the reaction mixture, the filtrate was evaporated to dryness on a Rotavapor. The residue was dissolved in diethyl ether and extracted with 5% potassium hydroxide and then washed with water several times. Dipropargyl ether of bisphenol A was obtained with quantitative yield after removal of the diethyl ether. The monomer was recrystallized from heptane (1 g. from 30 ml.) to yield large crystalline needles with m.p. of 85-86°C (m.p. 85-86°C)(1).

The other propargyl monomers including the monofunctional propargyl derivatives of para- and meth-cresol have been prepared in a similar manner. Their purity and structure has been proven by GC, m.p., infrared, ^{1}H- and ^{13}C-NMR spectroscopy.

Monomer Polymerization. 10 g. of monomer was dissolved in 40 ml. of dimethylformamide and added to a solution of 152 ml. of dimethylformamide, 48 ml. of pyridine, 0.8 g. of copper (I) chloride, and 1.6 ml. of N,N,N',N'-tetramethylethylenediamine. The solution was stirred and oxygen gas was bubbled through for 6 hours at room temperature. The polymer was precipitated by dropwise addition to 1500 ml of methanol, filtered and dried. Yield was 93%.

Several copolymers and the dimer of propargyl ether of isomeric cresols have been prepared in a similar manner.

Coatings Preparation. The UV radiation curing was carried out using a Radiant Curing UV Processor on coatings with different thickness from 0.2 to 2 mil (0.002" to 0.0002") with a Hanovia UV lamp system. The laboratory system utilized one 200 watt/inch, medium pressure mercury lamp. The lamp height above the coated substrate was 5.4 inches (13.7 cm.). A line speed of 50 ft./min. in air was used throughout this study. The thickness of the coatings was measured with a General Electric Magnetic Thickness Meter.

Conclusion

The initial monomers, dipropargyl ethers of bis(s)phenols and their polymers, are available in a one-step easy preparation in quantitative yield. They are suitable for industrial production in large scale and potentially available at a low price. The coatings based on these linear polymers require low energy for curing. Their radiation cure is completed for several seconds in the absence of a photosensitizer at room temperature, whereas thermal cure proceeds for half an hour at 100°C. The initial unoptimized properties of these coatings appear very attractive. They exhibit excellent adhesion to metal substrates, hardness, flexibility, impact resistance, good solvent resistance, low water absorption, and good corrosion protection. The initial results indicate that these polymers might be very attractive for corrosion protection, lithographic applications, etc.. Low VOC formulations, based on polymer solutions in (meth)-acrylate reactive diluents, produce excellent compatible, transparent coatings.

Acknowledgment

The authors thank Dr. John Graham from the Coatings Research Institute for many fruitful discussions and suggestions.

Literature Cited

1. Hay, A.S.; et al. Polymer Letters 1970, 8, 97.
2. Hay, A.S. U.S. Patent 3594175, 1971.
3. Picklesimer, L.G. U.S. Patent 4226800, 1980.
4. Feng, Yan; Dirlikov, S. Proc. 197th ACS Mtg.: Polym. Mat.: Sci. and Eng., 1989, 60, 618.

RECEIVED September 13, 1989

Chapter 16

Polymeric Coatings Containing Chlorendic Anhydride

Decomposition Products of Chlorendic Anhydride

John C. Graham and David J. Gloskey

Coatings Research Institute, Eastern Michigan University, Ypsilanti, MI 48197

Chlorendic anhydride (4,5,6,7,8,8-hexachloro-3a,4,7,7a-tetrahydro-4,7-methanisobenzofuran-1,3-dione, I, CAS registry number 115-27-5) has been used since 1964 in polyester coatings that can be cured under ultraviolet (UV) light. The presence of the chlorendic moiety in the polymeric backbone is observed to increase the curing rate of the polymer under these conditions, presumably because of the generation of Cl radicals. In our studies, we have photolyzed chlorendic anhydride with and without photoinitiators in acetone and dioxane and examined the products generated by GC/MS procedure. The participation of these products in the rate of UV initiated curing of chlorendic containing polymeric coatings is discussed.

Chlorendic anhydride (CAS Registry No. 115-27-5) has been used since 1964 in the synthesis of UV curable coatings (1-12). In previous studies designed to evaluate the effect of chlorendic based diluents on the rate of cure and physical properties of ultraviolet (UV) curable coatings, two chlorendic based reactive diluents (I and II) were synthesized and substituted in place of conventional reactive diluents in acrylate based UV curable formulations (13). The results, which are reproduced elsewhere (13), show that the addition of the chlorendic diluents (I and II) increased the rate of cure, mar resistance and physical properties of UV cured coatings. Although the enhanced rate of UV cure observed with chlorendate based coatings has been ascribed to the cleavage of carbon-chlorine bonds leading to an increase in the concentration of free radicals (12,13), no experimental results

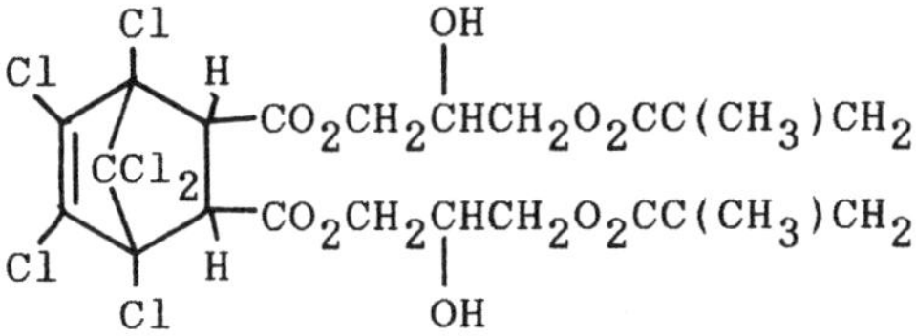

Structure I:
Cl₂C=C(Cl)–C(Cl)= ... (norbornene with CCl₂ bridge, Cl, H substituents)
–CO₂CH₂CHCH₂O₂CC(CH₃)CH₂
       |
       OH
–CO₂CH₂CHCH₂O₂CC(CH₃)CH₂
       |
       OH

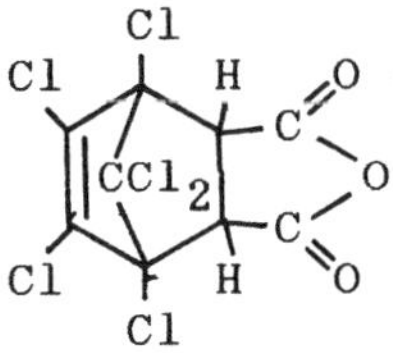

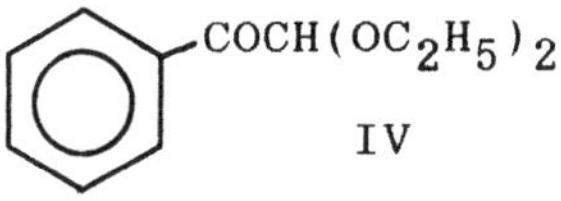

I

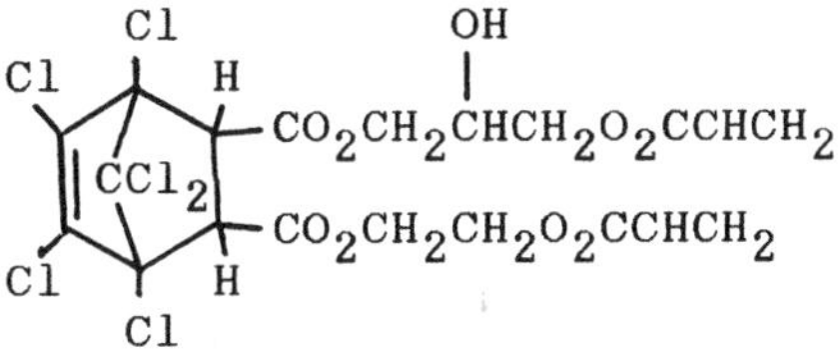

Structure II:
–CO₂CH₂CHCH₂O₂CCHCH₂
       |
       OH
–CO₂CH₂CH₂O₂CCHCH₂

II

Structure III: chlorendic anhydride

Structure IV: C₆H₅–COCH(OC₂H₅)₂

addressing this issue have been published, to date. In these studies, we examined the photolysis of chlorendic anhydride (III) in acetone and dioxane as solvents using diethoxyacetophenone (IV) as the photoinitiator and GC/MS techniques to identify the photolysis products.

Experimental

Raw Materials. Chlorendic anhydride (III), dioxane, acetone, and diethoxyacetophenone (IV), were obtained from commercial sources and used, except when otherwise indicated, without further purification.

Photolysis Conditions. Molecular sieves were added to the solvents acetone and 1,4-dioxane, then dry nitrogen was bubbled through the solvents to remove any dissolved oxygen. Finally the solvents were filtered to remove the molecular sieves. Chlorendic anhydride (III) (lit m.p. 235-239°C) was recrystallized from hot chlorobenzene, rinsed with heptane and dried at 105°C. Diethoxyacetophenone (IV) was used as received from the manufacturer.

The photolysis reactions were carried out using an Ace Glass photochemical reaction vessel (model #7841-03), quartz immersion well (model #785-25), Conrad-Hanovia power supply (model #7830-54), and Conrad-Hanovia immersion lamp (model #7825-32) rated at 200 watts, 115-130 volts, and 1.9 amps.

The reaction solution was placed in the photochemical reaction vessel and purged with dry nitrogen. The purge continued throughout the duration of the experiment (6 hours or 24 hours). Volatiles were collected by passing the purge flow through a trap immersed in a Dewar flask filled with liquid nitrogen. The solution was gently stirred with a magnetic stirring bar and tap water was used to cool the medium pressure mercury lamp. The temperature of the solution was maintained at approximately room temperature during the course of the reaction.

After the reaction, samples were taken from both the reaction vessel and the trap and combined for GC/MS analysis.

Instrumentation and Equipment. GC/MS analysis was performed using a Finnigan 9610 gas chromatograph equipped with a capillary column, at a head pressure of 10 psi, with a temperature ramp set at $50^{\circ}C$ for two minutes, then a $20^{\circ}C/min$ increase to $275^{\circ}C$ and held at $275^{\circ}C$ until all the compounds had eluted from the column. No detector (i.e. FID nor TCD) was coupled to the Finnigan 4500 Mass Spectrometer (the mass spectrometer acted as the detector). An electron impact ionization source with an electron energy of 70 eV and ionization temperature of $150^{\circ}C$ was used to fragment the compounds as they eluted from the column.

Results and Discussion

The 24 hour photolysis of chlorendic anhydride (III) in acetone in the presence of diethoxyacetophene (IV) yielded at least twelve measurable products as observed by the GC/MS procedure (Table I). Some of the products observed, i.e. 4-methyl-4-hydroxy-2-pentanone, biacetal, 4-methyl-3-pentene-2-one, ethyl acetate, 3-methyl-3-hydroxy-2-butanone, 4-methyl-2-pentanone, and 2,4-pentanedione, are clearly derived from the photochemically induced decomposition of acetone and/or diethoxyacetophenone.

Although the compounds and preferred modes of decomposition/recombination vary with the overall systems, diethoxyacetophenone in acetone exhibits primarily Norrish Type I cleavage as evidenced by the formation of benzaldehyde, ethyl benzoate, acetophenone and benzil. Although acetalydehyde may indicate Norrish Type II cleavage is observed, it is likely that this compound is derived from the solvent and not the photoinitiator (Scheme 1).

However, in the presence of chlorendic anhydride, the decomposition of diethoxyacetophenone occurs only by Norrish I type cleavage, yielding benzoyl chloride, acetophenone and benzil. No Norrish Type II cleavage products were observed. The 24 hour photolysis of chlorendic anhydride in acetone using diethoxyacetophenone as the photoinitiator shows carbon-chlorine homolytic bond cleavage and the generation of chlorendic radicals as evidenced by the appearance of benzoyl chloride, chloro-

Table I

Summary and Comparison of Products Found After Photolysis Using Acetone as the Solvent

Acetone	Diethoxy-acetophenone/ acetone	CA/acetone	Diethoxy-acetophenone/ CA/acetone
4-methyl-4-hydroxy-2-pentanone	benzaldehyde	4-methyl-4-hydroxy-2-pentanone	4-methyl-3-penten-2-one
2-pentanone	ethylbenzoate	acetic acid	benzoyl chloride
biacetal	acetophenone	3-methyl-3-hydroxy-2-butanone	4-methyl-2-pentanone
acetaldehyde	benzil		
	acetic acid	2-butanone	3-methyl-3-hydroxy-2-butanone
carbon dioxide	acetaldehyde	biacetal	
	2-butanone		4-methyl-4-hydroxy-2-pentanone
propane	2,4-pentane-dione	ethyl acetate	
	4-methyl-4-hydroxy-2-pentanone	2,4-pentane-dione	chloroacetone
		4-methyl-2-pentanone	2,4-pentane-dione
	ethyl acetate		biacetal
	3-methyl-3-hydroxy-2-butanone	4-methyl-3-pentene-2-one	ethyl acetate
	diethoxy-acetophenone	unknown chlorendic compound	unknown chlorendic compound
			unknown chlorendic compound
		chlorendic anhydride	chlorendic anhydride

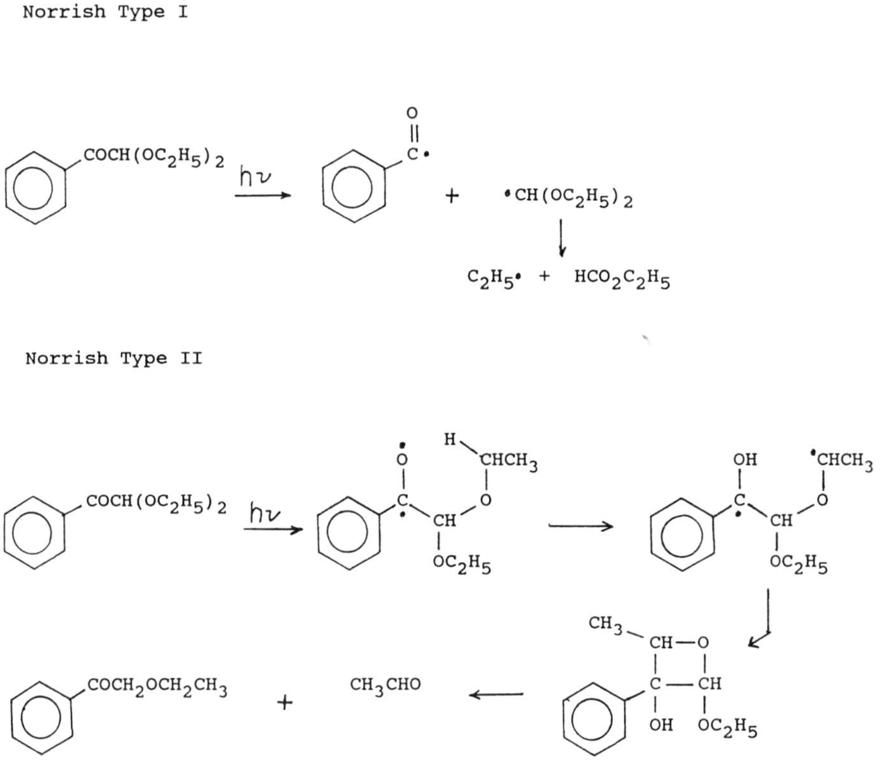

Norrish Type I

Norrish Type II

Scheme 1. Photolytic decomposition pathways of diethoxyacetophenone.

acetone and two unknown chlorendic compounds where chlorine atoms have been replaced by hydrogen atoms. Since only one chlorendic compound was observed in the product mixture obtained from the 24 hour photolysis of chlorendic anhydride in acetone, it is proposed that the presence of diethoxyacetophone in acetone as the solvent, enhances carbon-chlorine bond cleavage. The UV spectra of chlorendic anhydride shows absorption maxima at 272 and 286 NM which closely corresponds to the UV spectral emission of the lamp indicating that chlorendic anhydride is capable of homolytic chain cleavage in the absence of photoinitiators.

Using dioxane as the solvent (Table II), a similar situation is observed with the decomposition of diethoxyacetophenone occuring by Norrish Type I cleavage as evidenced by the formation of ethyl formate, benzoyl chloride and acetophenone with no evidence for Norrish Type II cleavage. Evidence for hydrogen abstraction from the dioxane molecule is observed in the generation of dioxane dimers and chlorodioxane. Although the formation of chlorodioxane and benzoyl chloride can be justified on the basis of a homolytic displacement mechanism, the presence of at least three chlorendic based molecules where one or two chlorines have been lost from chlorendic anhydride and replaced by hydrogen atoms clearly indicates homolytic cleavage and the generation of chlorine radicals. Similarly, chlorendic compounds were observed in the photolysis of chlorendic anhydride in dioxane and it appears that diethoxyacetophenone may have less influence in the generation of products in dioxane than in acetone.

Unlike chlorendic anhydride where the MS shows a fragmentation pattern involving the sequential loss of Cl (m/e = 35), CO (m/e = 28), CO_2 (m/e = 44), and C_2H_2 (m/e = 26) as well as the reverse Diel-Alder from the molecular ion peak (m/e = 368) accounting for the higher molecular weight fragments (Scheme 2), the unknown chlorendic compounds show fragmentation patterns consistent with the loss of chlorine from the starting material and replacement by hydrogen atoms in both solvents used. A typical compound observed in the case where dioxane is used as the solvent shows the MS fragmentation pattern outlined in Scheme 3.

Conclusions

Our findings clearly show that chlorendic anhydride and presumably polymers containing the chlorendate group are capable, under UV light, of undergoing multiple carbon-chlorine bond cleavage generating chlorine and carbon radicals. The presence of these radicals, by increasing the concentration of free radicals in the system, are capable of increasing the rate of reaction and crosslinking of UV curable coatings containing the chlorendic

Table II

Summmary and Comparison of Products
Found After Photolysis
Using Dioxane as the Solvent

Diethoxy-acetophenone/ Dioxane	CA/Dioxane	Diethoxy-acetophenone/ CA/Dioxane
ethyl formate	hydroxydioxane	ethyl formate
ethyl dioxane	chlorodioxane	hydroxydioxane
diethoxy-methane	dioxanyl dioxane	chlorodioxane
benzoic acid	(2 isomers)	benzoyl chloride
1-phenyl-2-ethoxyethanone	CA − 3 Cl + 2H	dioxanyl dioxane
	CA − 2 Cl + H	(2 isomers)
benzaldehyde	CA − Cl + H	acetophenone
	chlorendic anhydride	CA − Cl + H
		CA − Cl + H
		CA − 2 Cl + H
		chlorendic anhydride

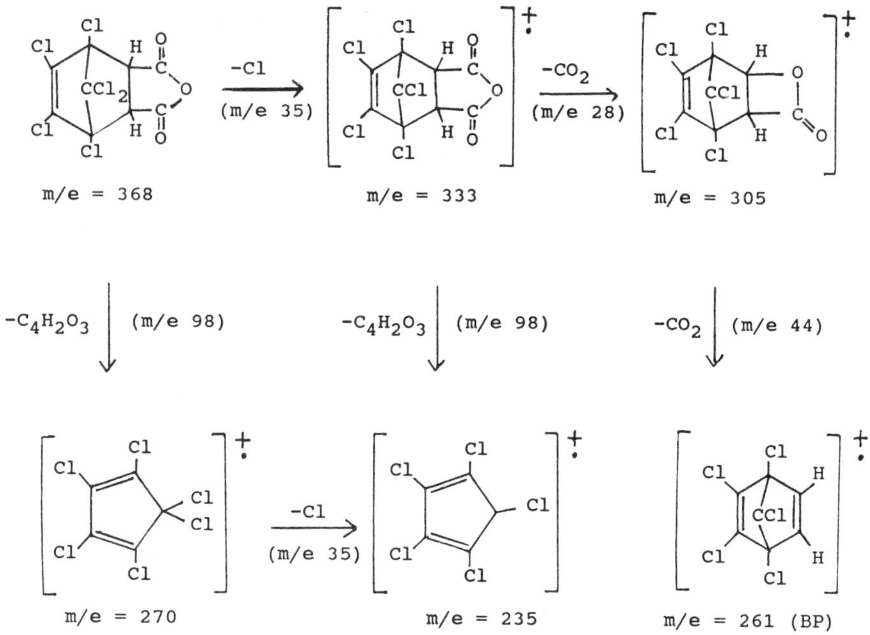

Scheme 2. MS Fragmentation Pattern for Chlorendic Anhydride

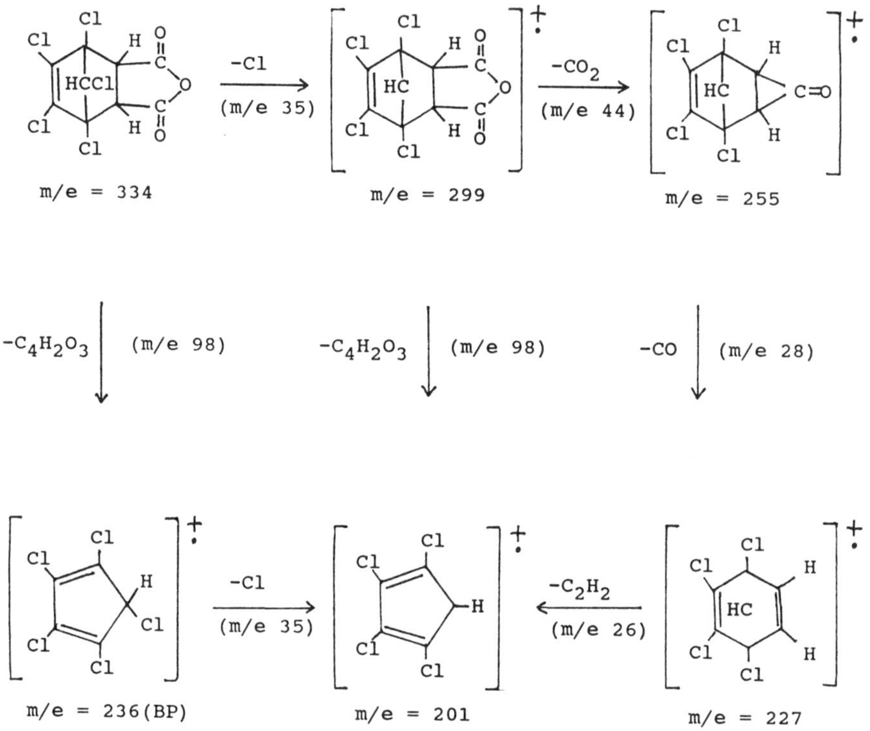

Scheme 3. MS Analysis of One of the Chlorendic Based Photoproducts When Dioxane is Used as the Solvent

moiety. This will result in much harder coatings and faster rate of cure (cross-linking) as observed by ourselves and others.

Acknowledgment

The authors are grateful to Velsicol Chemical Corporation for their support of this work.

References

1. Gloskey, D.J., M.S. Thesis, submitted to Eastern Michigan University, 1989.
2. Hoch, Bellet U.S. Patent 3 157 709, 1964.
3. Robitschek U.S. Patent 3 249 565, 1966.
4. Nass, G. I.; et. al. U.S. Patent 3 551 311, 1970.
5. Sun Chemical Br. Patent 1 241 824, 1971.
6. Schupan, I., et al.; Z. Naturforsch 1972, 27 b, 147-156.
7. Shiraishi, M., et al. Jpn. Kokai 73 62 711, 1973.
8. Nishikubo, T. et al. Jpn. Kokai 74 02 601, 1974.
9. Przezdziecki, W. M., et al. Br. Patent 1 356 390, 1974.
10. O'Brien Corp. Br. Patent 1 375 177, 1974.
11. Parker, G. M., et al. U.S. Patent 3 785 849, 1974.
12. Wagner, H. M., et al. Res. Discl. 1975, 134, 19-21.
13. Graham, J.C. and Gloskey, D.J., Polym. Paint Colour J., 1987, 177, 4188.
14. Laws, A; J. Oil Colour Chem. Assoc., 1976, 59(6), 206-9.
15. Kang, U. G. et al. U.S. Patent 4 591 522, 1986.

RECEIVED September 13, 1989

Chapter 17

Photochemical Approaches to Ordered Polymers

Michael A. Meador, Mahmoud Abdulaziz, and Mary Ann B. Meador

Polymers Branch, Materials Division, NASA Lewis Research Center, Cleveland, OH 44135

>The photocyclization of o-benzyloxyphenyl ketone chromophores provides an efficient, high yield route to the synthesis of 2,3-diphenylbenzofurans. The synthesis and solution photochemistry of a series of polymers containing this chromophore is described. The photocuring of these polymers is a potential new approach to the synthesis of highly conjugated polymers based upon a p-phenylene bisbenzofuran repeat unit.

Highly ordered aromatic and heteroaromatic polymers, such as polyimides and the poly(benzobisazoles)[1], PBT, PBI, and PBO, have been the subject of a great deal of interest over the past three decades. The superb mechanical properties and thermal oxidative stability of these polymers have attracted a great deal of interest leading to, in some cases, the commercialization of these materials. However, recent work with these systems has uncovered some impressive electronic and nonlinear optical characteristics. Studies have shown that electrochemical doping of PBT films raises their electrical conductivities from semiconductor levels to as high as 20 ohm^{-1} cm^{-1} [2]. Recently, Prasad and others [3] have reported that highly ordered films of PBO and PBI possess outstanding third-order nonlinear optical properties.

A high degree of aromaticity and extended electronic conjugation confer these desirable properties, but highly aromatic polymers are generally intractable and difficult to process. In many cases the synthesis and processing of highly aromatic polymers

This chapter not subject to U.S. copyright
Published 1990 American Chemical Society

requires the use of fairly corrosive solvents. High yields of PBT and PBO are only possible if the condensation polymerization is carried out in polyphosphoric acid.

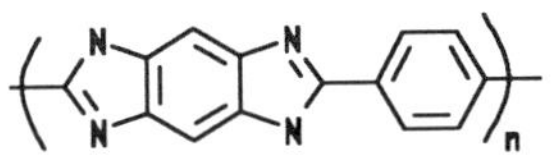

PBI

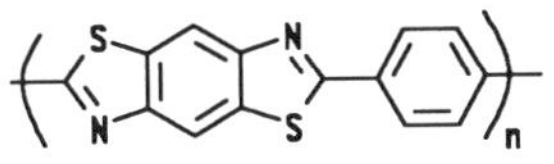

PBT

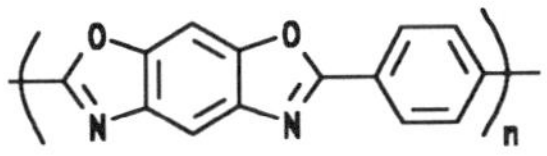

PBO

Kevlar, PBO, PBI, and PBT form lyotropic liquid crystalline solutions at high concentration. Processing under these conditions can lead to highly ordered fibers and films. Solutions having high polymer concentrations are only possible in strong acids, such as polyphosphoric acid, methanesulfonic acid, and trifluoromethanesulfonic acid. These solvents are highly corrosive and difficult to remove from the polymer once it has been processed into a film, coating, or fiber. The lack of solubility of these polymers in typical organic solvents prohibits processing in these media. Recent work by Jenekhe [4] has addressed this problem through the use of Lewis acids as a means of increasing the solubility PBT, PBO, and PBI in aprotic organic solvents. However, these techniques add an additional processing step, i.e., the removal of the complexed Lewis acids from the polymer once processing is complete.

To take full advantage of the beneficial properties of these highly aromatic polymers, there is a clear need to develop new polymers and synthetic strategies which favor processability. Polymer photochemistry provides such an approach. Advances in photcurable polymers have made possible the fabrication of a variety of shapes from films and fibers to dental fillings. Photochemical processes offer the added advantage that they are typically faster and more energy efficient than conventional thermal or chemical processes. Furthermore, some photochemical processes present routes to new compounds which could not otherwise be prepared via ground state techniques.

Some of these reactions could provide a high yield, less severe route to similar heteroaromatic polymers. One such reaction is the photocyclization of o-benzyloxyphenyl ketones, **1**, to the corresponding 2-hydroxy-2,3-dihydrobenzofuran, **3** [5,6] (Scheme I). This reaction occurs from the ketone triplet and involves formation of a 1,5-biradical intermediate, **2**, via an initial δ-hydrogen abstraction. Biradical **2** has three mechanistic pathways available to it - reabstraction of the alkoxy hydrogen by the benzyloxy radical to give starting ketone (path a), rearrangement of the biradical followed by oxidation to a 1,2-diacyl benzene (path b), or cyclization of the biradical to the desired dihydrobenzofuran, **3** (path c).

For o-benzyloxy-benzophenone (R=Ph), cyclization of the biradical occurs exclusively [6] to afford the desired 2,3-diphenyl-2-hydroxy-2,3-dihydrobenzofuran as a mixture of two diastereomeric products (Z/E = 8) in overall quantitative yield. The process is highly efficient, the quantum yield for photocyclization of o-benzyloxy-benzophenone is 0.95. Both photoproducts are readily dehydrated by refluxing in benzene containing a trace amount of hydrochloric acid to afford 2,3-diphenylbenzofuran in 100% yield [7]. Thus, this scheme constitutes a highly efficient, two-step synthesis of phenyl substituted benzofurans.

Application of this photocyclization to polymers, such as **4**, should afford an efficient photochemical route to the synthesis of ordered polymer systems based upon a phenyl substituted benzofuran repeat unit (Scheme II). Due to a fair degree of molecular flexibility, photopolymer **4** should be soluble in a variety of organic solvents. Photocyclization of **4** could then be carried out either in solution or the solid state (films, fibers, or coatings). This paper describes the photochemistry of low molecular weight formulations of **4** (n= 1, 5, 10) and that of an appropriate model compound, 1,4-bis(2-benzyloxy-benzoyl)benzene, **4a**.

<u>Experimental</u>

All NMR spectra (^{1}H, ^{13}C, and CPMAS) were recorded with a Bruker AM-300 spectrometer. Solution spectra were recorded in $CDCl_3$ and referenced to tetramethylsilane. CPMAS spectra were run with a high power solids attachment with a probe equipped with a dual air bearing

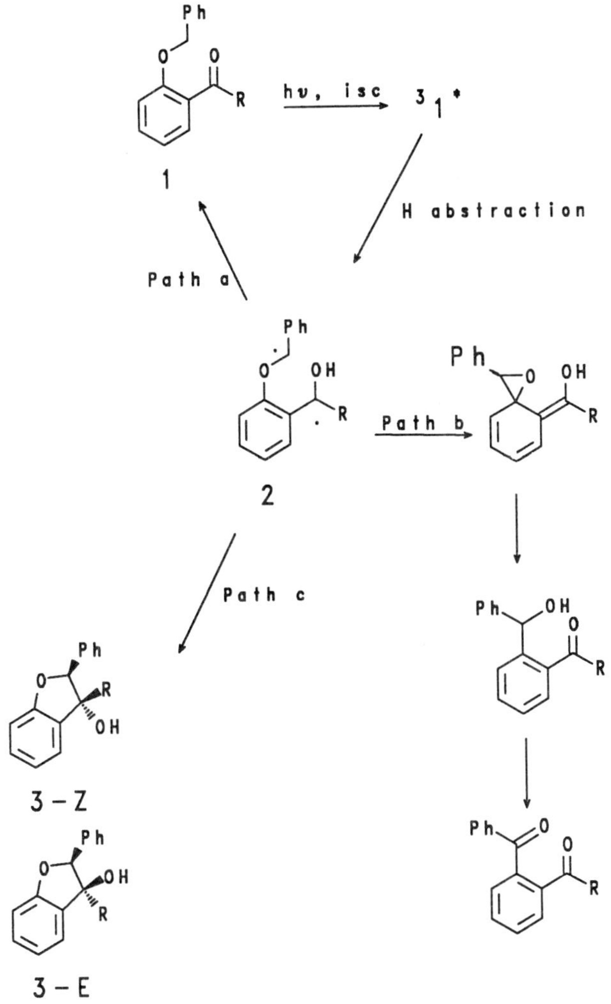
Ph
O
O
R
hν, isc
$^{3}1^{*}$
1
H abstraction
Path a
Ph
O
OH
R
Path b
Ph
O
OH
R
2
Path c
Ph
OH
O
R
Ph
O
R
OH
3 – Z
Ph
O
OH
R
3 – E
Ph
O
O
R

Scheme I

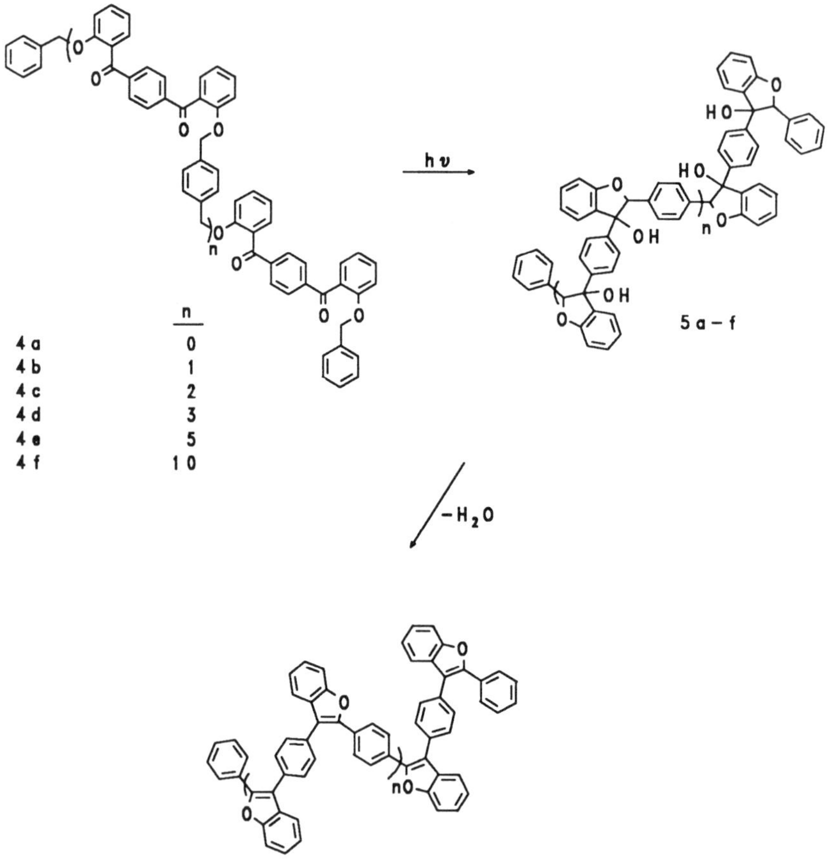

	n
4a	0
4b	1
4c	2
4d	3
4e	5
4f	10

Scheme II

capable of magic angle spinning at rates up to 5 kHz, and tunable over the frequency range from ^{14}N to ^{63}Cu. CPMAS spectra were referenced to the carbonyl resonance of glycine. Infrared spectra (KBr pellet) were recorded on a Perkin Elmer Model 1700 FTIR. Ultraviolet-visible absorption spectra were obtained on a Varian Model 2390 Spectrophotometer.

Thermal analysis of polymers was performed on DuPont Model 910 Differential Scanning Calorimeter and a Perkin Elmer Model TGS-2 Thermal Gravimetric Analyzer using a heating rate of 10°C/min. Both systems were interfaced to an Omnitherm Data Station for experiment control, and data acquisition and processing.

GPC analyses were conducted using a Waters Liquid Chromatograph equipped with three Ultrastyrogel columns (100, 500, 1000 A) in series. Tetrahydrofuran was used as the mobile phase, and the output monitored by uv absorption at 280 nm. The detector was interfaced to an IBM PC-AT equipped with a Dynamic Solutions Maxima chromatography software package.

All photochemical reactions were performed in benzene solutions under nitrogen using light from a 450 Watt Hanovia medium pressure Hg lamp filtered through Pyrex.

Benzyl bromide, tetra-N-butylammonium bromide, α α'-dibromo-p-xylene, 2-bromoanisole, terephthalonitrile, and boron tribromide were purchased from Aldrich Chemical Company and used without further purification. Tetrahydrofuran (THF) was freshly distilled under nitrogen from potassium-benzophenone.

1,4-Bis(2-Hydroxybenzoyl)benzene (BHB).

1,4-Bis(2-anisoyl)-benzene (BAB) was prepared in 72% yield from the reaction of 2-anisoyl magnesium bromide (2.1 equivalents) with terephthalonitrile, followed by acid hydrolysis of the imine.

A solution of boron tribromide (10.0 mL, 100 mmole) in 100 mL dichloromethane was added dropwise to a stirred solution of BAB (9.2g, 26.6 mmole) in 200 mL dichloromethane at -78°C under nitrogen. Once the addition was complete, the reaction mixture was allowed to warm to room temperature over 3h. The resulting solution was then poured into 300 mL distilled H_2O. The aqueous layer was separated and extracted (3x200 mL) with chloroform. The organics were combined and washed (3x200 mL) with an aqueous KOH solution (5%, 3x200 mL). The base extracts were combined, neutralized with conc. HCl, and extracted with chloroform (3x200 mL). The chloroform extracts were combined, dried ($MgSO_4$), and the solvent removed to provide crude BHB as a pale yellow solid. Recrystallization from chloroform/ethanol afforded 6.2g (73%) of pure BHB as bright yellow crystals, mp 216-218°C.

^{1}H-NMR δ 6.68-7.56 (m,5H), 7.76 (s,2H), 11.8 (s,1H).

^{13}C-NMR δ 118.3, 118.48, 118.5, 128.5, 132.9, 136.5, 140.3, 163.0, 200.3.

1,4-Bis(2-Benzyloxybenzoyl)benzene (4a). BHB (1.0g, 3.14 mmole), K_2CO_3 (Fisher, 0.91g, 6.6 mmole), and benzyl bromide (0.8 mL, 6.6 mmole) were combined in 100 mL dioxane and refluxed with stirring under nitrogen for 18h. The solution was then allowed to cool to room temperature and poured into 300 mL distilled H_2O. The resulting mixture was extracted (3x200 mL) with chloroform. The organic extracts were combined, extracted with 150 mL of a 5% aq. KOH solution, and dried ($MgSO_4$). The solvent was removed under vacuum to provide 1.4g (89%) of crude **4a** as a cream colored solid. Pure **4a** was obtained by recrystallization from ethanol, mp 172-174°C.
^1H NMR δ 4.96(s, 2H), 6.84-7.92 (m, 11H).
^{13}C NMR δ 70.21, 112.9, 121.0, 126.7, 127.7, 128.3, 128.8, 129.3, 130.1, 130.3, 132.5, 132.8, 136.1, 136.7, 141.5, 156.7, 196.1.

Typical Procedure for Synthesis of Polymers, 4. A solution of BHB (3.18g, 10.0 mmole) and 25 mL of 2N aq. KOH solution was stirred under nitrogen at room temperature for 30min. A solution of α, α'-di-bromo-p-xylene (2.20g, 8.33 mmole), benzyl bromide (0.57g, 3.33 mmole), and 0.1g tetra-N-butylammonium bromide in 25 mL 1,2-dichloroethane was added and the resulting mixture refluxed under nitrogen for 18h with vigorous stirring. The reaction mixture, which had gone from deep red to colorless, was then cooled to room temperature and 30 mL hexane was added. The resulting precipitate was then filtered and washed with 100 mL distilled H_2O, and dried under vacuum at 80°C for 18h to yield 4.0g (94%) of **4f** as a cream colored solid.

Results and Discussion

Synthesis and Characterization. Compounds **4b-f** were synthesized as depicted in Scheme III. Typical yields from this approach were on the order of 90-100%.

Presence of any unreacted 2-hydroxybenzoyl moieties in the final polymer would act as a uv cutoff filter, and thereby diminish the photoreactivity of the system. Thus, it was important to assure that each of the polymers prepared was free of any unreacted hydroxyl groups. ^1H NMR of $CDCl_3$ solutions of **4a-f** revealed no detectable free -OH groups (the hydroxyl proton of the starting material appears as a sharp singlet at 11.8 ppm).

UV spectra of **4a-f** show a strong (π,π^*) absorption at 266 nm ($\epsilon = 10^4$ to 10^5 l mole^{-1}cm^{-1}) and an (n,π^*) transition as a shoulder at 300nm ($\epsilon = 10^4$ l mole^{-1}cm^{-1}). Note that these spectra show only slight tailing at wavelengths greater than 350nm, indicating the absence of unreacted o-hydroxybenzoyl chromphores. A comparison of the UV spectra of THF solutions of **4a and f** and BBB is given in Figure 1.

Infrared spectra of **4a-f** are essentially identical and consistent with the proposed structure, a representative spectrum is shown in Figure 2. Key features of the spectrum include a strong carbonyl

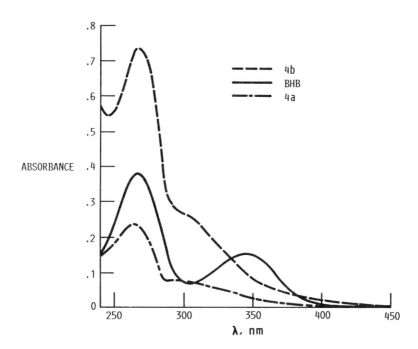

Figure 1. UV spectra of **4a,b** and BHB in tetrahydrofuran.

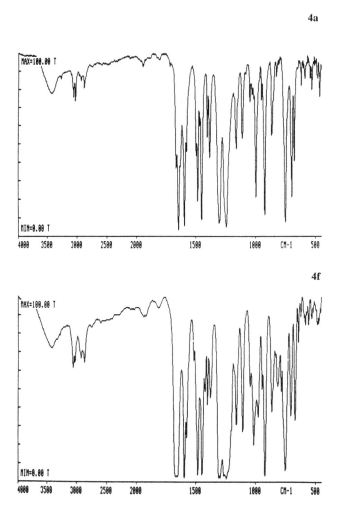

Figure 2. A comparison of the ir spectra (KBr pellet) of **4a** and **4f**.

stretch at 1651 cm^{-1}, aromatic C-H stretch at 3065 and 3033 cm^{-1}, saturated C-H stretches at 2927 and 2872 cm^{-1}, strong aromatic C-C ring stretches at 1598 and 1485 cm^{-1}, a strong band at 1449 cm^{-1} due to -CH$_2$-O- scissoring, and a strong assymetric C-O-C stretch at 1242 cm^{-1}.

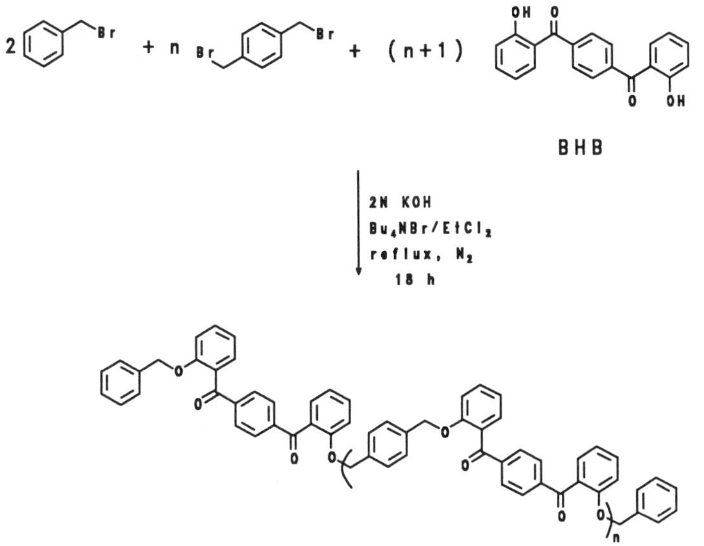

Scheme III

A comparison of typical ^{1}H and ^{13}C spectra of polymer and **4a** is depicted in Figure 3. As would be expected, spectra for both compounds are similar. Resonances for the benzyloxy methylene protons appear as a singlet at 4.96 ppm in the ^{1}H spectrum of **4a**, and as a singlet at 4.98 and minor resonances from 4.70-4.97ppm for those protons in **4f**. The ^{13}C NMR spectra of these compounds show carbonyl resonances at 196.1ppm for **4a**, and 195.1 and 196.4ppm for **4f**. Benzyloxy carbons appear at 70.21ppm in the ^{13}C spectrum of **4a**, and at 69.85 and 73.13ppm in that of **4f**.

Photochemistry. Photolysis of a benzene solution (.01M) of **4a** produced a quantitative yield of the desired photocyclized product, **5a**,

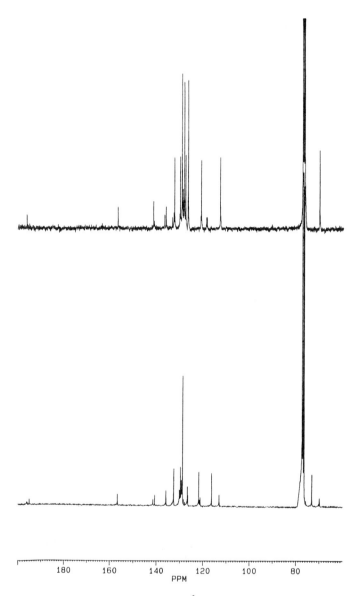

Figure 3. A comparison of the ^{1}H spectra of 4a (top) and 4f (bottom). <u>Continued on next page</u>.

17. MEADOR ET AL. *Photochemical Approaches to Ordered Polymers* 231

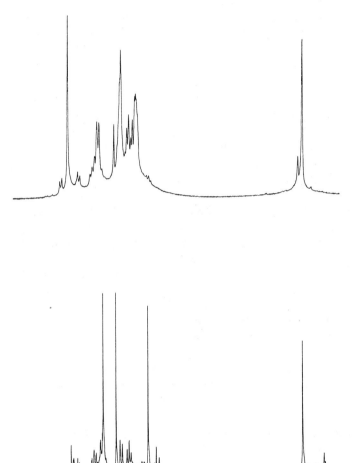

Figure 3. <u>Continued</u>. A comparison of the ^{13}C spectra of 4a (top) and 4f (bottom).

as a mixture of diastereomers. The ^{1}H and ^{13}C spectra of **5a** are shown in Figure 4. The absence of peaks at 5.69 ppm in the ^{1}H spectrum and of the carbonyl peak and the benzyloxy carbon peak in the ^{13}C spectrum at 196.1 and 70.21 ppm, respectively, indicate complete conversion of both o-benzyloxy-benzoyl moieties. The presence of two singlets at 5.66 and 5.67ppm in the ^{1}H spectrum and peaks at 83.49, 95.33 and 95.45 ppm in the ^{13}C NMR of the photoproduct substantiate the formation of a dihydrobenzofuran. The absence of a carbonyl peak at 1651 cm^{-1}, formation of a strong -OH band at 3543 cm^{-1}, and a shift of the -C-O-C- stretch from 1242 cm^{-1} to 1255 cm^{-1} further corroborates the proposed structure. Thus, all spectral evidence is consistent with the formation of **5a** and is in good agreeement with that previously reported for 2,3-diphenyl-2,3-dihydro-benzofuran-2-ol [7].

While the photocylization of **4a** proceeded cleanly to afford only the desired dihydrobenzofuran products, irradiation of **4b-f** did not always give complete conversion of each o-benzyloxybenzoyl chromophore. The best conversions were realized in low molecular weight oligomers. For example, photolysis of **4b** resulted in complete photocyclization to afford **5b**. ^{1}H and ^{13}C NMR spectra of the photoproduct (Figure 4) are consistent with proposed structure. A broad peak at 5.69 ppm in the ^{1}H spectrum as well as peaks at 84 and 95 ppm in the ^{13}C spectrum of the photoproduct confirm the formation of the desired dihydrobenzofuranol **5b**. Examination of the ir spectrum of the photoproduct shows no carbonyl resonance at 1657 cm^{-1}, a shift of the -C-O-C- stretch from 1242 cm^{-1} to 1257 cm^{-1}, and formation of a strong -OH stretch at 3447 cm^{-1} consistent with the formation of **5b**.

Complete photocyclization of higher molecular weight compounds **4e** and **4f**, was not realized, perhaps due to the insolubility of partially photocyclized **4e** and **f** in benzene. ^{13}C NMR analysis of these photoproducts was performed in the solid state. ^{13}C CPMAS spectra of **4f** before and after photolysis are shown in Figure 5. There is a noticable decrease in the intensity of the carbonyl peak at 185 ppm and a broadening of the benzyloxy carbon resonance at 50 ppm. The ir spectrum of irradiated **4f** shows the presence of unreacted carbonyl and benzyloxy groups in addition to peaks at 3447 and 1257 cm^{-1} indicating that some photocyclization has occured.

The spectral data gathered on these partially photocyclized systems does not indicate the presence of any products of competitive side reactions (Scheme IV). Rearrangements, similar to a photo-Fries, have been reported for substituted anisoles [8]. The same

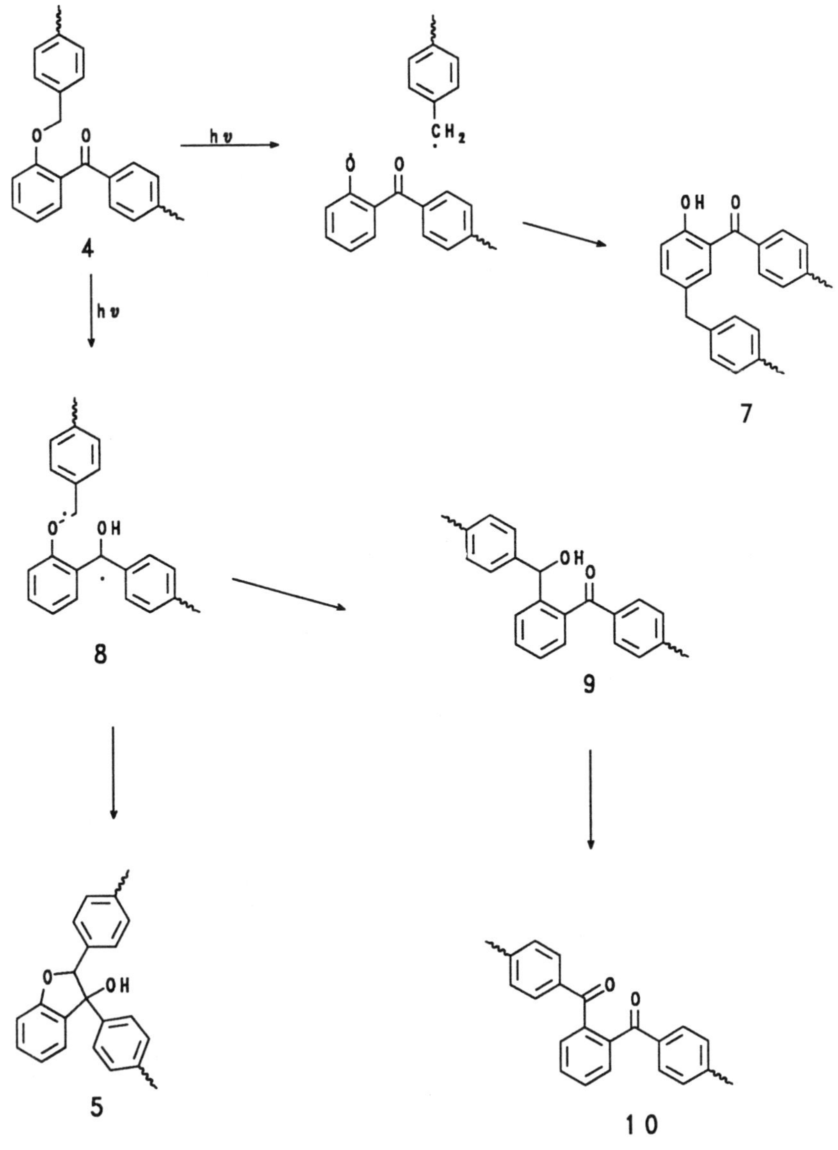

Scheme IV

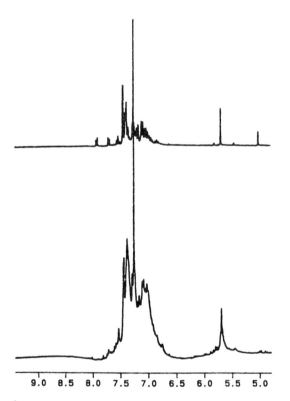

Figure 4. ^{1}H spectra of 5a (top) and 5b (bottom). <u>Continued on next page</u>.

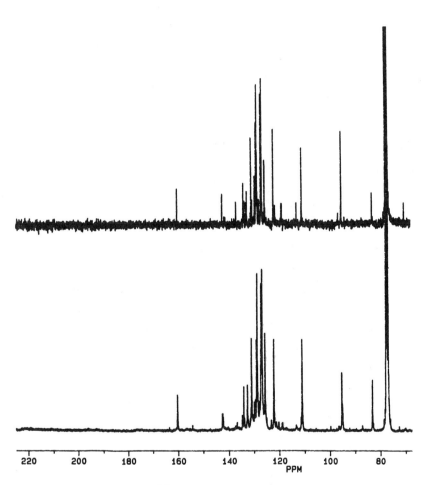

Figure 4. Continued. ^{13}C spectra of 5a (top) and 5b (bottom).

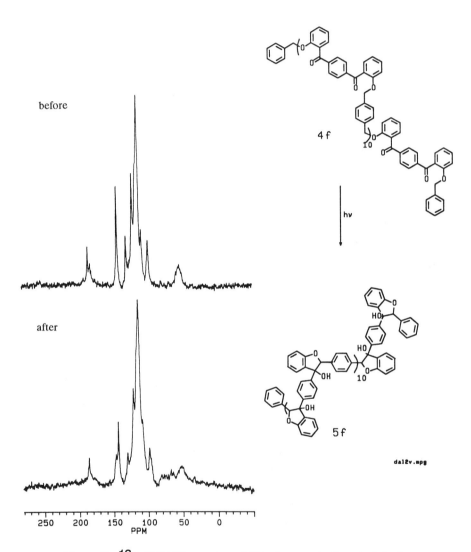

Figure 5. ^{13}C CPMAS spectra of **4f** before and after photolysis.

rearrangement of **4** would produce compounds such as **7**. The ir spectrum of **7** should have a carbonyl stretch at 1620 cm^{-1} as is observed for other o-hydroxyphenyl ketones, such as BHB. Examination of the ir spectrum of irradiated **4f** shows no such shift of the carbonyl stretch at 1661 cm^{-1}.

A rearrangement of biradical **8**, similar to that described in Scheme I for simple o-benzyloxyphenyl ketones, would produce o-benzoylbenzhydrol **9** which might oxidize to o-dibenzoylbenzene **10**. This would result in a slight shift of the carbonyl band in the ir spectrum, as well as a broadening of the benzyloxy resonances in the CPMAS spectrum. However, such a rearrangement should not result in the loss of the carbonyl resonance at 185ppm since the rearranged structure still has a benzoyl group. In fact, if oxidation of **9** occurred, an increase of this carbonyl peak would be expected due to the additional benzoyl moiety.

Conclusions

Thus, we have demonstrated that photocylization in low molecular weight polymers containing o-benzyloxybenzoyl moieties is possible. In those instances where complete photocyclization did not occur, the insolubility of partially photocyclized polymer may have prevented complete reaction. We are currently investigating ways of maintaining solubility in these systems as well as examining this photocylization in other polymers.

Acknowledgments

The authors wish to thank Mr. Frank Danko and Mr. Terry Kacik of Sverdrup Technologies, Inc. for assistance in chemical analysis of polymers and photoproducts.

Literature Cited

1. Wolfe, J.R.; Arnold, F.E. Macromolecules 1981, **14**, 909; Wolfe, J.R.; Loo, B.H.; Arnold, F.E. Macromolecules 1981, **14**, 915.
2. De Pra, P.A.; Gaudiello, J.G.; Marks, T.J. Macromolecules 1988, **21**, 2295.
3. Teng, C.C ; Garito, A.F. Nonlinear Optics, SPIE Proc., 1986; Rao, D.N.; Swiatkiewicz, J.; Chopra, P.; Ghoshal, S.K.; Prasad,P.N. Appl. Phys. Lett. 1986, **48, 1187**.
4. Jenkhe, S.A.; Johnson, P.O.; Agrawi, A.K. Proceedings of the ACS Div. of Polymeric Materials: Science and Engineering 1989, **60**, 404.
5. Pappas, S.P.; Pappas, B.C.; Blackwell, J.E., Jr J. Org. Chem. 1967, **32**, 3066.
6. Wagner, P.J.; Meador, M.A.; Scaiano, J.C. J. Amer. Chem. Soc. 1984, **106**, 7988.
7. Meador, M.A. Ph.D. Thesis, Michigan State University, 1983.
8. Houser, J.J.; Chen, M-C, Wang, S.S. J. Org. Chem. 1974, **39**, 1387.

RECEIVED October 2, 1989

PROPERTIES OF RADIATION-CURED MATERIALS

Properties of Radiation-Cured Materials

One of the critical aspects in the development of radiation cured systems for specific applications is the utilization of various methods of testing curing rates and final film properties. By generation and interpretation of physical, spectroscopic and thermal data it is possible to specify changes in radiation curable materials which can lead to improvement in performance. In choosing a particular parameter to measure, it must be established that the property which is assessed is relevant to the application for which the cured film is designed.

In Chapter 17, the authors present what proves to be an excellent example of the wealth of information which can be extracted on photocurable systems (in this case adhesives) by combining IR spectroscopy, calorimetry, mechanical spectroscopy, and thermomechanical analysis. This chapter provides one of the few examples in the literature of a time-temperature superposition analysis of a photocured material. For anyone attempting to conduct a thorough comprehensive study of photocuring using a variety of complimentary techniques to assess cure, this chapter is highly recommended.

Chapters 18 and 19 are concerned with evaluation of final physical properties such as tensile strength, modulus, elongation, hardness, adhesion, etc. of photocured acrylated urethane coatings. In one case (Chapter 19), cure rates were also measured. The film properties were correlated with the structural components, i.e. polyol type and molecular weight, biscarbamate chemical structure, and functionality and type of reactive diluents. The variation in properties detected points out the need for careful consideration of structure-property considerations when formulating radiation curable coatings.

Chapter 20 deals with the topic of enthalpy relaxation in photocured acrylated epoxy films which can result in marked physical changes. The results indicate that film relaxation can play a major role in establishing the final strength, hardness and adhesion properties of radiation cured films and must be considered when formulating the initial coatings. In addition, the rate of cure and resulting crosslink density/extent of polymerization may affect such relaxation phenomenon.

Chapters 21 and 22 are excellent examples of using non-conventional analytical methods to understand photocuring phenomena associated with a specific lithographic printing application. In Chapter 21, use of a photographic sensitometric method was used to develop complex time-temperature-transformation diagrams for a photocurable microencapsulated system based on a trifunctional acrylate. The technique was employed successfully to described reciprocity failure which hampers the use of this system over a wide intensity/temperature range. The results in Chapter 22 establish the procedure of using microcolumn membranes to simulate the restricted environment of microencapsulated photocurable systems. Image based sensitometric measurements, similar to those employed in Chapter 21, were used to confirm the utility of microcolumn membranes. Finally, results on storage modulus, loss modulus, Tg, and image density were correlated to provide a complete description of the photocurable systems in the microcolumns.

Chapter 18

Comparison of Thermal, Mechanical, and Spectroscopic Techniques for Characterization of Radiation-Cured Adhesives

G. M. Allen and K. F. Drain

Formulated Systems Group, Ciba—Geigy Corporation, Madison Heights, MI 48071

> Radiation cure adhesives are becoming increasingly important for structural material applications. In order to obtain optimum performance and process efficiency, it is necessary to analyze these materials using several techniques. Thin film applications have been successfully characterized by traditional methods such as infrared spectroscopy and thermal analysis. This investigation includes comparison of traditional methods and mechanical spectroscopy for characterization of structural adhesive applications. In addition, mechanical spectroscopy provides viscoelastic data dependent on structure property relationships.

Radiation cure materials are becoming increasingly important for structural adhesive applications. In the formulation of these adhesives great care is taken using structure property relationships to choose monomer, oligomer, photoinitiator and adhesion promoter ingredients. However, one of the most significant factors in determining the performance of a UV curable adhesive is the intensity and spectral emission of the source, and this is often beyond the control of the adhesive formulator (1). Accordingly it is important to be able to characterize an adhesive in terms of its degree of cure under different conditions of irradiation. The rate and extent of cure of thin films has been effectively monitored by infrared spectroscopy (2). This technique has limitations when used to study UV curable adhesives as curing does not occur uniformly in the adhesive because of the variation in the percentage of incident radiation absorbed with depth of penetration (3). The extent of this limitation may be further understood if we appreciate that UV adhesives are commonly used in bondline thicknesses ranging from 5-125 mil. More appropriate than infrared spectroscopy for thick specimen evaluation is photo differential scanning calorimetry, which measures residual photoactivity in the cured bulk adhesive to determine percent conversion. In this paper we compare IR spectroscopy and photo-DSC as techniques for monitoring the cure efficiency of UV curable adhesives. The establishment of the optimum cure conditions for an adhesive is of chief importance in insuring that reproducible mechanical properties are achieved and an appropriate joint design load established. This paper also describes the correlation between percent conversion and physical properties as measured by mechanical spectroscopy and proposes that modulus information can be used in a manner similar to DSC and IR. In addition, properties relating to performance of completely cured adhesives, such as glass transition temperature and coefficient of thermal expansion, as determined by mechanical spectroscopy are also considered.

The adhesive of this study has been designed for an automotive 'under-hood' application. Accordingly, it was important to understand the behavior of the adhesive over a wide frequency (engine RPM) and temperature range. Time-temperature superposition allows characterization beyond the frequency range of our instrumentation (4). More significantly, it is hoped to use time-temperature superposition data to explore the molecular weight implications of incomplete polymerization.

Experimentation

Sample Preparation. Samples of an acrylated epoxy adhesive were prepared as rectangular solids (length 50mm, width 12mm, and thickness 1.75mm) using an aluminum mold (Figure 1). Liquid adhesive was weighed into the mold and irradiated with an ultraviolet lamp system (Fusion UV Curing System, F450, Lamp Series 4D). The exposure time was varied to achieve several degrees of cure (0-100%). Samples were removed from the mold and analyzed without further preparation for all experiments except time-temperature superposition studies. Cure conditions are listed in Table I.

Thermomechanical Analysis. A thermomechanical analyzer (Perkin-Elmer TMS 2) was used to measure glass transition temperature (T_g) and coefficient of thermal expansion (α) of completely cured samples in the penetration and expansion modes respectively. Experimental conditions are listed in Table I.

Photo Differential Scanning Calorimetry. A modified differential scanning calorimeter (Perkin-Elmer DSC 2C) was used to determine percent conversion for various irradiation times. This was calculated from the residual polymerization energy remaining after sample irradiation.

The photo-DSC was assembled with a low pressure mercury lamp system (Oriel lamp 6035) mounted in the draft shield. The sample holder enclosure cover was machined to include ports containing quartz windows (1 cm dia.) located directly above the sample/reference holder compartments (Figure 2). The DSC was operated in the isothermal mode with a nitrogen purge. Experimental conditions are listed in Table I.

Infrared Spectroscopy. Fourier transform infrared spectroscopy (Perkin-Elmer 1750) was used to measure surface characteristics of samples of various percent conversion. An attenuated total reflectance accessory with KRS-5 crystal (45°) provided sufficient absorp-

UV SOURCE

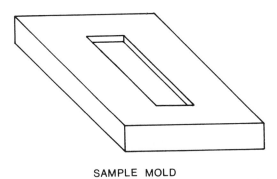

SAMPLE MOLD

Figure 1. Sample preparation technique.

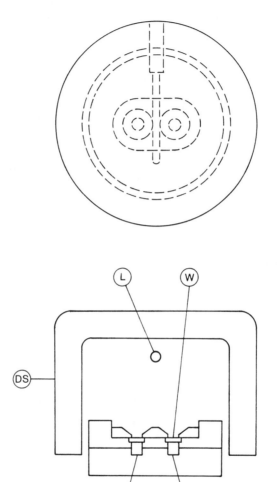

Figure 2. Photocalorimeter sample holder assembly; S) sample cell, R) reference cell, DS) draft shield, L) UV lamp, W) quartz window.

tion for this purpose. Area of absorption bands at 830 and 810 cm^{-1} was monitored for each sample. Experimental conditions are listed in Table I.

Mechanical Spectroscopy. The dynamic mechanical properties of samples representing 0-100% cure were measured by shear rheometry (Rheometrics RDA 700). All determinations were performed below the sample critical strain value. Solid and liquid samples were examined using torsion rectangular and parallel plate fixtures respectively. Glass transition temperature and coefficient of thermal expansion of completely cured samples were measured by temperature scanning at 1, 10, and 100 rad/sec frequency. The mechanical properties relating to percent conversion were obtained by frequency scanning at constant temperature (25°C). Time-temperature superposition studies were performed on fully cured samples using isothermal frequency scans (25-125°C). These experiments were conducted on samples exposed to a thermal soak (20 min. @ 120°C) prior to measurement to ensure consistent thermal history. Experimental conditions are listed in Table I.

TABLE I
EXPERIMENTAL CONDITIONS

Sample Preparation		
Sample/Lamp Distance:	32 cm	
Intensity:	100%	
Duration:	0, 0.2, 1.0, 10.0, 100.0 sec	
Thermomechanical Analysis		
Heat Rate:	10°C/min	
Gas:	He 50 cc/min	
Load:	10.0g (penetration mode)	
	0 (expansion mode)	
Photo Differential Scanning Calorimetry		
Temperature:	25°C, isothermal	
Gas:	N_2 25 cc/min	
Radiant Power:	1.36 mW/cm^2	
Infrared Spectroscopy		
Resolution:	2 cm^{-1}	
Number of Scans:	64	
Gain:	4	
Mechanical Spectroscopy	Temperature Scan	Frequency Scan
Geometry	torsion rectangular	torsion rectangular
Temperature	0-150°C	25°C
Temperature Rate	2°C/min	---
Frequency	1, 10, 100 rad/sec	0.1-100 rad/sec
Strain	0.1%	0.1%
Auto Tension	on	off
Correlate Delay	---	1 sec

Results and Discussion

The adhesive formulation used in this study was comprised of a mixture of acrylated epoxy oligomers and diluent acrylate and vinyl monomers.

The percent conversion of bulk adhesive was evaluated using photo differential scanning calorimetry. Values were established by comparison of residual polymerization energy remaining after sample irradiation to total energy of unpolymerized adhesive. Thermograms representing 0-100% conversion are shown in Figure 3.

Infrared analysis of adhesive samples representing various percent conversion was achieved using reflectance spectroscopy. The transmission mode was ineffective due to sample thickness and opacity. Limitation of the reflectance technique for measurement of bulk adhesive polymerization was apparent on examination of the sample surface nearest the UV source which indicates complete cure (irrespective of irradiation time for $t > 0$ sec). Consequently, the sample surface furthest from the UV source was used to indicate conversion. Area ratio of absorption bands at 830 and 810 cm^{-1} was calculated for this determination. The 810 cm^{-1} band intensity, attributable to CH_2 deformation of carbon double bond functionality, is directly proportional to conversion. Spectra of samples representing 0-100% conversion are shown in Figure 4.

The mechanical spectra of fully cured adhesive measured by shear rheometry, are shown in Figures 5 and 6. These figures illustrate how modulus (loss and storage) and tangent delta vary with temperature and frequency. To evaluate the variation of mechanical properties with respect to percent conversion of an adhesive containing low boiling monomers, as in the case with the adhesive of this study, it was decided to assess only property/frequency response.

Storage modulus, loss modulus and tangent delta were monitored as a function of frequency for samples representing various irradiation times as in the IR and photocalorimetry experiments. Only the results of the storage modulus are reported here as this is the most significant parameter for the design engineer (Figure 7).

Of the techniques utilized, photocalorimetry is the method of choice to give a quantitative measurement of percent conversion. Using photo-DSC to define a cure response/irradiation time relationship makes it possible to plot mechanical or infrared response versus percent conversion. The results are listed in Table II and illustrated in Figure 8.

TABLE II
POLYMERIZATION

Irradiation Time	Differential Scanning Calorimetry Enthalpy	Polymerization Percent	Infrared Spectroscopy Area Ratio	Mechanical Spectroscopy Storage Modulus
0 sec	42.8 cal/g	0%	3.19	5.37 E0 dyne/cm^2
0.2	7.7	82	3.70	1.83 E9
1.01	1.1	97	9.81	2.84
10.0	0.4	99	12.00	7.22
100.0	0	100	14.00	9.95

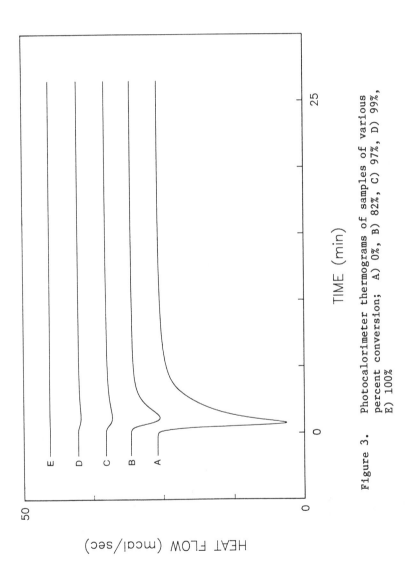

Figure 3. Photocalorimeter thermograms of samples of various percent conversion; A) 0%, B) 82%, C) 97%, D) 99%, E) 100%

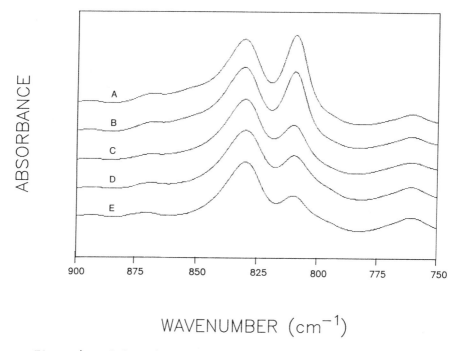

Figure 4. Infrared spectra of samples of various percent conversion; A) 0%, B) 82%, C) 97%, D) 99%, E) 100%

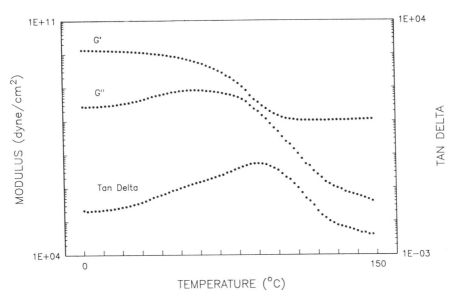

Figure 5. Temperature dependent mechanical spectrum of completely cured adhesive.

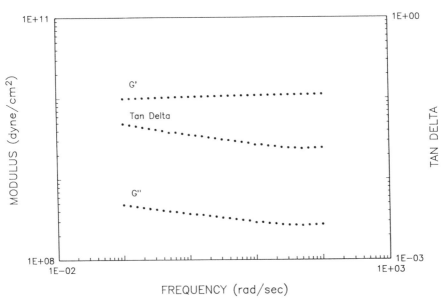

Figure 6. Frequency dependent mechanical spectrum of completely cured adhesive.

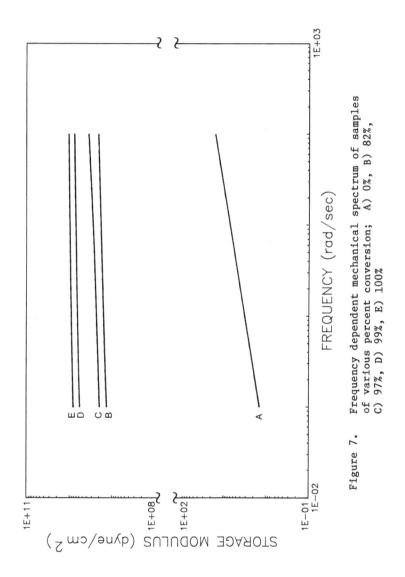

Figure 7. Frequency dependent mechanical spectrum of samples of various percent conversion; A) 0%, B) 82%, C) 97%, D) 99%, E) 100%

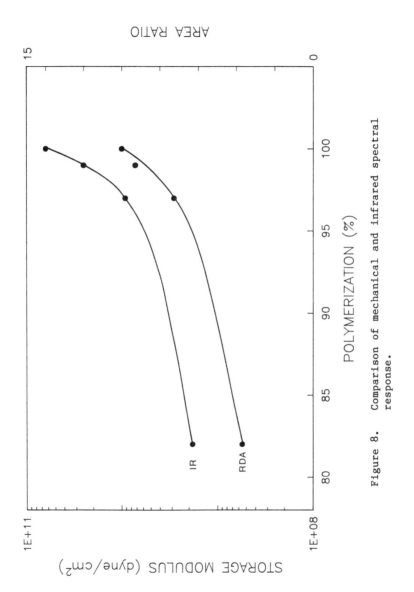

Figure 8. Comparison of mechanical and infrared spectral response.

Since infrared analysis perturbs approximately the first 5 microns of the polymer surface (representative of < 0.5% of the specimen thickness) a non-linear response is not surprising. The variation of the intensity of incident radiation with depth of penetration being the major contributing factor.

While mechanical spectroscopy is a measurement of the bulk specimen, a nonlinear response is also possible here because of plasticization and crosslinking effects as described below. Modulus increase may result from many causes including increasing crosslink density through polymerization of pendant double bonds of multifunctional monomers, or elimination of the plasticization effect of unreacted monomer/resin. Earlier work by Kloosterber and Lippits (5) suggests that even though photoinitiator radicals are short lived, polyacrylate radicals may be trapped in partially polymerized networks and this may provide the initiating species on continued irradiation. A further source of modulus increase may be through stress relaxation of the specimen as it gets heated (on continued irradiation) above its glass transition temperature, where the initial stress results from vitrification before complete conversion is reached. Identification of the elements within the adhesive contributing to this behavior will be the subject of future work.

Properties relating to performance of completely cured adhesive were determined by mechanical spectroscopy and thermomechanical analysis. Measurement of glass transition temperature and coefficient of thermal expansion was obtained from temperature scanning. Tg is reported as tangent delta maximum from mechanical spectra. The measurement of Tg by mechanical spectroscopy is frequency dependent and emphasizes the importance of appropriate selection of experimental conditions for comparison to other techniques. Results are listed in Table III.

TABLE III

COEFFICIENT OF THERMAL EXPANSION AND GLASS TRANSITION TEMPERATURE

	Thermomechanical Analysis	Mechanical Spectroscopy		
		1 rad/sec	10 rad/sec	100 rad/sec
T_g	78°C	78°C	84°C	89°C
α_G (0-50°C)	6.6 E-5/°C	6.6 E-5/°C	6.8 E-5/°C	6.8 E-5/°C
α_L (100-150°C)	1.8 E-4/°C	1.3 E-4/°C	1.3 E-4/°C	1.3 E-4/°C

As indicated above a goal of the next phase of this work is to study the molecular implications of incomplete polymerization. Wu (6) has proposed that this can be done by studying the terminal plateau modulus data for a series of test specimens. Accordingly, we attempted in this phase to show that the mechanical spectroscopy data generated from the adhesive specimens could be subjected to time-temperature superposition to yield complete modulus mapping over an extended frequency scale. Figure 9 illustrates the isothermal frequency scans obtained for time-temperature superposition. A master curve was constructed at reference temperature (T_o=75°C) approximately equal to the glass transition temperature (T_g.)

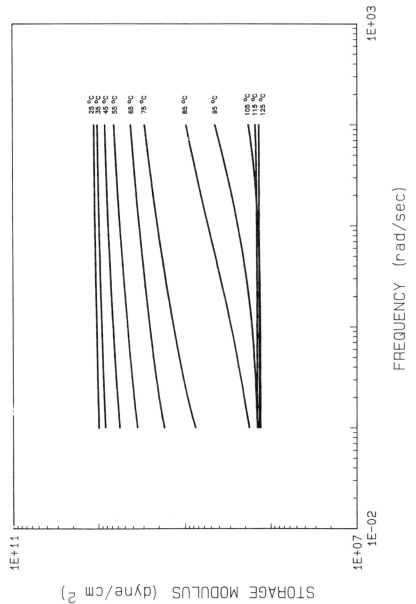

Figure 9. Frequency dependent mechanical spectra of completely cured adhesive at several temperatures.

Horizontal shift factors (a_t) were determined empirically. No vertical shift factor (To/T) was incorporated. The empirical shift factors were used with Arrhenius-type [log a_t = ($\Delta H/2.303R$)/(1/T-1/To)] and WLF-type [log a_t = $^-C_1$ (T-To)/(C_2 + T-To)] equations. A plot of T-To/log a_t versus T-To yielded WLF constants $C°_1$ and $C°_2$. The resultant values corresponding to the glassy zone (25-75°C) were used with the WLF-type equation to calculate horizontal shift factors. The agreement between empirical and calculated shift factors, as a function of temperature, is shown in Figure 10. This correlation indicates that the temperature dependence of a_t has an acceptable form consistent with WLF theory. Consequently, at least one criterion for the application of reduced variables to this adhesive system is fulfilled. A plot of log a_t versus 1/T (°K) also yielded a linear relationship consistent with Arrhenius theory and was used to calculate an activation energy. This data and WLF constants are listed in Table IV. The master curve of storage modulus resulting from empirical shift factors is shown in Figure 11. This composite curve defines the glassy, transition, and rubbery zones at constant temperature as a function of frequency and closely resembles the storage modulus as a function of temperature shown in Figure 5. The acceptable fit confirms the validity of time temperature superposition for modulus mapping of the cured adhesive polymer of this study.

TABLE IV

ACTIVATION ENERGY [1] and WLF CONSTANTS [2]

Enthaly	$C°_1$	$C°_2$
449 Kj/mol	19.88	133.86

1 calculate for temperature range 25-125°C
2 calculate for temperature range 25-75°C

Conclusions

While this paper reports only preliminary findings, it does illustrate the usefulness of photocalorimetry to define optimum cure conditions for UV curable adhesives. In addition, once the mechanical spectrum of fully cured adhesive has been mapped, mechanical spectroscopy can be used to monitor cure efficiency. In this paper we have not explored the molecular weight implications of incomplete polymerization. Preliminary evaluation of loss and storage modulus data would suggest that time-temperature superposition may be necessary to evaluate molecular weight/degree of cure relationships and terminal, plateau, and transition zones (4).

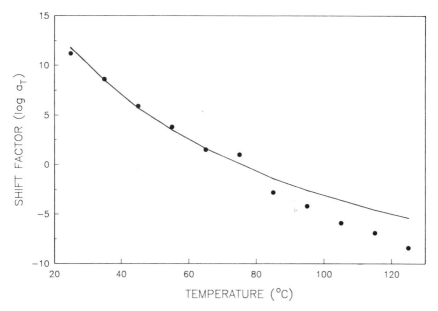

Figure 10. Temperature dependence of the shift factor; points are empirical, curve is theoretical.

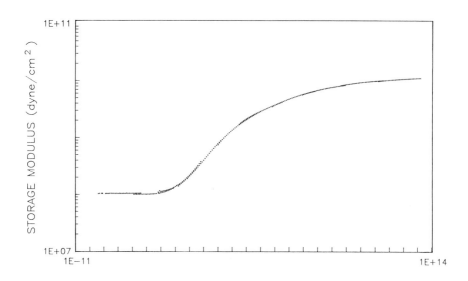

Figure 11. Frequency dependent master curve of completely cured adhesive resulting from time-temperature superposition.

REFERENCES

1 Drain, K. F.; Gatechair, L. R. Short Course on Fundamentals of Adhesion, State University of New York, 1988.

2 Kosnik, J. F.; Schweri, R. J. Proc. Radtech, 1988.

3 Gatechair, L. R. Proc. Radtech, 1988.

4 Ferry, J. D., Viscoelastic Properties of Polymers, John Wiley and Sons, New York, NY 1980, p 264.

5 Kloosterboer, J. G.; Lippits, G. J. M. Polym Preprints, 1985, 26, 351.

6 Wu, S. Polym. Eng. Sci, 1988, 25, 122.

RECEIVED September 13, 1989

Chapter 19

Mechanical Properties of UV-Cured Coatings Containing Multifunctional Acrylates

G. K. Noren, J. M. Zimmerman, J. J. Krajewski, and T. E. Bishop

DeSoto Inc., P.O. Box 5030, Des Plaines, IL 60017

> Multifunctional (>3) urethane acrylate oligomers were synthesized from HEA, IPDI, various diols and two trifunctional branching monomers. The variation of diol structure, diol molecular weight, type of functionality used on the branching monomer and overall composition of the oligomer was studied to establish structure-property relationships for the oligomers. Formulation of the oligomers with photoinitiators and monofunctional reactive diluents gave low viscosity coating materials which could be applied to various substrates and cured with UV radiation. From the stress-strain behavior of the cured films it was concluded that the molecular weight of the diol reactant was the most significant variable in the control of the mechanical properties. In agreement with results on difunctional oligomers, higher molecular weight diols produced lower tensile strength and modulus while giving increased elongation. Selection of the diluent monomer also contributed significantly to adjusting the mechanical properties of the cured films.

In addition to traditional applications such as graphic arts, wood finishing and floor coatings, the use of ultra-violet (UV) and electron beam cured materials has gained significant acceptance in emerging technologies such as coatings for thermally sensitive substrates, magnetic media, printed circuits and optical devices. The advantages of the use of radiation curable materials include efficient space utilization, fast line speeds, no solvent emissions and reduced energy consumption. Although several mechanisms exist to effect UV-curing, the most common technology involves the free radical polymerization of materials containing acrylic or methacrylic functionality. In general, this type of UV-curable coating is formulated to contain a photoinitiator, a reactive diluent and an acrylate functional oligomer.

Variation of the components of a coating formulation can produce improvements in the performance of the coating. Recent technology for the photoinitiator component has emerged which gives higher effective radical yields and faster rates of cure.(1) Many factors must be considered in the choice of reactive diluents used in a UV-curable coating formulation.(2-4) Each diluent has a different level of effectiveness in reducing viscosity and improving application properties.(5,6) Changes in reactive diluent will result in changes in final product performance such as: hardness, cure speed, tensile strength, elongation and modulus.(6) A mixture of several diluents is sometimes beneficial in achieving the optimum results. The oligomer portion of a UV-curable coating has a major effect on both the application properties of the formulated coating and the final film properties. Commercially available oligomers include urethane acrylates, epoxy acrylates, polyester acrylates, polyether acrylates and acrylated acrylic resins. Acrylate functionality is preferred to methacrylate functionality due to its faster cure speed. Little or no effect on the physical properties of cured films has been observed when changing the type of reactive functionality from acrylate to methacrylate.(7,8)

The urethane acrylates have gained wide industrial acceptance because they impart the toughness, abrasion resistance and elastomeric properties inherent in polyurethanes to the coating. The conventional synthesis of urethane acrylates, which consists of end-capping a polyester or polyether polyol with a diisocyanate followed by reaction with a hydroxyalkyl acrylate, has been reviewed.(9,10) Structure-property studies that have been undertaken with urethane acrylate oligomers have used methacrylate functionality, pure oligomer systems or high molecular weight oligomers cast from solution.(2,11-14) Other studies have used mainly difunctional oligomers with film branching introduced through the use of commercially available multifunctional acrylate diluents such as trimethylolpropane triacrylate (TMPTA) (15) or have been cured using electron beam radiation.(11,16,17) Our interest in radiation curable coatings led us to study complex multifunctional urethane acrylate oligomers in formulated coatings to determine if similar structure-property behavior could be observed.

Experimental

All reactants and reactive diluents were used as obtained from the manufacturer without further purification. All viscosities were measured on a Brookfield Viscometer (Model RV) at 25°C.

The oligomers were all prepared in a similar manner. The preparations were carried out in a 1 liter 4-necked round bottom flask fitted with a stirrer, thermometer, addition funnel, dry air sparge and a thermostatically controlled heating mantle or a room temperature water bath.

In each case, a solution of the desired diol in 2-hydroxyethyl acrylate was charged to the addition funnel and the isophorone diisocyanate, inhibitor and urethane catalyst were charged to the reaction flask with the water bath in place. The addition took one hour and after a two hour hold at room temperature, the appropriate trifunctional branching agent (TBA) was added. When the amine functional TBA was used, a 50 : 50 solution of the TBA in the proprietary reactive diluent was prepared and added over a 2 hour period at room temperature. When the hydroxy functional TBA was used a 50 : 50 solution of the TBA in the proprietary reactive diluent was added over a period of 20 minutes at 60°C. In each case, the reaction was then held at 60°C until the isocyanate absorption in the infra-red spectrum at 2250 cm^{-1} had completely disappeared.

The coatings were formulated from the above oligomer solutions by adding the required amount of photoinitiator and enough additional proprietary diluent to adjust the viscosity to 3000-10,000 cps at 25°C. The final oligomer concentration was generally 42-55 percent. Films were made by drawing the coating down on a 6 inch by 12 inch glass plate using a 6 inch, 3 mil bird bar. The films were cured at 3.5 Joules/sq-cm using two medium pressure, electrodeless mercury lamps with a Fusion Systems Corporation Radiation Curing Unit.

Mechanical properties were obtained on 4 inch by 0.5 inch strips of cured film using an Instron Model 4201 with the initial jaw separation distance of 2 inch at a strain rate of 1"/min.

Results and Discussion

Synthesis and Formulation. A conventional two step process, which is shown in Scheme I, was used for the synthesis of the multifunctional urethane acrylate oligomers. Step 1 involved the reaction of IPDI with a blend of HEA and the desired diol. The choice of IPDI as the diisocyanate was based on the differential reactivity of the NCO groups (15,18) which theoretically allowed the oligomer molecule to be built up in a step-wise fashion to control the molecular weight at its lowest level and thus the viscosity of the final oligomer. The question of which NCO group exhibits greater reactivity (19) will not be addressed in this paper, but for simplicity in representing oligomer structures it will be assumed that the primary NCO reacts first.(18) Step 2 involved the reaction of the NCO terminated mixture from step 1 with a suitable trifunctional branching agent (TBA). Either a hydroxyl functional or an amine functional TBA was used.

From Scheme I it is evident that four variables relating to oligomer structure could be investigated. They are: diol structure, diol molecular weight, functional group on the TBA, and the overall composition. A fifth variable, diisocyanate structure, was held constant (IPDI). Although there is some indication that diisocyanate structure may be related to crosslinking reactivity (20), it has been reported to have very little effect on modulus, tensile strength or elongation in a comparison between IPDI and TDI.(15)

It was desired to control the oligomer molecular weight within the range of 1500 to 3000 in order to maintain practical viscosities at 100% non-volatile material. Theoretical number average molecular weights were calculated based on the materials charged using the Macosko equations.(21) A typical GPC curve is shown in Figure 1, but actual molecular weights were not determined. An additional variable was evaluated by formulating the coatings with various reactive diluents. Vinyl caprolactam (VC), isobornyl acrylate (IBOA), dicyclopentadienyl acrylate (DCPA), and Diluent A which is a monofunctional acrylate containing a urethane group were all investigated. Diluents are generally used to adjust viscosity but

Step 1

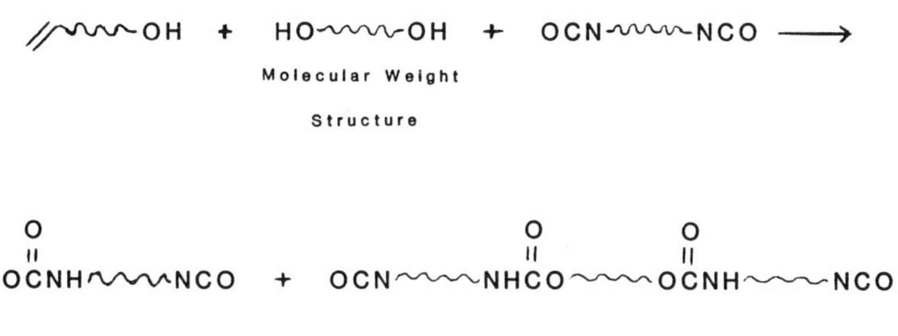

//∼∼∼OH + HO∼∼∼OH + OCN∼∼∼NCO ⟶
      Molecular Weight
         Structure

⌐OCNH∼∼∼NCO + OCN∼∼∼NHCO∼∼∼OCNH∼∼∼NCO
         ‖   ‖   ‖
         O   O   O

Product 1

NOTE: Due to the unequal reactivity of the NCO groups of IPDI virtually no diacrylate is formed in Step 1.

Step 2

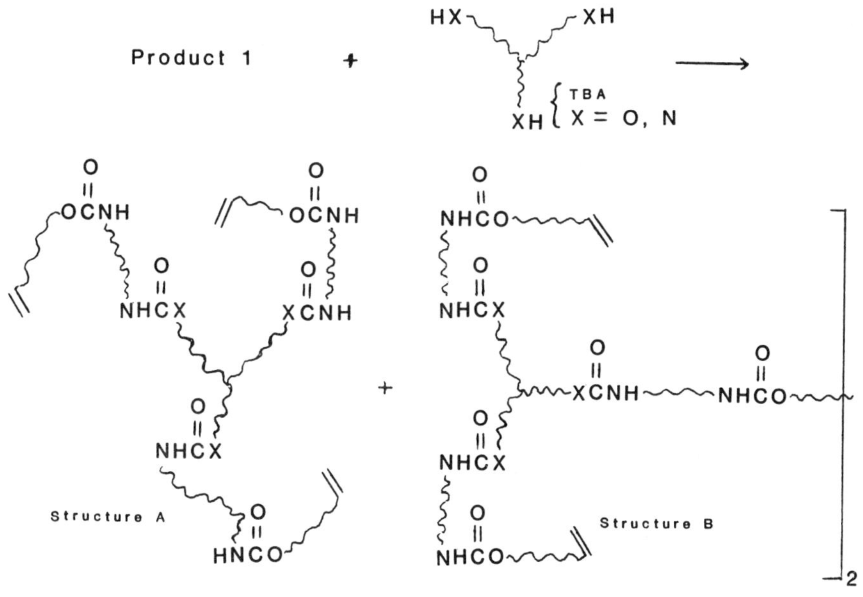

Product 1 + HX∼⌐∼XH
                 ⟩ {TBA
              XH⌐ X = O, N   ⟶

Structure A + Structure B

Scheme I

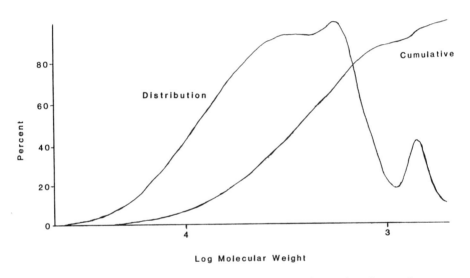

Figure 1. Representative GPC curve for multifunctional acrylates.

can help in property development such as adhesion, cure rate and hardness.(9)

Effect of Diol Structure. Model UV-curable oligomers were synthesized containing various diol structures. The diol structure was varied to contain ether, ester or carbonate linkages in an otherwise linear carbon chain while the molecular weight of the diol was maintained between 480-880. These oligomers were prepared using the amine trifunctional branching agent (ATBA), and, therefore, contained urea linkages in addition to urethane linkages. The total percent -CONH- in this set of oligomers varied from 15.7% to 17.0% based on diisocyanate charged. The theoretical molecular weight varied only slightly from a low of 1846 to a high of 1973, and the double bond functionality was 3.175+/-0.005. Oligomer compositions are shown in Table I. The viscosities of the oligomers ranged from 380,000-640,000 cps at $25^{o}C$. These high viscosities were attributed to H-bonding from the urea linkages. The order was Polycarbonate (PC) = Polycaprolactone/medium MW (PCL/MMW) > Polycaprolactone/low MW (PCL/LMW) > Polytetramethylene glycol (PTMG) = Polyester (PE) > Polypropylene glycol (PPG) which does correlate roughly with the molecular weights of the truly linear diols even in this narrow molecular weight range. The exception is the PPG which contains methyl side chains.

The oligomers shown in Table I were formulated into UV-curable coatings which are shown in Table II. The viscosities of the formulated coatings ranged from 3100 to 8090 cps at $25^{o}C$ and did not correlate with either the viscosities of the parent oligomers or the molecular weight of the diol. However, the oligomer containing PPG did again give the lowest viscosity. The higher viscosity of Formulation F (8090 cps) was probably due to the reduced amount of VC which was used as will be discussed later. The physical properties of 3 mil films cured at 3.5 Joules/sq-cm are also given in Table II. The order for tensile strength and modulus decreased according to the series PE > PTMG > PC > PCL/LMW > PCL/MMW > PPG. There was no correlation between these physical properties and the percentage of the -CONH- moiety or the molecular weight over this narrow range, therefore, the differences have been attributed to the structural features of the diol. This order is only slightly different from that reported for strictly linear difunctional oligomers.(11,16,17) The percent elongation decreased according to the series PCL/MMW > PPG > PC > PTMG > PCL/LMW > PE which, with the exception of the PC, correlates well with the molecular weight of the diol segment. The higher the molecular weight of the diol the greater the distance between crosslinks and thus the greater the elongation.

The Effect of Diol Molecular Weight. The molecular weight of the diol component was varied from 108-2040. The low molecular weight diol was 1,6-hexanediol while all the others were polycaprolactone based diols. These oligomers were prepared using a hydroxyl functional branching agent and thus contained all urethane linkages. Oligomer compositions are shown in Table III. The viscosities of the oligomers ranged from 134,000-386,000 cps and for the whole series could not be directly correlated with the urethane content of the oligomer or with the molecular weight of the diol component.

Table I. The Effect of Diol Structure: Composition and Properties of Oligomers

REACTANT (Structural Variable)	Segment MW	I	II	III	IV	V	VI
Polytetramethylene Glycol	640	6.7					
Polyester	480		5.0				
Polycarbonate	880			8.9			
Polypropylene Glycol	785				8.0		
Polycaprolactone (Moderate MW)	820					8.4	
Polycaprolactone (Low MW)	520						5.6
Isophorone Diisocyanate		27.6	27.5	26.5	26.8	26.8	28.1
Hydroxyethyl Acrylate		12.5	12.5	12.1	12.2	12.2	12.8
Amine Functional TBA		17.7	17.6	17.0	17.2	17.2	18.0
Urethane Catalyst		0.07	0.08	0.07	0.07	0.07	0.07
Free Radical Inhibitor		0.02	0.02	0.02	0.02	0.02	0.02
Proprietary Reactive Diluent		35.4	37.3	35.4	35.8	35.4	35.5
Brookfield Viscosity(cps)		500,000	500,000	640,000	380,000	640,000	632,000
RPM		1	5	5	10	5	5
Spindle		6	7	7	7	7	7
NCO/OH Equivalent Ratio		1.00	1.00	1.00	1.00	1.00	1.00
Theoretical Mn		1894	1846	1973	1936	1948	1857
Double Bond Equivalent Weight		598	581	621	612	615	586
Double Bond Functionality		3.17	3.18	3.18	3.17	3.17	3.17
Percent -CONH- Linkage		16.6	17.0	15.7	16.2	16.1	16.6

Table II. The Effect of Diol Structure: Coating Formulations and Physical Properties of the Films

COMPONENT	FORMULATION					
	A	B	C	D	E	F
Oligomer I	43.3					
Oligomer II		42.0				
Oligomer III			43.3			
Oligomer IV				43.0		
Oligomer V					43.3	
Oligomer VI						45.1
Proprietary Reactive Diluent*	43.7	45.0	43.7	44.0	43.7	45.9
Vinyl Caprolactam	10.0	10.0	10.0	10.0	10.0	5.0
Photoinitiator I	3.0	3.0	3.0	3.0	3.0	4.0
Photoinitiator II						
Brookfield Viscosity(cps)	5110	5430	7020	3100	4900	8090
RPM	100	100	100	100	100	100
Spindle	6	6	6	6	6	6
Tensile Strength(psi)	4654	5282	3639	3049	3086	3320
Modulus(kpsi)	139	146	101	79	82	91
Percent Elongation	54	10	57	72	73	52

* Includes proprietary reactive diluent used as a solvent during oligomer synthesis.

Table III. The Effect of Diol Molecular Weight: Composition and Properties of the Oligomers

REACTANT	OLIGOMER NUMBER			
	VII	VIII	IX	X
1,6-Hexanediol	1.5			
Low MW Polycaprolactone		6.5		
Moderate MW Polycaprolactone			9.9	
High MW Polycaprolactone				22.3
Isophorone diisocyanate	35.9	32.8	31.4	28.7
Hydroxyethyl Acrylate	16.2	15.9	14.2	13.1
Hydroxy Functional TBA	15.4	14.2	13.5	12.7
Urethane Catalyst	0.08	0.08	0.08	0.07
Free Radical Inhibitor	0.02	0.02	0.02	0.02
Proprietary Reactive Diluent	30.9	30.4	30.8	23.2
Brookfield Viscosity(cps)	386,000	134,000	147,000	200,000
RPM	10	20	20	10
Spindle	7	7	7	7
Molecular Weight of Diol Segment	108	525	830	2040
NCO/OH Equivalent Ratio	1.00	1.00	1.00	0.99
Theoretical Mn	1579	1694	1760	2106
Double Bond Equivalent Weight	495	533	562	682
Double Bond Functionality	3.19	3.18	3.13	3.09
Percent -CONH- Linkage	20.1	18.6	17.7	14.5

However, for the pure polycaprolactone based oligomers the viscosity increased as the molecular weight increased and the urethane content decreased.

The effect of the molecular weight of the diol segment on the physical properties of the cured films was evaluated using the oligomers shown in Table III which were formulated into the coatings shown in Table IV. The viscosities of the formulated coatings ranged from 3290 to 8340 cps. The lower viscosity of Formulation G was attributed to the presence of VC which appears to be a better solvent for the oligomers as will be discussed later. Neither urethane content nor segment molecular weight appeared to correlate with the viscosity of these formulations. As the segment molecular weight increased both the tensile strength and modulus of the films decreased while percent elongation increased. This result indicates that these properties are mainly dependent on the distance between crosslinks an observation which correlates with results reported for systems having double bond functionalities of two.(11,15)

The Effect of the Functional Group on the TBA. A comparison of Oligomers V and VI from Table I and Oligomers VIII and IX from Table III allows the evaluation of the effect of the urea vs the urethane linkages obtained from the TBA. Overall the viscosities of the urea oligomers are higher than the urethanes (H-bonding). This is also true for the specific examples using the low molecular weight PCL (VI vs VIII) and the moderate molecular weight PCL (V vs IX) based oligomers.

The effect of varying the branching linkage of the oligomer from urea to urethane can be obtained by comparison of Formulations K,E,H, and F which constitute a simple 2 x 2 factorial design with the variables of urea vs urethane and diol molecular weight (520 vs 820).

```
                TS = 3086 psi                    TS = 3796 psi
    MW = 820    Mod =   82 kpsi  (E)     (K)     Mod =   60 kpsi
        ↑       %E  =   73                       %E  =   93
        │
        ↓       TS = 3320 psi                    TS = 3468 psi
    MW = 520    Mod =   91 kpsi  (F)     (H)     Mod =   89 kpsi
                %E  =   52                       %E  =   63

                         UREA  ←──────→  URETHANE
```

From a visual analysis of the factorial design it can be seen that tensile strength and percent elongation increase in moving toward Formulation K or higher molecular weight while modulus increases in moving toward Formulation F or lower molecular weight and urea.

The Effect of Overall Composition. An important consideration in the design of oligomer structures is the variation of the molar ratio of the monofunctional HEA : diol : TBA while maintaining the NCO/OH ratio equal to 1. This was analyzed using Carothers' Equation for stoichiometric systems as a simple approximation. From this analysis f(av) must be less than 2 to obtain non-gelled oligomers. This requires the HEA : TBA molar ratio to be greater than 1, and increasing the amount of TBA will result in gellation.

This ratio also controls the final double bond functionality of the oligomer; the higher the desired double bond functionality the closer the HEA : TBA ratio will be to 1. It should be noted that at a constant HEA : TBA ratio both the molecular weight and the equivalent weight per double bond increase by adding diol. Two additional materials were synthesized to study the general composition and functionality of oligomers. They are summarized in Table V and can be compared with Oligomer IX in Table III. Significantly, the viscosity of these oligomers increased as the HEA : TBA ratio decreased indicating that viscosity was directly related to the TBA concentration. Although the actual oligomer concentration is slightly higher for Oligomer XII, this difference should not significantly affect this result.

The effect of varying the oligomer composition is shown in Table VI and can be compared with Formulation I from Table IV. These formulations were made at the same concentration of double bonds by using additional diluent as a consequence of the high viscosity obtained with Oligomer XII. As functionality increases from 3 to 4 the tensile strength only increases about 300 psi while lowering the molecular weight from about 3000 to about 1700 results in an increase of about 800–1000 psi. Increasing the molecular weight results in a decrease in modulus and an increase in percent elongation. Generally, we see very little difference in these results.

The Effect of Reactive Diluent. Four monofunctional reactive diluents were evaluated using Oligomer I in Formulation A (Table VII). The viscosity of these formulations showed a marked effect due to the diluent and increased in the order VC << IBOA < DCPDA = Diluent A, indicating that VC was a better solvent in terms of reducing viscosity. Analysis of the mechanical property data showed that both the tensile strength and modulus decreased along the order: VC > DCPDA > IBOA > Diluent A. The percent elongation increased in the order: VC < DCPDA < IBOA < Diluent A. As the Tg's of homopolymers of DCPDA and IBOA are roughly equivalent, 95° and 94°C, respectively, and the Tg of VC is expected to be < 50°C, these effects do not appear to be the result of long homopolymer chains from the diluent monomer. Further analysis was not attempted.

Conclusions

Several multifunctional (>3) urethane acrylate oligomers were synthesized by a two step process that gave materials which could be readily formulated into UV-curable coatings. The molecular weight of the diol used in the oligomer synthesis had the greatest effect on the mechanical properties of UV-cured films prepared from the formulated coatings. The effects of diol structure and diol molecular weight were similar to those reported for strictly difunctional oligomers. The presence of urea groups in the oligomer tended to increase the modulus while decreasing the percent elongation and tensile strength. The diluent vinyl caprolactam produced the greatest reduction in viscosity for the formulated coatings.

Table IV. The Effect of Diol Molecular Weight: Coating Formulations and Physical Properties of the Cured Films

COMPONENT	FORMULATION			
	G	H	I	J
Oligomer VII	45.6			
Oligomer VIII		52.9		
Oilgomer IX			52.6	
Oligomer X				55.3
Proprietary Reactive Diluent*	45.4	43.1	43.4	40.7
Vinyl Caprolactam	5.0			
Photoinitiator II	4.0	4.0	4.0	4.0
Brookfield Viscosity(cps)	3290	7130	6530	8340
RPM	100	100	100	100
Spindle	6	6	6	6
Tensile Strength(psi)	6600	3468	3796	2283
Modulus(kpsi)	185	89	60	4
Percent Elongation	18	63	93	113

* Includes proprietary reactive diluent used as a solvent during oligomer synthesis.

Table V. The Effect of Oligomer Functionality: Composition and Properties of the Oligomers

REACTANT	OLIGOMER NUMBER	
	XI	XII
Moderate MW Polycaprolactone	23.9	19.3
Isophorone diisocyanate	26.1	31.7
Hydroxyethyl Acrylate	10.3	11.1
Hydroxy Functional TBA	8.7	14.2
Urethane Catalyst	0.08	0.06
Free Radical Inhibitor	0.02	0.02
Proprietary Reactive Diluent	30.8	23.6
Brookfield Viscosity(cps)	86,600	2,000,000
RPM	10	
Spindle	6	
NCO/OH Equivalent Ratio	1.00	0.99
Theoretical Mn	2353	3145
Double Bond Equivalent Weight	784	805
Double Bond Functionality	3.00	3.90
Percent -CONH- Linkage	14.6	16.0

Table VI. The Effect of Oligomer Functionality: Coating Formulations and Physical Properties of the Cured Films

COMPONENT	FORMULATION	
	K	L
Oligomer XI	54.7	
Oligomer XII		42.8
Proprietary Reactive Diluent*	36.3	58.2
Vinyl Caprolactam	5.0	5.0
Photoinitiator II	4.0	4.0
Brookfield Viscosity(cps)	9820	3680
RPM	100	100
Spindle	6	6
Tensile Strength(psi)	2611	2900
Modulus(kpsi)	8.5	11
Percent Elongation	136	146

* Includes Proprietary Reactive Diluent used as a solvent during oligomer synthesis.

Table VII. The Effect of Monomeric Reactive Diluent on the Physical Properties of Oligomer I in Formulation A

PHYSICAL PROPERTY	REACTIVE DILUENT			
	VC	DCPDA	IBOA	Diluent A*
Brookfield Viscosity(cps)	5110	9810	8460	9920
Tensile Strength(psi)	4654	3640	3452	3136
Modulus(kpsi)	139	105	79	21
Percent Elongation	54	68	83	109

* Diluent A is a monofunctional acrylate containing a urethane group.

Acknowledgment

The authors would like to thank DeSoto for the opportunity to publish this research. We also wish to acknowledge the Physical Testing Department for performing the Instron testing.

Literature Cited

1. Bassi, G. L. J. Radiat. Curing, July 1987, p. 18.
2. Speckhard, T. A.; Hwang, K. K. S.; Lin, S. B.; Tsay, S. Y.; Koshiba, M.; Ding, Y. S.; Cooper, S. L. J. Appl. Polymer Sci., 1985, 30, 647.
3. Thanawalla C. B.; Victor, J. G. J. Radiat. Curing, Oct. 1985, p. 2.
4. Van Neerbos, A. J. Oil Col. Chem. Assoc., 1978, 61, 241.
5. Rybny, C. B.; DeFazio, C. A.; Shahidi, J. K.; Trebellas, J. C.; Vona, J. A. J. Paint Technol., 1974, 46(596), 60.
6. Priola, A.; Renzi, F.; Cesca, S. J. Coatings Technol., 1983, 55(703), 63.
7. Larson, E. G.; Spencer, D. S.; Boettcher, T. E.; Melbauer, M. A.; Skarjune, R. P. Radiation Phys. Chem., 1987, 30(1), 11.
8. Christmas B. K.; Zey, E. G. Spec. Chem., 1988, 8(1), 24.
9. Martin, B. Radiation Curing, Aug. 1986, p. 4.
10. Miller, H. C. Radiation Curing, May 1984, p. 4.
11. Lin, S. B.; Tsay, S. Y.; Speckhard, T. A.; Hwang, K. K. S.; Jezerc, J. J.; Cooper, S. L. Chem. Eng. Commun., 1984, 30(3-5), 251.
12. Chiang, W. Y.; Chan, S. C. J. Appl. Polym. Sci., 1987, 34, 127.
13. Ando M.; Uryu, T. J. Appl. Polym. Sci., 1986, 33, 1793.
14. Li, C.; Nagarajan, R. M.; Chiang, C. C.; Cooper, S. L. Polym. Eng. Sci., 1986, 26(20), 1442.
15. Koshiba, M.; Hwang, K. K. S.; Foley, S. K.; Yarusso, D. J.; Cooper, S. L. J. Mater. Sci., 1982, 17(5), 1447.
16. Oraby, W.; Walsh, W. K. J. Appl. Polym. Sci., 1979, 23, 3227.
17. Oraby, W.; Walsh, W. K. J. Appl. Polym. Sci., 1979, 23, 3243.
18. Huls Technical Bulletin 22-ME 377-7.
19. Ono, H. K.; Jones, F. N.; Pappas, S. P. J. Polym. Sci.; Polym. Lett. Ed., 1985, 23, 509.
20. Wang, M.; Du, C.; Dong, L.; Ma, Z. Fushe Yanjiu Yu Gongyi Xuebao, 1986, 4(1), 15; Chem. Abstr. 1987, 107, 116720s
21. Macosko, C. W.; Miller, D. R. Macromolecules, 1976, 9, 199.

RECEIVED September 13, 1989

Chapter 20

Structure—Performance Relationships of Urethane Acrylates

JoAnn A. McConnell and F. Kurt Willard

Radcure Specialties Inc., 9800 Bluegrass Parkway, Jeffersontown, KY 40299

> Acrylated urethanes are an important class of commercial radiation-curable oligomers. Industrial applications of these materials cover a wide range, including binders for magnetic media, vehicles for inks, and coatings for vinyl floor tiles, optical fibers, and paper. The compositions, and therefore, the properties of the acrylated urethanes are varied in order to meet the performance criteria of the different end uses. Properties of various acrylated urethanes will be discussed as they relate to structure.

Radiation curing technology is rapidly expanding into numerous, commercial applications. Due to their fast cure response, acrylate functional raw materials dominate the industry. A wide range of acrylated monomers and oligomers is available in order to meet the various application requirements.

Acrylated oligomers can be divided into three main classes: polyester acrylates, epoxy acrylates, and urethane acrylates. Urethane acrylates are an especially important commercial class of oligomers. Applications include binders for magnetic media, vehicles for inks, and coatings for vinyl floor tiles, optical fibers, and paper. Each application has certain performance requirements that must be met by the urethane oligomer. This paper will discuss the effect of urethane acrylate composition on end properties.

URETHANE ACRYLATES

A generic urethane acrylate is shown in Figure 1. It is formed by the reaction of a diisocyanate with a polyol and a hydroxy alkyl acrylate. Two types of substructures are found in urethanes, hard and soft segments. The hard segment is derived from the diisocyanate and urethane linkages. The soft segment imparts flexibility to the urethane acrylate, and is derived from the polyol.

By varying R, D, and P in Figure 1, a multitude of compositions, and therefore properties, can be obtained. In this study R was limited to hydroxyethyl acrylate (HEA) and hydroxypropyl acrylate (HPA), and no differentiation was made between the two.

Three diisocyanates (D) were studied: toluene diisocyanate (TDI), isophorone diisocyanate (IPDI), and dicylcohexylmethane-4,4'-diisocyanate (HMDI). Weatherability is one consideration in the choice of a diisocyanate. Oligomers based on aliphatic diisocyanates form coatings that weather well, whereas coatings based on aromatic diisocyanates yellow on exposure.[1]

The possibilities for P, the polyol, are virtually unlimited. Besides varying the functionality (di-, tri-, or tetra-ol), a number of different backbones can be employed: polyether, polyester, polycaprolactone, or polycarbonate. The substructure and molecular weight can also be changed. Only di- and tri-hydroxy polyethers and polyesters were used in this study.

Besides varying the nature of R, D, and P, the relative ratios of these three components can also be changed. These changes in stoichiometry allow molecular weight, degree of unsaturation, and urethane content to be varied.

One other compositional variable that must be considered in urethane design is illustrated in Figures 1 and 2. The urethane acrylate shown in Figure 1 is the theoretical reaction product of 4 eq. of diisocyanate with 2 eq. of HEA/HPA and 2 eq. of polyol. In practice, three different types of urethanes are always obtained from this reaction[2] (see Figure 2). The distribution of these three products is dependent on stoichiometry.

To summarize, Table I lists the ways that the composition of urethane acrylates can be altered, and the properties that may be affected by these changes. Each of these properties will be discussed as they relate to composition.

TABLE I

COMPOSITIONAL VARIABLES AND END PROPERTIES OF URETHANE ACRYLATES

Compositional Variables	End Properties
polyol type	viscosity
diisocyanate type	cure speed
functionality	tensile strength
molecular weight	tensile modulus
& distribution	elongation
urethane content	hardness
	adhesion
	flexibility
	solvent resistance

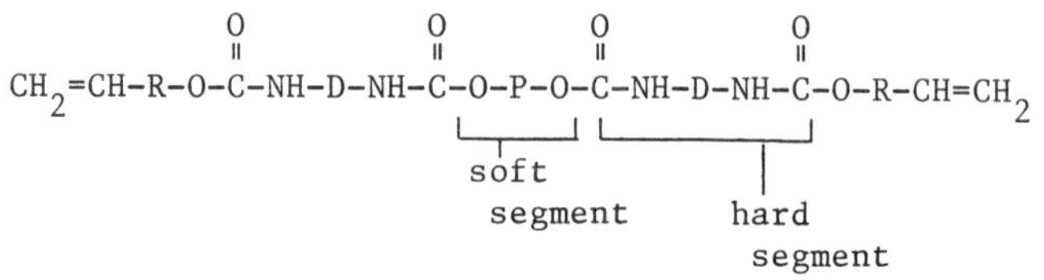

$$CH_2=CH-R-O-\overset{O}{\underset{\|}{C}}-NH-D-NH-\overset{O}{\underset{\|}{C}}-O-P-O-\overset{O}{\underset{\|}{C}}-NH-D-NH-\overset{O}{\underset{\|}{C}}-O-R-CH=CH_2$$

soft segment   hard segment

Figure 1. Generic urethane acrylate in which R = hydroxy alkyl acrylate backbone, D = diisocyanate backbone, and P = polyol backbone.

$$CH_2=CH-R-O-\overset{O}{\underset{\|}{C}}-NH-D-NH-\overset{O}{\underset{\|}{C}}-O-R-CH=CH_2$$

capped diisocyanate

$$CH_2=CH-R-O-\overset{O}{\underset{\|}{C}}-NH-D-NH-\overset{O}{\underset{\|}{C}}-O-P-O-\overset{O}{\underset{\|}{C}}-NH-D-NH-\overset{O}{\underset{\|}{C}}-O-R-CH=CH_2$$

urethane acrylate

$$CH_2=CH-R-O-\overset{O}{\underset{\|}{C}}-NH-D-NH-\overset{O}{\underset{\|}{C}}-O-P-O-\overset{O}{\underset{\|}{C}}-NH$$

$$CH_2=CH-R-O-\overset{O}{\underset{\|}{C}}-NH-D-NH-\overset{O}{\underset{\|}{C}}-O-P-O-\overset{O}{\underset{\|}{C}}-NH \quad D$$

extended urethane acrylate

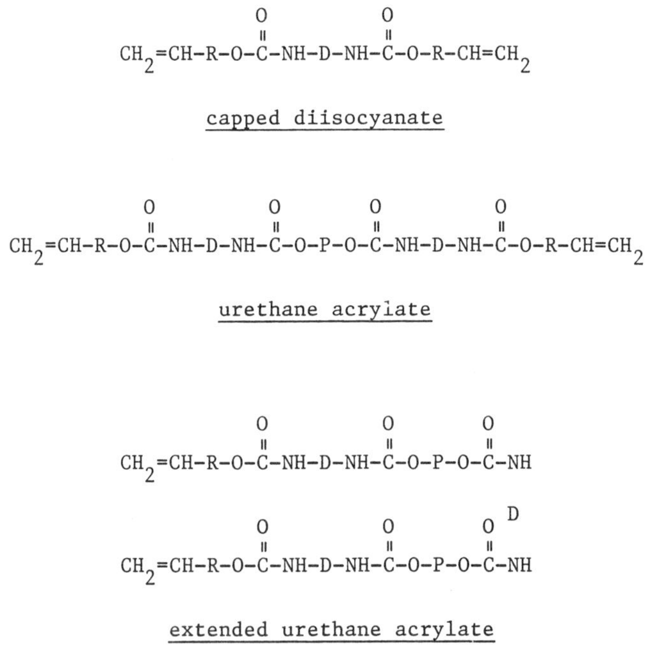

Figure 2. Experimental reaction products of urethane formation in which R = hydroxy alkyl acrylate backbone, D = diisocyanate backbone, and P = polyol backbone.

EXPERIMENTAL

The acrylated urethanes were synthesized by adding HEA or HPA to a diisocyanate in the presence of a catalyst and polymerization inhibitor. After a suitable hold time, polyol was added, and the reaction held for completion (<0.2% NCO). In some cases, the order of addition was reversed. Catalyst, inhibitor, hold times, and temperatures varied with the urethane composition.

Viscosities were measured using a Brookfield viscometer with a Thermosel attachment. Stress-strain properties were determined on 5-10 mil films of the homopolymerized oligomer using an Instron Tensile Tester. The films were obtained from warm resin containing 4 pph photoinitiator, Irgacure 184.

Cure speed and coating properties were obtained on the following formulation: 70% acrylated urethane, 30% 1,6-hexanediol diacrylate (Ebecryl HDODA), and 4 pph Irgacure 184. The formulations were drawn down on Parker Bonderite 40 steel panels using a #6 Mayer rod, ~0.5 mil films. These were cured in air using two 300 watt/inch mercury vapor electrodeless lamps at the maximum belt speed that gave a tack-free film. These films were then tested in general accordance with the ASTM methods for hardness (pencil hardness, ASTM D3363), flexibility (conical bend, ASTM D522-60), and adhesion (cross-hatch, ASTM D3359-83). Solvent resistance (MEK double rubs) was also measured.

EFFECT OF COMPOSITION ON PROPERTIES

VISCOSITY. The oligomer is primarily responsible for the basic coating properties. Viscosity is therefore an important parameter since it determines the amount of oligomer that can be formulated into the final product. All of the variables listed in Table I affect the oligomer viscosity.

The use of a symmetrical diisocyanate (HMDI), in which the two NCO groups possess the same reactivity, will produce an oligomer with a higher viscosity than one that contains an unsymmetrical diisocyanate (TDI or IPDI). This is due to the presence of more extended urethane acrylate (Figure 2) in HMDI based oligomers.[2] Table II clearly shows this trend.

Acrylate functionality, varied by choice of polyol, also has a profound effect on viscosity (see Table III). The increase in viscosity with functionality can be attributed to both the increased branching and the lower weight per double bond of the oligomer. A lower weight per double bond corresponds to a lower weight per urethane linkage or hard segment. These hard segments are not flexible and tend to associate with each other, increasing the viscosity.[2]

Table IV further illustrates the effect of weight per urethane linkage, as well as that of oligomer molecular weight. If molecular weight and weight per urethane linkage are increased by choice of polyol, without changing the stoichiometry, a decrease in viscosity is observed (compositions G, F, E, K). In this case, the effect of urethane content dominates.

On the other hand, if a change in stoichiometry leads to increased molecular weight and weight per urethane linkage, the

TABLE II
EFFECT OF DIISOCYANATE ON VISCOSITY AND CURE SPEED

Composition[1]	Diisocyanate	Viscosity (cps @ 150°F)	Cure Speed (fpm, 2 lamps)
A	IPDI	3,100	
	TDI	3,700	
	HMDI	10,100	
B	IPDI	22,000	
	TDI	23,000	
	HMDI	80,000	
C	IPDI	19,000	4p @ 15*
	TDI	24,000	25
D	IPDI	17,000	2p @ 15
	TDI	11,000	70

*tacky
[1] Within each composition, R, P, and the stoichiometry are identical.

TABLE III
EFFECT OF ACRYLATE FUNCTIONALITY ON VISCOSITY AND CURE SPEED

Comp[1]	Functionality (th)	Mol Wt (th)	Wt/Double Bond (th)	Viscosity (cps @ 150°F)	Cure Speed (fpm, 2 lamps)
E	2.06	1600	777	2,500	40
	2.3	1500	652	6,000	40
F	2.0	1800	900	9,000	60
	2.4	2300	965	24,000	100
	2.4	1700	715	27,000	-
	2.6	1700	650	51,000	25

[1] Composition. Within each composition, R, D, and the stoichiometry are identical; P is chemically the same, and is a mixture of diol and triol, only the ratio of diol to triol changes.

TABLE IV
EFFECT OF MOLECULAR WEIGHT AND/OR URETHANE CONTENT
ON VISCOSITY AND CURE SPEED

Comp[1]	Mol Wt (th)	Wt/Urethane Linkage (th)	Stoichiometry[2]	Viscosity (cps @ 150°F)	Cure Speed (fpm, 2 lamps)
G	1200	300	4:2:2	23,000	-
	1400	350	4:2:2	15,500	2p @ 15
	1500	375	4:2:2	11,000	4p @ 15
F	1500	320	4:2:2	107,000	-
	1700	375	4:2:2	27,000	-
	2300	525	4:2:2	24,000	-
E	1600	400	4:2:2	2,500	40
	2500	625	4:2:2	1,300	30
H	1200	300	4:2:2	3,500	-
	1900	317	6:2:4	8,000	-
	2600	325	8:2:6	13,500	-
I	2600	650	4:2:2	1,200	15
	7000	875	8:2:6	10,600	4p @ 15*
J	1500	375	4:2:2	4,000	40
	1800	450	4:2:2	2,600	60
	2300	380	6:2:4	33,000	70
K	1550	305	4:2:2	172,000	15
	1700	340	4:2:2	106,500	30

*tacky
[1]Composition. Within those compositions with the same stoichiometry, R and D are identical, and P has the same chemical stucture and functionality, but with different molecular weights. Only the stoichiometry changes for the other compositions: R, D, and P are the same.
[2]eq NCO:eq HEA/HPA:eq polyol

viscosity increases (compositions H, I, J). This increase can be attributed to a change in the distribution of the products shown in Figure 2. More of the extended urethane acrylate, and less or no capped diisocyanate are obtained as the stoichiometry changes from 4:2:2 to 6:2:4 to 8:2:6. The extended urethane acrylate contributes greatly to the viscosity of the oligomer, which can be attenuated somewhat by the capped diisocyanate.[2] Therefore, the increase in molecular weight and concurrent change in distribution override the decrease in urethane content, and viscosity increases.

Predictably, different polyols give oligomers with different viscosities. Table V lists some examples. The same trend in viscosity is seen in compositions L and M.

TABLE V
EFFECT OF POLYOL TYPE ON VISCOSITY AND CURE SPEED

Composition[1]	Polyol	Mol Wt (th)	Viscosity (cps @ 150°F)	Cure Speed (fpm, 2 lamps)
L	Polyester 1	1600	25,000	25
	Polyester 2	1400	11,000	70
	Branched Polyether	1600	2,700	40
M	Polyester 1	1700	19,000	4p @ 15*
	Polyester 2	1500	12,600	4p @ 15
	Branched Polyether	2700	850	4p @ 15*
N	Linear Polyether	1800	2,600	60
	Branched Polyether	1900	8,000	2p @ 15

*tacky
[1]Within each composition, R, D, and the stoichiometry are identical; P is difunctional.

CURE SPEED. Cure speed is very important in those applications that require high output. TDI based urethanes cure faster than IPDI based urethanes (Table II). Variations in the polyol also impact cure speed. Linear polyether linkages in a polyol have previously been shown to enhance cure speed.[1,2] This effect is illustrated in Table V, composition N.

Unexpectedly, cure speed does not increase with increased acrylate functionality (Table III). In composition E, no change in cure speed occurs with a change in functionality (2.06-2.3). In composition F, cure speed actually decreases with a decrease in the weight per double bond (965-650). Composition F is based on a linear polyether, which as mentioned earlier, enhances cure speed. Also, there is a correlation between weight percent of polyether in the oligomer and cure speed. It appears that cure speed is more dependent on the amount of polyether in the oligomer than on the functionality of the oligomer.

As molecular weight and weight per urethane linkage increase, the cure speed of those oligomers that do not contain a linear polyether decreases (Table IV, compositions G, E, I). This is primarily a dilution effect. The cure speed of those urethanes that contain a polyether correlates, once again, with the weight percent of polyether in the oligomer (Table IV, compositions J, K).

STRESS-STRAIN PROPERTIES. Tensile strength, modulus, and elongation of urethane acrylates are important considerations in the design of an oligomer for a specific end use. The structures of the urethanes can be manipulated in order to achieve the desired stress-strain properties. Diisocyanate and polyol can be varied to produce concurrent changes in tensile, modulus, and elongation (Tables VI and VII). Generally, TDI based oligomers give higher tensile and lower modulus and elongation than IPDI based ones. The less flexible structure of TDI contributes to these effects.

Decreases in molecular weight, weight per urethane linkage, and weight per double bond all give increases in tensile strength and modulus and decreases in elongation (Tables VIII and IX). These effects are all related to crosslink density and to the ratio of hard and soft segments in the molecule.[2] More hard segments give higher tensile and modulus and lower elongation. The converse holds for more or larger soft segments.

TABLE VI
EFFECT OF DIISOCYANATE ON STRESS-STRAIN PROPERTIES

Composition[1]	Diisocyanate	Tensile (psi)	Modulus (psi)	Elongation (%)
O	IPDI	4800	8.1×10^5	0.6
	TDI	4950	7.5×10^5	0.7
C	IPDI	2200	8.4×10^3	135
	TDI	2800	8.1×10^3	120
D	IPDI	4100	2.5×10^4	125
	TDI	4750	7.8×10^3	120

[1] Within each composition, R, P, and the stoichiometry are identical.

TABLE VII
EFFECT OF POLYOL TYPE ON STRESS-STRAIN PROPERTIES

Composition[1]	Polyol	Mol Wt (th)	Tensile (psi)	Modulus (psi)	Elongation (%)
L	Polyester 1	1600	2800	8.1×10^3	120
	Polyester 2	1400	4750	7.8×10^3	120
	Branched Polyether	1600	600	4.8×10^3	65
M	Polyester 1	1700	2200	8.4×10^3	135
	Polyester 2	1500	1400	6.8×10^3	120

[1] Within each composition, R, D, and the stoichiometry are identical; P is difunctional.

TABLE VIII
EFFECT OF ACRYLATE FUNCTIONALITY ON STRESS-STRAIN PROPERTIES

Composition[1]	Functionality (th)	Mol Wt (th)	Wt/Double Bond (th)	Tensile (psi)	Modulus (psi)	Elongation (%)
F	2.4	2300	965	1900	6.9×10^3	100
	2.6	1700	650	3900	1.0×10^5	65
E	2.06	1600	777	560	2.9×10^3	80
	2.3	1500	652	1070	6.7×10^3	70

[1]Within each composition, R, D, and the stoichiometry are identical; P is chemically the same, and is a mixture of diol and triol, only the ratio of diol to triol changes.

TABLE IX
EFFECT OF MOLECULAR WEIGHT AND/OR URETHANE CONTENT ON STRESS-STRAIN PROPERTIES

Composition[1]	Mol Wt (th)	Wt/Urethane Linkage (th)	Tensile (psi)	Modulus (psi)	Elongation (%)
G	1200	300	5900	2.0×10^5	100
	1400	350	5400	3.6×10^4	130
	1500	375	2900	1.1×10^4	130
E	1600	400	600	5.0×10^3	60
	2500	625	300	7.1×10^2	80
P	5000	830	150	4.2×10^2	75
	7000	875	90	2.4×10^2	85
H	1200	300	4800	2.0×10^5	60
	1900	317	3600	6.1×10^4	105
	2600	325	3200	1.8×10^4	150
J	1000	305	8200	3.4×10^5	5
	1500	375	3050	3.2×10^4	90
	2300	380	2350	5.3×10^3	130

[1]Within those compositions with the same stoichiometry, R and D are identical, and P has the same chemical structure and functionality, but with different molecular weights. Only the stoichiometry changes for the other compositions: R, D, and P are the same.

COATING PROPERTIES. Coating properties such as hardness, flexibility, adhesion, and solvent resistance can be impacted by changing the oligomer or the formulation. One formulation was used in this study, and variations in the oligomer were investigated as a means to change the coating properties. Because the oligomer is diluted in the formulation, the effect of the oligomer on final coating properties is abated.

From Table X, the IPDI based oligomers have better solvent resistance than the TDI based ones. IPDI based oligomers also appear to exhibit better adhesion. Other properties are not affected by diisocyanate choice. Polyester based materials seem to give better flexibility and solvent resistance than polyether based urethanes (Table XI). The branched polyether in composition N gives better adhesion and solvent resistance than the linear polyether. No other generalities can be drawn from these data.

TABLE X
EFFECT OF DIISOCYANATE ON COATING PROPERTIES

Composition[1]	Diisocyanate	Pencil Hardness	Conical Bend (in)	Adhesion (%)	Solvent Resist (MEK D.R.)
C	IPDI	2H	0	75	75
	TDI	3H	0	0	50
D	IPDI	2H	0	5	>200
	TDI	2H	0	0	145

[1]Within each composition, R, P, and the stoichiometry are identical.

The data in Table XII show no correlation between weight per double bond and coating properties. However, solvent resistance does increase with functionality. This is due to an increase in the crosslink density of the cured film. No correlation is seen for adhesion and molecular weight or weight per urethane linkage (Table XIII). However, there is a slight trend of decreased solvent resistance with increased molecular weight and weight per urethane. Softness and flexibility show a definite increase with increase in molecular weight and weight per urethane linkage. The trends noted in Table XIII are due to the ratio of hard and soft segments in the urethane acrylate. More hard segments and/or shorter soft segments give better solvent resistance, harder surfaces, and less flexible coatings.[2]

As mentioned earlier, the effect of the oligomer on final coating properties is abated due to dilution in the formulation. Therefore, the differences in performance are more subtle, and may not be defined adequately by the properties measured in this paper. Tests specific to a certain application may be needed to better define oligomer performance.

TABLE XI
EFFECT OF POLYOL TYPE ON COATING PROPERTIES

Comp[1]	Polyol	Mol Wt (th)	Pencil Hardness	Conical Bend (in)	Adhesion (%)	Sol. Resist. (MEK D.R.)
L	Polyester 1	1600	3H	0	0	50
	Polyester 2	1400	2H	0	0	145
	Branched Polyether	1600	3H	1/4	0	10
M	Polyester 1	1700	2H	0	75	75
	Polyester 2	1500	2H	0	2	>200
	Branched Polyether	2700	HB	1/2	25	5
N	Linear Polyether	1800	HB	0	5	40
	Branched Polyether	1900	2H	0	75	>200

[1]Within each composition, R, D, and the stoichiometry are identical; P is difunctional.

TABLE XII
EFFECT OF ACRYLATE FUNCTIONALITY ON COATING PROPERTIES

Composition[1] (th. func.)	Mol Wt (th)	Wt/Double Bond (th)	Pencil Hardness	Conical Bend (in)	Adhesion (%)	Sol. Resist. (MEK D.R.)
F (2.0)	1800	900	HB	1/4	0	90
(2.4)	2300	965	B	0	10	>200
(2.6)	1700	650	F	0	10	>200
E (2.06)	1600	777	3H	1/4	0	10
(2.3)	1500	652	2H	0	0	50

[1]Within each composition, R, D, and the stoichiometry are identical; P is chemically the same, and is a mixture of diol and triol, only the ratio of diol to triol changes.

TABLE XIII
EFFECT OF MOLECULAR WEIGHT AND/OR URETHANE CONTENT
ON COATING PROPERTIES

Comp[1]	Mol Wt (th)	Wt/Urethane Linkage (th)	Pencil Hardness	Conical Bend(in)	Adhesion (%)	Sol. Resist. (MEK D.R.)
D	1400	350	2H	0	10	>200
	1500	375	2H	0	2	>200
E	1600	400	3H	1/4	0	10
	2500	625	HB	1/4	0	5
P	3000	750	HB	1/2	25	4
	5000	830	2B	1/4	50	2
	7000	875	4B	1/4	95	2
H	1200	300	2H	3	75	>200
	1900	317	2H	0	75	>200
	2600	325	H	0	40	180
I	2600	650	HB	3/4	50	5
	7000	875	2B	1/4	5	3
J	1500	375	H	0	2	100
	2300	380	HB	0	0	145

[1]Composition. Within those compositions with the same stoichiometry, R and D are identical, and P has the same chemical structure and functionality, but with different molecular weights. Only the stoichiometry changes for the other compositions: R, D, and P are the same.

CONCLUSION

The composition of urethane acrylates can be varied in order to achieve certain end properties required by the intended application. The structure-property relationships discussed in this paper can provide guidance in the intelligent design of this class of radiation cure oligomer.

LITERATURE CITED

1. H. C. Miller, Rad. Curing, 11(2), 4 (1984).
2. B. Martin, Rad. Curing, 13(3), 5 (1986).

RECEIVED September 13, 1989

Chapter 21

Enthalpy Relaxation in UV-Cured Epoxy Coatings

Geoffrey A. Russell and W. Eugene Skiens

Optical Data, Inc., 9400 Southwest Gemini Drive, Beaverton, OR 97005-7159

Development of structural ordering upon annealing is a well-recognized phenomenon in glassy polymers. It has significant effects on modulus, impact resistance and other mechanical properties. Structural ordering can occur in thin films as well as in bulk materials. In this study it has been observed in UV-cured acrylated epoxy coatings. It has been characterized by quantitative differential thermal analysis (DSC) and by thermomechanical analysis (TMA). The extent of structural relaxation is a function of cure dose, thermal history and type and level of comonomer. Because all of the formulations studied exhibited broad glass transitions beginning near ambient temperature, significant structural ordering occurred at room temperature. Relaxation rates decrease with increasing cure dose or upon addition of mobile multifunctional crosslinkers. Based upon these observations, it appears that structural ordering plays a significant role in the aging of UV-cured coatings.

It is frequently observed that the properties of UV-cured films and coatings change as a function of time after cure. This process can be accelerated by a brief thermal treatment. In the case of cationically-cured systems this phenomenon is generally attributed to the reaction kinetics of the system. The initiation step is photochemically-activated and is very rapid. The propagation step is slower, and requires time and/or thermal activation to proceed to completion. Since there are few chain termination reactions which affect cationic systems, cure can proceed for extended periods after UV exposure, as long as there are still reactive species present.[1,2] In the case of free radical-initiated systems, the basis for this post-cure change in properties is less clear. Both the initiation and propagation steps proceed rapidly, even at ambient temperature. Furthermore, free radicals generally are short-lived species, and are subject to a number of reactions with atmospheric species which consume them. Therefore, the changes in physical properties observed in acrylated epoxies and other free radical-initiated coating systems must be accounted for by some other mechanism. Since the changes in properties occur at ambient temperature, it is unlikely that a chemical mechanism will account for them. Enthalpy relaxation, a common phenomenon in bulk amorphous glasses, appears to be the principal mechanism of post-cure aging in free radical-initiated coatings.

Enthalpy Relaxation in Bulk Polymeric Glasses

Structural relaxation has been observed in a wide variety of organic, inorganic and metallic glasses.[3-13] It gives rise to systematic changes in such mechanical properties as modulus, impact strength and heat capacity as a function of the thermal history of the material.[4,5,12,13] The basis for structural relaxation is the nonequilibrium nature of the glassy state due to supercooling effects observed at any finite cooling rate. Structural relaxation has been shown to be a nonlinear, nonexponential process[10] due to the fact that the enthalpy of a glass is a function of the entire thermal history of the glass rather than a state variable which is independent of the path by which the glass was formed. Differential scanning calorimetry (DSC) is a widely-used method for studying structural relaxation in polymeric glasses.[3,5-9] It is convenient in that proper calibration of the DSC instrument allows direct determination of the heat capacity of the sample as a function of temperature. Also, the sample can be annealed for varying periods in the DSC pan and re-analyzed conveniently. The extent of structural relaxation is determined by the area of the "excess enthalpy" peak which appears as the annealed sample is first heated from below Tg through the glass transition region. This is shown schematically in Figure 1. The excess enthalpy peak may appear below, in the middle or at the end of the glass transition region, depending upon the specific material system. Its area is a function of the entire thermal history of the glass after it is quenched from the rubbery state. As structural relaxation occurs, impact strength decreases[12] and modulus increases.[3] While the work reported to date has dealt with structural ordering in bulk polymeric glasses, the same structural relaxation processes ought to occur in amorphous glassy films as well, and should have a significant effect on their properties.

Experimental

All the formulations used in our experiments utilized a diacrylated Bisphenol A epoxy, Novacure 3700 (Interez, Inc.), as the base resin. Four multifunctional acrylates were used to modify cure behavior and physical properties. They were: trimethylolpropane triacrylate (TMPTA), polybutanediol diacrylate (PBDDA), pentaerythritol triacrylate (PETA) and dipentaerythritol triacrylate (DPEPA). The base resin, TMPTA and PETA were supplied by Interez, Inc. PBDDA was obtained from Alcolac Chemical. DPEPA was supplied by Sartomer. All were commercial grades, and were used without further purification. The photoinitiator used for all experiments was Darocur 1664 (EM Industries), a mixture of aromatic ketones whose absorption spectrum is shown in Figure 2. It was chosen because it is well-matched to the emission spectrum[14] of the Fusion "V" lamp which was used as the source of irradiation. The formulations used are summarized in Table I.

Formulations were prepared by dissolving the base resin in 100 phr of 2-butanone, then adding the photoinitiator and any monomers to the resin solution. Films were cast from the resin solution onto 75 um polyethylene terephthalate (PET) sheets, dried in a convection oven at 60°C for 15 minutes to remove residual solvent, and cured. A LESCO C612 conveyor system with Fusion Systems F450-10 irradiator was used for curing. A 300 W/inch Fusion "V" lamp was used in all experiments. Cure dose was varied by changing belt speed at constant irradiator power.

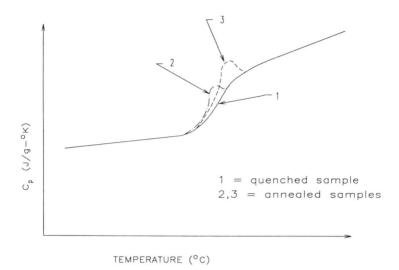

FIGURE 1: Schematic representation of DSC curves for quenched and annealed samples showing structural relaxation effects.

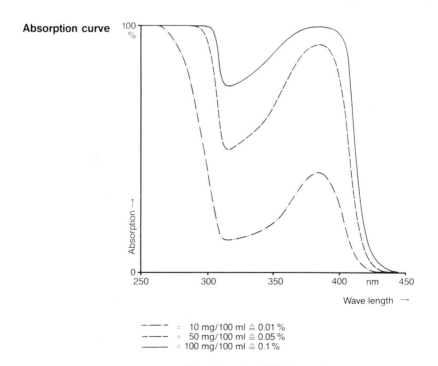

FIGURE 2: Absorption spectrum of Darocure 1664 photoinitiator.

After irradiation, samples were peeled from the PET substrates and analyzed by quantitative differential thermal analysis (DSC) and thermomechanical analysis (TMA) using a Mettler TA3000 thermal analysis system with DSC30 and TMA40 modules.

TABLE I: Formulation Parameters

Epoxy Acrylate(1)	Comonomer Type(2)	Level	Photoinitiator(3) Level	Sample Designation
Novacure 3700 100phr	None	-	2phr	N37D2
	None	-	5phr	N37D5
	TMPTA	15phr	2phr	N37T15D2
	TMPTA	30phr	2phr	N37T30D2
	PBDDA	15phr	2phr	N37PB15D2
	PBDDA	30phr	2phr	N37PB30D2
	PETA	15phr	2phr	N37PE15D2
	PETA	30phr	2phr	N37PE30D2
	DPEPA	15phr	2phr	N37DP15D2
	DPEPA	30phr	2phr	N37DP30D2

(1) Bisphenol A epoxy diacrylate (Interez, Inc.)
(2) TMPTA = Trimethylolpropane triacrylate;
    PBDDA = Polybutanediol diacrylate;
    PETA = ppentaerythritol triacrylate;
    DPEPA = dipentaerythritol pentaacrylate
(3) Photoinitiator = Darocure 1664 (EM Industries)

Samples for FTIR analysis were cast from dilute 2-butanone solution onto KBr plates, then dried for 15 minutes at 60°C in a convection oven. The absorption spectra of the unexposed films were recorded from 4000-600 cm-1 using a Nicolet 5DXB FTIR spectrometer. Spectra were obtained at 2 cm-1 resolution using 128 co-added scans per sample. Each sample was then irradiated at 18 m/min and then re-analyzed. This process was repeated until a total of six irradiations was accumulated.

Percent extractibles were determined by peeling cured films from the PET substrate, heating each sample in a convection oven for 30 minutes at 60°C to remove volatiles, and weighing the dried film. Each film was then placed in a scintillation vial and extracted with 2-butanone for 30 minutes on a wrist-action shaker. Films were removed intact from the solvent and dried to constant weight at 60°C in vacuo, then re-weighed to determine weight loss.

Results and Discussion

A single pass through the UV cure system converted all the films to a tack-free state. DSC thermograms were obtained on samples cut from each film to determine the effect of cure dose and composition on Tg and other properties. The DSC thermogram for Sample N37D2, the base resin without added comonomer, is shown in Figure 3(a). The sample exhibits a broad glass transition with an onset at 30°C. At 41°C an additional endotherm is observed, superposed on the glass transition. The endotherm is complete at 63°C. The glass transition is complete at 90°C. When the sample is quenched and immediately reheated, a broad glass transition is observed, extending from 28°C to 92°C, but no endotherm is observed, as seen in Figure 3(b). If the sample is allowed to anneal at 20°C for 25 hours, the endotherm is again observed in the DSC thermogram, but its area is significantly

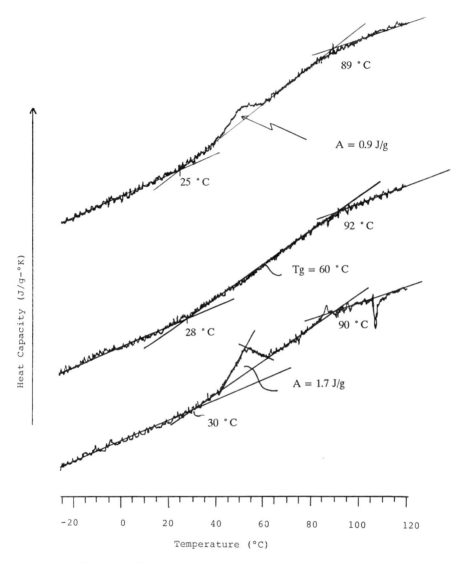

FIGURE 3: DSC thermograms of quenched and annealed samples of N36D2 cured 2 X @ 12m/min.

reduced, as seen in Figure 3(c). Similar behavior is exhibited by all the compositions tested. DSC data for Tg onset, midpoint and final temperature are summarized in Table II, along with the area of the endotherm observed during the first heating of each sample. All samples were annealed at 20°C for approximately one week prior to the first DSC analysis.

TABLE II: Summary of DSC Results on Cured Films

FORMULATION	CURED 1 X @ 18m/min					Cured 2 X @ 12 m/min				
	Tg (°C)	ΔTg (°C)	ΔHex (J/g)	Tg (°C)	ΔTg (°C)	Tg (°C)	ΔTg (°C)	ΔHex (J/g)	Tg (°C)	ΔTg (°C)
N37D2	42	58	3.4	50	48	60	60	1.7	60	64
N37T15D2	48	69	3.0	43	53	59	67	1.3	59	57
N37T30D2	46	74	2.7	40	58	65	82	1.7	64	78
N37PB15D2	49	74	2.2	47	58	53	61	1.2	62	90
N37PB30D2	47	60	1.9	43	50	50	71	1.6	49	79
N37PE15D2	48	59	2.3	44	57	61	72	1.4	62	72
N37PE30D2	41	66	2.2	39	66	68	79	1.9	68	91
N37DP15D2	43	54	2.7	44	74	54	59	1.3	54	80
N37DP30D2	42	72	2.4	35	56	47	37	1.0	57	34
	First Heating			Second Htg		First Heating			Second Htg	

The breadth of the glass transition can be explained by the immiscibility of the acrylic and epoxy components in the film. Since the two components would phase separate if they were not covalently linked, their glass transition reflects the relaxation spectra of the acrylic component and the epoxy component. The endotherm observed in annealed samples reflects the structural rearrangement which is occurring in the glassy state. This type of endothermic transition is frequently observed in aged or annealed samples of amorphous polymeric glasses.[3,5,6] It has been termed an "excess enthalpy" peak, and is caused by structural ordering and densification within the amorphous glass. The presence in the acrylated epoxy films of the acrylic component, with a relaxation onset at or below ambient temperature, provides a mechanism for structural relaxation at ambient conditions.

In order to determine the rate of structural relaxation in the cured films, Sample N37D2 was annealed for varying periods at 20°C and then re-analyzed by DSC. The area of the excess enthalpy peak was determined by a simple geometric integration technique. A plot of the excess enthalpy, Hex, versus the reciprocal of the annealing time was constructed, as shown in Figure 4. Extrapolation to infinite time ($t^{-1} = 0$) predicts that Sample N37D2 would have a Hex of 1.8-1.9 J/g at equilibrium, with a $t_{1/2}$ for structural relaxation of 25 hours at 20°C. Examination of the data in Table II shows that Hex decreases with increasing cure dose. At low UV dose, the addition of comonomer causes a decrease in the excess enthalpy peak. At higher UV dose, addition of comonomer has little or no effect on Hex.

Thermomechanical analysis (TMA) is a very useful technique for determination of the properties of coatings and thin films. In many cases it can be used to determine the Tg of a film which is too thin or too adherant to the substrate to permit analysis by DSC. However, interpretation of TMA results can be complicated by the effects of sample curl during curing coupled with volume relaxation effects. This is illustrated by the TMA thermogram of Sample N37T15D2, cured 2X at 12 m/min, which is shown in Figure 5. Instead of a monotonic increase in thermal expansion upon heating, the as-cured sample shown in Figure 5(a) exhibits a net decrease in thickness beginning at 27°C and continuing until 90°C. Above 90°C the rate of thermal expansion

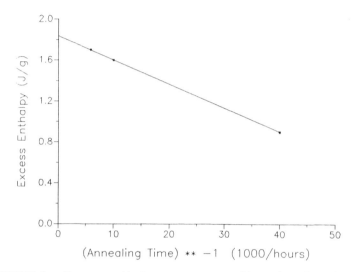

FIGURE 4: Excess enthalpy versus annealing time for N37D2

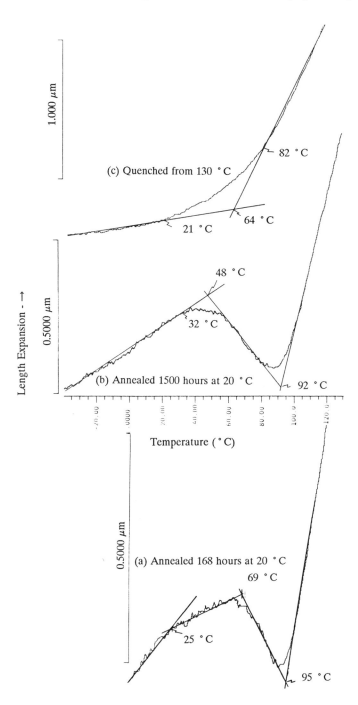

FIGURE 5: TMA thermograms of N37T15D2 cured 2 X @ 12m/min

increases and a net expansion occurs. The sample annealed 1500 hours at 20°C exhibits similar behavior, as shown in Figure 5(b). However, when the sample shown in Figure 5(b) is quenched from 130°C and immediately reheated, no contraction is observed, as shown in Figure 5(c). The Tg determined by the intersection of the glassy state and rubbery state portions of the thermogram is 67°C for the quenched sample. This is slightly higher than the value of 59°C determined by DSC. However, the range of temperature over which the contraction occurs in the annealed samples agrees well with the range over which enthalpy relaxation is observed by DSC. While quantitative estimation of enthalpy relaxation is not possible from the TMA thermogram, Tg can be determined from the thermogram of annealed thin film specimens. Care must be exercised in interpretation of the thermograms of annealed samples, however, as structural relaxation will be superposed on thermal expansion once the sample is heated above the onset of the glass transition.

Since both UV dose and the composition of the film had an effect on the degree of structural relaxation observed, it was important to determine the effect of these variables on the extent of cure as well. The degree of cure was determined by two methods: FTIR analysis to determine the loss of vinyl unsaturation as a function of composition and UV dose, and solvent extraction to determine the extent of crosslinking in the cured film.

The vinyl unsaturation peak at 810 cm^{-1} was used as a measure of the extent of reaction.[15-18] Vinyl group disappearance was also correlated with changes in the intensity of the acrylate H2C=CH-stretching in the 1620-1640 cm^{-1} region. (17) FTIR spectra were obtained as a function of UV dose for samples N37D2, N37D5 and N37D30D2. The spectrum of each unexposed sample was stored and used for all subsequent analyses. Each sample was then exposed by passing it through the UV source at 18 m/min, and a new FTIR spectrum was obtained. Spectra were obtained after 1, 2, 4 and 6 passes through the source. The spectrum of each exposed sample was then subtracted from the spectrum of the unexposed sample. A typical difference spectrum is shown in Figure 6. The peak at 810 cm^{-1} in the difference spectrum was then integrated using the software supplied with the spectrometer. Percent conversion was then calculated as:

$$\% \text{ conversion} = \frac{[A810cm^{-1}, n=0] - [A810cm^{-1}, n=1,2,4,6]}{[A810cm^{-1}, n=0]}$$

Results are summarized in Figure 7. In all cases, increased UV exposure led to increased conversion over the dose range studied. The rate of conversion and the maximum extent of conversion varied significantly among samples. Sample N37D2 had the lowest cure rate and lowest extent of conversion (40%). Increasing the photoinitiator level to 5 phr (Sample N37D5) increased cure rate and increased the extent of conversion to 50%. Addition of 30 phr of TMPTA to a formulation with 2 phr of photoinitiator was even more effective in increasing the cure rate and the final extent of conversion (57%). It is interesting to note, however, that the maximum degree of conversion observed for any sample was approximately 60%. There are two possible explanations for this:

1. the samples contain some species in addition to the vinyl group which has an absorbance at 810 cm^{-1}; or
2. as curing proceeds a significant number of the unsaturated species in the system are immobilized and are unable to react further.

The fact that the absorbance of the acrylate group in the C=C stretching region parallels changes in the vinyl group absorbance

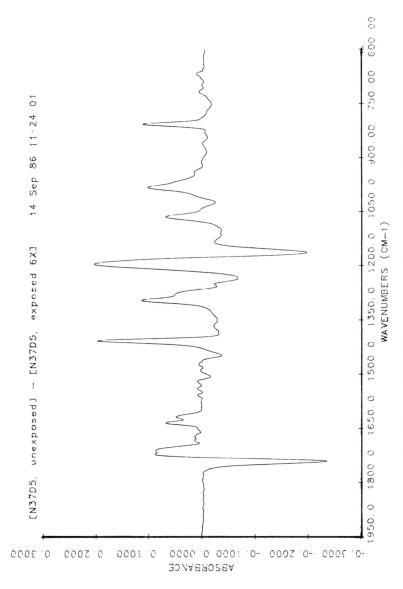

FIGURE 6: FTIR difference spectrum, Sample N37D5

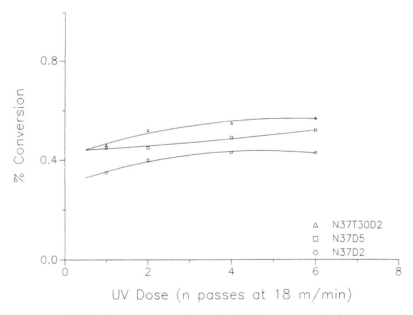

FIGURE 7: % Conversion (by FTIR) versus UV dose

tends to discount the former explanation. Recent work on polymerization of methyl methacrylate to high degrees of conversion in undiluted systems tends to support the latter explanation.[19] In the case of bulk polymerization of methyl methacrylate at 0°C, a rapid increase in polymerization was observed as the viscosity of the system increased (the well-known Trommsdorff effect). However, as the reaction proceeded to higher degrees of conversion (approximately 70%), monomer mobility in the system decreased, polymerization rate decreases, and complete conversion was never obtained. In UV-curing systems we might expect to see similar behavior, especially when multifunctional monomers are used. As soon as one of the unsaturated groups is incorporated into a growing chain, the probability that the other unsaturated groups in the molecule will encounter either another chain radical or an active photoinitiator fragment are reduced. As cure proceeds, the fraction of unsaturated species remaining which are capable of diffusion over significant distances becomes vanishingly small. Even if they encounter an active initiator fragment, the probability that they can participate in a reaction with another monomer group is thus decreased, and the reaction stops well short of complete conversion. This may account for the free radicals observed in cured acrylated epoxy film in a recent study.[15]

Data for percent extractibles versus cure dose are summarized in Table III for Sample N37D2 and N37T30D2. The maximum weight loss observed was 2%, even at low cure dose and low photoinitiator concentration. This indicates that the samples are highly crosslinked, even though their FTIR spectra indicate that they contain significant residual unsaturation. Loss of mobility of the reactive groups during cure effectively limits the degree of conversion, even though essentially all of the low molecular weight species have been incorporated into the network.

TABLE III: Butanone Extractibles versus Cure Dose for Acrylated Epoxy Films

Formulation	UV Dose (1)	% Extracted (2)
N37D2	1 X @ 18m/min	0.8%
	1 X @ 12m/min	1.6%
	2 X @ 12m/min	0.8%
	2 X @ 12m/min	0.5%
N37T30D2	1 X @ 18m/min	2.0%
	1 X @ 12m/min	1.1%
	2 X @ 12m/min	2.0%

(1) Films were cured using Fusion F450-10 irradiator with "V" lamp
(2) 30 minute extraction with 2-butanone

Conclusions

Acrylated epoxy systems undergo significant changes in thermomechanical properties on standing at room temperature. Other studies[15,20] have shown that free radicals are very long-lived in UV-cured systems kept under inert atmospheres, and that polymerization can continue for extended periods due to the low mobility of the residual unsaturated species at higher conversions. In this study, however, the films were exposed to air immediately after irradiation, which should effectively quench the residue free radicals. Therefore, these changes in properties cannot be fully accounted for by changes in unsaturation or chemical structure, since they are thermally reversible. The most probable source of the changes observed in this study is structural relaxation in the cured films. The broad glass transitions exhibited by all samples indicate a degree of incompatibility between the acrylic and epoxy components of the system studied. The broad relaxation, which exhibits an onset at or near ambient temperature, provides a mechanism by which annealing can occur at ambient temperature, even though the midpoint of the Tg is 50-70°C.

The maximum extent of acrylic group reaction observed after cure was approximately 60%, even though the films contained less than 2% extractible material. This indicates that a significant fraction of the acrylic groups are sterically hindered and unable to be incorporated into active network chains. Instead they act as additional chain ends, increasing free volume within the cured film. This provides an additional driving force for structural relaxation during annealing.

The ability of acrylated epoxy films to undergo structural relaxation at ambient conditions has significant implications on such film properties as impact strength, adhesion, hardness and abrasion resistance. In comparing properties among formulations, it is important that all the films or coatings tested be subjected to the same thermal history prior to analysis. While quenched samples will yield the most consistent results, their properties will differ significantly from those of annealed or aged films.

Acknowledgments

The authors wish to thank Mr. Robert G. Marx and Ms. Laura T. Onstott for their help in preparation of formulations and in performing FTIR and thermal analysis measurements.

References

1. Gaube, H.G. RadCure '86: Conf. Proc., 15-27(1986)
2. Lapin, S.C. RadTech '88: Conf. Proc., 395(1988)
3. Tribone, J.J.; O'Reilly, J.M.; Greener, J. Macromolecules, 19, 1732(1986)
4. Kovacs, A.J. Fortschr. Hochpolym.-Forsch. 3, 394(1963)
5. Marshall, A.S.; Petrie, S.E.B. J.Appl.Phys. 46, 4223(1975)
6. Berens, A.R.; Hodge, I.M. Macromolecules 15, 756(1982)
7. Hodge, I.M.; Berens, A.R. Macromolecules 15, 762(1982)
8. Hodge, I.M.; Huvard, G.S. Macromolecules 16, 371(1983)
9. Hodge, I.M. Macromolecules 16, 898(1983)
10. Hodge, I.M. Macromolecules 20, 2897(1987)
11. Privalko, V.P.; Demchenko, S.S.; Lipatov, Y.S. Macromolecules 19, 901(1986)
12. Krzewski, R.J.; Labovitz, M.; Sieglaff, C.L., pp.67-76 in "Structure and Properties of Amorphous Polymers", A.G. Walton (Ed.), Studies in Physical and Theoretical Chemistry, Volume 10, Elsevier Scientific Publishing Company, Amsterdam, 1980.
13. Flick, J.R.; Petrie, S.E.B., pp.145-171, ibid.
14. Levine, L.S.; Ury, M.B. RadCure '86: Conf. Proc., 1-1 (1986)
15. Decker, C.; Moussa, K. Polymeric Material Science and Engineering: Proceedings, 55, 552 (1986)
16. Kosnik, F.J.; Schweri, R.J. RadCure '86: Conf. Proc., 9-21(1986)
17. Bellamy, L.J. "Alkenes", Chapter 3 in The Infrared Spectra of Complex Molecules, Volume One, Third Edition, Chapman and Hall, London, 1975.
18. Anderson, D.G. RadTech '88: Conf. Proc., 513(1988)
19. Sack, R.; Schulz, G.V.; Meyerhoff, G. Macromolecules 21, 3345(1988)
20. Kloosterboor, J.G., "Advances in Polymer Science", Volume 84, "Polymers in Electronics, pp 46-59 (1988).

RECEIVED September 13, 1989

Chapter 22

Temperature Effect on the Phase Transformation of UV-Curable Systems

L. Feldman[1] and T. C. Ward[2]

[1]Mead Imaging, 3385 Newmark Drive, Miamisburg, OH 45342
[2]Department of Chemistry and Polymer Materials, Virginia Polytechnic Institute and State University, Blacksburg, VA 24061

> An extension has been made to the time-temperature-transformation (TTT) cure diagram for studying thermosetting systems. Traditionally, the TTT diagrams are obtained using torsion pendulum, DMTA, DSC, or similar instruments. By the use of a technique familiar in photographic science, this work demonstrates that such phase transition boundaries may be revealed by sensitometric methods. The D-log E curve is a measure of color development in response to exposure. When evaluated, this curve reveals information about the phase transformation. For crosslinking acrylate polymerizations, the TTT clearly shows the details of physical and chemical processes. These changes are brought about through a radiation induced chemical reaction in the microencapsulated multifunctional acrylate system.

The time-temperature-transformation (TTT) state diagram common to material science can be successfully applied to thermosets to establish a cure path which leads to the desired performance. It can also be a very useful tool in developing a fundamental understanding of the cure process and the nature of the solid state (1-5). Several techniques can be used to define the boundaries of the material transformation from fluid to solid through gelation due to an observable change in macroscopic properties. Thermomechanical methods (TP, TBA, DMTA), dilatometry, and calorimetry are widely used. Such techniques are similarly useful in the study of radiation-curable systems, but require special adaptation for irradiation, or are used after irradiation. Fluorescent labeling, although useful, sometimes leads to an unclear interpretation (6,7).

An imaging system based on photopolymerizable microcapsules has recently been developed (8-10). The microcapsules (Figure 1) contain photoinitiators, acrylic monomers, and a colorless dye precursor.

The capsule wall is transparent to UV, but provides a substantial barrier to atmospheric oxygen. The capsules are usually coated on a flexible substrate. Upon exposure, the monomer hardens to a degree dependent on the intensity of the light. When pressure is applied to the capsules, the monomer which has not been immobilized can be transferred onto a special image developing paper (Figure 2).

Color density (D) of the final image depends on the proportionality between exposure (E) level and the change in physical characteristics which control the dye delivery mechanism (11,12). This proportionality can be estimated from D-log E curves which are familiar in photographic science. These curves are generated by exposing media through a transparent optical density mask, which modulates light intensity (I) in logarithmic increments. In accordance with reciprocity law ($E = I \times t$), results of a single exposure can be extrapolated to the logarithmic time scale when applicable (13).

A typical sensitometric curve which exemplifies capsule internal phase transformation is represented in Figure 3. The highest density is produced by unexposed fluid material, while completely exposed regions do not release any dye precursor at all. Intermediate densities correspond to a gradual change in physico-structural state. By assigning the first transition on the curve to the beginning of gelation and the second transition to vitrification, a phase diagram can be generated over a broad temperature range.

In this work several formulations were studied in the temperature range between $(-60)°$ and $(100)°C$. Experimental data and a phenomenological interpretation in the analogy with thermosets are provided.

Experimental

Photo-crosslinkable monomer containing predissolved initiator, colorless dye precursor, and other necessary ingredients was emulsified in water and then encapsulated. The capsule slurry was machine coated on paper and dried to form a uniform layer of approximately 12 μm thick. The average capsule size was estimated to be 8 μm by Coulter Counter.

A Vivitar Model VP-1 xenon flash lamp with a preset exposure time of 1/500 s, and a desk lamp with two F15T8 black light fluorescent bulbs were used as exposure devices in two experiments. In both cases, the distance from the light source to the sample plane was 10 cm. The light intensity was modulated using a 30 step transparent wedge of 0.1 optical density increments. Before each exposure, the sample and the mask were thermally equilibrated for 5 min at a designated temperature.

The exposed capsule donor was coupled with the receiver containing a color developing component, and was pressure developed with a set of steel laboratory pressure rollers.

Optical reflection color density was measured using a Macbeth model 914 filter densitometer. For simplicity, step-wedge numbers were used to represent logarithmic increments on the intensity axis. Phase diagrams were constructed by plotting the step number corresponding to a transition on the sensitometric curve against the exposure temperature.

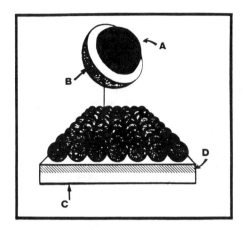

Figure 1. Schematic representation of the microencapsulated imaging system: A-microcapsule internal phase, B-transparent shell, C-substrate, D-dye developer layer.

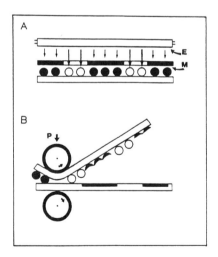

Figure 2. Exposure and image development in the microencapsulated system: E-exposure device, M-photographic mask, P-pair of pressure rollers.

Results and Discussion

Data from experiments with the flash lamp exposure demonstrate that the radiation cure path may be highly affected by the temperature. These results are presented in Figure 4, where three major regions of density levels can be observed: A, of the maximum density (Dmax); B, where densities are below Dmax; and C, in which dye transfer is not detectable.

Although the material in region C possesses the physical properties of a solid, it contains a significant amount of extractable monomer (11). Dielectric thermal scans indicate that this monomer has been completely immobilized. Therefore, we assume that the minimum density line in Figure 4 defines vitrification conditions. The gel phase and the boundary of its formation corresponds to region B, where gradual density change correlates with the amount of unreacted monomer disappearance. The highest density region may be divided into three parts. One of them (A1) is obviously below the glass transition temperature of the reactants in their initial state ($T_{g,0}$). Photopolymerization was not initiated below this temperature. The area A2, below the horizontal line, corresponds to insufficient light intensity, where dye delivery is not yet controlled by polymerization. The third region, A3, represents a temperature dependent, high intensity reciprocity failure, manifested as a reversal in the density vs. step number curve (Figure 5). As displayed on the diagram, the critical temperatures, $T_{g,0}$; $_{gel}T_g$ (coincidental with $T_{g,0}$ in this case); and $T_{g,\infty}$, define the processing window as well as the aging behavior. As in thermosets these values are very system specific and are predetermined by the monomer type, the chemistry of the initiators, and often by the physico-chemical characteristics of the reagents added.

A microencapsulated system containing hexaaryl bis-imidizole dimer, isopropylthioxanthone (Quanticure ITX), and TMPTA (trimethylolpropane triacrylate) provides a representative example. Its low sensitivity to exposure (UV lamp, 4 s) is accentuated with a very narrow processing temperature range between (-10) and (35)°C (Figure 6-B). This range significantly broadens (Figure 6-A) with addition of viscosity reducing agents, such as PMA and MHT (2-heptylthio-5-mercapto-1,3,4-thiodiazole). As a result, $T_{g,0}$ shifts toward its lowest end (-47)°C, where T_g of neat TMPTA is usually observed. Such an effect would not be noticed if experiments were not done in a substantial temperature range. For example, DPHPA/ 1,6-HDDA / ketocoumarin systems with and without 2,6-N,N-tetra methylaniline (TMA) are not very different in their performance at 19°C (see sensitometric curves in Figure 7). The dissimilarities become evident though, at lower temperatures (Figures 8, 9). This and other observations indicate a dual role of hydrogen donors and reactive diluents in modifying both the system photosensitivity and the thermal response.

It may be shown from the phase diagrams that the intensity span between gelation and vitrification can also be affected by additives. For example, the ability of Quanticure EPD to change the vitrification path is revealed in experiments with GPTA monomer. This can be observed through comparison of Figures 10 and 11. The

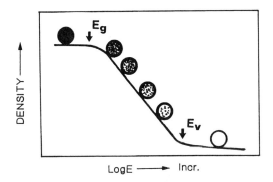

Figure 3. Sensitometric response of the microencapsulated acrylate system.

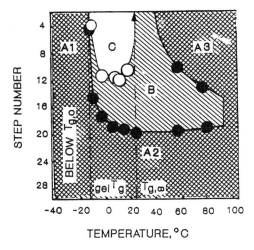

Figure 4. Phase diagram based on the color density developed after exposure with flash lamp. Solid circles-gelation, open circles-vitrification profiles.

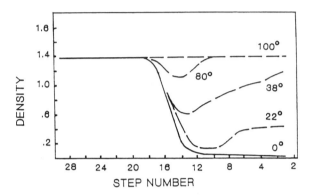

Figure 5. Color density vs. photographic step number (D-logI) for different exposure temperatures. Corresponds to phase diagram in Figure 4.

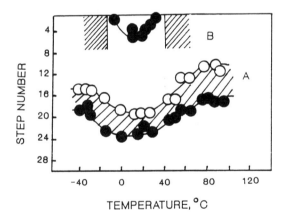

Figure 6. Phase transformation profiles of TMPTA / Quanticure ITX containing hexaaryl bis-imidizole dimer with (A) and without (B) PMA and MHT. (Solid circles-gelation, open circles-vitrification).

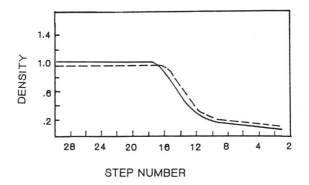

Figure 7. Sensitometric curves obtained at 19°C for DPHPA / 1,6-HDDA / ketocoumarin system with (solid line) and without (dashed line) TMA.

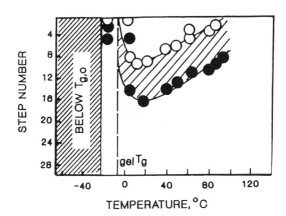

Figure 8. Phase transformation profiles of DPHPA / 1,6-HDDA / ketocoumarin system. (Solid circles-gelation, open circles-vitrification).

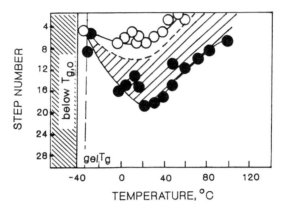

Figure 9. Phase transformation profiles of DPHPA / 1,6-HDDA / ketocoumarin system with TMA. (Solid circles-gelation, open circles-vitrification).

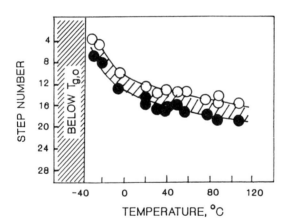

Figure 10. Phase transformation profiles of GPTA / Irgacure 907 / Quanticure BMS system. (Solid circles-gelation, open circles-vitrification).

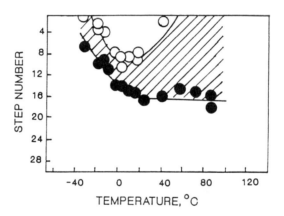

Figure 11. Phase transformation profiles of GPTA / Quanticure EPD / Quanticure BMS system. (Solid circles-gelation, open circles-vitrification).

first graph represents the typical behavior of compositions with Irgacure 907 and Quanticure BMS as the photoinitiation system. Note the very broad temperature range (low $T_{g,o}$ and high, $T_{g\infty}$), the smooth, unusual improvements in photographic sensitivity with an increase in temperature, and the very tightly controlled dynamic range. Gelation and vitrification profiles are very parallel. When Irgacure 907 is replaced by Quanticure EPD (Figure 11), the gelation path is not affected. At the same time, the vitrification curve exhibits a sharp minimum at approximately $10^{\circ}C$. A completely different behavior is displayed past this minimum, where temperature can reduce toe speed in a significant manner. At $50^{\circ}C$ and above, capsule internal phase does not vitrify under the experimental exposure conditions. Some other examples seem to confirm that EPD mainly affects the post-gelation process by inducing characteristic minimum ($T^{o}min$) on the vitrification profile.

Conclusions

In analogy with thermosets, the cure path of photo-hardenable monomers can be described phenomenologically using phase transformation diagrams. The example of microencapsulated imaging systems demonstrates a strong dependence of the cure path on the temperature. Color development methods coupled with sensitometry allow easy data collection methods. With this technique, arbitrary boundaries between the fluid, soft gel, and the vitrified solid can be deduced from the D-log intensity curve. Experimental results demonstrate that such observations provide additional insights on the specific contributions of each reactant in photocurable systems. Although the dye labeling method in these experiments was used to obtain color images, the technique may be extended to colorless compositions, where other optical instruments can be used to read the contrast induced by the material's phase transformation (14,15). Encapsulation of the test material is not necessary, in that the sample may be prepared in a microcolumn system format (16).

Literature Cited

1. Gillham, J. K. In Development in Polymer Characterization-3; Dawkins, J. V., Ed.; Applied Science: Englewood, NJ, 1982; pp 159-227.
2. Enns, J. B.; Gillham, J. K. J. Appl. Polym. Sci. 1983, 28, 2567-2591; 2831-2846.
3. Gillham, J. K.; Enns, J. B. SPSE Proc.,Fall Meeting of the American Chemical Society, Los Angeles, Ca; American Chemical Society: Washington, DC, 1988; Vol.59, pp 851-858.
4. Wisanrakkit, G.; Gillham, J. K. ibid.; pp 969-974.
5. Schiraldi, A.; Pezzati, E.; Baldini, P. Thermochimica Acta 1987, 120, 315-323.
6. Levy, R .L.; Ames, D. P. Polym. Sci. Technol. (Plenum) 1984,29, 245-255.
7. Sung, Ch. S. P.; Chin, I.-J.; Yu, W.-Ch. Macromolecules 1985, 18, 1510-1512 .
8. Sanders, F. W. U.S. Patent 4,399,209, 1983.
9. Sanders, F. W. U.S. Patent 4,565,137, 1985.
10. Diamond, S. Electron. Imaging 1984, 35 .

11. Arney, J. S.; Dowler, J. A. J. Imaging Sci. 1988, 32, 3, 125-128.
12. Arney, J. S. Paper Summaries, SPSE's 36th Annual Conference, San Francisco, CA, 1983.
13. Todd, M. and Zakia, R. In Neblette's Handbook of Photography and Reprography; Sturge, J. M., Ed.; Van Nostrand Reinhold: N.Y., N.Y., 1977; pp 175, 176, 173.
14. Murray, R.D. ibid., p 437.
15. Volkova, M. M.; Bel'kovsky, I. M.; Golikov, I. V.; Semyannikov, V. A.; Mogilevich, M. M.; Indeykin, E. A. Vysokomol. Soed 1987, XXIX, 3, 435-440.
16. Feldman, L.; Cage, M. R.; Shi, D. J.; R. C. Liang Paper Summaries, SPSE's 42nd Annual Conference, Boston, MA, 1989; SPSE: Springfield, VA, 1989; pp 400-403.

RECEIVED September 13, 1989

Chapter 23

Microcolumn Imaging

Simulation of the Microcapsule Imaging System

L. Feldman, M. R. Cage, D. J. Shi, T. K. Kiser, and R. C. Liang

Mead Imaging, 3385 Newmark Drive, Miamisburg, OH 45342

New methods of imaging require ingenious delivery systems. One such system is Cylithography, in which microcapsules are used to regulate the delivery of dyes and acrylic monomers. It produces continuous tone, full-color, high-resolution images at high speed with panchromatic light (1-3). A novel experimental technique was developed through the use of highly constrained microporous supports that simulates the imaging mechanism involved in Cylithography while avoiding the time-consuming encapsulation steps. Inert films containing microcolumns of 0.4-8.0 μm diameter and approximately 10.0 μm depth were filled with a photosensitive composition consisting of a leuco dye, photoinitiator, and multifunctional acrylic monomer, and then exposed through a mask. Various development techniques can be used such as compression, extraction, treatment with dye materials, etc. This versatile technique is suitable for different applications that include image reproduction, printing, and use as a research tool. This work demonstrates that microcolumn imaging mimics photographic properties of the microencapsulated system and can be used for a detailed study of the image formation mechanism.

The microencapsulated acrylate process of imaging is a unique application of radiation curing technology. In this process, the photohardenable composition is contained within a microcapsule wall and consists of a multifunctional acrylic monomer with photoinitiators and leuco dye precursors dissolved in it. When a thin coating of microencapsulated media on paper is irradiated, a latent image is formed as a pattern of various extents of cure. The final color image is developed by compressing the capsule donor sheet together with a receiver sheet, which contains a color-developing agent. Usually pressure rollers are used for this purpose. The color density (D) of the final image is controlled by the amount of leuco dye delivered with monomer. The amount of transferable monomer, in turn, is regulated by the change in the mechanical properties of the microcapsule ($\underline{4}$). This relationship determines the sensitometric response of the system as shown in the characteristic D-log E curve (Figure 1). The maximum density (Dmax) is produced by the dye delivered with fluid monomer from soft, unreacted capsules. Upon depletion of oxygen to some critical level with exposure dose E_1, which determines photographic sensitivity, gelation is initiated. Changes in the material state in response to exposure beyond this point are reflected by the color density decrease in the dynamic portion of the curve. E_2 marks another threshold region, where monomer becomes immobilized inside the microcapsule and minimal color density (Dmin) is produced.

The critical parameters of this process depend on the interplay of the exposure conditions and the properties of the encapsulated materials, including initiator efficiency, the type of monomer, and contributions of co-reactants to the chemical and physical characteristics of the capsule internal phase.

It is desirable to study and evaluate the effect of each constituent in the capsule internal phase on the system performance with a minimum number of variables that may be introduced during encapsulation. A simple method that employs microporous membranes with discrete vertical channels is proposed for this purpose. The channels are filled with the photosensitive composition and both planes of the film are tightly sealed using two covers, one of which should be transparent. The film thickness and pore size can be selected to approximate the capsule coating thickness and single capsule size, respectively. Similar to the microencapsulated system, dye transfer is controlled by light-induced crosslinking polymerization and accomplished through compression (Figure 2). Prints developed by this technique are comparable to those obtained with microcapsules. Using color density vs. exposure characteristic curves, sensitometric properties of the different formulations can be evaluated and compared.

Experimental work reported below shows that such tests can be used to predict microencapsulated system behavior and its imaging characteristics. It also demonstrates that the microcolumn imaging system provides a convenient vehicle for observations with SEM and other analytical methods, including GC, IR and TA.

Experimental

Sample Preparation. Photohardenable formulations based on TMPTA in combination with other reagents including photoinitiators (Quanticure ITX or 7-diethylamino-3-cinnamoyl coumarin, depending on the specific experiment) and colorless dye precursors developable upon contact with acidic resin coated paper were used as a test composition.

In one case such compositions were encapsulated and machine coated on a flexible substrate. In the second set of tests, the identical formulations were applied to microcolumn type membranes (manufactured by Nuclepore). Transparent polyester plastic sheets 25 μm thick were used to cover both sides of these samples after the application step. Laboratory steel pressure rollers that applied

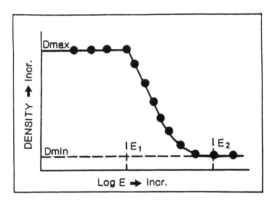

Figure 1. Typical sensitometric curve obtained with microencapsulated media via pressure development.

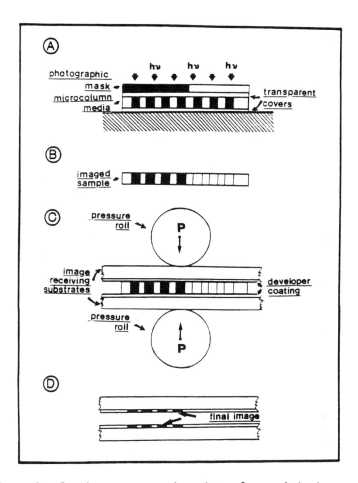

Figure 2. Imaging process using microcolumn substrate.
A-exposure, B-imaged sample, C-pressure development, D-final images on the receiver.

1000 N of force per cm of roller length were used to remove the excess liquid from the membrane and provide a tight, uniform seal.

Samples were exposed to the light through a photographic step tablet with 0.1 optical density increments. This provided the necessary range of exposure levels at a single exposure time (5). Specific details on the exposure time and light sources used are given separately in the discussion.

Both covers were then removed from the exposed microporous samples before the samples were placed on the receiver sheet and developed. Microencapsulated media does not need to be covered during exposure.

To obtain a final image both microencapsulated and microcolumn media were similarly developed using identical pressure rollers, receiver, and glossing step at 150°C for 1 min.

Characterization. Optical reflection densities on the final image were measured with a Macbeth model 914 filter densitometer. Photographic sensitivities were estimated from the threshold intensity values in relative units corresponding to the shoulder and toe regions on the D-log I curves.

Microscopic observations were done on a Phillips High Resolution Scanning Electron Microscope. To reveal solid material inside the microcolumn, a freeze-fracture technique was used after samples were washed with methanol and dried under normal conditions.

The amount of extractable monomer was estimated using GC (HP 5890 Model with a FID detector and a phenylmethylsilicon fused silica 10 m capillary column). Before extraction, FT-IR spectra were obtained using an IBM IR-32 Model with a DTGS detector. The percentage conversion of double bonds was calculated from the 1640 cm^{-1} (6) peak height.

To estimate the exposure effect on the physical characteristics, dynamic mechanical thermal analysis was employed. In this case samples were prepared without the application of pressure rollers. Instead, a squeegee was used to remove the excess liquid and preserve the monomer as a continuous layer on the surface of the substrate. Thermal scans at 10°C/min and a fixed frequency of 1Hz were done using a Du Pont DMA 983 instrument.

Results and Discussion

Reciprocity behavior is an important characteristic of photographic materials. Its failure limits practical use because of the unpredictable, nonlinear response to a proportional change in exposure (5). In this regard microcolumn material shows superior results. Figure 3 demonstrates a reciprocity curve for the system with Quanticure ITX, exposed on the sensitometer with a combination of "black" and "day light" (General Electric) fluorescent bulbs at time intervals between 4 s and 2.5 h. The shoulder speeds on this plot correspond to a constant value of total exposure (log E=log I+ log t). According to Arney (7) the rate of diffusion of oxygen into the system is negligible. Apparently polyester covers provide an excellent barrier. Microcapsules, on the other hand, need a perfect wall (7).

A further comparison of the microcolumn and microcapsules show that their properties are parallel in many respects. They both possess a similar spectral sensitivity profile with a maximum in the same region. An example is given in Figure 4 for 7-diethylamino-3-cinnamoyl coumarin / TMPTA / CVL leuco dye samples exposed for 256 s with a Xenon 1000 W arc lamp and monochrometer. This evaluation was performed for a typical initiator concentration of 0.20%, which is very close to the optimal range (Figure 5), where maximum sensitivity is reached in both systems. Based on these observations a suggestion can be made about high optical densities, indicating that microcolumn and microencapsulated systems may be classified as bulk-response systems according to Thommes and Webers (8).

Although an exact explanation of the almost linear displacement in the sensitivity between these two systems is not yet available, the observations show that this difference is affected by the properties of the matrix. For example, among clear materials polycarbonate based membranes are not very effective for the broad band and UV exposures, probably because of the light absorption in the matrix. Tinted microcolumn membranes and translucent materials with interconnected pores produce even lower sensitivities. Polyester was found to be the least attenuating of those materials considered. This demonstrates that geometry and optical properties of the support are very important.

Effect of the pore diameter on the sensitivity and Dmax was also noticed (Figure 6). However, such an effect may be connected to a difference in thickness, which was also measured. Lower sensitivity and a higher Dmax were observed with thicker samples (Figure 7). Due to the direct correlation between membrane thickness and microcolumn depth, these results provide additional evidence of the bulk-responsiveness in the microcolumn imaging system.

It has been discussed (9) that the color density in the case of the microencapsulated system is a continuous function of the change in morphology of encapsulated material. This was not directly observed using microscopy methods, but deduced from other experiments. Using the microcolumn technique, such observations were made. They indicate a good correlation with the proposed model. Samples corresponding to different positions on the D vs. log I curve (Figure 8) were viewed with SEM after preparation as described above and sputter coating with gold. The photomicrographs in Figure 9 represent two different stages of curing. It appears that, similar to microcapsules, each microcolumn images in an analog fashion, and that the log exposure range of the system reflects the log exposure range intrinsic to an individual microcolumn.

The amount of extractable monomer and unreacted double bonds as a function of exposure were estimated using methods described above on the ITX system embedded in PC microcolumn membrane and exposed for 128 s at the illumination $I_0 = 1.29$ mJ/cm^{2}s in the band width of 325-410 nm. In Figure 10 such data are compared to the color density change on the transferred image. A rather high conversion was determined in the sample exposed to the full intensity level available.

A rough estimate of the double bonds consumed per reacted monomer can be made, as demonstrated in Figure 11, where experimental data are compared to theoretical limits. The ratio between 1.0 and 1.5, corresponding to the conversion in monomer up to 70% is agreeable with observations made on a microencapsulated system (10).

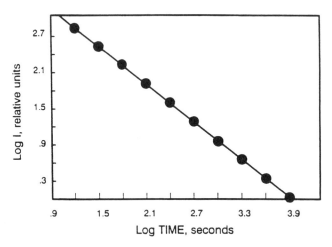

Figure 3. Reciprocity behavior of the microcolumn imaging system.

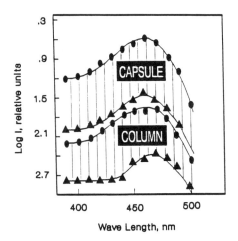

Figure 4. Spectral sensitivity profiles of identical formulations in microcapsules and microcolumn. Ovals-shoulder speed, triangles-toe speed.

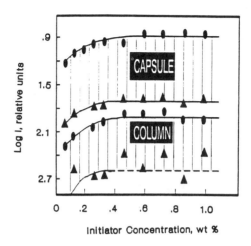

Figure 5. Photographic sensitivity vs. initiator concentration. Ovals-shoulder, triangles-toe. Monochrometer exposure at 470 nm for 64 s.

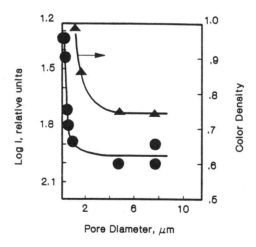

Figure 6. Photographic shoulder speed and final image color density vs. pore diameter for PC microcolumn membranes.

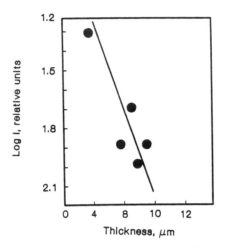

Figure 7. Photographic shoulder speed vs. thickness of membrane. For comparison, the first data point in the left top corner was obtained without membrane or any spacer.

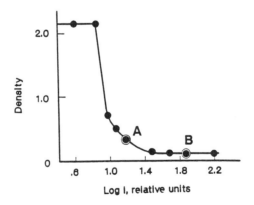

Figure 8. D-log I curve of the microcolumn sample used for examination with SEM.

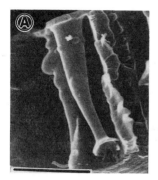
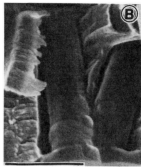

Figure 9. SEM images of polymer formed inside the microcolumn at corresponding intensity levels indicated by A and B in Figure 8.

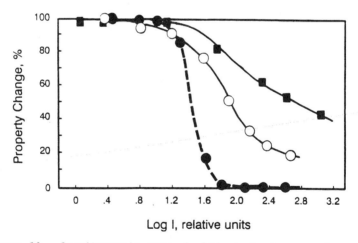

Figure 10. Sensitometric curve (solid circles), percentage of residual vinyl groups (rectangles), and unreacted monomer (open circles) vs. log intensity.

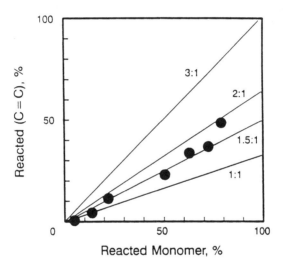

Figure 11. Vinyl groups conversion (by FT-IR) vs. monomer conversion (by GC analysis).

Final image density is controlled by the amount of fluid monomer and the rigidity of the polymer formed. Figure 12 shows the relationship between fractional conversion of monomer and fractional change in density. This is similar to results obtained for microcapsules, where the direct proportionality between image density and elastic deformation of the substrate was considered (10).

Changes in visco-elastic characteristics of the microcolumn samples can be observed experimentally with DMA. Figure 13 (A and B) demonstrates the results for storage and loss modulus, respectively. Numerals on the curves indicate log I of each exposure at a constant exposure time (128 s). Corresponding values representing the interrelationship among modulus change, T_g, and image density are shown in the following figures. According to this method, T_g of the unreacted material ($T_{g,0}$) is $(-47)°C$. It reaches $106°C$ upon exposure to the maximum intensity used. A sudden jump from $(-26.5)°C$ to $(+41.4)°C$ is indicative of a sharp rheological transition in the middle of the log intensity range (1.4-1.5). It corresponds to the toe on the sensitometric curve (Figure 14) and may represent vitrification. Dye transfer is completely restricted from the material with a T_g exceeding room temperature. Although the chemical reaction threshold may not be reached, the physical properties component becomes a controlling factor manifesting high efficiency of the amplification mechanism relied upon in this process.

Because image transfer is accomplished at room temperature it is interesting to compare the modulus at this temperature with color density. This is done in Figure 15, where logarithmic increments of these two characteristics are plotted vs. log I. A fairly good correlation is produced by this method.

It may be concluded from what has been observed that the sensitometric evaluation based on the color density detects both transitions - from liquid to gel and from soft gel to vitrified solid - most precisely, while other techniques reveal information on the continuous polymerization process.

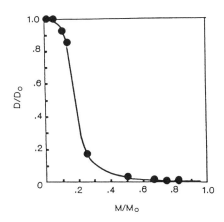

Figure 12. Fractional dye control vs. fractional conversion of monomer in the microcolumn imaging system.

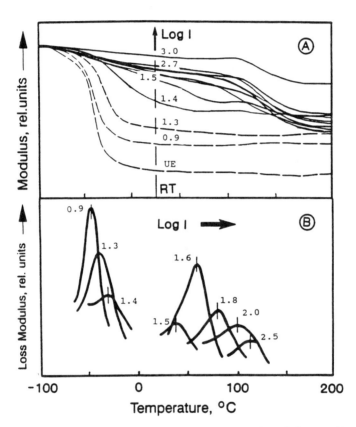

Figure 13. Storage modulus (A) and loss modulus (B) DMA diagrams obtained on exposed microcolumn samples.

23. FELDMAN ET AL. *Microcolumn Imaging* 321

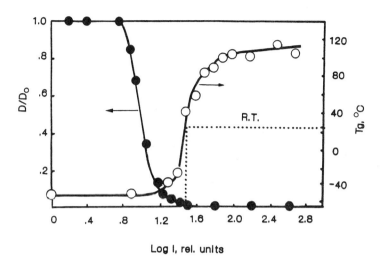

Figure 14. T_g (open circles) and fractional density change as a function of exposure.

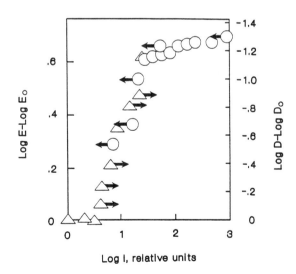

Figure 15. Superimposed data corresponding to image density (circles) and modulus (triangles) change vs. log intensity.

Conclusions

Photographic behavior of the microcolumn imaging system parallels the microencapsulated system in many respects. Microcolumn provides a convenient sampling format for application of the multi-technique approaches in addressing the interdependence of the imaging characteristics, curing conditions, and the structure of the polymer formed.

Acknowledgments

The authors would like to acknowledge J.S. Arney (Mead Imaging) for guidance, M.A. Zumbrum and S.R. McCartney (Virginia Polytechnic Institute and State University, Blacksburg, VA) for microscopy work, P.C. Adair and S.R. Buczynski (Mead Imaging) for help with microencapsulated media.

Literature Cited

1. Sanders, F. W. U.S. Patent 4,399,203, 1983.
2. Diamond, A. S. Electron. Imaging 1984, 35.
3. Rastogi, A. K.; Wright, R. F. Presented at the Electronic Imaging Devices and Systems '89 Symposium, SPIE/SPSE, Los Angeles, CA, Jan.1989.
4. Arney, J. S. Paper Summaries, SPSE's 36th Annual Conference, San Francisco, CA, 1983.
5. Todd, M.; Zakia, R. In Neblette's Handbook of Photography and Reprography; Sturge, J. M., Ed., Van Nostrand Reinhold: NY,NY, 1977; pp 125,173.
6. Van Oosterhout, A. C. J.; Van Neebros, A. J. Radiat. Curing 1982, 9, 19-33.
7. Arney, J. S. J. Imaging Sci. 1987, 31, 1, 27-30.
8. Thommes, G. A.; Webers, V. J. J. Imaging Sci. 1985, 29, 3, 112-116.
9. Arney, J. S.; Dowler, J. A. J. Imaging Sci. 1988, 32, 3, 125-128.
10. Arney, J. S. J. Imaging Sci. 1989, 33, 1, 1-6.

RECEIVED October 19, 1989

PHOTODEGRADATION OF RADIATION-CURED FILMS

Photodegradation of Radiation-Cured Films

One of the problems which continues to confront the polymer industry is the degradative failure of coatings and plastics upon exposure to ultraviolet radiation. This results in significants decreases in the long-term durability of materials which are designed for use where exposure to sunlight is expected. The problem of ultraviolet stability is particularly acute in radiation cured coatings where additives such as photoinitiators may accelerate the degradation process.

The two papers in this section (Chapters 23 and 24) explore the light stability of photocured (Chapters 23 and 24) and electron beam (Chapter 24) cured coatings. In Chapter 23, the loss in surface hardness of bisphenol-A-epichlorohydrin based epoxy-acrylate photocured films has been investigated by detailed product studies of photolyzed model compounds in solution. By careful analysis of products the major pathways for photodecomposition were shown to involve cleavage/rearrangement processes. As well, a semi-quinone type product was postulated though direct evidence awaits future analysis. The results in Chapter 24 deal directly with the photooxidation of UV and EB cured acrylate networks formed from amine terminated acrylates. By employing a direct Michael synthesis of acrylated amines and subsequent UV or EB curing in the presence and absence, respectively, of photoinitiator the authors have illustrate the critical role of photosensitizers such as benzophenone in the photostability of UV-cured acrylated networks. Of particular note is the unexpected increase in photolability for certain of the EB cured films. Obviously, the mechanism of curing greatly influences the final long-term stability of radiation-cured materials!

Chapter 24

Photolysis Studies of Bisphenol-A-Based Model Compounds

Effect of Decomposition Products on the UV Stability of Bisphenol-A-Based Epoxy Coatings

John C. Graham, Dennis J. Gaber, Yifang Liu, and Pravin K. Kukkala

Coatings Research Institute, Eastern Michigan University, Ypsilanti, MI 48197

> During the UV initiated curing of bisphenol-A (4,4-isopropylidenediphenol) based coatings, it was observed that the surface hardness reached a maximum and thereafter decreased to less than 1/2 of the peak hardness level with prolonged exposure to UV radiation. In order to understand the chemistry that contributes to this decrease in surface hardness, model bisphenol-A compounds were synthesized and exposed to UV radiation using a medium pressure Hanovia lamp and the decomposition products analyzed by GC/MS procedures. In addition to Photo-Fries rearrangement reactions and bond scissions, primarily at the aromatic ether oxygen-carbon bond, we also observed isopropylidene cleavage to form biphenyl compounds.

Numerous studies have shown that bisphenol-A-epichlorohydrin (BAE) condensation polymer coatings are prone to degradation on prolonged exposure to UV light (1-5). Gaber's (3) work on photoinitiator performance in UV curable coatings using BAE condensates with reactive acrylate diluents such as β-carbonyl ethylacrylate (β-CEA) or isobornyl acrylate (IBA), showed that the hardness of the coating initially increased with time on exposure to UV light, then decreased. Pappas, et al. (4) found that the fluorescence from BAE type polymeric films rapidly decreased on continued excitation due to resonance energy transfer from excited singlet states of the aromatic ring resulting in the formation of photoproduct(s). Timpe, et al. (5) attributed the photophysical and photochemical properties of BAE condensation polymers to the Ar-O chromophore. Major photolysis products of model compounds such as 1,3-diphenoxy-2-

propanol were phenol and substituted phenols and the photochemical reactions were found to be similar to those of other aromatic ethers (6).

Grassie, et al. (7) conducted an extensive study on the photodegradation of bisphenol-A diglycidyl ether and compared the photodegradation products with those observed in thermal degradation. Many of the products formed were found to be identical in both cases, especially the phenols and derivatives of phenol. The principal differences were the appearance of hydrogen in the photoreaction, but not in the thermal reaction and the formation of a series of aromatic hydrocarbons such as xylenes. Mechanisms were outlined in which non-phenolic aromatic products (xylenes, ethylbenzene, styrenes, etc.) are formed by the scission of the phenyl-oxygen bond.

Gaber (8) also found that surface softening of epoxy-acrylate polymeric films on prolonged exposure to UV light is independent of the type of initiator or reactive diluent used and their respective concentrations (Figure 1). However, softening was not observed with polyester-acrylate films or urethane-acrylate films (Figure 2). In order to rationalize the above results, it was apparent that the contribution of the molecular structure and the possible side reactions of the epoxy segment of the polymer chain should be more closely examined under UV light. A typical BAE polymer segment follows:

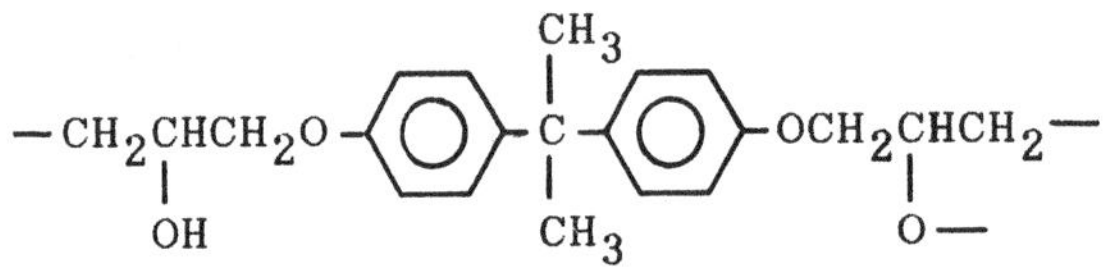

$$-CH_2CHCH_2O-\phi-C(CH_3)_2-\phi-OCH_2CHCH_2-$$
$$\quad\quad\quad |\quad\quad\quad\quad\quad\quad\quad\quad\quad\quad\quad\quad |$$
$$\quad\quad\quad OH\quad\quad\quad\quad\quad\quad\quad\quad\quad\quad\quad O-$$

For the purpose of studying the above polymer segment, model compounds (I, II, and III) were synthesized representing the various sections of the BAE polymer as shown below:

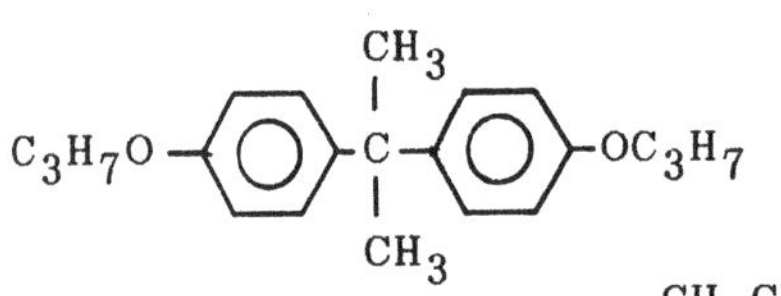

$$C_3H_7O-\phi-C(CH_3)_2-\phi-OC_3H_7 \quad\quad (I)$$

$$CH_3CHCH_2O-\phi-C(CH_3)_2-\phi-OCH_2CHCH_3 \quad\quad (II)$$
$$\quad\quad |\quad\quad\quad\quad\quad\quad\quad\quad\quad\quad\quad\quad\quad\quad\quad |$$
$$\quad\quad OH\quad\quad\quad\quad\quad\quad\quad\quad\quad\quad\quad\quad\quad\quad OH$$

$$CH_3CO_2-\phi-C(CH_3)_2-\phi-O_2CCH_3 \quad\quad (III)$$

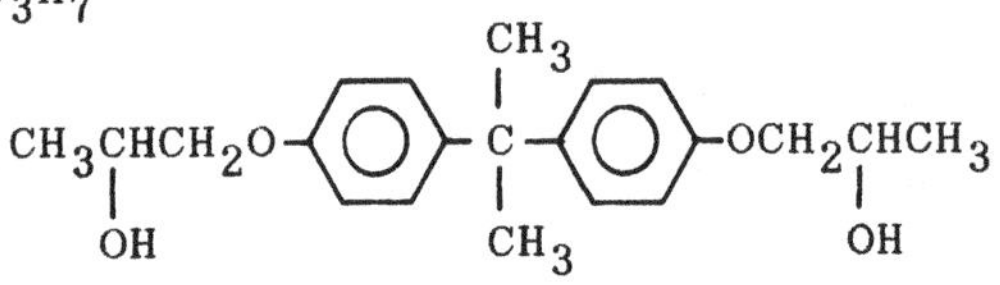

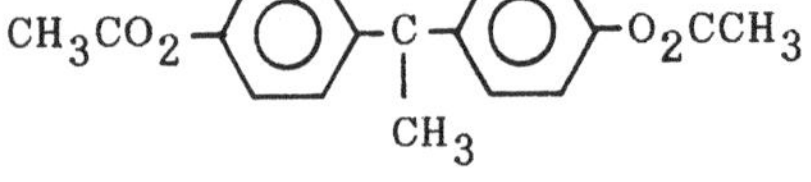

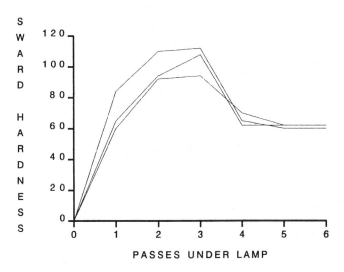

Figure 1. Cure curves for a typical epoxy acrylate varying the amount of reactive diluent.

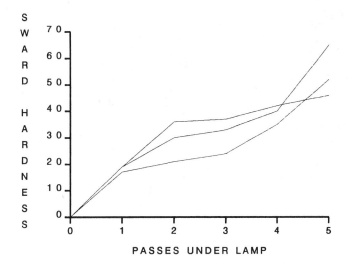

Figure 2. Cure curves for a typical polyester acrylate where the thickness is varied.

These compounds were dissolved in dioxane in the presence of a UV photoinitiator and exposed to a medium pressure mercury lamp. The products were analyzed by GC/MS techniques(9).

Attempts to use diglycidyl bisphenol A in this study were unsuccessful since this compound did not resolve well under our GC conditions.

Experimental

The experimental section is divided into three parts: (1) the preparation of model compounds, (2) their photolysis conditions under UV light, and (3) instrumentation. All chemicals except the model compounds were obtained from commercial sources and used without further purification.

Preparation Of Model Compounds

Bis(n-propyl ether) of Bisphenol-A (I). The sodium salt of recrystallized bisphenol A (0.4 mole) was dissolved in 750 mL water containing NaOH (2.4 moles), to which n-propyl bromide (1.6 moles) and tetrabutylammonium bromide (0.03 moles) (10) were added and kept under vigorous agitation at 38°C for 72 h under a nitrogen blanket. 200 mL of methylene chloride was then added and agitated for 12 h. The organic layer was then separated and evaporated, the residue was dissolved in ether, washed with 5% NaOH solution and water three times, passed through $MgSO_4$ and finally the ether was evaporated and the residue kept in a vacuum oven for 24 h at 50°C to obtain the propyl ether of bisphenol A (I). I is a colorless liquid (lit. bp. 200-202°C at 3mm) (11).

The compound was identified by FTIR which showed peaks at 1048 cm^{-1} (the alkyl aromatic ether peak), and by absence of a broad OH peak at 3715 cm^{-1}, by NMR peaks at δ = 0.99 (CH_3), δ =1.73 (CH_2) and δ =3.87(O-CH_2) and GC/MS showing the molecular weight as 312 as expected. The yield was 70% and purity determined by GC and GPC analysis was approximately 99%.

Bis(2-hydroxypropyl ether) of Bisphenol-A (II). A mixture of the sodium salt of pure bisphenol-A (0.25 moles), 1-bromo-2-propanol (1 mole) and tetrabutylammonium bromide (0.019 moles) (6) in 600 mL of water containing NaOH (1.5 moles) was vigorously agitated at 50°C for 96 h under N_2 blanket. 500 mL of methylene chloride was then added and agitated for 24 h. The organic layer was separated. The remaining work-up was similar to that used in the synthesis of compound I.

The product obtained was initially a light brown colored liquid which crystallized in 48 h. It was decolorized and crystallized from isopropanol. Pale yellow crystals of II were obtained (m.p. 65-67°C).

II was identified by FTIR by the appearance of the alkyl-aromatic ether peak at 1048cm^{-1}; by NMR peaks at δ=3.83 (-O-C$\underline{H}_2$), δ = 3.9 (aliphatic O$\underline{H}$) and δ = 4.0 (-C$\underline{H}$-); and GC/MS showed the molecular weight to be 344 as expected. The yield was 93% and the purity determined by GC and GPC analysis was >99%.

Bis(methyl ester) of Bisphenol-A (III). A mixture of bisphenol-A (0.5 moles), acetic anhydride (2 moles) and anhydrous sodium acetate (0.6 moles) was refluxed for 12 h. The product precipitated on cooling. The cooled reaction mixture was stirred with 400 mL of water and filtered. The product was dissolved in ether. The remaining work-up was similar to that used in the synthesis of compound I. White crystals were formed from ether overnight, m.p. 91.5 - 92.5°C (lit. mp.: 80°C) (9).

The product was identified by FTIR peak analysis, i.e. the appearance of a strong ester peak at 1752 cm^{-1}; NMR analysis, i.e. δ = 2.2 (COC$\underline{H}_3$). The GC/MS showed the molecular weight to be 312 as expected. The yield was 95%. No impurities were observed by GC analysis.

Photolysis of Model Compounds

Photolysis studies were carried out using 5% solutions of the model compounds in anhydrous 1,4-dioxane in the presence of a photoinitiator (3% by weight of the model compound). The photoinitiators used were diethoxyacetophenone (DEAP) and 1-hydroxy-1-cyclohexylphenyl ketone (HCPK). The following sequence of experiments were carried out:

- model compound + 1,4-dioxane
- model compound + 3% DEAP + 1,4-dioxane
- model compound + 3% HCPK + 1,4-dioxane

Samples were removed at 6, 12 and 24 h. Any volatiles trapped at -70°C in a Dewar flask were also analyzed.

Instrumentation

The preparation of the model compounds was monitored using an IBM FTIR IR44. NMR studies were conducted on a Brucker 360 MHz FT-NMR. GC/MS studies were accomplished using a Finnigan 9610 GC equipped with a 30 m DB-5 capillary column coupled to a Finnigan 4500 Mass Spectrometer which acted as the detector. Hewlett-Packard's gas chromatograph #5890 with a 30 m DB-5 capillary column was used for GC analysis. GPC analysis were performed using a M-45 Waters Associate pump and R.I. detector coupled to a Hewlett-Packard HPLC 1040A Chemwork station.

Irradiation studies were conducted under a nitrogen blanket in a 250 ml Ace Glass Co. photochemical reactor (model #7844-04) equipped with a stopcock at the bottom and a Conrad-Hanovia immersion lamp (#7825-32) rated at 200 watts, 115-130 volts and 1.9 amps. Volatiles were trapped using cold traps and Dewar condensers in series.

Results and Discussion

The reduction in the surface hardness of epoxy-acrylate (UV-cured) films on prolonged exposure to UV light (Figure 1) has been attributed to photo-chemically induced cleavage of chemical bonds in the BAE segments of the polymer chain (8). Although thermally, the initial cleavage of the bonds in epoxy systems is said to occur at the C-CH$_3$ (isopropylidene) group (1), in the case of photochemical degradation, Norrish-type reactions have been observed as the main primary photochemical reactions in similar compounds (10).

We have conducted photochemical studies on model compounds I, II, and III and obtained evidence for the formation of photochemical products involving degradation and rearrangements from GC and GC/MS analysis (9).

Photolysis of Model Compound I. The bis-n-propyl ether of bisphenol-A (I) is a colorless liquid and easily miscible in 1,4-dioxane. Figure 3 shows the GC as drawn by the computer and Table I shows the various significant peaks that appeared in the GC spectrum and their corresponding m/e ratios for all three photolysis runs at various intervals of time. The proposed photodegradation pathways are shown in Scheme 1.

After photolysis, the gas chromatograph of the irradiated samples showed six significant compounds including starting material (Figure 3). The retention time of the largest fraction in all cases corresponds to the undissociated I (m/e=312). There were numerous other photoproducts formed to a lesser degree but identification was beyond the range of the instruments.

The dissociation pattern of I in the mass spectrophotometer showed that the isopropylidene -CH$_3$ fragmented first yielding the base peak (m/e=297), followed by cleavage of the ether linkage with the loss of the n-propyl group (-C$_3$H$_6$) to yield a phenol (m/e=255). This was followed by the loss of another -C$_3$H$_6$ to produce a biphenol (m/e=213) (Table I).

Mass spectral studies of the products after photolysis showed that the largest fraction other than undissociated I exhibited a molecular weight of 354, i.e. 42 mass units higher than starting compound I.

An increase in molecular weight can occur through rearrangement followed by combination. In this case, the 42-mass unit increase can arise if a -C$_3$H$_6$ species combines with the starting compound I. A possible pathway for the generation of the increased molecular weight compound (IV) is shown in Scheme I. Accordingly, the ether linkage cleaves to generate n-propyl radicals and aryloxy radicals. Statistically, since two n-propyl radicals can

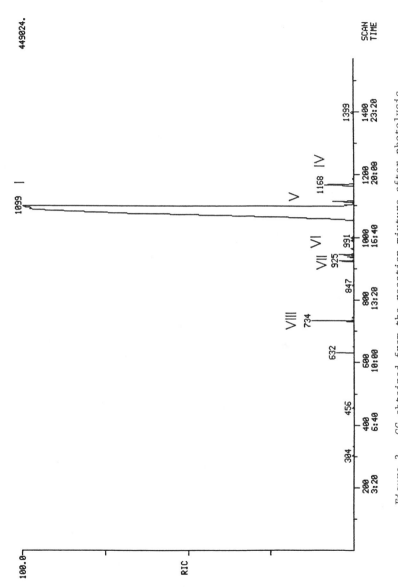

Figure 3. GC obtained from the reaction mixture after photolysis of model compound I.

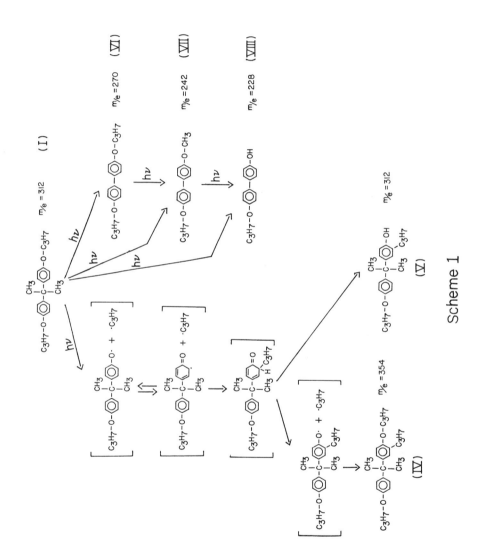

Scheme 1

TABLE I

MASS SPECTRAL ANALYSIS OF THE REACTION MIXTURE
AFTER PHOTOLYSIS OF MODEL COMPOUND I

m/e ratio	I	IV	V	VI	VII	VIII	Bisphenol A
354		MI					
339		X					
312	MI		MI				
297	BP	X	X				
270				MI			
255	X	X	X				
242				X	MI		
239		X					
228			X	X	X	MI	MI
213	X		X				BP
211		X					
199				X	X		
186						BP	
185				BP	X		
171				X	X	X	
153				X	X	X	
119	X	X	X				X
43	X	BP	BP	X	X	X	
41	X	X	X	X	BP	X	X
39							X

Legend: MI = molecular ion peak
BP = base peak
X = appearance of an intense peak

be formed for every parent radical, the concentration of n-propyl radicals will be higher than the parent radical by a factor greater than one. In order to achieve a MW equal to 354, product (IV) is assumed to possess three propyl groups, as indicated in Scheme 1. In addition, compound IV can possess two isomeric structures (IVa and IVb) as shown below:

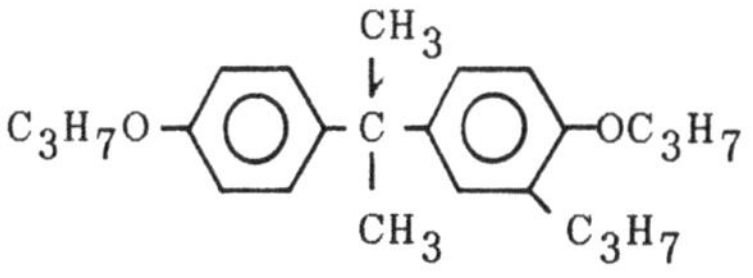

$$C_3H_7O-\phi-C(CH_3)(CH_3)-\phi(C_3H_7)-OC_3H_7 \quad\quad C_3H_7O-\phi-C(CH_3)(CH_3)-\phi(C_3H_7)(C_3H_7)-OH$$

IVa 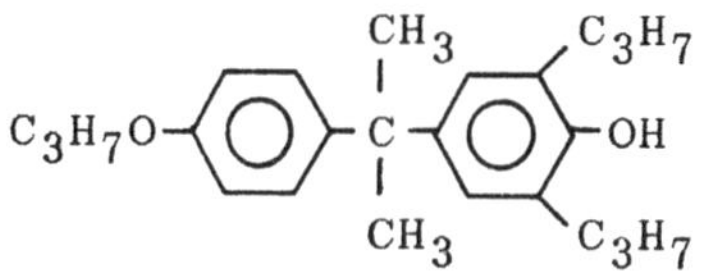   IVb

One other major stable photoproduct obtained after the photolysis of I exhibited a molecular ion peak at 312. Since it possesses the same molecular weight of the starting compound I, it could be either an isomer of I or a new compound arising via a Photo-Fries rearrangement of the n-propyl moiety (V). Differences in the MS spectra of I and the new compound indicate that rearrangement of the n-propyl group has occurred. This can result in two isomers, Va and Vb, assuming one or two n-propyl groups are rearranged as shown below:

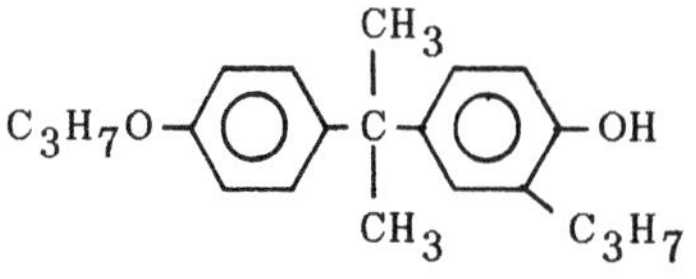

$$C_3H_7O-\phi-C(CH_3)(CH_3)-\phi(C_3H_7)-OH \quad\quad HO-\phi(H_7C_3)-C(CH_3)(CH_3)-\phi(C_3H_7)-OH$$

Va 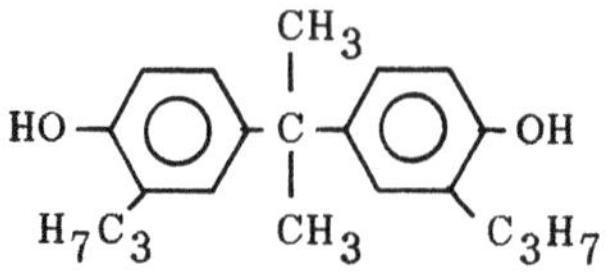   Vb

The formation of V can also arise from the degradation of the higher molecular weight photoproduct IV with the loss of one n-propyl moiety (attached to the ether oxygen) from IVa followed by hydrogen abstraction. The loss of the n-propyl moiety from the ortho position in IVb can also lead to Va, but this is unlikely as the C-C bond energy is higher than the C-O bond energy of the ether.

Other than the formation of compound V and the higher molecular weight compound (IV), three other stable photoproducts (VI, VII, and VIII) are formed in minor amounts. These products possess molecular ion peaks at 270, 242, and 228 respectively (Scheme 1 and Table I).

GC/MS analysis show that compounds VI, VII, and VIII appear to have been generated by the loss of the isopropylidene group in I, IV and/or V, resulting in biphenol type ethers. The proposed structure for photoproduct VI (m/e = 270) is shown in Scheme 1.

Compound VII whose molecular weight is 242, may be formed from compound VI by the loss of a $\cdot C_2H_4$ radical (m/e = 28) from the alkyl side chain as shown in Scheme 1. The fragmentation pattern of VII is similar to compound VI indicating that compound VII may have been derived from compound VI during photolysis (Table I).

Compound VIII, whose molecular ion peak appears at 228 could have been formed by the loss of the $\cdot CH_2$ moiety (m/e = 14) from VII, and/or from the cleavage of a n-propyl moiety from VI. VIII does not appear to possess the isopropylidene group as evidenced by large differences in the fragmentation patterns of VIII and bisphenol A. Although the structures indicate only one isomer other isomers are possible.

The presence of the biphenol group in compounds VI, VII, AND VIII as opposed to the bisphenol A group is based on a number of factors:
1. MS fragmentation patterns of VI, VII and VIII do not match that of bisphenol A. Also, the presence of bisphenol A as a photoproduct has not been detected.
2. The appearance of an intense peak at m/e = 119 signifies the presence of a styryl moiety (which typically appears in bisphenol A type structures). This appears in the MS spectra of IV and V, but is absent in the MS spectra of compounds VI, VII, and VIII.
3. Consideration of the bond energy for the aryl-isopropylidene bond (83 kcal/mole) in comparison to the ether bond (86 kcal/mole), indicates that such cleavages are possible.

Peaks appearing in the computer generated GC below a retention time of 650 seconds represent decomposition products from the photoinitiator (in this case, DEAP) and the solvent (dioxane) as seen in Figure 3. These are primarily a minor photoproduct (m/e = 190) (632 secs), a DEAP fragment (456 secs), and a 1,4-dioxane fragment (304 secs).

Photolysis of Model Compound II. Compound II, the bis-2-hydroxy propyl ether of bisphenol A, represents a section of a BAE polymer. Figure 4 shows the GC peaks as drawn by the computer and Table II shows the significant peaks that appeared in the GC/MS spectrum and their corresponding m/e ratios. The proposed photodegradation pattern is shown in Scheme 2.

After photolysis, the gas chromatograph showed only three significant peaks at 1571 secs, 1151 secs, and 815 secs; the retention time of the largest fraction, as expected, corresponded to unchanged II (m/e=312).

GC/MS data for photoproduct IX shows the molecular ion peak at 286, which can be obtained by cleavage of the ether oxygen-carbon linkage with the loss of an isopropylene alcohol moiety [$CH_2=C(OH)-CH_3$ (m/e = 58)] followed by regeneration of the aromatic ring, as shown in Scheme 2. The MS fragmentation pattern can be easily correlated to the proposed structure of the compound (IX). The peak at m/e = 119 indicating the presence of the isopropylidene group also supports the structure as proposed. This

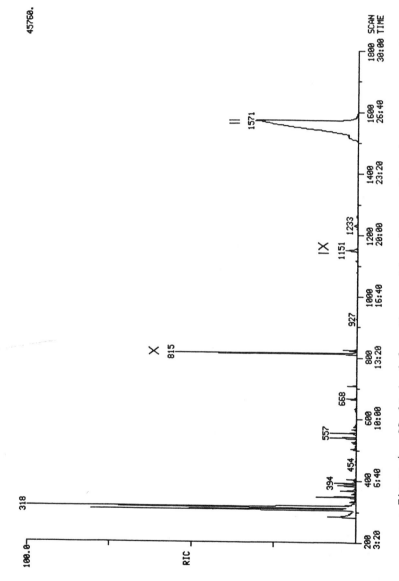

Figure 4. GC obtained from the reaction mixture after photolysis of model compound II.

TABLE II

MASS SPECTRAL ANALYSIS OF THE REACTION
MIXTURE AFTER PHOTOLYSIS
OF MODEL COMPOUND II

m/e ratio	II	IX	X
344	MI		
329	BP		
286		MI	
271	X	BP	
244			MI
213	X	X	
186			BP
171			X
153			X
119	X	X	
45	X		X
41	X	X	X

Legend: MI = molecular ion peak
       BP = base peak
       X = appearance of an intense peak

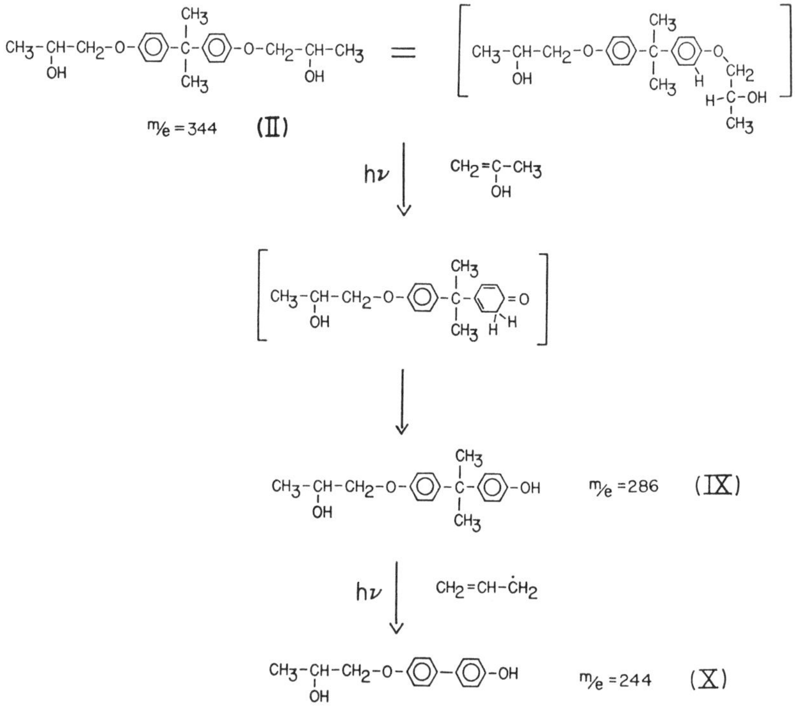

Scheme 2

product (IX) appeared to form in larger quantities in the presence of a photoinitiator.

The largest photoproduct fraction formed (X) had a retention time of around 815 seconds. In comparison to other photoproducts from either I or III, the retention time of this product is short and the quantity formed is large. The MS spectra showed the molecular ion peak at 244. From Scheme 2, it can be seen in the case of photoproduct X (m/e=244), if cleavage occurs at the isopropylidene group, a biphenyl is generated. As stated earlier when studying the photodegradation pattern of model compound I, the isopropylidene group can break away and after H^+ abstraction and rearrangement, form a stable allylic radical (m/e=41). This photoproduct (X) is formed in quantities larger than the other photoproducts, even in the absence of photoinitiators.

The MS analysis shows excellent correlation between the proposed structure for X and its fragmentation pattern. For example, in the ionization chamber, the loss of an isopropylene alcohol moiety (m/e = 58) leads to the base peak at m/e = 186. This corresponds to a p,p'-dihydroxybiphenyl molecule. One other significant feature from the MS spectra (Table II), is the absence of the peak at m/e=119 corresponding to the styrene molecule typically observed with bisphenol A compounds. The same argument as applied to photoproducts VI, VII, and VIII arising from model compound I, is applied here in assigning the biphenyl structure to X.

Peaks appearing in the GC (Figure 4) below 815 seconds represent decomposition products from the photoinitiator and the solvent. These are primarily photoinitiator fragments (668 secs), dimer of 1,4-dioxane (557 secs), and 1,4-dioxane (394 secs).

Photolysis of Model Compound III. The gas chromatograph (Figure 5) shows the presence of four major compounds in the reaction mixture after 24 hours irradiation of model compound III in dioxane. The retention time of the largest fraction corresponds to unchanged III, with the maximum amount of decomposition being less than 20%. There are numerous other photo-products formed to a far lesser degree, but identification of these is beyond the range of the instrument. The photodegradation pattern is shown in Scheme 3, and the GC/MS significant peak analysis is shown in Table III.

Mass spectral studies have shown that the largest fraction other than III has a molecular weight of 312. This compound (XI) appears to be formed by a photochemical rearrangement; most likely, a Photo-Fries arrangement of III (13), as shown in Scheme 3. It is difficult to predict at this stage whether the rearrangement occurs at only one of the aromatic centers or both, i.e. to form a monophenol (XI) or bisphenol (XII).

The largest stable photodegraded product, XIII has a molecular weight of 270, which can arise by the loss of

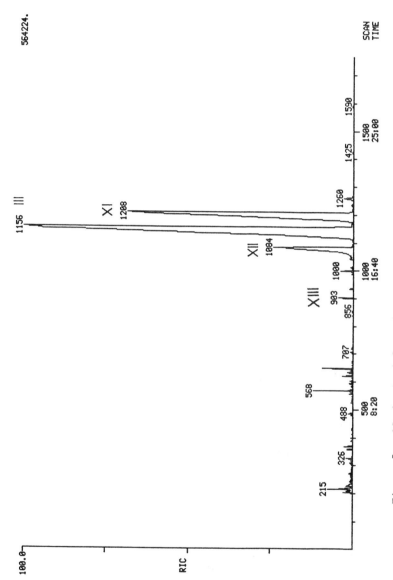

Figure 5. GC obtained from the reaction mixture after photolysis of model compound III.

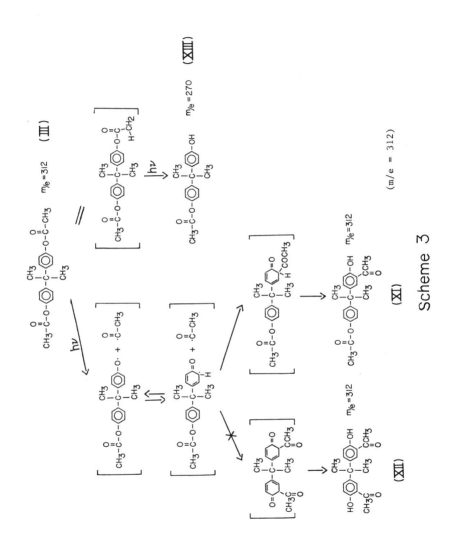

Scheme 3

TABLE III

MASS SPECTRAL ANALYSIS OF THE REACTION MIXTURE AFTER
PHOTOLYSIS OF MODEL COMPOUND III

m/e ratio	III	XI isomer	XIII	XIV
312	MI	MI		
297	X	X		
270	X	X	MI	
268				MI
255	X	BP	X	
253				X
228	X		X	
226				X
213	BP		BP	
211				BP
119	X	X	X	
117				X
43	X	X	X	X

Legend: MI = molecular ion peak
BP = base peak
X = appearance of an intense peak

a $COCH_2$ (m/e = 42) moiety from any of three compounds (III, XI, and XII) although loss from III is most likely (Scheme 3). The other stable photoproduct (XIV) obtained in detectable amount has a molecular weight of 268. This could be formed either from dissociated product XIII or directly from III, XI and/or XII. The GC/MS analysis shows that the fragmentation pattern of XIV is very similar to that of XIII except that the fragmentation pattern of XIV is two units lower than the fragmentation pattern of XIII.

The structure of XIV is difficult to elucidate but may be formed by the combination of the loss of the $-COCH_2$ moiety from and Photo-Fries rearrangement of III to yield most probably a semi-quinone type of structure. If this is the case, then it appears that Photo-Fries rearrangements compete with photodegradation in these processes. Allen, et al. (15,16) have attributed the yellowing of BAE films under UV light to the formation of quinone structures, as shown below (XV), by photo-sensitized oxidation of the bisphenol A unit.

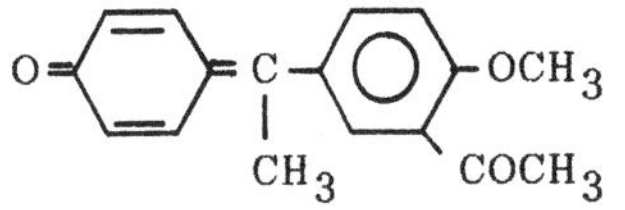

In our case, the initial colorless reaction mixture turns deep yellow on exposure to UV light, which may indicate the formation of such chromophoric structures.

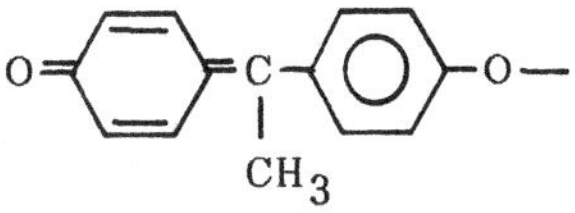

The products that appear in the GC below a retention time of 650 seconds are due to the decomposition and combination of photoinitiator and dioxane (Figure 5). These are primarily photoinitiator (568 secs), dimer of 1,4-dioxane (488 secs), and 1,4-dioxane (215 secs).

The effect of the presence of photoinitiators, in general, was not significant enough to observe any major differences in the photoproducts obtained though it could have participated in generating some of the unexpected photoproducts. However, one minor effect was observed, i.e. in the presence of the photoinitiators, the amount of photoproducts formed increased to a slight extent in comparison to photolysis runs that did not contain photoinitiator. This may be explained on the basis of the chromophoric nature of the initiators which are capable of acting as energy reservoirs or reaction sites for the generation of photoproducts.

Conclusions

Some of the interesting conclusions that can be drawn from photodegradation products of model compounds I, II, and III are as follows:
1. Rearrangement reactions such as the Photo-Fries re-

arrangement appear to be significant reactions during the photolysis of these model compounds. As expected, the extent of rearrangement appears to be dependent on the type of group attached to the bisphenol A group i.e. the more sensitive the group to UV radiation, such as in the case of the ester group in III, the greater the extent of rearrangement.
2. Bond scissions occur primarily at the aromatic ether oxygen-carbon bond. This degradation reaction appears to compete with the rearrangement reactions and may occur simultaneously.
3. Surprisingly, we observe that the isopropylidene group cleaves to form biphenyl type photoproducts, a fact which has not been observed by other researchers. Since the aromatic ring-oxygen combination is a strongly absorbing chromophore, the substituents linked to the oxygen atom and the aromatic ring may determine the extent of formation of allylic radical from the isopropylidene group, especially during photodegradation reactions.
4. The presence of solvent cages appears to be an important element in the photo-rearrangement and photodegradation reactions.
5. In contrast to Grassie's (7) studies, we did not detect any scission of the phenyl-oxygen bond nor formation of styrene.

Certainly chain cleavage reactions as well as rearrangements are involved in the photolysis of bisphenol A polymers which explains Gabers results (8) where a softening and loss of strength was observed in UV cured epoxy coatings on exposure to UV light. Conversely, the formation of quinone type compounds which can function as inhibitors during a photoinitiated free-radical curing reaction could lower the cure rate and/or reduce the hardness. Chain cleavage and the formation of quinone structures can be used to explain the instability and yellowing of epoxy coatings on exposure to weathering.

Acknowledgments

This research was supported in part by Alcolac.

References

1. Patterson-Jones, J.C.; J. Appl. Polym. Sci., 1965, 9, 3195.
2. Lin, S.C.; Balkin, B.J.; Pearce, E.M.; J. Polym. Sci., Polym. Chem. Ed., 1979, 17 3121.
3. Gaber, D.J.; J. Coatings Techn., 1986, 58, 743, 12.
4. Pappas, S.P.; Gatechair, L.R.; Breskman, E.L.; Fischer, R.M.; in "Photodegradation and Photostabilization of Coatings", ACS Symposium Series #151, Ed. Pappas, S.P. and Winslow, F.H., 1981, p109.
5. Timpe, H.J.; Garcia, C.; Pappas, S.P.; Gatechair, L.R.; Breskman, E.L.; Fischer, R.M.; Polym. Photochem., 1985, 6, 41.

6. Kohler, G.; Kittel, G.; Getoff, N.; <u>J. Photochem.</u>, 1982, <u>18</u>, 19.
7. Grassie, N.; Guy, M.I.; Tennent, N.H.; <u>Polym. Degrad. Stab.</u>, 1985 <u>13</u>, 249.
8. Gaber, D.J.; M.S. Thesis, Eastern Michigan University, April, 1986.
9. Liu, Yifang, M.S. Thesis submitted to Eastern Michigan University, 1989.
10. McKillop, A.; Fiaud, J.C.; Hug, R.P., <u>Tetrahedron</u>, 1974, <u>30</u>, 1379.
11. Corson, B.B.; Heintzelman, W.J.; Schwartzman, L.H.; Tiefenthal, H.E.; Lokken, R.J.; Nickels, J.E.; Atwood, G.R.; Pavlik, F.J.; <u>J. Org. Chem.</u>, 1958, <u>23</u>, 544.
12. Yohe, G.R.; Vitcha, J.F.; <u>J. Amer. Chem. Soc.</u>, 1935, <u>57</u>, 2259.
13. Cicchetti, O.; <u>Adv. Polym. Sci.</u>, 1970, <u>7</u>, 70.
14. Sander, Hedaya, Trecker; <u>J. Amer. Chem. Soc.</u>, 1968, <u>90</u>, 7249.
15. Allen, N.S.; Binkley, J.P.; Parsons, B.J.; Philips, G.O.; Tennant, N.J.; <u>Polym. Photochem.</u>, 1982, <u>2</u>, 97.
16. Allen, N.S.; Robinson, P.J.; White, N.J.; Swales, D.W.; <u>Polym. Degrad. Stability</u>, 1987, <u>19</u>, 147.

RECEIVED September 13, 1989

Chapter 25

Photooxidative Stability and Photoyellowing of Electron-Beam- and UV-Cured Multifunctional Amine-Terminated Diacrylates

Norman S. Allen[1], Peter J. Robinson[1], Roy Clancy[2], and Nicholas J. White[3]

[1]Department of Chemistry, Faculty of Science and Engineering, John Dalton Building, Manchester Polytechnic, Chester Street, Manchester M1 5GD, United Kingdom
[2]Radiation Curing, Harcros Chemicals (UK) Ltd., Silk Street, Eccles, Manchester, United Kingdom
[3]Howson-Algraphy Ltd., Coal Road, Leeds, United Kingdom

> A number of dialkylamine-terminated diacrylate monomers are prepared by a Michael addition reaction and their photooxidative stability and photoyellowing are studied after ultraviolet (uv) and electron-beam (EB) curing by infra-red and uv absorption spectroscopy. Photooxidative stability of the resin is controlled more by the structure of the triacrylate monomer used to prepare the dialkylamine diacrylate monomer, while the structure of the latter markedly influences the growth and decay of photoyellowing. Monomer terminated with di-n-butylamine groups gives the greatest amount of photoyellowing whilst alkylethanolamines give the least (as measured by second-order derivative uv absorption spectroscopy at 275 nm). The former, however, is photobleached more rapidly than the latter. UV cured resins exhibit a more facile photoyellowing than the same electron-beam cured systems, and this is associated with hydrogen atom abstraction and oxidation of the alkylamine group by the residual photoinitiator. It is suggested that the primary process of photoyellowing is oxidation of the methylene groups adjacent to the nitrogen, giving rise to unstable hydroperoxides which dissociate to give α,β-unsaturated carbonyl groups.

A number of studies have been carried out to investigate the photo-oxidation and photoyellowing phenomena which occur in various multifunctional acrylate systems (1-16). The complex formulations of radiation curable resins complicate our understanding of the mechanisms involved. For example, high concentrations of photoinitiators in uv curable resins will often dominate the subsequent photoreactions and contribute significantly to photoyellowing (3-7). With regard to photooxidative stability, hydrogen abstracting type photoinitiators such as benzophenone are more detrimental than the photofragmenting types such as benzoin ethers (2-7). Other factors include the use of tertiary amines as co-synergists, which also contribute toward the photoyellowing phenomenon and photostability of the resin, and these may be free or reacted with acrylate groups in the resin. For example in earlier studies (14,15) we found that whilst the use of tertiary amines contributed significantly toward the subsequent photoyellowing of resins, they nevertheless markedly improved their photooxidative stability. The yellowing was associated with hydroperoxidation of the amine, resulting in the formation of conjugated unsaturated carbonyl products which absorb strongly in the uv region with a maximum at 275 nm. The latter on the otherhand was associated with ability of the amine to scavenge oxygen effectively and operate as an effective sacrificial site for hydrogen-atom abstraction reactions, thus protecting the resin structure. These observations were also extended to a diethylamine-terminated diacrylate resin, where it was found that hydroperoxides are the key initiators of the photoyellowing phenomenon but not in photooxidation. Owing to the complex nature of uv cured resin films, it has proved difficult to elucidate the nature of the species and mechanisms responsible for inducing and controlling the photooxidative stability and photoyellowing of di- and triacrylate resin films. To this end we have examined the post-cured photooxidative stability and photoyellowing of a number of different dialkylamine-terminated diacrylate resins prepared via a Michael addition reaction of the corresponding secondary amine with different multifunctional monomers. The monomers were cured under both uv and EB irradiation conditions.

Experimental

Materials. Samples of three multifunctional monomers namely glycerol propoxylate triacrylate (GPTA), trimethylolpropane ethoxylate triacrylate (TMPETA) and trimethylolpropane triacrylate (TMPTA) were supplied by Harcros Chemicals Limited, Manchester, UK. Benzophenone was obtained from the Union Carbide Company, USA and used as supplied. Dimethylamine, morpholine, diethylamine, dicyclohexylamine, di-n-butylamine and diethanolamine were obtained from the Aldrich Company, UK and used as supplied. N-methylethanolamine, N-ethylethanolamine and N-methylethylamine were all obtained from Fluka Chemicals, UK and used as supplied. Carbon tetrachloride (Analar grade) was obtained from Fisons Limited, UK.

Samples for infra-red analysis were coated onto aluminised glass and/or cardboard using wire wound applicator bars giving an eventual film thickness in this work of 12 microns. Samples of monomer were also coated onto the surface of quartz flats (1 x 4 cm) for uv absorption spectroscopic analysis.

Ultraviolet Curing Equipment. The equipment used during this study was a Mini-cure apparatus available from Primarc (Jigs and Lamps) Limited, UK. The equipment consists of a medium pressure mercury arc lamp mounted over a conveyor system, the lamp being located at one of the foci of an elliptically shaped polished aluminium reflector with the conveyor arranged to pass through the other focus. This arrangement ensured that the material being irradiated received the maximum radiation dose possible from the lamp as it passed beneath it. The conveyor belt was adjusted to give a belt speed of 0.5 m s^{-1} and the total rating of the lamp used was 80 W cm^{-1}. The benzophenone (0.5% w/w) was incorporated into the monomers at 60°C with vigorous stirring.

Amine Acrylate Synthesis. During this study a number of amine acrylates and model compounds were synthesised. The syntheses were based on the Michael type addition of a secondary amine to an activated acrylate double bond. Our procedure is as follows.

The acrylate compound was charged into a 250cm³ four necked flask equipped with a condenser, stirrer, thermometer and dropping funnel. The required amount of amine was added slowly from the dropping funnel with stirring. An immediate exotherm was observed. Periodically, a sample of the reaction mixture was removed for titrimetric analysis as outlined below. When the reaction was complete and the correct amine value obtained the amine acrylate was steam stripped to remove unreacted secondary amine. Water (5% w/w) was added, and the reaction vessel fitted with an air bleed and a distillation head leading to a trap and a high vacuum system. The temperature of the reaction mixture was raised to 40°C and the water stripped off under a reduced pressure of 13 mm Hg. Remaining traces of water were removed at 75°C. Stripping was considered complete when less than 0.2% w/w water was found to be present Karl-Fischer titration. Analytical details may be obtained elsewhere (17).

Total Amine Value (TAV). The following TAV values (mg KOH/g) were obtained for the resins prepared. Diethylamine with GPTA (111.9), morpholine with GPTA (108.8), dicyclohexylamine with GPTA (92.1), di-n-butylamine with GPTA (100.7), N-ethylethanolamine with GPTA (108.5), N-methylethanolamine with GPTA (111.5), dimethylamine with GPTA (118.6) and diethylamine with TMPETA (111.9) and TMPTA (152).

Electron-Beam Curing. The equipment used in th s study was an Electro-curtain curing unit from Energy Sciences Inc., Massachusetts, USA and was located at Lankro Chemicals Limited, Manchester, UK. The accelerating voltage operating range of the equipment was between 150 and 175 kV with a beam current range from 0 to 10 mA. The dose range of the equipment was 35 kilorads to 35 megarads and the nominal dose rate was 1×10^8 rads s^{-1} at a beam current of 5 mA. The sample to be cured was passed under the electron beam curtain via a conveyor system and a single pass dose in MRads depended on the following relationship :

$$\text{Single pass dose} = \frac{Ib \times C}{Sc}$$

where Ib = beam current in mA
Sc = conveyor speed in ft min^{-1}
C = a machine dependant constant and was 70 for the equipment used

During this work a single pass dose of 4 MRads was employed under nitrogen to avoid the inhibiting effect of oxygen.

Photooxidation. Film samples were irradiated using a Microscal unit (Microscal Limited, London) utilising a 500 W high pressure mercury/ tungsten lamp (wavelengths 300 nm) with an operating temperature of 50°C and ambient relative humidity.

Infra-red Measurements. Infra-red measurements were carried out using a Perkin-Elmer Model 1420 ratio recording spectrometer linked to a Perkin-Elmer Model 3600 data station. Film samples were analysed using a specular reflectance attachment and oxidation rates were monitored using a hydroxyl (3450 cm^{-1}) and alkyl (2878-2889 cm^{-1}) band index :

$$\text{Oxidation index} = \frac{\text{Absorbance of Monitored Band}}{\text{Absorbance of Reference Band}}$$

The reference band (2940 cm^{-1}) was used to compensate for film thickness variation and is independent of cure conditions and photooxidation time. An absorption band at 1612 cm^{-1} was also monitored and is discussed further below.

Absorption Spectroscopy. Second derivative absorption spectra were recorded using a Perkin-Elmer Model 554 absorption spectrophotometer. Film samples were analysed by transmission on optically transparent quartz flats.

Results and Discussion

Initial experiments were undertaken with morpholine, dicyclohexylamine and dibutylamine terminated GPTA systems as well as the diethylamine sample which has already been investigated in previous work (14,15). The effect of Microscal irradiation on the increase in hydroxyl index of these amine acrylates is illustrated by the data in Table I. This shows that the nature of the terminal amine functionality has a marked effect on the hydroxyl index profile during irradiation. The following order of increasing hydroxyl index was observed relative to amine functionality :

dibutyl $\rangle$ morpholine $\rangle$ diethyl $\rangle$ dicyclohexyl

Table I. Hydroxyl Index Values during Irradiation in a Microscal Unit for UV Cured (5% w/w Benzophenone) GPT (12μm thick) after Reaction with Various Amines

Amine	Hydroxyl Index Values		
	100h	200h	300h
Diethylamine	0.070	0.094	0.120
Morpholine	0.120	0.150	0.170
Dicyclohexylamine	0.045	0.070	0.091
Di-n-butylamine	0.210	0.279	0.327

The improved stability of the dicyclohexylamine terminated system over the diethylamine terminated one may be related to the reduced number of C-H bonds alpha to the nitrogen in the former while the reduced stability of the morpholine terminated system may be due to the presence of C-H bonds adjacent to the ether oxygen. The latter would act as additional proton donation sites to benzophenone during exciplex interaction on irradiation. The relative stability of the dibutylamine terminated system is more difficult to rationalise. It is possible that the bulky nature of the butyl groups is hindering efficient exciplex interaction between this the terminal amine group and benzophenone thus allowing the residual benzophenone to attack the polyether backbone of the parent polymer. If this was the case, however, one would have expected the dicyclohexylamine system to show a similar effect.

During irradiation all of these amine acrylates exhibited a rapid loss of the residual benzophenone photoinitiator along with the production of a new uv absorption band at 275 nm as detected by second-order derivative uv spectroscopy. The effect of Microscal irradiation on the growth of this band at 275 nm is illustrated in Figure 1. The most interesting result is that all of these amine acrylates exhibit photoyellowing and the growth of a new band in their uv spectra at 275 nm. Therefore the nature of the chromophore responsible for the new absorption must be the same in all cases as deduced earlier (14,15). The data shows that all of the amine acrylates exhibit an initial rapid increase in the level of the absorbance at 275 nm during the early stages of irradiation. The rapid increase in this uv absorption band also coincided with a rapid initial photoyellowing in all of the samples.

The second most interesting feature of this data was the relative rates of photobleaching of the absorption at 275 nm for each amine acrylate. The dicyclohexylamine, morpholine and dibutylamine terminated samples all exhibited a maximum in the growth of the uv band within the first 20 hours of irradiation while the diethylamine terminated sample exhibited a maximum after 50 hours of irradiation. The dibutylamine sample has the next fastest rate of photobleaching which is complete after 90 hours of irradiation. The morpholine and diethylamine systems photobleached at comparable and very much slower rates with the chromophore persisting at over 300 hours of irradiation. These results suggest that although the nature of the amine functionality has little or no effect on the rate of formation of this chromophore, it does have a dramatic effect on its rate of photobleaching. The nature of the substitution of this chromophore, as governed by the amine structure from which it is formed has an apparently dramatic effect on its photostability as exhibited by the different rates of photobleaching. It is possible that the photobleaching may be sensitised by different species in each system which are produced during the curing process or in the early stages of irradiation.

This uv curing investigation was extended to evaluate a wider range of amine acrylates which had been EB cured. Under these conditions there are no complications due to the presence of photoinitiator. The additional amine acrylates investigated included those derived from N-ethyl ethanolamine, N-methyl ethanolamine and dimethylamine. During this study additional features were monitored from the infra-red data.

The effect of Microscal irradiation on the increase in hydroxyl index for these electron-beam cured amine acrylates is shown by the data in Table II.

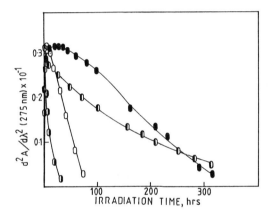

Figure 1. Change in intensity of the second derivative uv absorption band at 275 nm during Microscal irradiation for uv cured (5% w/w benzophenone) (4μ m thick) GPT after reaction with (●) diethylamine, (◐) morpholine, (◑) dicyclohexylamine and (○) di-n-butylamine.

Table II. Increase in Hydroxyl Index Values during Irradiation in a Microscal Unit of EB Cured (4MRad) GPT (12μm thick) after Reaction with Various Amines

Amine	Hydroxyl Index Values		
	100h	200h	300h
Diethylamine	0.127	0.164	0.185
Morpholine	0.052	0.082	0.109
Dicyclohexylamine	0.027	0.052	0.080
Dimethylamine	0.100	0.145	0.176
N-ethylethanolamine	0.024	0.035	0.052
N-methylethanolamine	0.012	0.021	0.032
Di-n-butylamine	0.119	0.166	0.194

The general order of increasing stability for these amine acrylates is as follows :

> dibutyl = diethylamine = dimethylamine ⟩ morpholine ⟩
> dicyclohexylamine ⟩ N-ethyl ethanolamine ⟩
> N-methyl ethanolamine

The dibutylamine system is much more stable after EB curing than after uv curing while the morpholino system has comparable stability whereas the diethylamino system appears to be less stable after EB curing. Clearly the nature of the curing process has a marked and complicated effect on the oxidation processes in these systems. One would expect the uv cured samples to be in general less stable than the EB cured samples during the initial stages of irradiation. An interesting point relating to the hydroxyl index data is that of the four most stable amine acrylates three of them have amine functionality which contains oxygen either as ether or as a hydroxyl group. This additional functionality may be exerting a stabilising effect in these systems. The dicyclohexylamine may be inducing increased stability due to the reduced number of abstractable hydrogen atoms alpha to the nitrogen in this amine. The three least stable amine acrylates namely; diethylamine, dibutylamine and dimethylamine all have the same number of abstractable hydrogen atoms alpha to nitrogen in their structure.

The effect of Microscal irradiation on the growth of the uv absorption band at 275 nm for these EB cured amine acrylates is illustrated in Figure 2. Comparing this data to that obtained for the uv cured samples earlier it can be seen that the growth profile is considerably different with the EB cured sample. As with the uv cured samples all of the EB cured samples photoyellowed to a greater or lesser extent and exhibited the growth of the band at 275 nm in their uv spectra. Thus, all of these amine acrylates produce the same or similar chromophore and photoyellowing phenomenon. The general order of photoyellowing was found to decrease as follows :

> dibutyl ⟩ diethyl ⟩ morpholine ⟩ dimethyl = dicyclohexyl
> = N-ethyl ethanolamine = N-methyl ethanolamine

The last three all exhibit very similar photoyellowing

25. ALLEN ET AL. *Photooxidative Stability and Photoyellowing* 353

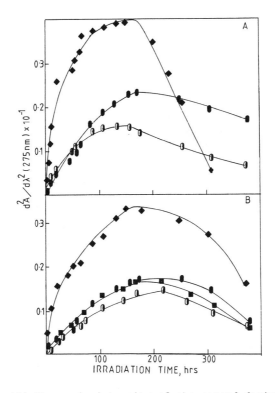

Figure 2. (A) Change in intensity of the second derivative uv absorption band at 275 nm during irradiation in a Microscal unit of EB cured (4MRad) GPT (4 μm thick) after reaction with (◆) di-n-butylamine, (●) morpholine, and (○) dicyclohexylamine and (B) after reaction with (◆) N-diethylamine (○) N-ethylethanolamine (■) N-methylethanolamine and (●) dimethylamine.

properties. This trend also follows closely the relative rates of increase of the observed uv band at 275 nm. The overall maximum intensity of this uv band is governed by the nature of the amine functionality as is the irradiation time required to reach the uv absorption maximum. The rate of decomposition of the uv absorbance is also amine structure dependent with the relative order being similar to that observed for the rate of formation.

It may have been expected that the systems which exhibited the greatest production of the chromophore responsible for the yellowing would have done so due to its low rate of photobleaching whereas the opposite is observed the amine acrylates which exhibited the fastest rate of growth in the uv band also exhibited the greatest overall increase in this band as well as the greatest photodecomposition rate. The intensity of the observed band at 275 nm, although measured in relative terms, cannot be related directly to the concentration of the chromophore in these systems. One reason for this is that although the nature of the chromophore is undoubtedly the same, the type of substitution and therefore the extinction coefficient may differ depending on the amine structure. The initial rates in the uv band formation are slower in the EB cured samples compared to the uv cured samples because no residual photoinitiator is present in the former systems which would sensitise the oxidation of the terminal amino group. The rate of photoyellowing as well as the extent of photoyellowing is clearly dependent on amine structure. Also, as for the hydroxyl index results there is a tendency to greater stability for the oxygen amine acrylates with the exception of the dicyclohexylamine system which suggests that for simple dialkyl amines two methylene protons alpha to the nitrogen are required for photoyellowing. The oxygen containing amine acrylates may be providing some specific antioxidant activity or reduced susceptibility to degradation under these irradiation conditions. It must be remembered that all of these amine acrylates contain terminal amine groups which have at least one of the alkyl substituent groups in common.

Thus, all of these amine acrylates have a carbon atom alpha to nitrogen which has two methylene protons. This type of site may be the site of oxidation which is common to all of these amine acrylates producing the same photoyellowing chromophore in each to a lesser or greater extent depending on the nature of the alkyl groups. Another point worth noting about the uv data is that some of these amine acrylates exhibit different rates of photobleaching depending on the curing process. The most notable example of this is seen with the dicyclohexylamine based amine acrylate where the uv band is photobleached rapidly in the uv cured sample and at a much slower rate in the EB cured sample. This may be due to the formation of a chromophore with different types of substitution depending on the curing process. Under uv curing conditions specific sites on the amino functionality are attacked by benzophenone. These sites are usually the methylene protons alpha to the nitrogen atom. They will also be initiating sites for the polymerisation and therefore their substitution will increase as they attack an acrylate double bond. There will also be an increase in the level of tertiary carbon atoms alpha to the nitrogen in the uv cured dicyclohexylamine amine acrylate. Under EB curing conditions the initiating radical sites are produced in a less specific manner and thus the level of tertiary carbons alpha to nitrogen will be less than in the uv cured sample. It is this possible difference in substitution as a consequence of the curing process which may account for the observed differences in data. However, this curing condition phenomenon does not occur with all of the amine acrylates.

Two additional areas of the infra-red spectra were monitored for the EB cured samples. These were the C-H and carbonyl stretching regions. These amine acrylates exhibit similar bands in the C-H stretching region which are associated with the terminal amine functionality. The exact frequency of absorption was not common to the various amine acrylates used. The band situated at 2978 cm^{-1} exhibited the typical increase in intensity associated with the amine substitution but was shifted to slightly higher or lower wavenumber depending on the nature of the substitution. The additional band which is observed in amine acrylates but not in the unmodified multifunctional monomer material is situated at 2813 cm^{-1} for the diethylamine system but again the exact frequency of absorption was different in the other amine acrylates with two bands being observed for the dimethylamine system.

The effect of Microscal irradiation on the rate of decrease of these alkyl bands, expressed as an index, is shown by the data in Tables III and IV for the EB cured samples. It is interesting to note that the dicyclohexylamine sample did not exhibit this additional band at 2813 cm^{-1} and produced a complex C-H stretching profile which made it impossible to monitor this sample. The fact that all of these amine acrylates exhibit a decrease in amine alkyl band intensity during irradiation indicates that the amine group is the preferential site of oxidation. This is confirmed by the fact that the intensity of the band situated at 2875 cm^{-1}, which is associated with the polymer backbone, remained unchanged during irradiation. This observation is consistent with the relatively low levels of degradation observed with the hydroxyl index data. The rate of decrease of the amine alkyl band situated at 2813 cm^{-1} depends on the amine structure with the general order for decreasing rate loss of this band being :

dimethyl (2774 cm^{-1}) $>$ dimethyl (2823 cm^{-1}) $>$ diethyl (2813 cm^{-1}) $>$ N-methyl ethanolamine (2810 cm^{-1}) $>$ dibutyl (2812 cm^{-1}) $>$ morpholine (2815 cm^{-1}) = N-ethyl ethanolamine (2818 cm^{-1})

Table III. Decrease in Alkyl Band Index (2813 cm^{-1}) during Irradiation in a Microscal Unit of EB Cured (4MRad) GPT (12μm thick) after Reaction with Various Amines

Amine	Alkyl Band Index Values	
	100h	200h
Diethylamine	0.734	0.634
Morpholine	0.874	0.771
Dimethylamine (2873 cm^{-1})	0.734	0.622
Dimethylamine (2774 cm^{-1})	0.637	0.515
N-ethylethanolamine	0.885	0.773
N-methylethanolamine	0.775	0.639
Di-n-butylamine	0.783	0.755

Table IV. Decrease in Alkyl Band Index (2976 cm^{-1}) during Irradiation in a Microscal Unit of EB Cured (4MRad) GPT (12 μm thick) after Reaction with Various Amines

Amine	Alkyl Band Index Values	
	100h	200h
Diethylamine	0.885	0.856
Morpholine	0.989	0.980
Dimethylamine	0.974	0.967
N-ethylethanolamine	0.970	0.956
N-methylethanolamine	0.977	0.965
Di-n-butylamine	0.962	0.940

The rates of loss do not correlate with the rates of increase in the hydroxyl index for these systems or with the photoyellowing and uv phenomena. This indicates that the mechanism of oxidation of the amine functionality is complex. An additional feature of interest is the profile of this band loss as related to terminal amine structure. In the simple alkyl substituted systems namely diethyl, dimethyl and dibutyl there is an initial rapid loss followed by a slowing down of the rate of decrease on prolonged irradiation. However, with the oxygen containing systems namely morpholine, N-ethyl ethanolamine and N-methyl ethanolamine the rate of decrease is linear with irradiation time. Evidently, there must be two different mechanisms in operation here. The loss of these infra-red absorption bands may be due to a process of dealkylation via oxidation (see reactions later). The band situated at 2978 cm^{-1} exhibited similar behaviour although the extent of the loss was less than that observed with the previous band and the order of rate loss was marginally different.

The above data clearly indicates that altering the nature of the terminal amine functionality of these amine acrylates has a profound effect on their photodegradation profiles. All exhibit the same photoyellowing phenomenon with a significant reduction being observed with the dicyclohexylamine and two alkanolamine systems. These three, along with the morpholine system were clearly the most photostable. Furthermore, all of these amine acrylates were significantly more photostable than the parent monomer GPTA as found in previous work (14-16).

Apart from varying the functionality of the terminal amine in these amine acrylates it is also possible to vary the structure of the parent monomer and investigate its effect on the photoyellowing and photodegradation behaviour of the derived amine acrylate. Two new amine acrylates were prepared using diethylamine and the parent monomers TMPETA and TMPTA. These amine acrylates were investigated in conjunction with that prepared with GPTA and irradiated in a Microscal unit.

The effect of the Microscal irradiation on the growth of the uv band at 275 nm for these amine acrylates is illustrated in Figure 3. It is interesting to note that, despite the difference in the multifunctional monomer structure, the extent and growth profile of the uv absorption is identical for each amine acrylate. This indicates that the photoyellowing is independent of the multifunctional monomer

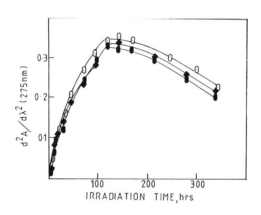

Figure 3. Change in intensity of second-derivative uv absorption band at 275 nm during Microscal irradiation of EB cured (4MRad, 4μm thick) (○) GPT, (◆) TMPETA and (●) TMPTA after reaction with diethylamine.

structure and is associated with oxidation of the terminal amine functionality. Considering the infra-red data the effect of irradiation on the increase in hydroxyl index is shown by the data in Table V. The rate of increase was found to depend on the amine acrylate structure with the following order of increasing stability :

$$\text{TMPETA} > \text{GPTA} > \text{TMPTA}$$

Table V. Increase in Hydroxyl Index Values during Irradiation in a Microscal Unit of EB Cured (4MRad) Resins (12 m thick) after Reaction with Diethylamine

Resin	Hydroxyl Index Values		
	100h	200h	300h
GPT	0.078	0.100	0.131
TMPETA	0.141	0.187	0.219
TMPTA	0.047	0.069	0.090

The same relative order was found for the decrease in the 2812 and 2878 cm^{-1} band indices as shown by the data in Tables VI and VII. The infra-red data shows that the photooxidative stability of the amine acrylates depends on the nature of the polymer backbone. The order of decreasing stability is linked directly to the level of methylene protons alpha to the ether oxygen in each system. With the diethylamine derivative of TMPTA there is no ether functionality and this is the most photostable. The amine acrylate based on GPTA contains three alpha methylene protons adjacent to an ether oxygen. The amine acrylate based on TMPETA contains four alpha methylene protons adjacent to an ether oxygen.

Table VI. Decrease in Alkyl Band Index (2813 cm^{-1}) during Irradiation in a Microscal Unit of EB Cured (4MRad) Resins (4μm thick) after Reaction with Diethylamine

Resin	Alkyl Band Index Values	
	100h	200h
GPT	0.740	0.663
TMPETA	0.703	0.630
TMPTA	0.889	0.826

Table VII. Decrease in Alkyl Band Index (2976 cm^{-1}) during Irradiation in a Microscal Unit of EB Cured (4MRad) Resins (4μm thick) after Reaction with Diethylamine

Resin	Alkyl Band Index Values	
	100h	200h
GPT	0.885	0.852
TMPETA	0.837	0.796
TMPTA	0.955	0.941

These methylene protons are relatively labile and can be abstracted easily by alkoxy and peroxy radicals which may be present in an oxidising polymer matrix. Thus, it is not surprising that the TMPTA based amine acrylate is the most stable. It is interesting to note that the bands lost in the infra-red spectra are associated with the amine functionality and their rate of loss by photooxidation is dependent on the polymer backbone structure. It is clear that the nature of the polymer backbone controls the oxidation of the amine functionality but not the photoyellowing phenomenon.

Conclusions

The structure of the amine acrylates used in this study after curing is likely to be very complex and this makes it difficult, if not impossible, to evaluate clearly the nature and mechanism of the photooxidation and photoyellowing observed in these systems. The infra-red data clearly indicates the loss of amine alkyl functionality resulting in amine or amide formation. The carbonyl absorption reported previously (14,15) at 1680 cm^{-1} indicates the latter. The band situated at 1612 cm^{-1}, also reported previously (14,15), may be due to imine, enamine, vinyl ether, amine or an , -unsaturated carbonyl species. If it was due to imine or enamine it would be easy to explain its loss due to hydrolysis during irradiation. However, the exact assignment remains difficult.

A number of the findings above found coloured product formation during the oxidation of amine containing systems with the coincidence of a uv absorption band at 275 nm. The results indicate that the chromophore responsible is primarily an α,β-unsaturated carbonyl species associated with the tertiary amine functionality.

The structure of the alkylamino group is important in controlling the photooxidative stability and photoyellowing of the cured resin film. The alkyl group adjacent to the nitrogen atom is the prime site for oxidative attack particularly at an alkyl methylene adjacent to an oxygen atom. In general, a di-n-butylamine diacrylate resin gave the most photoyellowing but with the fastest photobleaching while N-alkylethanolamines gave the least photoyellowing but a slow photobleaching. Ultraviolet cured coatings gave rise to more rapid photoyellowing than electron beam cured systems and this is associated with facile hydrogen atom abstraction and subsequent hyroperoxidation of the alkylamino groups in the former case by the residual benzophenone photoinitiator.

Literature Cited

1. Gatechair, L. R., in "UV Curing : Science and Technology" Vol II, (S.P. Pappas, (Ed)), Technology Marketing Corporation, Norwalk, CT., USA, 1985.
2. Tu, R. S., Soc. Man. Eng., Techn. Pap., 1980, FC80-579.
3. Puglisi, J and Vigeant, F., J. Rad. Curing, 1980, 7, 31.
4. Filipescu, N and Minn, F. L., J. Am. Chem. Soc., 1968, 90, 1544.
5. Hult, A., Yuan, Y. Y. and Ranby, B., Polym. Deg. & Stabil., 1984, 8, 241.
6. Gismondi, T. E., J. Rad. Curing., 1984, 11, 14.
7. Schmid, S. R., J. Rad. Curing., 1984, 11, 19.

8. Pappas, S. P., Gatechair, L. R., Breskman, E. L., Fischer, R. M., Klein, U. K. A., in "Photodegradation and Photostabilization of Coatings," Pappas, S. P.; Winslow, F. H., Eds.; ACS Symposium Series 151; American Chemical Society: Washington, DC, 1981; pp 109–116.
9. Timpe, H. J., Garcia, C., S. P. Pappas, Gatechair, L. R., Breskman, E. L. and Fischer, R. M., Polym. Photochem., 1984, 4, 1
10. Breskman, E. L. and Pappas, S. P., J. Coat. Techn., 1976, 48, 34.
11. Decker, C. and Bendaikha, T., Polym. Prep., 1984, 25, 42.
12. Decker, C. and Bendaikha, T., J. Rad. Curing, 1984, 11, 6.
13. Decker, C., Fizet, M. and Faure, J., J. Org. Coat. Plast. Chem., 1980, 42, 710.
14. Allen, N. S., Robinson, P. J., White, N. J. and Skelhorne, G. G., Eur. Polym. J., 1984, 20, 13.
15. Allen, N. S., Robinson, P. J., White, N. J. and Skelhorne, G. G., Eur. Polym. J., 1985, 21, 107.
16. Allen, N. S., Robinson, P. J., White, N. J. and Swales, D. W., Polym. Deg. & Stab., 1987, 19, 147.
17. White, N. J., PhD Thesis, "Photooxidative Stability of UV and EB cured Acrylates", Manchester Polytechnic, UK, (CNAA), 1988.

RECEIVED September 13, 1989

RADIATION CURING OF CATIONIC POLYMERIZATION

Radiation Curing of Cationic Polymerization

With the widespread interest in use of electron beams and UV light to initiate the polymerization of a variety of resins which proceed by a cationic propagation mechanism, there is a continuing effort to develop new applications and new monomers/oligomers which are specifically targeted for use with onium salt initiators. The papers presented in this section describe successful attempts to expand the use of photo and electron beam curable cationic systems.

The first three chapters (25-27) in this section describe the preparation and subsequent curing/characterization of non-traditional oligomers/polymers. The oligomers, which are the subject of Chapter 25 represent the vinyl ether analogs of acrylated urethanes which are commonly used in conventional radical-type photocurable systems. Such oligomers are extremely important since they provide the opportunity to develop cationic radiation curable formulations which possesses the final physical properties which to date have been enjoyed by traditional photocurable films cured by a free-radical process. The electron-beam and photoinitiated cure rates of the vinyl ether urethanes are reported to be quite high, even in the presence of oxygen.

Chapters 26 and 27 describe in detail the synthesis, characterization, and polymerization profiles of a relatively new series of multifunctional epoxy-silicone monomers and oligomers. In one case, use of an iodoniun salt initiator (Chapter 26), which is highly compatible with the silicone-based systems, resulted in films with excellent physical properties. One of the benefits of epoxy-silicones is their ability to increase the cure rate of conventional organic diepoxides when incorporated as a blend. The results reported in Chapter 27 provide evidence for the extremely high cure rates of low molecular weight epoxy-silicone multifunctional monomers and oligomers. Particularly interesting is the correlation which is made between changes in the structure of the difunctional systems and the resulting cure rates. As well, the outstanding physical properties of the cured films attest to their potential impact on the radiation curing industry.

The final paper (Chapter 28) dealing with cationic photocurable systems is concerned with thick coatings containing silicone additives. By using a silicone elastomer dispersion containing a distribution of various filler sizes and an appropriate adhesion promoter, photocurable formulations produce encapsulants which exhibit good adhesion to ceramic surfaces and possess other desirable thermal characteristics.

Chapter 26

Radiation-Induced Cationic Curing of Vinyl-Ether-Functionalized Urethane Oligomers

Stephen C. Lapin

Allied—Signal Engineered Materials Research Center, 50 East Algonquin Road, Des Plaines, IL 60017

```
Vinyl ether terminated urethane (VEU) oligomers were
prepared by reacting hydroxy vinyl ethers with isocyanates.
These materials were chain extended with polyols and
combined with divinyl ether monomers to produce radiation
curable coating formulations. Cationic curing was induced
by UV or electron beam irradiation in the presence of a
triaryl sulfomium salt catalyst. Curing occured at high
speeds and produced coatings with desirable physical
properties.
```

Radiation curable coatings consist of monomers and oligomers which have functional groups in which polymerization may be induced by the presence of UV or electron beam radiation. The major types of systems that have been used include: (1) unsaturated polyesters, (2) thiol/enes, (3) acrylates, and (4) cationically cured epoxies. Each of these systems has its own distinct advantages and disadvantages. One chooses a system that is best suited to a particular application. Acrylates at present are by far the most widely used system. They offer an excellent combination of high reactivity and moderate costs. In addition, there are a wide variety of commercially available acrylated monomers and oligomers which allow the formulation of coatings with a wide range of physical properties.

Although acrylates are extremely useful and versatile materials, they also have some distinct disadvantages. One disadvantage is the growing concern with health hazards that are associated with the use of acrylates (1-4). Another disadvantage is that the radical polymerization of acrylates is inhibited by atmospheric oxygen (5,6).

A system that has recently been receiving attention is radiation induced cationic curing. The interest in cationic curing has been inspired by the development of onium salt catalysts. Strong acids are liberated when the coatings are irradiated in the presence of certain onium salts. These acids are capable of catalyzing cationic polymerization reactions (7). Onium salts have been used mostly in UV curable systems. However, it has been shown recently that they may also be used for electron beam induced curing (8,9).

Reprinted with permission
© 1988 Radtech International

Most of the work on cationic curing has dealt with the homopolymerization of epoxies (9,10,11). Epoxy coatings are well known for their superior adhesion, toughness, and chemical resistance. This makes radiation curable epoxies extremely useful in some demanding applications. Another advantage of epoxies is that the cationic curing reaction is not inhibited by atmospheric oxygen. The major disadvantage of epoxies is their slow cure speed. One reason for using radiation curing is to take advantage of line speeds which can operate concurrently with other high speed processing steps. Epoxies do not cure fast enough to make them useful as direct replacements for acrylates in most applications. Epoxies will continue to polymerize in a "dark cure" reaction long after they are irradiated. This can complicate the handling of materials in a production line. A "thermal bump" is often applied following the irradiation step in order to hasten the attainment of the final coating properties (9,10). The use of this additional thermal curing step defeats part of the purpose of radiation curing, and is generally considered to be unacceptable.

It has recently be shown that alcohols may be added to the epoxy based formulations. This increases the cure speed of the coating through a chain transfer polymerization reaction. The chain transfer processes produces coatings with reduced crosslink density. This generally leads to a reduction in the mechanical properties and the chemical resistance in the cured coating. It has been suggested that multifunctional alcohols may used as chain transfer agents to help offset the loss of crosslink density (12). However, much of the published data shows a sharp decrease in T_g and tensil modulus upon the addition of polyols (13).

In 1982, Crivello and co-workers published a report on the UV initiated cationic polymerization of vinyl ether monomers using onium salt catalysts(14). Vinyl ethers are among the most reactive monomers which polymerize by a cationic mechanism. The radiation induced cationic curing of vinyl ethers occurs much faster than the cationic curing of epoxy coatings. In fact, cure rates that are at least as fast as the free radical polymerization of acrylates can be achieved(8,14). A recent report indicates that the cationic polymerization of vinyl ethers can occur even in the presence of certain polar functional groups(15).

One way in which vinyl ether monomers may be used is as reactive diluents in epoxy systems. The addition of a vinyl ether greatly increases the cure speed of the system. The rapid change to a tack-free state appears to be the result of the polymerization of the vinyl ether component of the mixture. A "thermal bump" is still necessary to cure the epoxy and bring the coating to a fully cured state(16). There is little evidence for any copolymerization of the vinyl ether and the epoxy groups.

In order to take full advantage of the high cure speed of vinyl ethers, systems are needed which have a curing mechanism based totally on the polymerization of vinyl ether groups. A few vinyl ether monomers are now commercially available(17); however, there are currently no sources of vinyl ether functionalized oligomers. It is difficult to formulate coatings based on monomers alone. Higher molecular weight materials are often required. Oligomeric components have more desirable rheological properties and they allow greater control of the cured film properties. In this paper we report the development of vinyl ether terminated oligomers. These materials greatly widen the scope of potential applications of radiation induced cationic curing.

Introduction

Vinyl ethers may be synthesized by the base catalyzed reaction of alcohols with acetylene at elevated temperatures and pressures(18). Diols may also be used to produce difunctional vinyl ethers (Equation 1) This is most likely a reversible reaction therefore, a mixture

$$HO-R-OH + 2C_2H_2 \xrightarrow[200°C, press.]{KOH} CH_2=CH-O-R-O-CH=CH_2 + CH_2=CH-O-R-OH \quad (1)$$

of products is obtained(18). The mixture includes the divinyl ether along with a monoacetylated product and unreacted diol. The divinyl ether has a boiling point which is lower than the other products and may be separated from the mixture by fractional distillation. The distillation also serves to separate the divinyl ether monomer from the KOH and other reaction by-products which would inhibit cationic polymerization. This method has been used to produce divinyl ethers from butane diol, diethylene glycol, and triethylene glycol and 1,4-cyclohexane dimethanol(17). The method does not lend itself well, however, to the production of oligomeric vinyl ethers from higher molecular weight polyols. It is not practical to purify higher molecular products by distillation. In addition the backbone of the oligomer would be limited to structures which are stable under the acetylene reaction conditions (i.e., 3% KOH and 200°C).

Another method which may be used to synthesize oligomeric vinyl ethers is the reaction of chloroethyl vinyl ether with alcohols (Equation 2) (14,19). The reaction is aided by the addition of a

$$CH_2=CH-O-CH_2CH_2-Cl + ROH \xrightarrow[PTC]{KOH} CH_2=CH-O-CH_2CH_2-OR \quad (2)$$

phase transfer catalyst. While this is good method for the laboratory scale preparation of certain vinyl ethers, it does not appear to be commercially attractive. There are no commercial sources of chloroethyl vinyl ether in the United States, and as with the acetylene reaction, one may have difficulty separating the desired product from the reaction mixture on a commercial scale.

We have chosen to use hydroxy vinyl ethers ($CH_2=CH-O-R-OH$) as intermediates in the synthesis of vinyl ether functionalized oligomers. Hydroxy vinyl ethers are easily obtained as by-products of the synthesis of divinyl ether monomers (Equation 1). The hydroxy group may be reacted to form a link to the oligomer backbone while the vinyl ether moiety remains free to be polymerized in a radiation induced reaction at a later time.

This paper will focus on vinyl ether terminated urethane (VEU) oligomers(20,21). VEU oligomers may be prepared by reacting hydroxy vinyl ethers with compounds containing isocyanate groups (Equation 3). The alcohol group of the hydroxy vinyl ether reacts with the isocyanate to form the desired urethane linkage.

$$R'-NCO + HO-R-O-CH=CH_2 \longrightarrow R'-NH-\overset{\overset{O}{\|}}{C}-O-R-O-CH=CH_2 \quad (3)$$

One interesting feature of the production of VEU oligomers is the way that it may be tied to the initial synthesis of the hydroxyvinyl ether. The reaction of acetylene with the diol produces a mixture of the divinyl ether, the hydroxy vinyl ether and the unreacted diol (Equation 1). Careful fractional distillation can separate the hydroxyvinyl ether for use in VEU oligomers; however, it may actually be desirable to use the mixture of these products, incorporating the diol and the divinyl ether into the radiation curable formulation (Scheme I). The diol serves to chain extend the resin. Each hydroxyl group on the diol reacts with an isocyanate group forming a higher molecular weight material. The total concentration of hydroxy groups from the diol and the hydroxy vinyl ether can be determined and then combined with an equivalent concentration of isocyanate. The divinyl ether does not have free hydroxy groups, therefore it will not react with the diisocyanate. The divinyl ether serves as a diluent which lowers the viscosity of the mixture during the oligomer synthesis. When the mixture is finally cured, the divinyl ether copolymerizes with the VEU oligomer, effectively increasing the crosslink density of the film.

Experimental

Synthesis of Triethylene Glycol Monovinyl Ether (TEGMVE). Triethylene glycol (250 mL) and ground KOH (7.5 g) were added to a 500 mL, 3-neck round bottom flask equipped with a mechanical stirrer, reflux condenser and gas inlet tube. The apparatus was placed behind a protective shield in a fume hood. The outlet from the condenser was vented to the upper area of the hood. The mixture was heated to 190 ± 10°C while purging with nitrogen. After the temperature was stabilized, acetylene flow was begun (1.0 liter/min). The reaction was continued for 5 hours. The flask was fitted with a distillation head. The product (125 mL) was collected with a boiling range of 70 to 84°C at 0.3 torr. Gas chromatographic analysis indicated that the products were a mixture of triethylene glycol divinyl ether (14%), triethylene glycol monovinyl ether (78%), and unreacted triethylene glycol (7%). Redistillation through a 12 inch Vigreaux column gave an 85 mL fraction that contained 95% TEGMVE.

Synthesis of VEU Oligomers. A number of different VEU oligomers was synthesized. The following is a typical procedure: diphenylmethylene diisocyanate (25.58 g, 0.102 moles) was dissolved in 100 mL of dry methylene chloride in a 250 mL round bottom flask equipped with a reflux condenser and a magnetic stir bar under a nitrogen atmosphere. TEGMVE (36.02 g, 0.204 mols) was added to the solution along with 2 drops of dibutyltin dilaurate. The mixture began to reflux within 2 to 3 minutes without external heating. After reacting for about 30

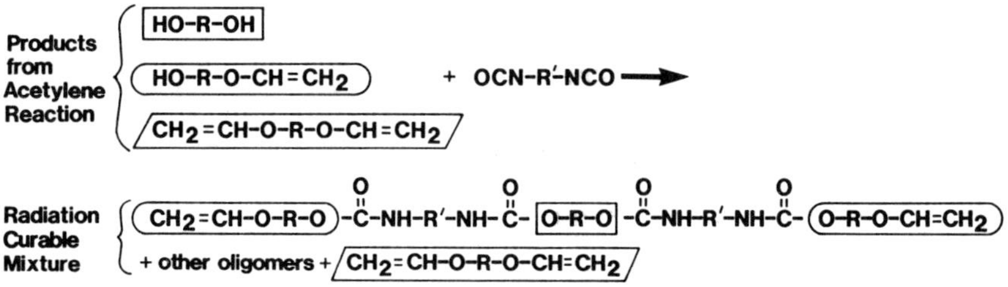

Products from Acetylene Reaction
$\left\{\begin{array}{l}\boxed{\text{HO-R-OH}} \\ (\text{HO-R-O-CH=CH}_2) \\ \diagup\text{CH}_2\text{=CH-O-R-O-CH=CH}_2\diagdown\end{array}\right.$ + OCN-R'-NCO ⟶

Radiation Curable Mixture
$\left\{\begin{array}{l}(\text{CH}_2\text{=CH-O-R-O})\overset{\overset{O}{\|}}{\text{C}}\text{-NH-R'-NH-}\overset{\overset{O}{\|}}{\text{C}}\text{-}\boxed{\text{O-R-O}}\text{-}\overset{\overset{O}{\|}}{\text{C}}\text{-NH-R'-NH-}\overset{\overset{O}{\|}}{\text{C}}\text{-}(\text{O-R-O-CH=CH}_2) \\ \text{+ other oligomers +}\diagup\text{CH}_2\text{=CH-O-R-O-CH=CH}_2\diagdown\end{array}\right.$

Scheme I

minutes the mixture began to cool. At regular intervals a few drops of the reaction mixture were removed for infrared analysis. After 5 hours, the -NCO band at 2250 cm^{-1} was no longer detectable. The solvent was removed in a vacuum and the resulting thick liquid product was analyzed by 1H and 13C NMR. The spectra were consistent with the expected product, I.

Many VEU oligomers were synthesized using triethylene glycol divinyl ether (TEGDVE) as a reaction solvent in place of methylene chloride. The TEGDVE was not separated from the products. It remained as part of the polymerizable mixture.

<u>Radiation Curing.</u> Samples to be irradiated were coated onto either Bonderite-40 treated steel test panels (Parker Chemical) or polyethylene coated paper board. An excess of the sample was placed at one end of the substrate and a #6 wire wound rod was drawn across the substrate with even pressure pushing excess material off the edge. This method produced coatings with a thickness of 6 to 12 um.

An RPC model QC-1202 processor was used for UV curing. The unit was equipped with two 12 inch medium pressure mercury arc lamps and a variable speed conveyor (50 to 500 ft/min). Only one lamp was used at a time in the testing (operated at 200 watts/in). The UV energy dosage was measured by passing a UV Process Supply Compact Radiometer under the lamp on the conveyor belt.

An Energy Sciences Electrocurtain model CB-150 equipped with a 15 cm linear cathode was used for EB curing. Electron energies of 160 KeV were employed. Samples were placed in a an aluminum tray on a variable speed conveyor (20- 235 ft/min) within the CB-150 unit. Irradiation occurred in a nitrogen atmosphere.

<u>Coating Evaluation</u> The coatings were evaluated within one hour after irradiation. The coatings were examined for solvent resistance using methyl ethyl ketone. The number of double rubs necessary to break through the coating was recorded. Reverse impact was measured on the steel panels using a Gardner impact tester according to ASTM Method D2794. The coating elongation was measured by bending the coated steel panel over a conical mandrel according to ASTM Method D522. Pencil hardness was measured according to ASTM D3363.

Results and Discussion

A number of different VEU oligomers may be synthesized from a variety of hydroxyvinyl ethers and isocyanate monomers or prepolymers. This paper will focus on oligomers synthesized from triethylene glycol monovinyl ether (TEGMVE) and diphenylmethane diisocyanate (MDI). Initially, the UV and EB curing parameters were determined using the simple 2:1 adduct of TEGMVE and MDI (I). The results were fairly representative of other VEU oligomers.

$$CH_2=CH-O-(CH_2CH_2O)_3-\overset{O}{\underset{\|}{C}}-NH-\hspace{-0.5em}\bigcirc\hspace{-0.5em}-CH_2-\hspace{-0.5em}\bigcirc\hspace{-0.5em}-NH-\overset{O}{\underset{\|}{C}}-O-(CH_2CH_2O)_3-CH=CH_2$$

I

The oligomer, I, was combined with a triarylsulfonium salt of hexafluorophosphate (General Electric UVE-1016), and then was coated onto the substrate. The curing properties were studied using UV and electron beam irradiation conditions.

Table 1 shows some of the physical properties of the coating as a function of the exposure to the UV lamp. A tack-free coating was obtained immediately upon irradiation at a energy dosages greater than 38 mj/cm^2. The coating was unaffected by 100 double rubs with methylethyl ketone (MEK) after being exposed at 60 mj/cm^2. Coatings cured at greater than 100 mj/cm^2 had a noticeable yellow discoloration immediately after irradiation, and turned a greenish color upon standing for 24 hrs. No additional changes in the coating properties were observed at higher UV exposure levels.

Table 1. UV Curing of VEU Oligomer I*

Energy Dose (mj/cm^2)	Pencil Hardness	MEK Rubs	Reverse Impact (in-lbs)	Elongation (%)
38	3B	60	>160	>50
44	HB	80	>160	>50
60	F	100	>160	>50
89	2H	>100	>160	>50
179	2H	>100	>160	>50
346	2H	>100	>160	>50

* 2 pph UVE-1016, nitrogen atmosphere

Table 2 shows the results of curing the same VEU oligomer (I), by electron beam irradiation. No curing was observed in the absence of the onium salt. In the presence of onium salt (2%), an energy dosage of 0.7 Mrads was required to produce a tack-free coating. A dosage of 4 Mrads produced a coating which was unaffected by 100 MEK rubs. The coating became noticeably discolored at dosages greater than 6 Mrads. These EB cure speeds were not particularly fast compared to typical acrylate based coatings. However, we have shown previously that much higher EB cure speeds may be obtained by using an iodonium salt in place of a sulfonium salt catalyst.8

Table 2. Electron Beam Curing of VEU I[a]

EB Dose (Mrads)	Pencil Hardness	MEK Rubs	Reverse Impact (in-lbs)	Elongation (%)
0.5	NC[b]	--	--	--
1.0	6B	4	>160	>50
2.0	2H	80	>160	>50
4.0	3H	>100	>160	>50
6.0	3H	>100	>160	>50
12.0	3H	>100	>160	>50

a. 2 pph UVE-1016, nitrogen atmosphere.
b. Coating remained tacky after irradiation.

The properties of the coatings shown in Tables 1 and 2 for UV and EB curing appear to be very similar. This suggests that similar polymer networks were created by the two irradiation methods. As expected, the coating obtained by curing this difunctional urethane oligomer was highly flexible. The coatings could not be ruptured by bending on a conical mandrel or by reverse impact up to 160 in-lbs. These flexural properties were attractive for a coating that also gave pencil hardness values of 2H to 3H.

A study was done to determine the effect of the sulfonium salt concentration on the cure speed of the VEU oligomer (I). A series of mixtures was prepared containing from 0.2% to 7.7% of UVE-1016. The energy dosages for both UV and electron beam irradiation were controlled by varying the sample conveyor speeds. The minimum energy dose which would produce a tack-free coating was recorded. Figure 1 shows the results for UV induced curing. The fastest cure speed occurred when the catalyst was present at a concentration of about 1% to 3%. The cure speed began to fall very rapidly at concentrations below 0.5%. A loss of cure speed also occurred at higher catalyst concentrations. This was likely due to high absorption of light at the surface of the coating by the sulfonium salt. This allowed little light to penetrate for curing underneath the surface. VEU oligomers derived from aromatic isocyanates will absorb light in competition with the onium salts. This is an important consideration in determining the optimum catalyst concentration.

Figure 2 shows the electron beam cure speed of the oligomer (I) as a function of the sulfonium salt concentration. Again the cure speed began to drop rapidly at concentrations below 0.5%. No significant gain in cure speed was achieved by increasing the catalyst concentration above about 3%. The electron beam easily penetrates the entire coating and is not selectively absorbed by the oligomer or the catalyst, therefore electron beam cured coatings are much less sensitive to catalyst concentration effects than UV cured coatings.

Properties such as tack and MEK rub resistance can give some indication of the degree of curing of a coating. It is often desirable however, to use spectropscopic methods in order to make a more quantitative determination. Free films of the radiation cured oligomer were analyzed by infrared transmittance. The spectra were compared to the unreacted starting material, I, (Figure 3). The vinylether has C=C absorption bands at 1635 and 1616 cm^{-1}(8). While the 1616 cm^{-1} band overlaps with bands arising the from the aromatic ring structure, the 1635 cm^{-1} band occurs in an unobstructed region of the spectrum. After UV irradiation at 100 mj/cm^2 or electron beam irradiation at 4 Mrads, the vinyl ether band at 1635 cm^{-1} was no longer detectable (Figure 3). These results indicate that radiation curing leads to a high conversion of the starting materials. The results do not necessarily indicate that all of the vinyl ether groups have polymerized. Other reactions such as hydrolysis or chain termination may occur.

A mixture of the products resulting from the reaction of acetylene with a diol may be used to synthesize a vinyl ether terminated oligomer (Scheme I). A VEU oligomer was prepared to illustrate this process. The concentrations of the reagents were adjusted so that equivalent amounts of hydroxy and isocyanate groups were reacted. Thus a mixture containing triethylene glycol divinyl

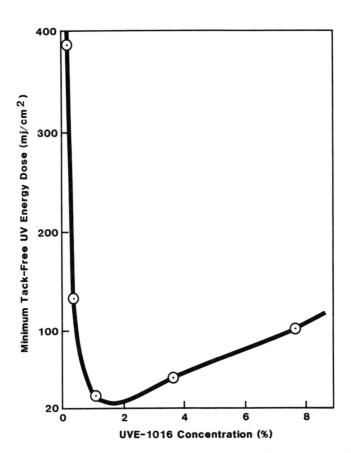

Figure 1. The effect of sulfonium salt concentration on the UV cure speed of oligomer I.

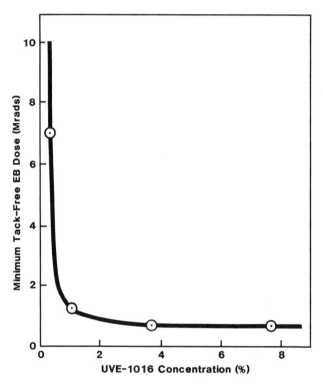

Figure 2. The effect of sulfonium salt concentration on the EB cure speed of oligomer I.

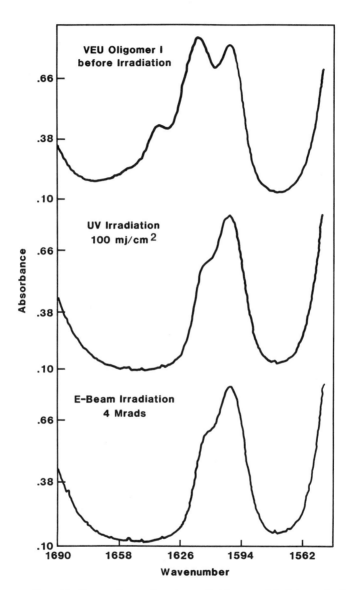

Figure 3. Infrared analysis of the radiation induced curing of oligomer I.

ether (40.0 g, 0.2 mol), triethylene glycol monovinyl ether (14.1g, 0.08 mol), and triethylene glycol (6.0 g, 0.04 mol) was reacted with diphenylmethane diisocyanate (20.0 g, 0.08 mol). The resulting VEU oligomer (II) was a clear thick liquid. GPC analysis showed a mixture of products with a range of molecular weights (Mn = 462, Mw = 1,464, including a contribution from the divinyl ether monomer, 202 g/mol). The oligomer (II) was combined with a sulfonium salt catalyst and cured by UV and electron beam irradiation. The results are shown in Table 3.

Table 3. Radiation Curing of VEU Oligomer II[a]

Min. Energy Dose for Tack-Free Cure	MEK Rubs	Pencil Hardness	Reverse Impact (in-lbs)	Elongation (%)
UV <40 mj/cm^2	>100[b]	3H[b]	>160[b]	>50[b]
EB 0.7 Mrads	>100[c]	3H[c]	>160[c]	>50[c]

a. 2 pph UVE-1016, nitrogen atmosphere
b. After UV curing at 75 mj/cm^2
c. After EB curing at 4.0 Mrads

The oligomer II was chain extended with triethylene glycol groups which would tend to produce softer coatings with a lower crosslink density. However, the mixture also contained 50% of the TEGDVE monomer which was incorporated during the oligomer synthesis. When the mixture was irradiated the divinyl ether monomer copolymerized with oligomeric components increasing the crosslink density of the cured film. The net results were that the coatings had similar properties to the coating produced from the monodisperse oligomer I. One important difference between oligomers I and II was that I gave coatings which were highly discolored while II produced coatings that were nearly colorless. This observation was likely due to the fact that oligomer I contained a much higher concentration of aromatic urethane groups relative to oligomer II. These groups appear to be at least partially responsible for coloration problems.

Multifunctional VEUs were prepared by incororating trimethylol propane (TMP) into the oligomer compositions. The oligomer (III) was the product of the reaction of TEGMVE and MDI with trimethylol propane (TMP) in a 3:3:1 mole ratio. The product was prepared using methylene chloride as a solvent which was later removed at reduced pressure. It was also prepared using TEGDVE as a solvent which remained in the mixture as a reactive diluent. In the absence of solvent the product was a colorless, low melting, amorphous solid. The simplest structure that would result from a combination of TEGMVE, MDI and TMP is shown below. GPC analysis indicated that the product (III) was actually a

$$CH_3CH_2C-[CH_2-O-\overset{O}{\underset{\|}{C}}-NH-\langle\bigcirc\rangle-CH_2-\langle\bigcirc\rangle-NH-\overset{O}{\underset{\|}{C}}-O-(CH_2CH_2O)_3-CH=CH_2]_3$$

III

mixture of several different oligomers (Figure 4). The molecular weight distribution and the viscosity were controlled by adjusting the stoichiometry of the reactants during the synthesis of the oligomer (Table 4). A 1:1 mole ratio of hydroxy groups to isocyanate groups was maintained in the three examples shown in Table 4. As expected, the viscosity and molecular weights of the products increased as the proportion of TMP in the mixture was increased.

Table 4. Effect of Reagent Stoichiometry on VEU Oligomer III

Mole Ratio TEGMVE:MDI:TMP	Viscosity* (cps)	Molecular Weight* Mn	Mw
3 : 3 : 1	1,130	1,273	3,888
3.5 : 4 : 1.5	3,140	1,472	4,861
4 : 5 : 2	8,380	1,577	8,160

* Including contribution from 50% added TEGDVE monomer (202 g/mol).

The viscosity of the oligomer (III) was also adjusted by varying the proportion of TEGDVE in the mixture (Figure 5). TEGDVE is an effective reactive diluent for the oligomer (III). Viscosities spanning more than three orders of magnitude were obtained by varying the proportion of TEGDVE from 20% to 80%. The mixtures of the VEU oligomer III and TEGDVE were combined with a triarylsulfonium salt (UVE-1016, 2 pph) and coated onto the substrate. The properties of the cured coatings are shown in Table 5.

Table 5. Radiation Curing of VEU Oligomer III[a]

TEGDVE (%)	Min Tack-Free Dose UV (mj/cm^2)	EB (Mrads)	Reverse Impact[b] (in-lbs)	Elongation[b] (%)
20	<40	0.7	>160	>50
40	<40	0.7	>160	>50
50	<40	0.7	>160	>50
70	<40	0.5	120	>50
80	<40	0.5	80	30
100	<40	0.3	60	6

a. 2 pph UVE-1016, nitrogen atmosphere.
b. After UV curing at 75 mj/cm^2

As the proportion of TEGDVE in the formulation was increased above 50% a decrease in the reverse impact and elongation properties were observed. This was expected due to the increase in crosslink density of the cured film. All of the mixtures gave coatings which showed good adhesion to untreated polyethylene substrate. Adhesion to the treated steel panels was relatively poor, but showed some improvement at higher monomer concentrations. Coatings produced from mixtures containing at least 50% TEGDVE monomer showed very little discoloration. Overall, the combination of the multifunctional VEU

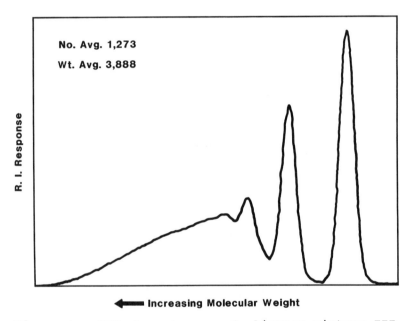

Figure 4. GPC chromatogram of oligomer mixture III.

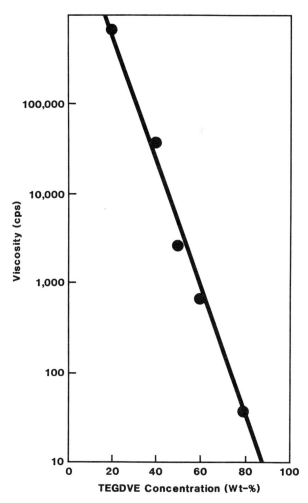

Figure 5. The effect of monomer on the viscosity of oligomer mixture III.

oligomer III and TEGDVE gave a desirable set of physical properties (Table 5). These properties compared very favorably with coatings produced by typical acrylated urethane formulations(22).

Cationic polymerization reactions are not inhibited by oxygen. This results in an enhancement of the cure speeds of vinyl ethers compared to acrylate based materials. A mixture of the VEU oligomer III and TEGDVE (50%) was cured side-by-side with some typical acrylate formulations. The samples were exposed to UV irradiation in a dry air atmosphere. The maximum conveyor speed that would produce a tack-free coating was recorded. The results are shown in Figure 6. There was a dramatic difference between the vinyl ether and acrylate based formulations. The vinyl ether reached a tack-free state at the maximum conveyor speed (500 ft/min., 40 mj/cm2) in the presence of less than 1% added photoinitiator. When the acrylate based formulations were cured at high speeds they typically formed a firm coating, but showed incomplete curing at the surface. This was due to oxygen inhibition of the free radical polymerization reaction. When the coatings were compared in a nitrogen atmosphere there were very few observable differences between the vinyl ether and acrylate based coatings. All of the formulations containing at least 1% photoinitiator, gave tack-free coatings at 500 ft/min (40 mj/cm2) in a nitrogen atmosphere. Oxygen inhibition of the surface cure in the acrylate based coatings was reduced slightly by increasing the photoinitiator concentration. Benzophenone/amine initiated acrylate systems are known to show reduced oxygen inhibition(23). This is clearly seen in Figure 6 at initiator concentrations of 2 to 3%. The cure speed obtained in the benzophenone/amine initiated system was improved; however, it still does not approach the cure speed obtained with the vinyl ether based coating.

Although oxygen does not appear to affect UV induced cationic polymerization, it does have an affect on electron beam induced cationic curing. We have previously suggested that the mechanism for the activation of the onium salts is different for UV and EB irradiation. The EB reaction mechanism involves an intermediate radical species which may be quenched by oxygen(8). The EB curing of the VEU oligomer III (containing 50% TEGDVE monomer and 2 pph UVE-1016 sulfonium salt) was compared in a dry nitrogen and a dry air atmosphere. A tack-free coating was obtained at 0.7 Mrads in nitrogen while a 1.8 Mrad dose was required in dry air. While this represents a substantial decrease in cure speed it is still well within acceptable limits of most commercial curing operations.

It is well known that cationic polymerization may be inhibited by nucleophilic species in the reaction mixture; however, vinyl ethers appear to be less sensitive than epoxies to presence of nucleophiles. The high reactivity of the vinyl ether group with the growing cationic chain is often favored over termination by a nucleophilic species. Water may act as a nucleophile and it has been reported that high atmospheric humidity conditions will inhibit the photocuring of epoxies(24). We have observed a similar effect on vinyl ether based coatings.

Table 6 shows the cure speed for the VEU oligomer III (with 50% TEGDVE) at various humidity levels. No differences were observed in the cured films after irradiation in dry nitrogen or in air at up to 34% relative humidity. At 61% relative humidity the cure speed was

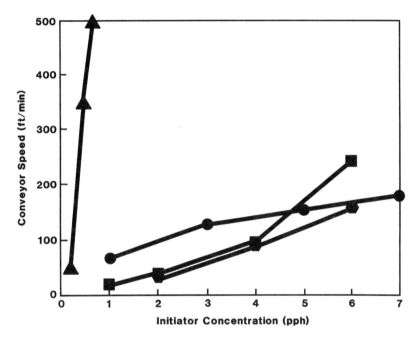

Figure 6. Maximum tack-free cure speeds. Samples UV cured (one 200 watt/in lamp) in dry air.
▲ = 50% oligomer III, 50% TEGDVE, UVE-1016
⬣ = 50% CMD 6700, 50% EO-TMPTA, Irgacure 184
■ = 50% Celrad 3700, 50% EO-TMPTA, Irgacure 184
● = 50% 3700, 50% EO-TMPTA, 2 pph MDEA, benzophenone

reduced to less than half of its original value. At 78% relative humidity no curing appeared to occur even at the lowest conveyor setting (50 ft/min., 350 mj/cm2). Although the curing was affected by atmospheric humidity, the formulation itself did not appear to be particularly sensitive to the addition of water. The VEU formulation used in Table 6 was spiked with water (5%). This mixture was easily cured at 500 ft/min (36 mj/cm2) in a nitrogen atmosphere to produce a coating which resisted greater than 100 double rubs with MEK.

Table 6. The Effect of Atmospheric Humidity on the UV Cure Speed of VEU Oligomer III [a]

Relative Humidity (%)	Max. Tack-Free Speed (ft/min)	MEK Rubs[b]
0 (dry N_2)	>500	>100
0 (dry air)	>500	>100
19	>500	>100
34	>500	>100
61	200	>100
69	150	>100
78	< 50	--

a. 50% TEGDVE, 2% UVE-1016, polyethylene substrate, one ~00 w/in lamp.
b. After curing at 100 ft/min.

It has been shown that the effect of humidity on cationic curing can be overcome by moderate increases in the temperature of the substrate (45°C). This increase in temperature is often achieved under commercial UV curing conditions due to the heat output of the lamps. This would tend to obscure the effects of high humidity(2͡5).

Conclusions

The combination of vinyl ether monomers and VEU oligomers along with onium salt catalysts provides a versatile new radiation curable coatings system. The coatings may be cured at high speeds using either UV or EB irradiation. The UV curing is not inhibited by oxygen although it can be affected by high atmospheric humidity.

Literature Cited

1. Leaf, D. A., Environmental Protection Agency, Presentation at Radcure 86, Baltimore, 1986.
2. Tu, R. S. In UV Curing: Science and Technology, Vol. II; Pappas, S. P., Ed.; Technology Marketing Corp.: Norwalk, Connecticut, 1985, p. 147.
3. Safe Handling and Use of UV/EB Curable Coatings, National Paint and Coatings Association: Washington, D.C., 1980.
4. Conning, D. M. Radcure 85: Technical Paper, Association for Finishing Processes: Basel, Switzerland, 1985; FC85-442.

5. Decker, C.; Bendaikha, T.; Fizet, M.; Faure, J. Radcure 85: Technical Paper, Association for Finishing Processes: Basel, Switzerland, 1985; FC85-432.
6. Decker, C.; Bendaikha, T. Europ. Polym. J. 1984, 20, 753.
7. For a review see: Crivello, J. V. Advances in Polymer Science, 1984, 62, 1.
8. Lapin, S. C. Radcure 86: Conference Proceedings, Association for Finishing Processes: Baltimore, Maryland, 1986; p. 15-15.
9. Watt, W. R. In UV Curing: Science and Technology, Vol. II, Pappas, S. P., Ed.; Technology Marketing Corp.: Norwalk, Connecticut, 1985; p. 247.
10. Koleske, J. V. Finishing Line, 2, (3), Association for Finishing Processes, 1986.
11. Gaube, H. G. Radcure 86: Conference Proceedings, Association for Finishing Processes: Baltimore, Maryland, 1986; p. 15-27.
12. Crivello, J. V.; Conlon, D. A.; Olsen, D. R.; Webb, K. K. Radcure Europe '87: Conference Prceedings, Association for Finishing Processes: Munich, 1987; p. 1-27.
13. Koleske, J. V. Radtech '88: Conference Proceedings, Radtech International: New Orleans, 1988; p. 353.
14. Crivello, J. V.; Lee, J. L.; Conlon, D. A. Radiation Curing VI: Conference Proceedings, Association for Finishing Processes: Chicago, 1982; p. 4-28.
15. Higashimura, T.; Aoshima, S.; Sawamoto, M. Makromol. Chem. Symp. 1986, 3, 99.
16. Dougherty, J. A.; Vara, F. J.; Anderson, L. R. Radcure 86: Conference Proceedings, Association for Finishing Processes: Baltimore, Maryland, 1986; p. 15-1.
17. GAF Corporation, Wayne, New Jersey.
18. Reppe, J. W. Acetylene Chemistry, I.G. Farbenindustrie A.A., Translated from German, Charles A. Meyer Co.: New York; 1949.
19. Gallucci, R. R.; Going, R. C. J. Org. Chem. 1983, 48, 342.
20. Lapin, S. C.; House, D. W. U.S. Patent 4,751,237, 1988.
21. Lapin, S. C. Radtech '88: Conference Proceedings, Radtech International: New Orleans, 1988; p. 395.
22. Christmas, B. K.; Zey, E. G. Radcure 86: Conference Proceedings, Association for Finishing Processes: Baltimore, Maryland, 1986; p. 14-53.
23. Brann, B. L.; Riggs, G. Radcure 86: Conference Proceedings, Association for Finishing Processes: Baltmore, Maryland, 1986; p. 4-57.
24. Watt, W. R. ACS Symposium Series: Epoxy Resin Chemistry, American Chemical Society: 1979, p. 18.
25. Hanrahan, B. D.; Manus, P; Eaton, R. F. Radtech '88: Conference Proceedings, Radtech International: New Orleans, 1988; p. 14.

RECEIVED September 13, 1989

Chapter 27

UV Cure of Epoxysiloxanes and Epoxysilicones

Richard P. Eckberg and Karen D. Riding

Silicone Products Division, General Electric Company, Waterford, NY 12188

Polydimethylsiloxanes, commonly referred to as silicones, are unique materials with a broad range of applications in industrial, consumer, and cosmetic products and processes. There are excellent opportunities for silicone suppliers to take advantage of growing market demand for radiation-curable silicone coatings used for release of pressure-sensitive adhesives, conformal coatings, and other diverse applications which combine unique silicone properties with low-temperature processing, high speed through-put, and the low energy and environmental impact of radiation processing.

Among the most successful of the new radiation-curable silicone coating systems are compositions incorporating cycloaliphatic epoxy-functional silicone polymers with compatible iodonium salt photocatalysts (1 - 5). The syntheses, structures, UV cure response, and other properties of these materials and some of their derivatives and monomeric homologs are discussed in the balance of this chapter. Their specific application for release coatings has been described in great detail elsewhere (5,6) and is not the subject of this paper.

A handy shorthand system of abbreviating silicone polymer structures has been developed by the industry and will be used herein. Chainstopper siloxy groups are designated "M" (monofunctional). Superscripts refer to organic functionality other than methyl, i.e.

$M = (CH_3)_3SiO_{1/2}$

$M^H = (CH_3)_2(H)SiO_{1/2}$

$M^E = $ O⟨cyclohexyl⟩CH$_2$CH$_2$Si(CH$_3)_2$O$_{1/2}$

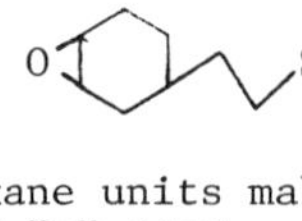

Siloxane units making up linear polysiloxane molecules are designated "D" (difunctional), and are superscripted in the same fashion as above; subscripts refer to chainlength of linear polymers.

$$D = \quad -\underset{\underset{CH_3}{|}}{\overset{\overset{CH_3}{|}}{Si}}O- \qquad D^H = \quad -\underset{\underset{CH_3}{|}}{\overset{\overset{H}{|}}{Si}}O- \qquad D^E = \quad -\overset{\overset{CH_3}{|}}{Si}O-\text{(cyclohexene epoxide)}$$

(Silicon in silicone polymers is always considered 4-coordinate).

<u>EPOXYSILICONES</u>

Platinum-catalysed reaction of olefin epoxides to SiH compounds, or hydrosilation, is an elegant and convenient means of sythesizing epoxy-functional silicones and siloxanes (<u>7</u>). The basic reaction is presented as equation 1:

$$\equiv SiH \; + \; \text{CH}_2=\text{CH-R-epoxide} \; \longrightarrow \; \equiv Si\text{-CH}_2\text{-CH}_2\text{-R-epoxide} \qquad (1)$$

Two olefin epoxides, allylglycidylether and 4-vinylcyclohexene-oxide (VCHO), are commercially available to the extent that large scale production of their siloxane derivatives is feasible:

$$\equiv SiH \; + \; \text{allyl glycidyl ether} \; \longrightarrow \; \equiv Si\text{-}\ldots \qquad (2)$$

$$\equiv SiH \; + \; \text{VCHO} \; \longrightarrow \; \equiv Si\text{-}\ldots \qquad (3)$$

Since extremely fast-curing UV-sensitive silicone materials are desired by the release coating industry, epoxysilicones derived from VCHO were chosen for development, since cationic UV cure of cycloalipliatic epoxies is known to be much faster than that of analogous glycidylethers (<u>8</u>). Two classes of cycloaliphatic epoxy containing materials have emerged from early research efforts: epoxysilicone polymers and epoxysiloxane monomers (or oligomers).

Epoxysilicone polymers suitable for radiation-cure release coating applications have to satisfy an important requirement in order to be useful: they must be coatable by three roll offset gravure (or other multiroll technique) coaters designed to provide complete coverage of substrate at low coat weights (typically 1 gram/meter^2 or less) without benefit of solvent. Since viscosity of such materials is constrained within 200-2000 centistoke (cstk), with 200-500 cstk preferred, the molecular weight of the silicone used must be tightly controlled, as well, and is normally between 5000-10,000 Daltons, or about 60 to 150 dimethylsiloxy (D) units.

Two methods to produce such polymers have been established: (a) hydrosilation addition of VCHO to an SiH-functional polymer pre-formed (by means of acid equilibration) to the desired molecular weight and SiH content and (b) basic equilibration of alkoxysilane-VCHO adducts with other siloxanes (<u>9</u>).

Since the trimethoxysilane

$$(MeO)_3 \; Si\text{-CH}_2\text{-CH}_2\text{-(cyclohexene epoxide)}$$

is the only such silane commercially available (as A-186, trademark of Union Carbide Co.), the former approach has proven to be the most versatile, and most readily translated to large scale production. Typical linear epoxysilicone polymers manufactured in this manner are represented below:

$$M^E D^E_5 D_{95} M^E \qquad\qquad MD^E_{10} D_{90} M$$

Polymer I   Polymer II

These epoxysilicones are similar in reactivity, cured film properties, and release characteristics when ultraviolet light - cured with iodonium photocatalysts (5). The cured release coatings are derived from unfilled low molecular weight dimethylsilicones, rich in highly reactive cycloaliphatic epoxy groups. The coatings are therefore brittle, highly crosslinked, easily abraded weak materials whose physical properties are unimportant so long as the coating is non-adherent to PSA's but anchors to substrates. Despite this poor property profile, the cationic UV-cure response of epoxysilicones is exceptionally fast compared to available organic epoxy resins. We therefore felt that cycloaliphatic epoxysiloxane monomers and oligomers might combine rapid cure response with improved physical properties permitting their use in applications other than paper release. Before discussion of results of that study, silicone-compatible iodonium salts and means of enhancing their solubility in epoxysilicone polymers will be described.

IODONIUM CATALYSTS FOR EPOXYSILICONES

Initial attempts to demonstrate UV-curability of epoxysilicones were unsuccessful because crystalline, high melting unsubstituted diphenyl iodonium salts and triphenylsulfonium salts were completely immiscible in silicone matrices without the use of solvents. Dimethyl silicones are non-polar materials; modifying the 'onium salts with nonpolar substituents to render them compatible with epoxysilicones proved to be the best approach to formulation of a workable system (1,4).

Two iodonium photocatalysts have been developed which are useful for UV-cure of epoxysilicones and epoxysiloxane monomers. The first to be described in the patent literature (1) is bis(dodecylphenyl) iodonium hexafluoroantimonate:

Catalyst A $[C_{12}H_{25}\text{-}\langle O \rangle\text{-}]_2 I^+ \; SbF_6^-$

This material is produced via synthesis derived from early work by Beringer et. al (10,11). The key to solubility of the ionic material lies in the nature of the dodecylbenzene used. Linear alkylate (detergent alkylate) dodecylbenzene is a mixture of at least two dozen isomers of several distinct compositions normally ranging from C_8H_{17}Ph to $C_{14}H_{29}$Ph. The bis (dodecylphenyl) iodonium salt derived from this mixture therefore includes over 400 separate compounds, so that the catalyst behaves like a supercooled fluid due to the freezing point depression phenomenon, and can therefore be dispersed in relatively nonpolar epoxysilicone media. (This catalyst remains immiscible in non-functional dimethylsilicones, however).

The nature of catalyst A, while rendering it compatible with reactive epoxysilicones, also makes it difficult to isolate and purify.

A crystalline, easily purified iodonium catalyst suitable for use in silicone media was subsequently developed, based in part on Koser et. al (12). (4-Octyloxyphenyl) phenyliodonium hexafluoroantimonate

$$\text{Catalyst B} \quad C_8H_{17}O-\langle O \rangle-\overset{+}{I}-\langle O \rangle \quad SbF_6^-$$

was found to be a low-melting solid capable of dissolution in low-polarity organic media and in some silicone fluids, although it is less soluble in epoxysilicones such as Polymer I or II than Catalyst A.

Both catalysts are deep UV-absorbers like other iodonium photoinitiators (8). The UV absorption spectrum of a .0001M solution of each species in methylene chloride is depicted in Figure 1. A bathochromic shift to higher wavelength absorption coupled with lower molecular weight results in Catalyst B being a more efficient photocatalyst than Catalyst A in those systems where both iodonium salts are freely miscible, as will be described in greater detail below.

The deep UV absorption characteristics of these catalysts requires deep UV lamp sources for optimum cure. All of the cure results presented herein were therefore obtained with an RPC Model 1202 QC Lab UV Processor equipped with 2 Hanovia medium pressure mercury vapor ultraviolet sources each capable of independent operation at 100, 200 or 300 watt/inch nominal power output and a conveyor which operates between 10 and 500 fpm. Actual power flux can therefore be widely varied per pass through the Processor as a function of total lamp output and conveyor speed. We have found that the microwave-fired Fusion Systems 'H' lamp is also well-suited for these iodonium/epoxysiloxane systems.

EPOXYSILICONE POLYMER MODIFICATIONS

Even though iodonium salts can be designed for compatibility in low polarity media, they are still ionic salts soluble only in certain organofunctional silicones. For example, while Catalyst A is miscible in Polymer II, it is only partially soluble in Polymer I; Catalyst B is immiscible in either silicone fluid. We therefore sought means by which reactive epoxysilicone fluids may be modified to permit dissolution of high concentrations of iodonium salts.

The polysiloxane backbone of an epoxysilicone polymer is undisturbed by many chemical processes, suggesting that reactions can be targeted to take place solely at reactive organofunctional groups attached to the silicone polymer, such as cycloaliphatic epoxy groups derived from VCHO. Oxiranes are readily esterified by carboxylic acids; this reaction is catalysed by tertiary amines. Neither species affects a dimethylsilicone. The reaction has been described by Coqueret et. al (13), and can be depicted as equation 4:

$$\equiv Si-R\overset{O}{\triangle} \;+\; HOOCR^1 \;\longrightarrow\; \equiv SiR\overset{OH}{\underset{}{\diagup}}CH_2OOCR^1 \qquad (4)$$

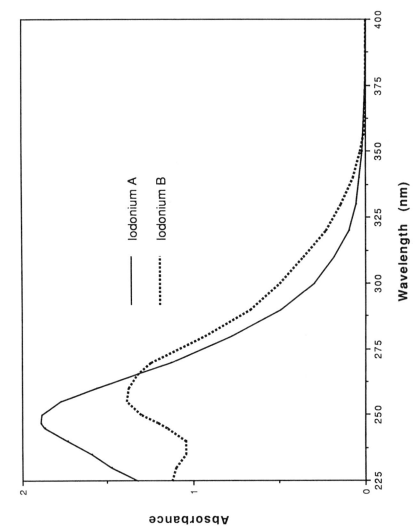

Figure 1. Iodonium salt ultraviolet absorbance spectra, 0.0001 Molar solutions in CH_2Cl_2.

where R = $-(CH_2)_3OCH_2^-$ and R^1 = CH_3. Analogous reactions have been taught in the patent literature, where epoxy-functional silicones are converted to hydroxyacrylate-functional polymers capable of EB- or UV-cure (14). Neither of the polymers described in these references, however, are capable of cationic-type UV cure in the presence of iodonium salt photocatalysts, because all oxirane groups present were converted into hydroxy-or di-esters. We therefore sought to determine if an epoxysilicone can be partially esterified to a more polar material without loss of so many oxirane groups that fast UV-cure response is lost.

A low viscosity hydride polymer of approximate structure $MD_3{}^HD_{20}M$ was selected as a good model structure on which to carry out this study. The first step was to react the hydride fluid with VCHO, creating a final product $MD_3{}^ED_{20}M$. A property comparison of these two polymers is depicted in Table I.

Table I. Comparison of Analogous Silicone Hydride and Silicone Epoxy Fluids

	$MD_3{}^HD_{20}M$	$MD_3{}^ED_{20}M$
$\overline{MW}$	1822	2194
25 Viscosity, cstk	18	70
$N_D{}^{25}$	1.4030	1.4245
EEW*	0	731

* Epoxy equivalent weight

The increase in viscosity and refractive index accompanying conversion of silicone hydrides to silicone epoxies is good indication of an increase in polarity. $MD_3{}^HD_{20}M$ is immiscible with Catalysts A and B; $MD_3{}^ED_{20}M$ readily accepts A but not B.

$MD_3{}^ED_{20}M$ was then reacted with acetic acid or benzoic acid according to the following method.

<u>Hydroxyester Synthesis:</u> $MD_3{}^ED_{20}M$ silicone polymer was dispersed in an equal weight of toluene. 0.1 wt % triethylamine was added, and varying amounts of benzoic acid or acetic acid were added, followed by reflux until titration with phenol red indicator showed that >95% of the carboxylic acid had been consumed. Toluene, triethylamine, and unreacted carboxylic acid were removed via vacuum strip to 150 C. Selected properties of polymer products are found in Table II.

Infrared and proton NMR spectra of all polymer reaction products were consistent with predicted hydroxyester structures. Viscosity build was about the same for analogous benzoic and acetic acid reaction products, suggesting that intermolecular hydrogen bonding caused by -hydroxl groups was responsible for the phenomenon. Observed changes in refractive index are consistent with observed refractive indices of phenyl-containing silicones.

Table II. $MD_3^E D_{20}M$ esterification products

Polymer	Carboxylic Acid	Mole % Oxirane Reacted	Polymer Viscosity, cstk	Polymer N_D^{25}	Catalyst Miscibility
3a	Acetic	50	970	1.4270	A only
3b	Acetic	20	220	1.4258	A only
3c	Benzoic	50	1000	1.4378	A and B
3d	Benzoic	20	250	1.4310	A and B

UV CURE OF MODIFIED EPOXYSILICONES

'Cure' of a liquid coating, whether by radiation or thermal means, is unavoidably subjective. For purposes of these investigations, cure is defined as minimum UV flux required to convert a 2 mil liquid film coated on polyethylene kraft substrates to a smear- and tack-free coating. UV flux for a given pass under the UV lamps in our RPC Processor was measured by an International Light Photometer using a model A309 Lightbug Accessory. Minimum cure conditions were qualitatively determined by manipulating lamp power and conveyer speed until the combination of fastest line speed and lowest UV lamp power for cure was ascertained. 0.5 Wt% photocatalyst A, $(C_{12}H_{25}Ph)_2ISbF_6$, was blended with the epoxysilicone fluids 30 minutes prior to exposure. Results are presented in Table III. (All cures were conducted in ambient atmosphere).

Table III. Cure Studies; Modified Epoxysilicones

Polymer	EEW*	UV Flux for Cure, MJ/cm^2
$MD_3^E D_{20}M$ (Control)	731	56
3a	1460	84
3b	914	65
3c	1460	425
3d	914	65

*calculated

Cure rates were roughly comparable (and quite fast) for the control $MD_3^E D_{20}M$ polymer and its 20% hydroxyester benzoate and acetate derivatives. Some cure loss was observed where ½ of the oxirane groups were replaced with hydroxyacetate, but the tenfold decrease in UV cure rate observed when ½ available oxirane is converted to hydroxybenzoate ester was surprising and unexpected. An explanation becomes obvious, however, on examination of the structure of these polymers:

Control = $MD_3^E D_{20}M$

3a = $MD_{1.5}^E D_{1.5}^{HA} D_{20}M$

3c = $MD_{1.5}^E D_{1.5}^{HB} D_{20}M$

where E= [structure: cyclohexane with epoxide]

HA= [structure: cyclohexyl with acetate ester]

HB= [structure: cyclohexyl with benzoate ester]

The PhCO moiety is a deep UV absorber commonly encountered in conventional free radical-type photosensitizers such as benzophenone and acetophenone derivatives. The hydroxybenzoate ester groups' ultraviolet absorption spectrum is in the same UV light range as that of the iodonium salt catalyst, so polymers 3c and 3d act as UV screens, absorbing UV light otherwise available for initiating acid generation by iodonium photocatalysts which in turn effects crosslinking. The UV absorption spectra of control and 3c polymers are presented in Figure 2, which clearly demonstrates this effect.

An interesting offshoot of the hydroxyester-modified epoxysilicone study was the synthesis of a self-sensitized acrylated epoxysilicone produced by treatment of $MD_3^E D_{20}M$ with equimolar quantities of benzoic and acrylic acids, resulting in a viscous fluid polymer incorporating hydroxybenzoate and hydroxyacrylate ester functions. 2 mil coatings of this composition were UV-cured to smear- and tack-free coatings when irradiated with 170 mJ/cm^2 UV flux in nitrogen without any added photoinitiator or sensitizer.

DIEPOXYSILOXANE MONOMERS AND OLIGOMERS

As we noted earlier in this chapter, UV cured films of the epoxysilicones described thus far are of limited utility because of poor tensile, elongation, and abrasion resistance properties. An intensive study of epoxysiloxane monomers and oligomers was carried out to determine if such materials, when UV cured with iodonium photocatalysts, can combine rapid UV cure response with physical properties similar to those reported for analogous organic epoxy compositions.

We prepared several different diepoxysiloxanes and diepoxysilanes, identities and selected properties of which are presented in Table IV.

The diepoxydisiloxane and polydimethylsiloxane homologs tabulated above may be designated $M^E M^E$, $M^E D_4 M^E$, $M^E D_{10} M^E$, and $M^E D_{16} M^E$, respectively, in accordance with the 'silicone shorthand' nomenclature. $M^E M^E$ was prepared by a previously known method (15); others by equilibration of $M^E M^E$ with octamethylcyclotetrasiloxane using tetramethyl ammonium hydroxide catalyst (9) or by platinum-

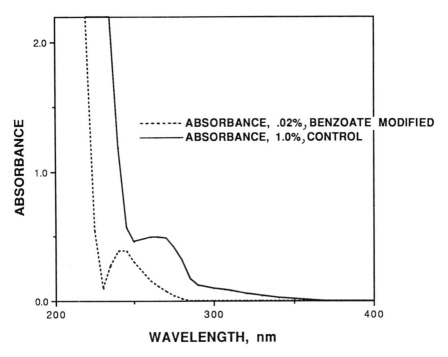

Figure 2. Epoxysilicone ultraviolet absorbance spectra (wt% concentration in CH_2Cl_2).

Table IV. Diepoxymonomers

$$\text{O}\diagdown\text{-CH}_2\text{-Si-R-Si-CH}_2\diagdown\text{O}$$

R	N_D^{25}	EEW*	Appearance
-O-	1.4726	191	Mobile, clear fluid, 70 cstk viscosity
-O(Si(CH$_3$)$_2$O)$_2$-	1.4475	273	Mobile fluid, 37.5 cstk
-O-(Si(CH$_3$)$_2$O)$_8$-	1.4312	495	Mobile fluid, 33.5 cstk
-O-(Si(CH$_3$)$_2$O)$_{14}$-	1.4235	717	Mobile fluid, 50 cstk
-CH$_2$CH$_2$-	------	197	Light brown solid, mp 41-43°C
-⌬-	1.5206	221	Clear, viscous fluid, @ 1250 cstk visc
-⌬-O-⌬-	1.5358	267	Clear, very viscous fluid, 10000 cstk

* Calculated

catalysed reactions of VCHO with the appropriate SiH-functional precursors, vis:

$$\text{HSi(CH}_3\text{)}_2\text{RSi(CH}_3\text{)}_2\text{(H)} + \text{VCHO} \longrightarrow \text{diepoxy product}$$

Yields were quantitative in each case; infrared spectra were consistent with expected products.

A cure rate study (as detailed previously) was carried out to assess UV cure response of the diepoxysiloxane model compounds, and to compare them with cure response of analogous diepoxy organic results. The UV flux required for cure of 2 mil coatings of these materials using both iodonium catalysts is displayed in Table V.

Two trends are readily discernable: the cycloaliphatic-terminated linear siloxane monomers and polymers are faster UV cure materials (with iodonium photo-catalysts) than the organic diepoxides tested. The siloxane-based Si-epoxy compounds also display faster cure response than those with non-siloxane bridges between silicon. The assymetric $(C_8H_{17}O-Ph)PhI^+$ salt is a more efficient catalyst than the $(C_{12}H_{25}Ph)_2I^+$ salt, particularly in the organic systems studied, except in $M^E D_x M^E$, x > 10, where it is not as miscible.

The remarkable cure speed of $M^E M^E$ suggested that it might serve as a cure accelerator for relatively sluggish, albeit widely used organic diepoxy analogs. $M^E M^E$ proved incompatible with Epon 825 BPA diglycidylether, but is freely miscible in all proportions with CY179. Cure response as a function of $M^E M^E$/CY179 ratio (o.5 wt. % iodonium salt B) is graphically depicted in Figure 3. As little as 20 wt % $M^E M^E$ doubles cure rate compared to 100% CY179.

Bulk UV cure of 20 mil sections of these diepoxides produced brittle, fragile films which fractured when attempts were made to determine tensile properties. DSC analyses of cured films determined approximate glass transition temperatures for these UV homopolymers, which are noted in Table VI.

Table V. Diepoxy cure; UV flux, mJ/cm^2

Compound	0.5% $(C_{12}H_{25}Ph)_2ISbF_6$	0.5% $(4-C_8H_{17}OPh)PhISbF_6$
$M^E M^E$	58	15
$M^E D_4 M^E$	78	29
$M^E D_{10} M^E$	80	67
$M^E D_{16} M^E$	80	105
E-SiMe$_2$-⟨O⟩-SiMe$_2$-E 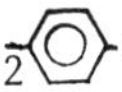	380	126
E-SiMe$_2$CH$_2$CH$_2$SiMe$_2$-E	90	60
E-SiMe$_2$-⟨O⟩-O-⟨O⟩-SiMe$_2$-E 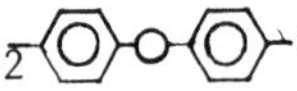	800	396
Limonenedioxide	160	65
CY 179 (Ciba-Geigy)	2000	370
Epon 825 (Shell)	4000	1460

where E = [structure]
Limonenedioxide = [structure]
CY179 = [structure]
Epon 825 = [structure]

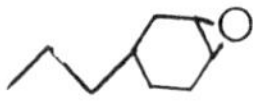

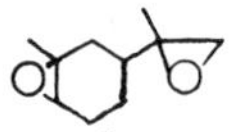

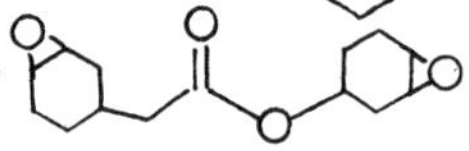

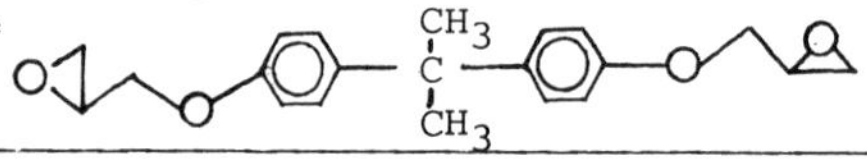

The high Tg observed for $M^E M^E$ and its analogs might reflect thermal cure taking place during the DSC measurements themselves; further analysis is required to resolve possible ambiguities. However, even relatively low molecular weight $M^E D_x M^E$ type polysiloxane oligomers' cured film properties resemble those of conventional cured dimethylsilicones, which retain their properties at very low temperature.

EPOXYSILOXANES AND POLYOLS

Alcohols, phenols, and polyols react with epoxies under cationic UV-cure conditions:

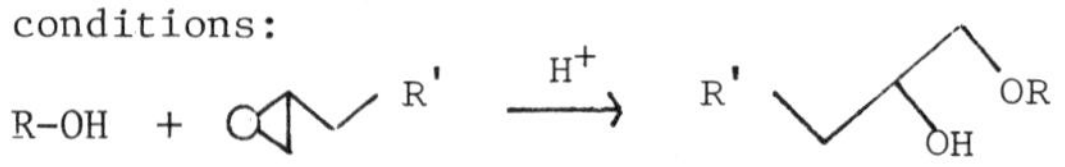

R-OH + [epoxide]-R' →(H$^+$) R'-[CH(OH)-CH$_2$-OR]

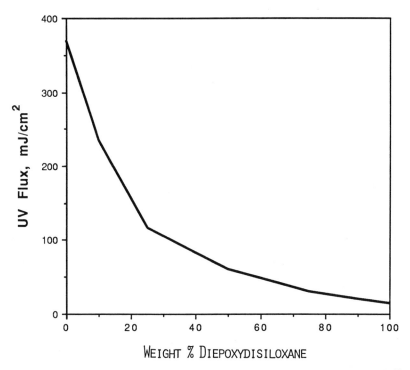

Figure 3. Ultraviolet flux required to cure 2 mil CY179/ $M^E M^E$ blends.

Table VI. Approximate Tg of UV-crosslinked Films *

Monomer	Tg, C
$M^E M^E$	165-195 (broad transition)
$M^E D_4 M^E$	30-70 " "
$M^E D_{10} M^E$	@-40 " "
$M^E D_{16} M^E$	@-100
E-SiMe$_2$-⌬-SiMe$_2$-E	200-215
E-SiMe$_2$CH$_2$CH$_2$SiMe$_2$-E	190-195
E-SiMe$_2$-⌬-O-⌬-SiMe$_2$-E	180-190
CY179	143
Epon 825	132

* 0.5 wt % $(C_{12}H_{25}Ph)_2 ISbF_6$, 2J/cm^2 UV flux

Numerous published reports describe how polyol or alcohol additives improve elongation, impact resistance, and general toughness of brittle, inflexible UV-cured cycloaliphatic epoxy resins (16,17). We endeavored to determine if film properties of these brittle UV-cured epoxysiloxanes can be improved by hydroxy-functional additives.

Initial studies were limited to the system $M^E D_x M^E$/2-ethyl-1,3-hexanediol (EHDO) which proved miscible when x = 0, 4, and 10, with 0.5 wt % iodonium catalyst B. We investigated the effect of EHDO on epoxysiloxane cure response, with results graphically depicted in Figure 4.

Polyols act as chain transfer agents which assist UV cure of cationic epoxy systems. Our results with EHDO and $M^E D_x M^E$ did not illustrate this effect, perhaps because all components of these blends are difunctional. We repeated this experiment by determining UV cure response of $M^E M^E$/1,2,6-trihydroxyhexane mixture using the same iodonium catalyst and concentration. These results are listed in Table VII.

Table VII. UV Cure Response, $M^E M^E$/Triol System

OH/Epoxy Mole Ratio	UV Flux for Cure, MJ/cm^2
1.00	212
0.75	70
0.5	32
0.25	30
0	15

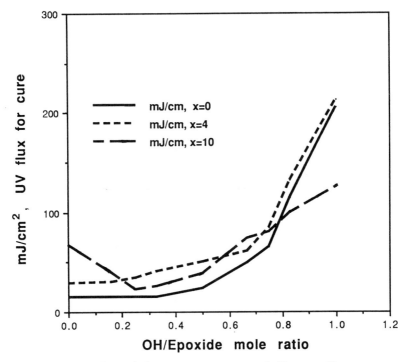

Figure 4. Ultraviolet cure response of diepoxysiloxanes, E-(SiMe$_2$O)$_x$-E, with 2-ethyl-1,3-hexanediol

Solubility of 1,2,6-trihydroxyhexane in $M^E M^E$ was limited, but this triol formed clear blends with $M^E M^E$ and octanol. An 8:1:1 (wt:wt:wt) mix of $M^E M^E$:triol:octanol cured to a tough, transparent 2 mil film on irradiation with 70 MJ/cm^2 UV energy in the presence of 0.5% iodonium salt B, suggesting that blends of various alcohols and polyols may be used with epoxysiloxanes or epoxysilicones to overcome solubility constraints.

Monofunctional alcohols may also be cured with epoxysiloxanes, although they act as chain terminators and slow UV cure as OH/oxirane ratios approach 1. This effect is illustrated for the $M^E M^E$/octanol system in Table VIII below:

Table VIII. UV Cure Response, $M^E M^E$/N-octanol

OH/oxirane mole ratio	UV flux for Cure, MJ/cm^2
1.0	No cure at 5000 MJ/cm^2
0.75	2700
0.5	140
0.25	70
0	15

Physical property measurement of thick section-UV cured epoxysiloxane/alcohol blends were difficult to obtain because many of the cured films (like unmodified epoxysilicones) proved too weak and brittle to test. Selected results are depicted in Table IX for 20 mil films UV cured with 0.5 wt% $(C_{12}H_{25}Ph)_2 ISbF_6$ by exposure to 2 J/cm^2 total UV flux.

Table IX. Cured Film Properties

Mole Ratio, OH/Oxirane	Peak Tensile, psi	Elongation at Break, %
EHDO/$M^E M^E$		
1.0	(very low)	100
0.8	111	62
0.6	643	29
0.4	(very high)	5
Octanol/$M^E M^E$		
0.5	17	14
0.4	36.5	22.5
0.33	130	26
0.25	159	23
EHDO/$M^E D_4 M^E$		
1.0	9	20
0.83	18	19
0.66	17	14.5
0.5	28	10.5
EHDO/$M^E D_{10} M^E$		
1.0	5	16
0.75	4	9
0.5	28	18

While most of these films are clearly low-modulus materials, they possess properties useful for such radiation-cure coating applications as conformal coatings, electronic encapsulent gels, or fiber optic coatings, particularly where such applications demand silicones' broad operational temperature range.

SUMMARY

Epoxysilicones ranging from low-molecular weight siloxane monomers to high molecular weight polymers have been developed for a wide range of radiation-curable coating applications. Modifications of iodonium photocatalysts and of epoxy-functional dimethyl silicones make the catalysts and resins mutually compatible, thus rendering these systems commercially feasible. Di-cycloaliphatic epoxy-terminated siloxane monomers and oligomers display extremely fast UV cure response and can be used to accelerate cure of organic epoxy resins. Physical properties of UV cured epoxysiloxanes are readily modified by co-cure with alcohols and polyols.

REFERENCES

1. R.P. Eckberg and R.W. LaRochelle, U.S. 4,279,717 (7/21/81).
2. R.P. Eckberg and R.W. LaRochelle, U.S. 4,421,904 (12/20/83).
3. R.P. Eckberg, U.S. 4,547,431 (10/15/85).
4. R.P. Eckberg, Radcure '84 Conference Proceedings, Atlanta, Georgia, pg. 2-1.
5. R.P. Eckberg, Radtech '88 North America Conference Proceedings, New Orleans, pp. 576-586.
6. Chemical & Engineering News, Nov 28, pg 32 (1988).
7. R.D. Mendecino, U.S. Pat. 4,046,930 (4/11/78).
8. J.V. Crivello, UV Curing: Science & Technology, (Ed. S. Peter Pappas), pp 24-77, Technology Marketing Corporation, (1978).
9. I. Yilgor and J.E. McGrath, Adv. Polymer Science, $\underline{86}$, 1-86 (1988) and references therein.
10. F.M. Beringer, M. Drexler, E.M. Gindler, and C.C. Lumpkin, J. Amer. Chem. Soc., $\underline{73}$, 2705 (1953).
11. F.M. Beringer, R.A Falk, M. Karmal, J. Lillien, G. Masullo, M. Mausner, and E. Sommer, J. Amer. Chem. Soc., $\underline{81}$,342 (1959).
12. G.F. Koser, R.H. Wettack, and C.S. Smith, J. Org. Chem., $\underline{45}$, 1543 (1980).
13. X. Coqueret, A. Lablache-Combier, and C. Loucheux, Eur. Polym. J., $\underline{24}$, 713-718 (1988).
14. R.P. Eckberg, U.S. 4,558,082 (12/16/85).
15. E.P. Plueddemann and G. Fanger, J. Amer. Chem. Soc., $\underline{81}$, 2632 (1959).
16. G.H. Smith, U.S. 4,256,828 (3/17/81).
17. J.V. Koleski, Radtech '88 North American Conference Proceedings, New Orleans, pp 353-371.

RECEIVED September 13, 1989

Chapter 28

UV Cure of Epoxy–Silicone Monomers

J. V. Crivello[1] and J. L. Lee

Corporate Research and Development Center, General Electric Company, Schenectady, NY 12301

Epoxy-functional silicone monomers are a new class of versatile monomers which are particularly attractive in their application to UV cationic curing. These monomers can be readily prepared by the platinum catalyzed hydrosilylation of Si-H containing compounds with epoxy compounds bearing vinyl groups. Novel epoxy monomers containing cyclic siloxane rings were prepared as well as multifunctional epoxy monomers with star and branched structures. Those monomers containing cyclohexylepoxy groups are characterized by their high rates of cationic photopolymerization. In addition, excellent cured film properties are obtained which make the new monomers attractive for potential applications in coatings.

As a consequence of their high cure and application speeds, essentially pollution-free operation, very low energy requirements and generally excellent properties, coatings prepared by photopolymerization techniques (UV curing) have made a substantial impact on the wood coating, metal decorating and printing industries. Early developments in this field centered about the photoinduced free radical polymerization of di and multifunctional acrylates and unsaturated polyesters. Still today, these materials remain the workhorses of this industry. While the bulk of the current research effort continues to be directed toward photoinduced free radical polymerizations, it is well recognized that ionic photopolymerizations also hold considerable promise in many application areas. Photoiniduced cationic polymerizations are particularly attractive because of the wealth of different chemical Pand physical properties which can potentially be realized through the polymerization of a wide variety of different vinyl as well as heterocyclic monomers. Further, photoinitiated cationic polymerizations have the advantage that they are not inhibited by

---

[1]Current address: Department of Chemistry, Rensselaer Polytechnic Institute, Troy, NY 12180–3590

oxygen and thus, may be carried out in air without the need for blanketing with an inert atmosphere to achieve rapid and complete polymerization (1).

Photoinitiators

The origin of our interest in photoinitiated cationic polymerization began with the discovery that certain onium salts, namely, diaryliodonium (**I**) and triarylsulfonium (**II**) salts, could rapidly and efficiently photoinitiate the polymerization of virtually all types of cationically polymerizable monomers (2-4).

$$Ar-I^+-Ar \qquad Ar-\underset{\underset{Ar}{|}}{S^+}-Ar$$
$$X^- \qquad\qquad X^-$$
$$\textbf{I} \qquad\qquad \textbf{II}$$

Where $X^- = BF_4^-, PF_6^-, AsF_6^-, SbF_6^-$

The photolysis of the above compounds results in the production of strong Brønsted acids which initiate cationic polymerization by direct protonation of the appropriate monomers. Over the past few years, we have successfully prepared a wide variety of different onium salts and have modified their structures for the purposes of tailoring their spectral absorption characteristics, enhancing their photoefficiency and changing their solubility. The ability of these compounds to be photosensitized at wavelengths both within the UV and visible regions of the spectrum adds a further dimension to the potential utility of these photoinitiators (5). Due to the above mentioned factors as well as to their commercialization by several companies, onium salts **I** and **II** are the most widely employed cationic photoinitiators in use today.

The Synthesis of Di, Tri and Tetrafunctional Epoxy-Silicone Monomers

As mentioned previously, the photoinitiated polymerization of almost any cationically polymerizable monomer can be carried out using onium salt photoinitiators **I** and **II**. However, among the most advantageous substrates for UV cationic polymerization are epoxide-containing monomers. The major reasons for this are as follows. Epoxide-based coatings are widely used in industry today and are noted for their outstanding chemical resistance and mechanical properties. Further, monomers containing the epoxide group are readily UV polymerized using onium salt photoinitiators (6). Accordingly, recent work in these laboratories has been directed to the preparation of new epoxy-containing monomers designed specifically for UV curing applications.

Silicon-containing epoxies with hydrolytically stable carbon-silicon bonds were first prepared by Pleuddeman by the addition of hydrogen functional silanes to epoxy compounds containing double bonds (7,8). We have employed this reaction extensively to prepare several different difunctional epoxy monomers as shown in Table I. An example of this reaction is given in equation 1 for the preparation of difunctional monomer **III**.

Table I

Characteristics of Silicon-Containing Epoxy Monomers

Compound	Notation	EEW*	Compound	Notation	EEW*
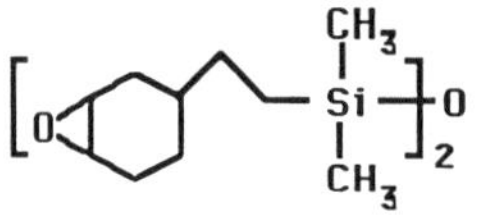 [epoxycyclohexyl-CH₂CH₂-Si(CH₃)₂-O]₂	III	191	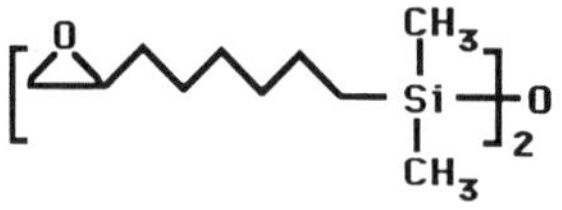 [epoxide-(CH₂)₄-Si(CH₃)₂-O]₂	VII	193
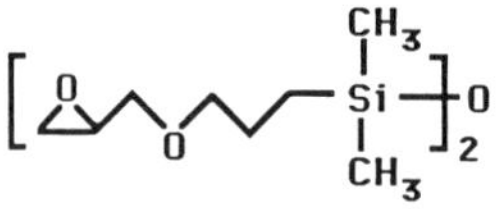 [glycidyloxy-CH₂CH₂-Si(CH₃)₂-O]₂	IV	181	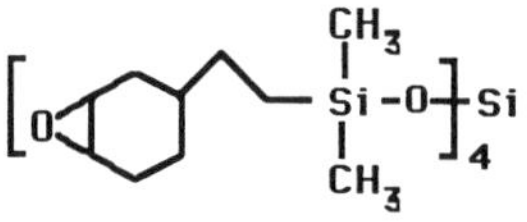 [epoxycyclohexyl-CH₂CH₂-Si(CH₃)₂-O-Si(CH₃)₂]₄	VIII	205
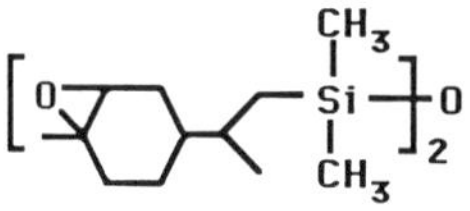 [epoxycyclohexyl-C(CH₃)₂-Si(CH₃)₂-O]₂	V	219	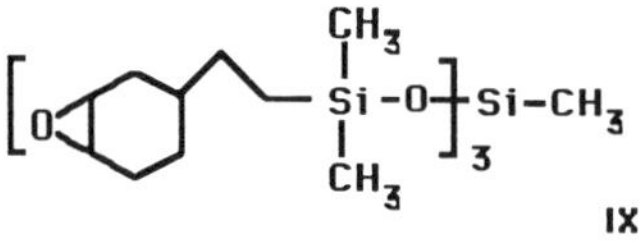		
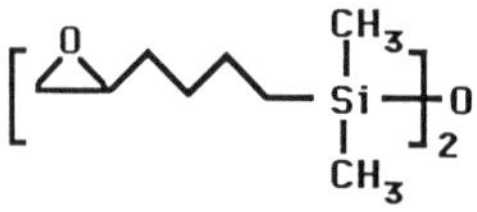 [epoxide-(CH₂)₃-Si(CH₃)₂-O]₂	VI	165	[epoxycyclohexyl-CH₂CH₂-Si(CH₃)₂-O-Si(CH₃)₂]₃-CH₃	IX	213

*Epoxy equivalent weight

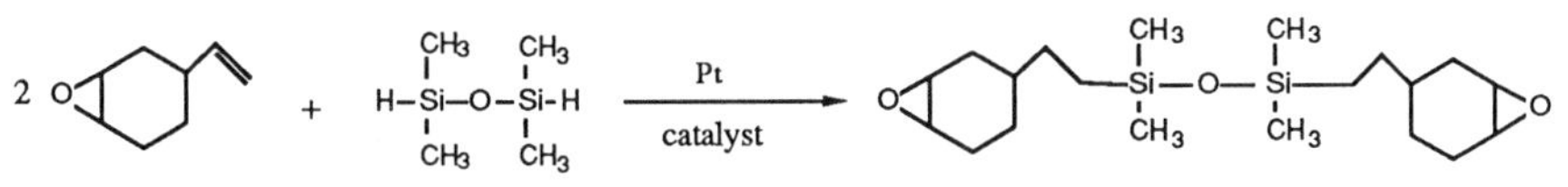

[Diagram: eq. 1 showing reaction of 2 vinylcyclohexene oxide + H-Si(CH₃)₂-O-Si(CH₃)₂-H with Pt catalyst yielding compound III]

eq. 1

The reactions proceed cleanly and quantitatively to give the desired epoxy functional silicones.

An interesting branched tetrafunctional epoxy-silicone monomer, **VIII**, can be readily prepared as shown in the following equation by the platinum catalyzed condensation of the tetrafunctional SI-H compound, tetrakis(dimethylsiloxy)silane, with 3-vinyl-7-bicyclo[4.1.0]heptane.

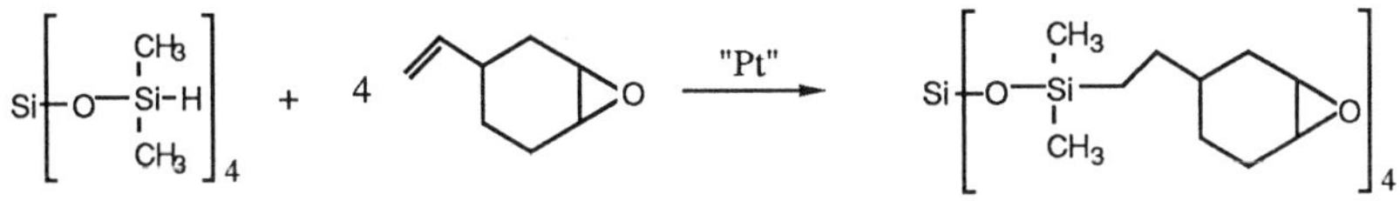

[Diagram: eq. 2 showing Si[-O-Si(CH₃)₂-H]₄ + 4 vinyl epoxide with "Pt" catalyst yielding compound VIII]

VIII eq. 2

In an analogous fashion, starting with methyltris(dimethylsiloxy)silane, the corresponding trifunctional epoxy monomer, **IX**, was prepared in quantitative yield. Similarly, a wide variety of complex resins containing Si-H groups and quaternary silicon are available within the silicones industry and can be applied to this chemistry.

The Preparation of Novel Cyclic Epoxy-Functional Siloxanes

The prospect of preparing compounds containing both epoxide rings and siloxane rings appeared to present some interesting possibilities for the synthesis of novel monomers with unusual properties. Starting with the commercially available 1,3,5,7-tetramethylcyclotetrasiloxane, it was possible to carry out a fourfold hydrosilylation reaction with various vinyl containing epoxides provided that the reaction was carried out under nitrogen and rigorously dry conditions. Equation 3 shows an example of this reaction.

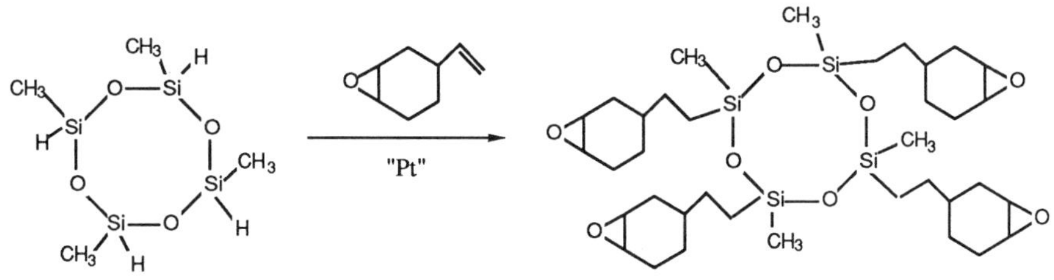

[Diagram: eq. 3 showing 1,3,5,7-tetramethylcyclotetrasiloxane + vinyl epoxide with "Pt" catalyst yielding compound X]

X eq. 3

Tetrafunctional cyclic epoxy-silicone monomer, **X**, was obtained as a mixture of stereo and regio isomers.

Using the synthetic route depicted in equation 4, the trifunctional cyclic epoxy-silicone monomer, **XIII**, shown was prepared.

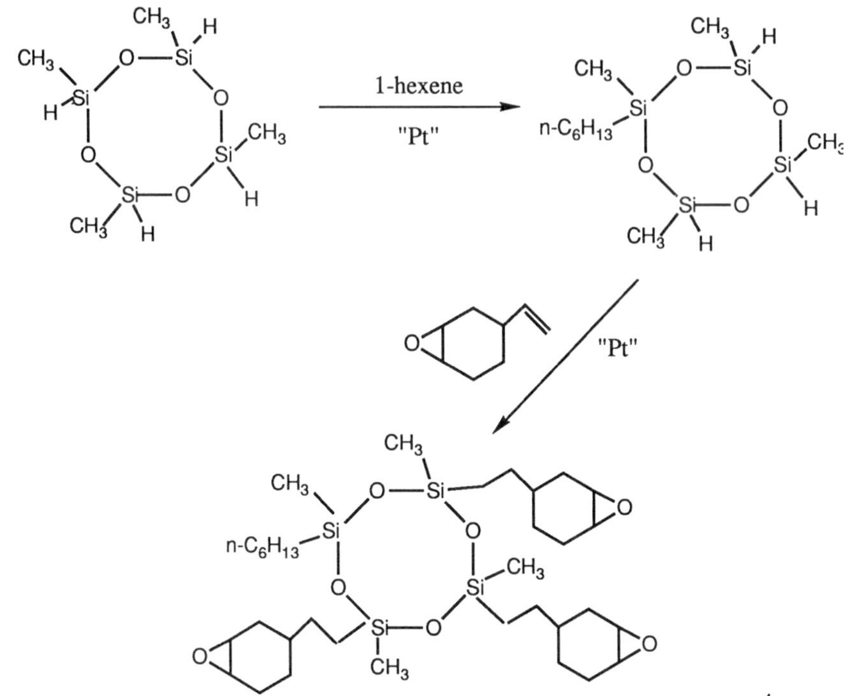

[Reaction scheme showing tetramethylcyclotetrasiloxane (with Si–H groups) reacting with 1-hexene in presence of "Pt" to give partially hexyl-substituted cyclic siloxane, then with 4-vinyl-1-cyclohexene epoxide and "Pt" catalyst to yield structure XIII.]

eq. 4

XIII

Poly(dimethylsiloxane) and poly(methylhydrogensiloxane) can be equilibrated in the presence of strong acids, such as trifluoromethanesulfonic acid to give cyclic componds. This is depicted in equation 5.

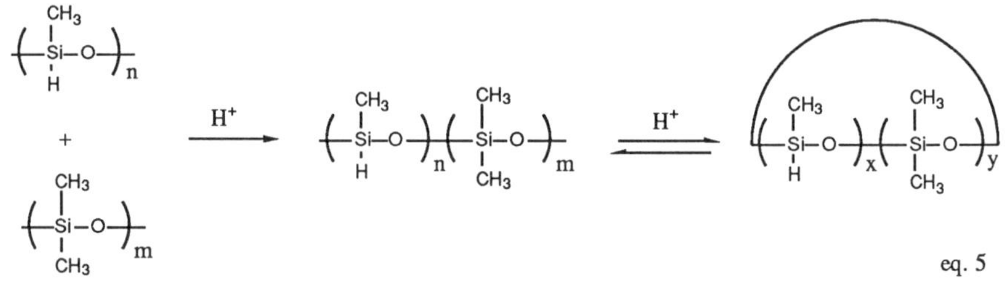

[Equation 5: poly(methylhydrogensiloxane) $(-Si(CH_3)(H)-O-)_n$ + poly(dimethylsiloxane) $(-Si(CH_3)_2-O-)_m$ →(H⁺) block copolymer $(-Si(CH_3)(H)-O-)_n(-Si(CH_3)_2-O-)_m$ ⇌(H⁺) cyclic product $(-Si(CH_3)(H)-O-)_x(-Si(CH_3)_2-O-)_y$]

eq. 5

Depending on the ratios of the two polymers used, one can produce equilibrium mixtures in which there are present as the major cyclic components six, eight and ten membered rings having one to five hydrogens attached per ring. These mixtures may be fractionated to give specific desired cyclic compound. However, in the usual case, an isomeric mixture of compounds of any given ring size will be obtained. For example, the above method was used for the synthesis of a cyclic difunctional epoxy-silicone monomer having an eight membered siloxane ring. This monomer actually consists of the two regio isomers shown below together with a number of related stereoisomers.

Shown in Table II are the structures of four novel cyclic epoxy-silicones which were prepared during the course of this work.

The Preparation of a,w-Epoxy-Functional Poly(dimethylsiloxane) Oligomers

A series of well characterized a,w-hydrogen difunctional polydimethylsiloxane oligomers were prepared as shown in equation 6 by the cationic ring opening polymerization of 2,2,4,4,6,6,8,8-octamethylcyclotetrasiloxane (D_4) in the presence of tetramethyldisiloxane as a chain stopper (10).

eq. 6

The platinum catalyzed condensation of the a,w-hydrogen difunctional polydimethylsiloxane oligomers with 3-vinyl-7-oxabicyclo[4.1.0]heptane proceeds smoothly and quantitatively. Under the above conditions, a,w-epoxy-functional polydimethylsiloxanes with n = 17, 41, 59 and 111 were prepared as colorless and odorless mobile oils.

DSC Characterization of Epoxy-Siloxane Monomers

To obtain qualitative and quantitative data concerning the reactivity of epoxy-siloxane monomers we employed differential scanning photocalorimetry (3,11). This is a general method for obtaining both qualitative and quantitative information on photopolymerizations. Specifically, the height of the exothermic peak gives a qualitative indication of the reactivity of the monomer, while the time from the opening of the shutter to the maximum of the exothermic peak wich relates to the time required to reach the maximum polymerization rate, gives a quantitative measure of the

Table II

Cyclic Silicon-Containing Epoxy Monomers

Compound	Notation	EEW*	Compound	Notation	EEW*
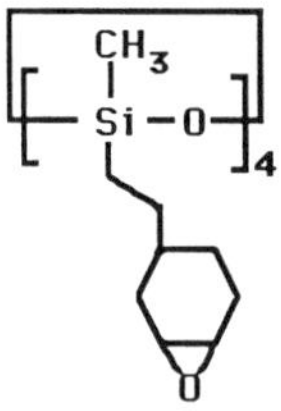 [structure: cyclohexene oxide-ethyl-Si(CH$_3$)-O]$_4$	X	184	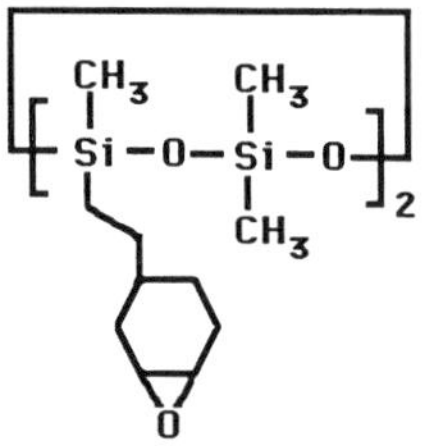 [structure: cyclohexene oxide-ethyl-Si(CH$_3$)-O-Si(CH$_3$)$_2$-O]$_2$	XII	247
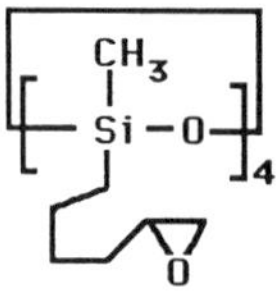 [structure: glycidyl-Si(CH$_3$)-O]$_4$	XI	158	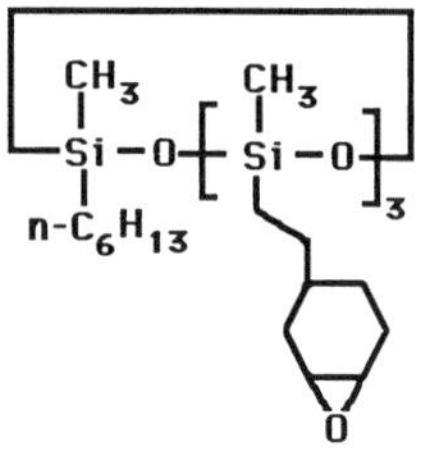 [structure: Si(CH$_3$)(n-C$_6$H$_{13}$)-O-Si(CH$_3$)(cyclohexene oxide-ethyl)-O]$_3$	XIII	232

*Epoxy equivalent weight

reactivity. Thirdly, the area under the exothermic peak gives a direct measure of the overall enthalpy of the polymerization and hence the conversion.

Figure 1 shows a composite of the differential photocalorimetry curves of several of the difunctional silicon-containing epoxy monomers given in Table I. Clearly, the most reactive of these monomers is III. The bisglycidyl ether IV is the least reactive, while monomer VI and monomer VII which is not shown in the figure are intermediate in their reactivity. This order of reactivity is similar to that which we have noted in an earlier publication for carbon based monomers, (3) i.e., those monomers containing cycloaliphatic moieties are more reactive than monomers containing glycidyl ether-type functional groups. Monomer V, also containing cycloaliphatic epoxy groups, is comparable in its reactivity to monomer III.

A comparison between difunctional monomer III and cyclic tetrafunctional monomer X is given in Figure 2. While both monomers are very reactive, some differences in their photocalorimetry curves can be discerned. The polymerization of III is slightly faster than that of X and is essentially complete within 3 minutes. The exceptional high reactivity of III and X were further confirmed by determining their tack-free times. When a 0.25 mole percent photoinitiator {(4-octyloxyphenyl)phenyliodonium SbF_6^-}(i.e. 0.25 moles photoinitiator/100 mol monomer) in the above two monomers was spread as 1 mil films, tack-free times of 500 ft/min were obtained using a single 300W medium pressure mercury arc lamp.

The differential photocalorimetric curves of four epoxy end-group functional poly(dimethylsiloxane) oligomers are given in Figure 3. It is interesting to note that, compared to monomer III (n =0), the longer chain compounds show a similar profile of their reactivities in cationic polymerization which are independent of their chain length. As one progresses from n = 0 to n = 111 in this series, the crosslinked polymers change from very hard and brittle in the case where n = 0, to soft and flexible (n = 17-59) and finally to elastomeric (n = 111).

Film Properties of Photopolymerized Epoxy-Silicone Monomers

Some preliminary properties of photocured films of several of the epoxy-silicone monomers described in this paper are shown in Table III. Excellent properties are obtained for these materials even at short irradiation doses. Most noteworthy are the very high glass transition temperatures which were obtained for the crosslinked polymers after an irradiation time of 5 seconds. High gel contents are noted in all cases for these materials after a 1 second irradiation. The hardness of the cured resins appears to be dependent on the degree of functionality (epoxy equivalent weight) of the respective epoxy-silicone monomer, with the highest hardness obtained for the cyclic tetrafunctional epoxy monomer X. In general, the new monomers exhibit excellent solvent resistance as measured by the number of methyl ethyl ketone double rubs. When cured, the monomers give clear, glossy smooth films which show a surprising degree of flexibility. Lastly, Figure 4 shows the thermogravimetric analysis curves in nitrogen and air for the photocrosslinked polymer derived from monomer III. This polymer is stable to 250°C in air and 350°C in nitrogen.

Conclusions

Epoxy-containing silicone monomers are a novel class of monomers which are very attractive as substrates for photopolymerizable coatings, inks, adhesives as well as other applications. Among the advantages which may be cited for these new monomers possess are: 1) they are easily prepared by simple, straightforward techniques 2) show outstandingly high cure rates and 3) give high quality films with excellent physical and chemical properties. Moreover, these monomers are freely miscible with other epoxy monomers and when added in modest amounts, substantially increase the rates of cationic photopolymerization which those epoxy monomers undergo. Such monomers also may be thermally cured using a wide variety of

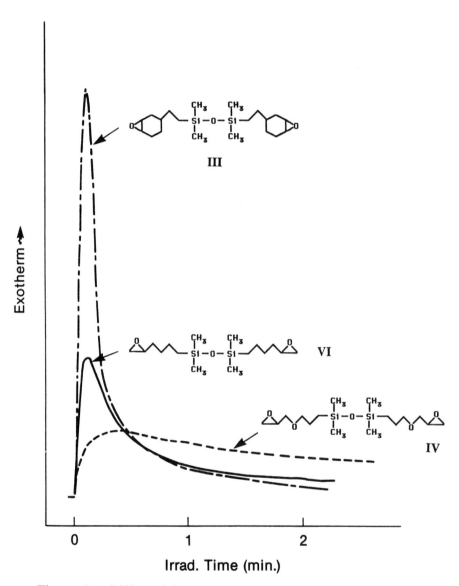

Figure 1. Differential scanning photocalorimeter UV cure response for various difunctional epoxy-silicone monomers using 0.5 mole % (4-octyloxyphenyl)phenyliodonium hexafluoroantimonate as photoinitiator.

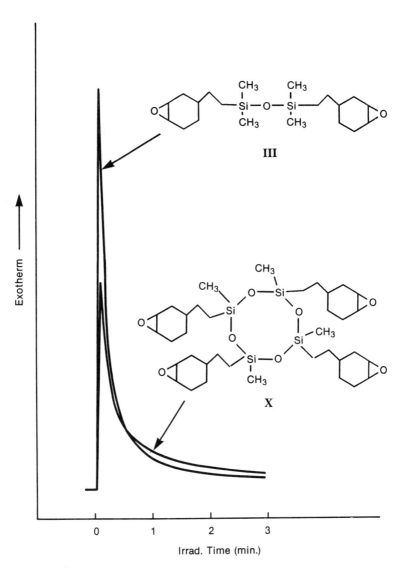

Figure 2. Differential scanning photocalorimeter curves for difunctional epoxy-silicone monomer III compared with tetrafunctional monomer X cured with 0.5 mole % (4-octyloxyphenyl)phenyliodonium hexafluoroantimonate as photoinitiator.

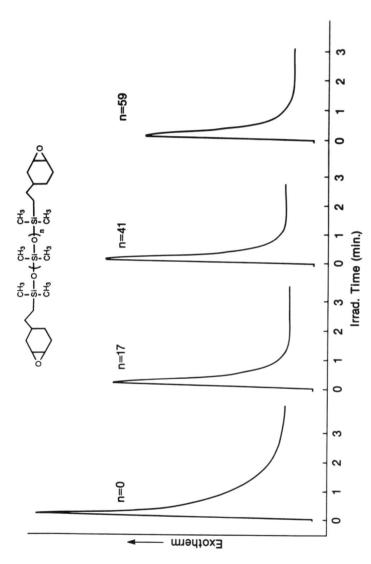

Figure 3. Comparison of the differential scanning photocalorimeter UV cure response of epoxy-terminated silicone oligomers with different chain lengths. The oligomers were cured using 0.5 mole % (4-octyloxyphenyl) phenyliodonium hexafluoroantimonate as photoinitiator.

Table III
Film Properties of UV Cured[†] Epoxy-Silicone Monomers

Compound	T_g [*] 5 sec. hν	Gel Fraction[†] 5 sec. hν (1 sec. hν)	P. Hardness 5 sec. hν (1 sec. hν)	Solv. Resist.[•] 5 sec. hν (1 sec. hν)
$\left[O\text{-}\bigcirc\text{-}CH_2\text{-}Si(CH_3)_2\text{-}O \right]_2$	187	99.4 (96.8)	H (H)	>750 (>750)
$\left[Si(CH_3)_2\text{-}O\text{-}Si(CH_3)(CH_2\text{-epoxycyclohexyl})\text{-}O \right]_2$	207	99.9 (93.1)	2H (H)	>750 (260-300)
$\left[Si(CH_3)(CH_2\text{-epoxycyclohexyl})\text{-}O \right]_4$	190	100 (99.8)	3-4H (2H)	>750 (>750)

[‡] 6 mil films cured with a GE H3T7 medium pressure mercury arc lamp using 0.25 mole % (4-octyloxyphenyl)phenyliodonium SbF_6^-. [*] Measured at 20°C/min. [†] Extracted with acetone. [•] Methyl ethyl ketone double rubs.

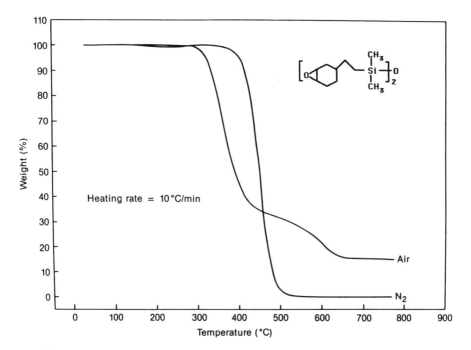

Figure 4. Thermogravimetric analysis curves in N_2 and air for monomer III polymerized by 5 seconds irradiation using 0.5 mole % (4-octyloxyphenyl)phenyliodonium hexafluoroantimonate as photoinitiator.

conventional epoxy curing agents. Ongoing studies are currently under way to explore these latter aspects of epoxy-functional silicone monomers.

Literature Cited

1. Sitek, F. Soc. Mfg. Eng. Radcure Europe '87 Conference Technical Paper FC 87-274 1987.
2. Crivello, J.V.; Lam, J.H.W. J. Polym. Sci., Polym. Symp. No. 56 1976, 383.
3. Crivello, J.V.; Lam, J.H.W.; Volante, C.N. J. Rad. Curing 1977, 4(3), 2.
4. Crivello, J.V.; Lam, J.H.W. J. Polym. Sci., Polym. Chem. Ed. 1979, 17(4), 2.
5. Crivello, J.V. Adv. in Polym. Sci. 1984, 62, 1.
6. Crivello, J.V.; Lam, J.H.W. In "ACS Symp. Ser. 114;" Bauer, R.S., Ed.; Am. Chem. Soc.: Washington, 1978, 1
7. Pleuddemann, E.P. Chem. Eng. Data, 1960, 5(1), 59.
8. Pleuddemann, E.P.; Fanger, G. J. Am. Chem. Soc. 1959, 81, 2632.
9. McGrath, J.E.; Yilgor, I. Adv. in Polym. Sci. 1988, 86,1.
10. Crivello, J.V.; Conlon, D.A.; Lee, J.L. J. Polym. Sci., Part A 1986, 24, 1197.
11. Moore, J.E. In "UV Curing Science and Technology;" Pappas, S.P., Ed.; Technology Marketing Corp.: Stamford, 1978, 1.

RECEIVED September 13, 1989

Chapter 29

UV-Induced Polymerization of Highly Filled Epoxy Resins in Microelectronics

H. Bayer and B. Lehner

Corporate Research and Development, Siemens AG, Otto-Hahn-Ring 6, D-8000 Munich 83, Federal Republic of Germany

High-performance epoxy resins for microelectronics can be UV-cured to a thickness of several millimeters by using sulfoniumsalt photochemistry. The matrix is built up by cationic crosslinking of cresolnovolak and cycloaliphatic epoxies. The Tg-values, as high as 150°C, are slightly affected by addition of á silicone based toughening agent. High filler contents in an at room temperature still processable liquid resin are achieved by combining particles of different size. Thus improvements concerning reaction shrinkage and coefficient of thermal expansion, as low as 20 ppm/K, are accomplished. CTE-values are recorded as displacement during dynamic mechanical testing. Also by DMTA information about cure state, glass transition and moduli is obtained. During the UV cure process the transmitted light is analyzed and the cure state at the bottom of the resin layer is monitored by measuring resistance and capacitance.
The resins may be used for potting and sealing in electronics, e.g. for globe topping in bare chip and wire technology. Therefore special efforts have been made to improve adhesion in thermal cycling.

Common high performance epoxy resins are cured at elevated temperatures or/and over a considerable period of time. Sufficiently reactive single component systems suffer from limited storage stability. Two-component systems have the disadvantages of necessitating mixing and degassing especially when highly filled with mineral fillers; the pot life of the mixture is limited. UV curing of epoxy resins (1-3) offers stable one component systems that cure reasonably fast and without additional heating so that there is the chance of inline compatibility with automatic production lines for large numbers of small pieces.

One example of such a highly automated process is bare chip and wire mounting on hybrid circuits where in many cases protection for chips and wire bonds is necessary (4). Protective coatings in the shape of globe tops may be soft-elastic down to lowest service temperatures and then preference would be given to silicone chemistry. Or they may be hard and of small thermal expansion up to highest service temperatures and then be a candidate for epoxy chemistry. In every case it is desirable to apply and cure drops up to a height of millimeters and to a diameter of about a centimeter within a few minutes. An additional advantage of UV-curing for globe top application is that drops retain their shape during curing process, because there is only a slight temperature increase before the resin gels.

UV curing in this application has been looked at sceptically (5), especially when the encapsulating drops consist of epoxy resins with a high content of mineral filler. We formulated some resins that show high glass transition temperatures and low coefficients of thermal expansion and can be cured by UV light to a thickness of several millimeters. Improved adhesion during thermal cycling has been achieved by using coupling agents and optimizing irradiation process.

Experimental

Materials and resin formulations. Commercial resins from Ciba Geigy (Araldit CY 179, cycloaliphatic diepoxide), Dow Chemical (Quatrex 3310, cresolnovolak glycidylether), and Union Carbide (Cyracure UVR 6200, cycloaliphatic epoxy diluent) were used without purification. The photoinitiator (Cyracure UVI 6974, aromatic sulfonium hexafluoroantimonate) and the adhesion promoter (Silane A 187, glycidoxypropyl-trimethoxy-silane) were obtained from Union Carbide. The toughening agent (Albidur EP 5220, silicone elastomer dispersion in epoxy resin) was provided by Hanse-Chemie. Flexibilizing epoxy resin (Edenol D 82, epoxidised soya bean oil) was obtained from Henkel.

Fused silica powders were supplied by Quarzwerke Frechen (Silbond FW 61 EST, medium particle size 25 μm, Silbond FW 600 EST, 4 μm, and VP 810-10/1, 1 μm). All fillers have been treated with glycidoxypropyl-trimethoxy-silane by the supplier and were used as received.

The formulations described in this paper were prepared using a dispersion disk and degassed in vacuum. The basic formulation of the resin matrix consisted of 33 parts of cycloaliphatic diepoxid, 33 parts of cresolnovolac epoxide, 33 parts of cycloaliphatic epoxy diluent, 0.5 parts of silane and 0.5 parts of photoinitiator (resin A). In addition fused silica 150 parts (resin B), 186 parts (resin C), 233 parts (resin D) was added. In the toughened mixture (resin E) the cycloaliphatic epoxide of resin D was substituted by 41.7 parts of a dispersion of 20% silicone elastomer in epoxy resin. In the flexibilized modification (resin F) 33 parts of epoxidised soya bean oil were added to resin D.

Viscosity measurements were done on a plate and cone viscosimeter (Epprecht Instruments) at 60 rpm and 25 $^{\circ}$C. The values are to be taken with care because of considerable deviation from

Newtonian behaviour and troublesome influence of the larger filler particles.

UV Curing and light monitoring. The UV source was a commercial metal-doped medium pressure Hg lamp equipped with a parabolic reflector. The distance from source to substrate could be varied from 10 to 40 cm. The photometric measurements were performed with a combination of a Jobin Yvon H 20 grating monochromator and a Hamamatsu R 928 photomultiplier driven by a Spectradata SD 90 data acquisition and monochromator control system.

Dynamic-mechanical testing. Dynamic-mechanical testing was done on a Polymer Labs DMTA spectrometer using specimens of about $2 \times 1 \times 20$ mm^3 in tensile mode with low strain and low pre-tension applying a sinusoidal force of 1 Hz. The displacement of the free sample ends was recorded and used to calculate the coefficient of thermal expansion.

UV Curing of Filled Epoxy Resins

The resin matrix. Mixtures of aliphatic and aromatic epoxy resins as used in this work almost totally absorb light below 300 nm. So only light of longer wavelength can be responsible for throughcure in the mm range. The photoinitiator efficiently absorbs up to a wavelength of 340 nm. For considerations on energy of effective radiation only this range of the lamp output has to be taken into account. Obviously the main part of the emitted spectrum is screened in the upper layers or only serves to warm up resin and substrate.

The inorganic filler. Many inorganic minerals are transparent for UV light. This is especially true for silica modifications. In addition the refraction indices of matrix and filler are not very different. So it is hardly revolutionary that silica fillers do not interfere with UV curing.

Spectrometry with transmitted light. One important condition for a successful application of a UV curable resin is the knowledge of the thickest layer that can be efficiently throughcured. As has been mentioned this implies the knowledge of the transmission or optical density of the resin referred to the emission spectrum of the UV source. Also the absorption (and concentration) of the photoinitiator has to be considered in the way that no excessive screening of the effective light takes place in the upper layers. Some considerations in this direction have been made in radical processes (6). Concerning our application, a premature surface cure might be of influence on the later performance of the drops because of internal stress.
We used a direct approach by placing a photometer looking permanently at the light source that effected the UV curing. Thus we got the possibility to compare spectrum and intensity of the incident light with the light that passes through real samples. Figure 1 shows the spectrum of our light source without sample and Figure 2 shows the light that passed 1 mm thick layers of resin A

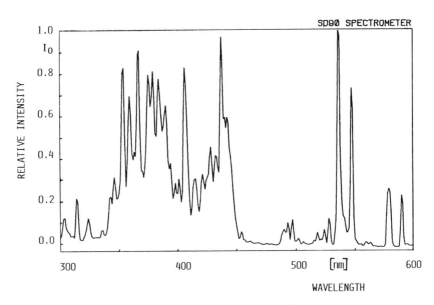

Figure 1. Spectrum of light source

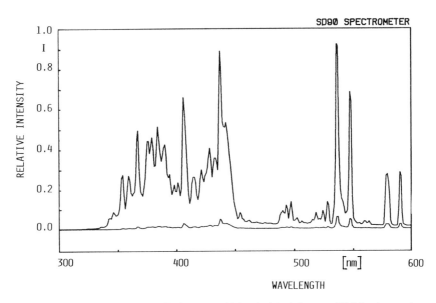

Figure 2. Spectra of transmitted light; unfilled resin (upper line), filled resin (lower line)

(unfilled) or resin C (filled). It is evident that part of the available light was absorbed by the resin matrix. The measured intensity of the light, that had passed through a 1 mm layer of filled resin and reached the monochromator slit is drastically lower when compared to the case of unfilled resin. A possible explanation is extensive light scattering; there was no longer a concentrated beam from the light source to the photometer. UV and visible light were attenuated to the same extent.
Figure 3 is a result of spectra arithmetics. Calculated is the optical density of resin A from measurements as described, with and without photoinitiator. We thus can distinguish between absorption of resin matrix and of photoinitiator.

Resistance and capacitance monitoring. A thick film metallic pattern on a ceramic substrate consisting of two combs separated by 100 µm was covered with a globe top. During irradiation capacitance and conductance were monitored with applied alternating current of 300 Hz and 3 kHz. As shown in Figure 4 capacitance and resistance reach a constant value after 3 min. respectively after 5 min.
 This simple test can be used to compare the cure speed of different globe tops and to determine the thickest possible layer to be throughcured.

Thermal postcure. DSC and DMTA measurements show that mere UV treatment is not sufficient in our case to achieve 100 % chemical conversion. There is a rest enthalpy of almost 10 % (7 J/g) and the temperature of the tan δ peak is 20 degrees below the final value. Prolonged storage at room temperature or short thermal postcure makes the resins show the final T_g of 140°C (Point of inflection in DSC), respectively 150°C (peak maximum of tan δ). A thermal postbake is often part of later process steps in manufacturing of hybrid circuits.

Thermomechanical properties of the filled resins

Modulus. Figure 5 presents dynamic-mechanical spectra of resins A and C. They show the anticipated influence of a reinforcing mineral filler to increase the modulus. The point of inflection in the modulus decrease during glass transition is shifted to slightly higher temperatures. The filler's influence on modulus of a thermoset resin is more pronounced at temperatures above Tg (7).

Coefficient of thermal expansion. One of our main concerns was to reduce the coefficient of thermal expansion and cure shrinkage. We therefore chose fused silica (CTE $\leq$ 0.5 ppm/K) as filler and optimized the particle size distribution to reach high contents in a mixture still flowing at room temperature.
 Several equations were established in literature (reviews are given in (8,9)) that predict lower CTEs than a simple equation of mixture when the continuous phase is of lower modulus and higher CTE than the dispersed phase. This is especially true if the dispersed phase does not consist of spheres but of rods, plates or fibers. When these are ordered, an additional complexity is to be expected because of the anisotropy of modulus and CTE (10). The

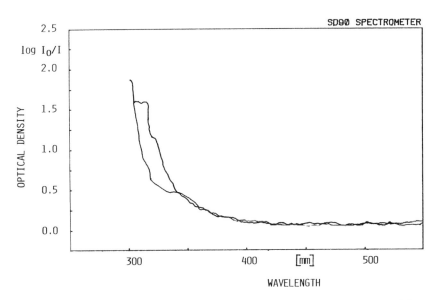

Figure 3. Optical density of resin A; with photoinitiator (upper line), without photoinitiator (lower line)

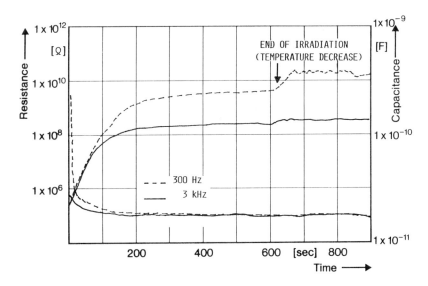

Figure 4. Resistance and capacitance during irradiation of resin D

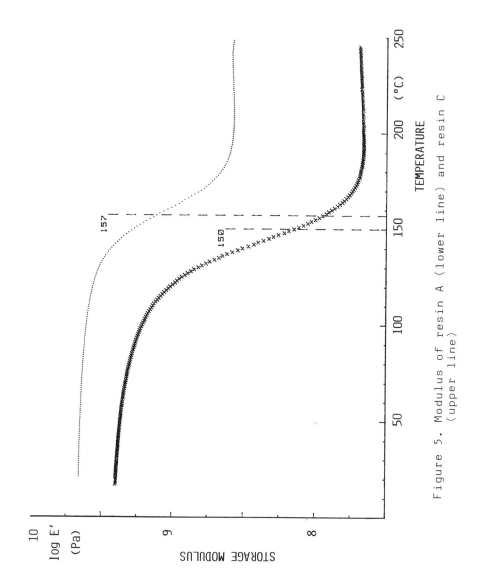

Figure 5. Modulus of resin A (lower line) and resin C (upper line)

fused silica powder that was used in our work consisted of splintered fragments and therefore the applicability of models is limited.

In our measurement of CTE the displacement of the sample's end was recorded during the dynamic mechanical temperature run. Figure 6 is the recording that corresponds to the run shown in Figure 5. Figure 7 shows the CTE curves that were calculated from these data. They are within the limits predicted by different theoretical (lower and upper bound) approaches. In Table I our CTE- values are listed. The rule of mixture would e.g. predict more than 30 ppm/K for resin C, we found 22 ppm/K.

With regard to particle size effects contradictory results have been published (11,12). Assuming that the filler particles change the properties of the resin in the immediate neighbourhood (13), an influence of particle size should be expected. A reduction of particle radius from 25 to 1 means a 1600 fold number of particles and a 300 fold surface area. Even if actually important aspects like agglomeration of filler particles and filler/matrix adhesion are neglected it seems almost impossible to deduce effective mechanical properties from constituent properties without knowing more about bondary layers.

Volume shrinkage. The volume shrinkage was determined by density increase. Density measurements were done with a pyknometer for liquid and a density scale for solid state. The volume shrinkage is 4.5% in case of the unfilled resin and decreases according to the decreasing volume portion of the reactive part of the mixture. The values are included in Table I.

Application and performance

Viscosity. As has been mentioned before the viscosity of a globe top resin should be suitable for dispenser applications. Viscosities of the investigated resins with increasing filler degree are listed in the last row of Table I. As filler we used a 80:18:2 mixture of fused silica types with medium particle sizes of 25, 4 and 1 μm. Further raising of filler content without increasing the viscosity should be possible by using different particle size distributions in optimized size ratios (14). The idea is that voids in any packing of regular spheres may be occupied by smaller spheres. Starting from our basic resin matrix filled with 70% of the 25 μm filler we substituted the latter successively with different portions of the 1 μm filler. Our results are shown in Figure 8. Up to 20% substitution there was no substantial increase of viscosity. An additional positive effect of this measure is an improvement in settling behaviour.

Flexibilization and toughening. The described resin matrix is highly crosslinked and rigid; this may be harmful for chips and wire bonds. For some applications it therefore seems necessary to introduce toughening and flexibilizing components. Tg should not be reduced significantly by these procedures. The results of two approaches are shown in Table II in comparison to unmodified resins.

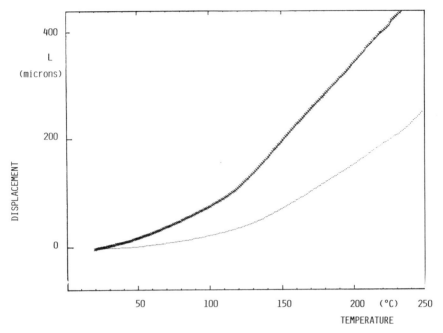

Figure 6. Displacement of the samples' ends; resin A (upper line), resin C (lower line)

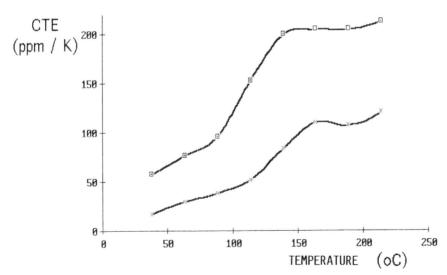

Figure 7. Coefficients of thermal expansion; resin A (upper line), resin C (lower line)

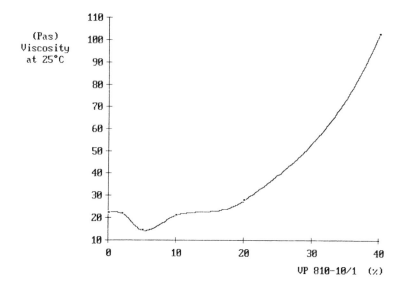

Figure 8. Influence of particle size distribution (increasing portion of fine filler)

Table I. Properties of resins with different filler contents

	resin A	resin B	resin C	resin D
Filler fraction (by weight)	0	0.60	0.65	0.70
Filler fraction (by volume)	0	0.437	0.490	0.547
Density of cured resin (g/cm3)	1.184	1.637	1.692	1.749
Thermal expansion				
25 - 75°C	65	27	22	19
175 -225°C (ppm/K)	208	119	108	95
Cure shrinkage (% vol)	4.5	2.5	2.3	2.0
Viscosity (Pas)	0.6	16	25	63

Table II. Properties of modified resins

	resin C	resin D	resin E	resin F
Filler fraction (by volume)	0.49	0.547	0.460	0.465
Glass transition (°C)	157	159	137	123
Thermal expansion				
25 - 75°C	22	19	27	25
175 -225°C (ppm/K)	108	95	102	105
Viscosity (Pas)	25	63	41	20

The best results were obtained with the silicone elastomer modification (resin E). Tg, CTE and viscosity are slightly affected but remain in acceptable ranges.

Addition of flexibilizing resin (resin F) clearly lowered Tg. Surprisingly globe tops comprising epoxidised soya bean oil damaged a ceramic substrate during thermal cycling (see below). This might be due to an antiplasticizing effect.

Formulations containing carboxyl-terminated butadiene/acrylo-nitrile elastomer (15) showed in our experiments very sluggish cure behaviour. In fact neither the initial mixture nor a prereacted adduct with epoxy resin gave more than a slight skin formation after prolonged irradiation.

Test performance. In electronic assemblies extensive accelerating tests are performed (16); the most important are listed in Table III.

For example a globe top has to withstand fast thermal cycling from 5 min/-55°C to 5 min/125°C for 100 or more times without showing cracks or loosing adhesion.
Table IV presents some of our results with drops of 3 mm height on aluminia ceramic. Numbers of cycles are listed when first delamination was detected by migration of ink into capillary delamination gaps. The second set of figures tells the numbers of cycles which led to total lift-off. It is clearly seen that adding glycidoxypropyl-trimethoxy-silane improves adhesion although the filler is already silane treated. There is an optimum in silane content; higher concentrations seem to act as dehesive.

Notable is the beneficial effect of curing with lower intensity of light, which meant in our case a greater distance between substrate and light source. Obviously optimization work is necessary for individual formulations, globe top geometry and irradiation equipment.

Conclusions

It has been demonstrated that thick layers of fused silica filled epoxy resins with processable viscosities can be UV cured in

Table III. Accelerated Testing in Microelectronics

High temperature tests	150°C/1000h, bias short exposure: 260°C/10s
Humidity test	85°C/85% rel.hum., bias
Pressure cooker stress test	121°C/24h, water vapour 2kp/cm^2,
Hydrolyzable Ions	detection from pressure cooker extract by ion chromatography
Temperature cycling	-55°C/5min, fast transport, 125°C/5min, 100 cycles

Table IV. Adhesion on alumina ceramic during thermal cycling (cycles are average values of 5 samples)

	1	2	3	4	5	6	7
Silane content (% weight)	0	0.02	0.04	0.08	0.20	0.40	1.00
Delamination (cycles) distance 11 cm	-	54	88	73	15	15	15
Delamination (cycles) distance 36 cm	89	243	246	331	241	83	35
Total lift-off (cycles) distance 11 cm	738	1462	1010	1022	818	934	1032
Total lift-off (cycles) distance 36 cm	1500	no lift off within 2000 cycles					

reasonable time with low cure shrinkage. By online photometry information is gained about absorption behaviour of photocurable resins. The throughcure can be monitored by impedance measurement. The cured resins show attractive coefficients of thermal expansion and good adhesion to ceramic substrates.

Yet one has to keep in mind that stress of an encapsulant on chip and bonds is reduced by accommodating CTE *and* by lowering elastic modulus (17). As these two properties are not independent and arbitrary, optimization work has to be done with resin matrix and filler reinforcement to keep stress within tolerable limits.

Acknowledgements

We thank D. Skudelny and W. Wolf from Quarzwerke Frechen for valuable discussions and many samples and UCC Deutschland for providing us with information and samples.

Literature Cited

(1) Pappas, S. P.; Pappas, B. C.; Gatechair, L. R.; Jilek, J. H.; Schnabel, W. Polymer Photochemistry 1984, 5, 1
(2) Crivello, J. V. Advances in Polymer Science 1984, 62, 2.
(3) Koleske, J. V. Polym. Paint Colour J. 1986, 76, 546
(4) Wong, C. P. Advances in Polymer Science 1988, 84, 63
(5) Long, L.E.; Chow, D. T. 29th National SAMPE Symposium 1984, 651
(6) Gatechair, L. R. Conference papers RADTECH'88, 1988, 28
(7) Lewis, T. B.; Nielsen, L. E. J. Appl. Polym. Sci. 1970, 14, 1449
(8) Schapery, R. A. J. Composite Materials 1968, 2, 380
(9) Holliday, L.; Robinson, J. J.of Material Science 1973, 8, 301
(10) Nielsen, L. E. J. Composite Materials 1967, 1, 100
(11) Pinheiro, M. DE F. F.; Rosenberg, H. M. J. Polym. Sci., Polym. Phys. Ed. 1980, 18, 217
(12) Feltham, S. J.; Yates, B.; Martin, R. J. J. Mater. Sci. 1982, 17, 2309
(13) Wang, T. T.,; Kwei, T. K. J. Polym. Sci. 1969, 7, 889

(14) Kuhlmann, H. W.; Wolter, F.; Mueller, E. R. ACS Org. Coat. Plast. Chem. 1963, 23, 449
(15) Drake, R. S.; Siebert, A. R. Polym. Sci. Technol. 1984, 29, 393
(16) Merrett, R. P.; Bryant, J. P.; Studd, R. Reliability Physics 1983, 21, 73
(17) S.Oizumi, N.Imamura, H.Tabata, H.Suzuki in Polymers for High Technology; Comstock, M. J., Ed.; ACS Symposium Series; American Chemical Society: Washington, DC, 1987; pp 537-546

RECEIVED September 13, 1989

LASER-INITIATED POLYMERIZATION

Laser-Initiated Polymerization

In the past few years, lasers, both pulsed and continuous, have been shown to be highly effective sources of ultraviolet light for initiating photopolymerization. In addition to their utility in investigation of photopolymerization mechanisms, their potential for use in photoresist and photoimaging applications provides the driving force for conducting basic studies of the kinetics of laser-initiated polymerization.

The two papers (Chapters 29 and 30) in this section describe recent studies of the photopolymerization of acrylate based systems using both pulsed and continuous lasers. In the first paper (Chapter 29), the percent conversion of a typical difunctional acrylate is measured by DSC as a function of the repetition rate and total photon dose of a pulsed excimer laser. The results indicate that the polymerization efficiency is highly dependent on the laser operating parameters. In the second paper, real time infrared spectroscopy is employed to assess the polymerization of acrylated systems induced by continuous (argon ion) and pulsed (nitrogen) lasers. The variety of monomers, functionalized oligomers, and photoinitiators included in this study clearly demonstrates the unique applicability of lasers to the investigation and processing of photopolymerizable systems.

Chapter 30

Laser-Initiated Polymerization of Multifunctional Acrylates

Repetition Rate Effects on Percent Conversion

Charles E. Hoyle and Martin A. Trapp

Department of Polymer Science, University of Southern Mississippi, Southern Station Box 10076, Hattiesburg, MS 39406—0076

>The pulsed laser-initiated polymerization of multifunctional acrylates has been investigated using a photocalorimeter. In the case where a limited number of pulses are delivered to the sample, the extent of conversion decreases as the laser repetition rate increases. However, for a large number of pulses, higher repetition rates do not lead to decreases in the overall degree of conversion.

The use of lasers to initiate the polymerization of both monofunctional and multifunctional monomers has been reported in a number of papers during the past decade. Decker (1) was the first to demonstrate that pulsed lasers could be effectively used to obtain relatively high degrees of polymerization for trimethylolpropane triacrylate. He showed that even for pulsed lasers which deliver up to gigawatts of peak power, polymerization could be effectively carried out over a wide range of conditions (1).

In recent reports (2-7), it has been shown that it is important to consider the effect of such laser operating parameters as pulse repetition rate on the polymerization kinetics. It was clearly demonstrated that pulsing the laser at narrow time intervals on the order of the lifetime of growing polymer radical chains resulted in a premature chain termination due to injection of small molecule "terminator" radicals into the system. In this paper we focus on the effect of pulse repetition rate on the polymerization of multifunctional acrylates, in particular 1,6-hexanediol diacrylate (HDODA) and trimethylolpropane triacrylate (TMPTA).

Experimental

1,6-hexanediol diacrylate (HDODA) (Aldrich), trimethylolpropane triacrylate (TMPTA) (Radcure Specialties) and hexyl acrylate (HA) (Scientific Polymer Products) were used as received. The photoinitiator, 2,2-dimethoxy-2-phenyl acetophenone (Irgacure 651, Ciba Geigy), was recrystallized several times from methanol. Two microliter samples were placed onto a DSC sample pan and an empty sample pan was used as the reference. Polymerization exotherms were monitored with a modified Perkin Elmer differential scanning calorimeter (DSC-1B) (8). The 2 second response time of the instrument results in a convoluted response function for the exotherm curves. Each sample was degassed for 10 minutes prior to irradiation. Copper tubing was used for all the nitrogen degassing lines. A 10 second baseline was collected before pulsing the laser. The laser source was a Lumonics Hyper-EX 440 operating with a xenon-fluorine fill gas with output at 351 nm (~15-ns per pulse fwhm). An IBM-AT was used to fire the laser and collect data from the DSC. Intensity measurements were accomplished using a Scientech Joule Meter (Model # 360401). The average power intensity was approximately 35 mJ/cm^2/pulse with no appreciable rolloff at the higher repetition rates.

Results and Discussion

Figure 1 shows the polymerization exotherms for the LIP of HDODA (2 wt percent photoinitiator) as a function of laser pulsing frequency (repetition rate ranging from 1 Hz to 80 Hz) for a total of 20 pulses delivered to the sample. It is quite obvious that the sample (curve a) exposed to a total of 20 pulses at a 1 second time interval between pulses (laser operation at 1 Hz) reaches a much higher total degree of polymerization, as indicated by the integrated area under the exotherm curve, than samples receiving pulses at shorter time intervals (curves b-f). The results for HDODA in Figure 1 were obtained by placing a neutral density filter of 3.0 in the laser path limiting the output of the laser to 0.035 mJ/cm^2/pulse. Figure 2 (curve a) shows a plot of percent conversion versus laser repetition rate for HDODA. In contrast to the results for HDODA, polymerization exotherms for hexyl acrylate (HA), also shown in Figure 2 (curve b), have integrated areas which are essentially independent of the laser repetition rate. [We should mention that at much higher pulse densities and photoinitiator concentrations, we would expect to see HA also display a dependence of percent conversion on the laser repetition rate (see reference 2)]. Furthermore, at any given repetition rate, the double bond percent conversion is significantly greater for HDODA than for HA despite the fact that the HA samples were exposed to the full output of the laser. The use of the full output of the laser was necessary for the HA samples since only very weak exotherms could be generated if a neutral density filter of 3.0 was used as in the case of HDODA. The results for HDODA and HA lead us to speculate that the very rapid polymerization rate of HDODA is influenced to a large extent by its difunctionality, i.e., two reactive acrylate groups. Indeed, the rate enhancement for bifunctional acrylates has been previously observed and postulated to arise from inhomogeneties and related consequences in the polymerizing medium (9-11). One question is left unanswered: why do longer time intervals (pulses greater than 50 msec apart--repetition rates less than 20 Hz) provide for enhanced conversion efficiency? A simple explanation can be offered. Since the radical lifetimes for difunctional monomers are very long relative to the lifetimes for

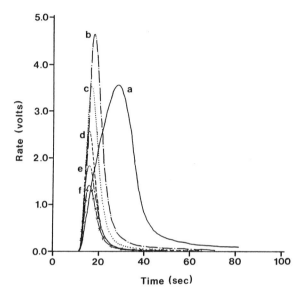

Figure 1: Plot of Exotherm Rate vs. Time for HDODA at 358 K, 2 wt% photoinitiator, neutral density filter 3.0 used, 20 pulses at different laser repetition rates:

(a) 1 Hz ———
(b) 5 Hz — — · —
(c) 10 Hz ············
(d) 20 Hz – – – –
(e) 40 Hz — - — -
(f) 80 Hz —X—X

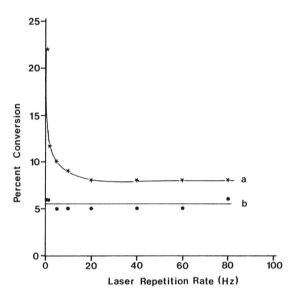

Figure 2: Plot of Percent Conversion vs. Laser Repetition Rate for (a) HDODA and (b) HA at 325 K (2 wt% photoinitiator, neutral density filter 3.0 used for HDODA, no neutral density filter used for HA, 20 pulses total).

polymer radicals generated from monofunctional monomers, premature injection of small molecule terminator molecules prior to or during the formation of gels (or microgels) greatly inhibits the polymerization, i.e., at high repetition rates (above 20 Hz) small molecule terminator radicals are injected into the polymerization medium and terminate the rapidly growing polymer kinetic chain. [A complete discussion of the premature termination process induced in pulsed laser-initiated polymerization appears in several recent publications (2-7)]. Another consequence of laser-initiated polymerization of HDODA is illustrated in Figure 3 which shows results for percent conversion versus repetition rate for a total number of pulses ranging from 20 to 3000. Upon initial inspection, it is seen that an increase in the total number of pulses delivered to the sample results in a dramatic rise in the percent conversion obtained at all repetition rates. From 50 to 3000 pulses, the percent conversion at higher repetition rates increases disproportionately. This result can be attributed to higher polymerization rates attained at the higher repetition rates which allows polymerization to actually precede the contraction process which accompanies crosslinking. Indeed, higher conversions at higher average powers using steady state lamp sources has previously been reported by Kloosterboer et al (4) and attributed to creation of a temporary excess free volume resulting in increased diffusion rates of reacting acrylate groups.

Figure 4 illustrates the effect of decreasing the photoinitiator concentration for a plot of the percent conversion versus laser repetition rate (shown for total pulses ranging from 20 to 3000 as in Figure 3 for direct comparison). The most obvious, but not necessarily the most important, effect of lowering the photoinitiator concentration is a decrease in the percent conversion for a given repetition rate. However, probably the most profound difference for a given pair of curves (curves a-f in Figure 3 versus the corresponding curves a-f in Figure 4) is realized when comparing the maximum percent conversion attained at low repetition rates to the percent conversion at high repetition rates. In each case in Figure 4, the difference in the conversion attained at lower repetition rates and the conversion attained at higher repetition rates is much less than the corresponding differences (for a given number of pulses) in Figure 3. For instance, the ratio of the maximum percent conversion at 1 Hz for 2.0 wt % photoinitiator for 20 pulses total (curve a, Figure 3) to the percent conversion at 40 Hz is 5.0 compared to a value of 1.6 for the same ratio (curve a, Figure 4) with 0.1 wt % photoinitiator. [In the latter case, the maximum percent conversion was obtained at 5 Hz and not 1 Hz]. These results can be attributed to the unique design of the laser-initiated polymerization experiment. Since the ability of the small molecule radicals produced at a given delay period to effect premature termination of growing polymer radicals depends on the photoinitiator concentration (see reference 11 for a complete discussion of this phenomenon), the systems with the highest photoinitiator concentration show the largest suppression in the conversion efficiency at higher repetition rates where premature chain termination is most effective.

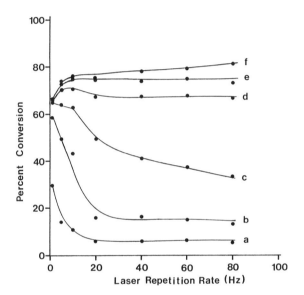

Figure 3: Plot of Percent Conversion vs. Laser Repetition Rate for HDODA at 358 K (2 wt% photoinitiator, neutral density filter 3.0 used) (a) 20 pulses, (b) 50 pulses, (c) 100 pulses, (d) 200 pulses, (e) 500 pulses, (f) 3000 pulses.

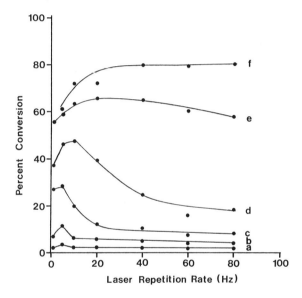

Figure 4: Plot of Percent Conversion vs. Laser Repetition Rate for HDODA at 358 K (0.1 wt% photoinitiator, neutral density filter 3.0 used) (a) 20 pulses, (b) 50 pulses, (c) 100 pulses, (d) 200 pulses, (e) 500 pulses, (f) 3000 pulses.

A second point should be noted for the curves in Figure 4. For curves a-e, unlike the corresponding systems in Figure 3 where the maximum conversion is obtained at 1 Hz, the maximum percent conversion occurs at 5 Hz (curves a-c), 10 Hz (curve d), and 20 Hz (curve e). This is due to a lack of effective premature termination of growing polymer chains, as previously discussed, which occurs when using lower photoinitiator concentrations. Apparently, if premature termination reactions did not occur, we would expect to observe higher extents of polymerization with increasing repetition rate for even low total numbers of pulses.

In summarizing the results in Figures 3 and 4, we note that two factors are operable which are opposite in their effect on the overall polymerization efficiency of laser-initiated polymerization of HDODA. The first factor is a result of efficient premature termination due to coupling of small molecule radicals produced by firing laser pulses at intervals less than 50 msec: this decreases the overall yield. The second factor is a consequence of carrying out the photopolymerization/photocrosslinking at higher overall light intensity: as the light intensity or overall average power of the laser increases an increase in the percent conversion is expected since polymerization precedes the conversion induced shrinkage of the film. At higher photoinitiator concentrations (Figure 3) the first factor dominates resulting in especially low percent conversions at high repetition rates (greater than 20 Hz). At a low photoinitiator concentration (Figure 4), the premature termination process is diminished and percent conversions at high repetition rates (while less than at low repetition rates) are only marginally lower than at low repetition rates.

An interesting comparison to the results in Figure 3 and 4 for HDODA is shown in Figure 5 for trimethylolpropane triacrylate (TMPTA). First, note that the percent conversion for TMPTA curve at 1 Hz for 20 pulses (curve a, Figure 5) is slightly greater than 20 percent, while for HDODA under similar experimental conditions it is almost 30 percent (curve a, Figure 3). Yet, at higher repetition rates the reverse is true, i.e., at 80 Hz the percent conversion for TMPTA is about 16 compared to 5 for HDODA. Apparently, for an identical photoinitiator concentration premature termination is greatly reduced for TMPTA at higher repetition rates. This is indicative of a lower termination rate for TMPTA, and/or an enhanced polymerization rate compared to HDODA. This premise is borne out by single pulse experiments for HDODA and TMPTA (13). Finally, TMPTA proceeds to approximately 50 percent conversion compared to about 80 percent for HDODA. This phenomenon is most assuredly a result of extensive crosslinking for trifunctional acrylate systems which prohibits attainment of high percentages of conversion.

Conclusions

The results in this paper demonstrate the utility of employing a pulsed-laser source to initiate and probe the kinetics of photopolymerization of

multifunctional acrylates. Specific results derived from this study are:
1. For low numbers of pulses delivered to a di- or trifunctional acrylate, higher degrees of conversion are obtained at lower pulse repetition rates.
2. Suppression of polymerization for the difunctional acrylate at higher repetition rates is enhanced at increased photoinitiator concentrations.

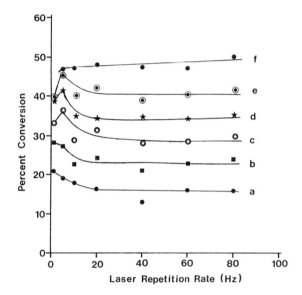

Figure 5: Plot of Percent Conversion vs. Laser Repetition Rate for TMPTA at 358 K, (2 wt% photoinitiator, neutral density filter 3.0 used) (a) 20 pulses, (b) 50 pulses, (c) 100 pulses, (d) 200 pulses, (e) 500 pulses, (f) 3000 pulses.

These results point to two processes, premature radical chain termination and film shrinkage, which compete in determining the ultimate polymerization conversion efficiency of multifunctional acrylates. It is obvious that critical attention must be paid to the pulse repetition rate, photoinitiator concentration, and acrylate functionality in developing any photopolymerizable system for laser-initiated polymerization. Future publications on laser-initiated polymerization of multifunctional acrylates will deal with monomer extraction of partially polymerized films, mechanical properties of laser polymerized films, and the kinetics of single-pulsed systems.

Acknowledgment

This research is supported by The National Science Foundation under Grant DMR 85-14424 (Polymers Program). Acknowledgement is also made to NSF for assistance in purchasing the laser system utilized in the course of this investigation (Grant CHE-8411829--Chemical Instrumentation Program).

Literature Cited

1. C. Decker, ACS Symposium Series 266; Materials for Microlithography: Radiation Sensitive Polymers, 9, 207 (1984).
2. C. E. Hoyle, M. A. Trapp, C. H. Chang, D. D. Latham and K. W. McLaughlin, Macromolecules, 22, 35 (1989).
3. C. E. Hoyle, M. A. Trapp and C. H. Chang, Polym. Mat. Sci Eng., 57, 579 (1987).
4. C. E. Hoyle, C. P. Chawla, P. Chattertion, M. Trapp, C. H. Chang and A. C. Griffin, Polymer Preprints, 29, (1), 518 (1988).
5. D. D. Latham, K. W. McLaughlin, C. E. Hoyle and M. A. Trapp, Polymer Preprints, 29, (2), 328 (1988).
6. C. E. Hoyle, C. H. Chang and M. A. Trapp, Macromolecules, accepted for publication.
7. C. E. Hoyle, M. A. Trapp and C. H. Chang, J. Polym. Sci., Polym. Chem Ed., 27, 1609 (1989).
8. C. E. Hoyle, R. D. Hensel and M. B. Grubb, J. Polym. Sci., Polym. Chem. Ed., 22, 1865 (1984).
9. J. G. Kloosterboer and G. J. M. Lippits, Journal of Imaging Science, 30, 177 (1986).
10. J. G. Kloosterboer, G. M. M. van de Hei, R. G. Gossink and G. C. M. Dortant, Polym. Comm., 25, 322 (1984).
11. J. G. Kloosterboer, G. M. M. van de Hei and H. M. J. Boots, Polym. Comm., 25, 354 (1984).
12. C. E. Hoyle and M. A. Trapp, J. Imaging Sci., accepted for publication.

RECEIVED October 5, 1989

Chapter 31

UV-Radiation- and Laser-Induced Polymerization of Acrylic Monomers

C. Decker and K. Moussa

Laboratoire de Photochimie Générale, Unité Associée au Centre National de Recherche Scientifique, Ecole Nationale Supérieure de Chimie, 3 rue Alfred Werner, 68200 Mulhouse Cedex, France

> The kinetics of ultrafast polymerization of acrylic monomers exposed to UV radiation or laser beams has been investigated by IR spectroscopy. An 8 fold increase of the cure speed was observed by using diphenoxybenzophenone as photoinitiator instead of benzophenone. The reactivity of polyurethane-acrylate or epoxy-acrylate systems was markedly improved by adding acrylic monomers that contain carbamate or oxazolidone groups and which impart both hardness and flexibility to the cured polymer. Time-resolved infrared spectroscopy was used to directly record the actual polymerization profile for reactions taking place within a fraction of a second upon UV or laser exposure. Comparison with other techniques of real-time analysis show the distinct advantages of this method for an accurate evaluation of the important kinetic parameters and of the dark polymerization which develops just after the irradiation.

Radiation curing has now become a well accepted technology which, owing to its distinct advantages, has found widespread industrial applications, going from the surface protection of materials by fast-drying coatings to the patterning of electronic components by insolubilization of photoresists (1-4). New polymer materials with tailor-made properties can thus be obtained instantly by simple exposure of a photosensitive resin to an intense source of UV radiation. This method of curing still suffers some inherent shortcomings which have somewhat restricted the fast growth that was expected to occur during the last decade.

Recent efforts have been directed towards an improvement of the performance of UV-curable systems, trying to overcome some of the most severe limitations by acting at different levels :
- increasing the resin sensitivity, with the development of more efficient photoinitiators and more reactive monomers ;
- lowering the residual unsaturation content of the UV-cured material ;
- developing new reactive diluents, less toxic and less irritating ;
- reducing the strong inhibition effect of atmospheric oxygen ;
- improving the properties of radiation cured polymers ;
- increasing the speed of cure by using more powerful radiation sources, like lasers.

In this paper we report some of the progress recently made in these various areas, concentrating on the light-induced polymerization of multiacrylic monomers which are today the most widely used UV-curable systems. We also describe here a new analytical method, based on IR spectroscopy, that permits the kinetics of photopolymerizations, which develop extensively in a fraction of a second, to be followed quantitatively and in real time.

EXPERIMENTAL SECTION

Materials. The photopolymerizable resin was made of 3 main compounds : (i) a photoinitiator that generates free radicals upon exposure to UV radiation ; (ii) a prepolymer end-capped with acrylate groups ; (iii) a reactive acrylic diluent to lower the viscosity of the resin. For most experiments, the selected photoinitiator was α,α' dimethoxyphenylacetophenone, DMPA (Irgacure 651 from Ciba-Geigy), because of its high initiation efficiency.

The performance of a new photoinitiator, diphenoxybenzophenone (DPB), has been evaluated and compared to benzophenone (BZP) and Irgacure 651. DPB was synthetized by phosgenation at 90°C of diphenylether in the presence of $AlCl_3$. A small amount of a tertiary amine, methyldiethanolamine (MDEA), had to be introduced in the formulation since, like for BZP, the radical production upon photolysis proceeds by hydrogen transfer from the donor molecule to the excited state of DPB.

Two types of functionalized prepolymer were used, either an aliphatic polyurethane-diacrylate (Actilane 20 from SNPE), or a diacrylate derivative from the glycidylether of bis-phenol A (Actilane 72 from SNPE), incorrectly called epoxy-acrylate. The reactive diluent consisted of mono, di or triacrylic monomers, namely, ethyldiethyleneglycol-acrylate (EDGA from Norsolor), ethylhexylacrylate (EHA from Norsolor), hexanediol-diacrylate (HDDA from UCB), tripropyleneglycol- diacrylate (TPGDA from UCB) and trimethylolpropane-triacrylate (TMPTA from UCB). Two new monomers recently developed by SNPE were also employed as reactive diluent : a carbamate-monoacrylate (Acticryl CL 960) and an oxazolidone- monoacrylate (Acticryl CL 959).

Typical resin formulations contained 2 to 5 % of photoinitiator and equal parts of the acrylic prepolymer and of the diluent. The resin was applied as a uniform layer of controlled thickness, between 12 and 36 µm, on a KBr disk with a calibrated wirewound applicator.

Irradiation. Samples were exposed for a short time to the radiation of a 2 kW medium pressure mercury lamp with a power output of 80 W per linear centimeter. The emitted light was focused by means of a semi-elliptical reflector on the sample, at which position the fluence rate was measured to be 500 mW cm^{-2} or 1.5×10^{-6} einstein $s^{-1}cm^{-2}$. A camera shutter was used to select a precise exposure time in the range of 2 to 100 ms. A less intense UV source (Philips HPK 125 W) was used for the kinetic investigation of the photopolymerization by real-time infra-red (RTIR) spectroscopy, with a fluence rate of 4 mW cm^{-2}.

The laser-curing experiments were performed with a continuous wave argon ion laser (Spectra Physics, Model 2000) tuned to its emission line at 363.8 nm. The radiant power at the sample position was measured to be in the range 30 to 200 mW cm^{-2}, depending on the selected laser output. Some photopolymerization experiments were carried out with a pulsed nitrogen laser (SOPRA, model 804 C) which emits at a wavelength of 337.1 nm. The instantaneous radiant power was calculated to be 500 kW cm^{-2}, based on the energy of a pulse (5 m J) and its duration (8 ns). The laser was operated in a multiple-pulse mode at a repetition rate between 2.5 and 40 Hz. An electronic shutter was used to select the desired number of pulses.

The fraction of incident light absorbed by the sample was determined by differential actinometry or from the absorbance of the coating at the wavelength of the relevant laser emission. Irradiations were carried out a room temperature, either in the presence of air or in a nitrogen-saturated reactor equipped with polyethylene windows which are transparent to both UV radiation and the IR analyzing beam.

Analysis. The extent of the polymerization process was evaluated quantitatively by IR spectroscopy (Perkin Elmer spectrophotometer, model 781) by monitoring the decrease of the sharp peak centered at 812 cm^{-1} (twisting vibration of the acrylic CH_2 = CH bond) which occurs after a given UV exposure. This analytical method has proved very valuable for measuring the polymerization rate of reactions which develop in the millisecond time scale (5), but it suffers the major disadvantage of requiring tedious point by point measurements.

A new method, RTIR spectroscopy, has recently been developed to study in real time the kinetics of ultra-fast photopolymerizations and has already been described (6). It was applied here to follow quantitatively the laser-induced curing of highly reactive acrylic photoresists. The method consists of exposing the sample simultaneously to the UV beam which induces the polymerization and to the analyzing IR beam, and monitoring continuously the resulting drop in the IR absorbance of the reactive double bond. Conversion *versus* time curves can thus be directly recorded for polymerizations developing extensively in 0.5 second or more. Faster polymerizations were followed by using a transient memory recorder ; the limiting factor is then the response of the IR detector, usually 30 ms.

The hardness of the cured film was evaluated by monitoring the damping of the oscillations of a pendulum (Persoz hardness) which is directly related to the softness of the sample. For a 35 µm thick UV-cured film, coated onto a glass plate, Persoz values typically range from 50 s for elastomeric materials to 300 s for hard and glassy polymers.

The film flexibility was evaluated by bending UV-cured coatings 180° around mandrels of decreasing diameters and measuring the diameter at which cracks first appear ; for a 3 mm diameter, the film undergoes a 30 % elongation. Highly flexible coatings were subjected to the most severe zero-T-bend test where the coated substrate was bent onto itself, with no spacer between the two halves. Samples passing this test without film cracking or crazing were marked 0 in the flexibility scale.

A NEW PHOTOINITIATOR : DIPHENOXYBENZOPHENONE

The efficiency of various substituted benzophenones for initiating the photopolymerization of multifunctional monomers in condensed phase was evaluated by IR spectroscopy from conversion *versus* exposure time kinetic curves. The best results were obtained with diphenoxy- benzophenone (DPB) associated to an hydrogen donor, like methyl- diethanolamine, for a one to one mixture of hexanediol-diacrylate (HDDA) and of a polyurethane-diacrylate (Actilane 20 from SNPE). Figure 1 shows some typical polymerization profiles obtained for a 30 µm thick film exposed, in the presence of air, to the UV radiation of a medium pressure mercury lamp at a light intensity of 1.5 × 10^{-6} einstein s^{-1} cm^{-2}.

At a photoinitiator concentration of 2 %, the reaction was found to develop 8 times faster with DPB than with benzophenone (BZP) and even more rapidly than with 2,2'-dimethoxyphenylacetophenone (Irgacure 651 from Ciba Geigy), one of the most efficient photoinitiators. An exposure time as short as 0.05 s proved to be sufficient to make polymerize more than 75 % of the acrylate functions. The higher initiator efficiency of DPB can be attributed to either a larger absorbance in the UV region or/and to a higher production of initiating radicals. In order to differentiate these two effects, the fraction (f) of the incident radiation absorbed by the sample was determined by actinometry ; it was found to be twice larger in DPB than in BZP formulations because of a substantial bathochromic shift of the DPB absorption spectrum toward longer wavelengths. From the observed 8 fold increase in the reactivity of DPB over BZP, it can be inferred that the remarkable performance of this novel photoinitiator not only results from a better capability for absorbing the radiation emitted by the UV source, but also from a higher intrinsic initiation efficiency.

The importance of this second factor can be assessed by evaluating the quantum yield of the polymerization (Φ_p) i.e. the number of acrylate functions polymerized per photon absorbed : Φ_p = 110 mol photon^{-1} for the BZP based formulation, and Φ_p = 530 mol photon^{-1} for the DPB one. For a given monomer (M), the variation of Φ_p are directly related to the variation in the initiation quantum yield Φ_i : $\Phi_p = \dfrac{k_p}{2k_t^{0.5}} [M].\Phi_i^{0.5}$.

The five fold increase of Φ_p observed when BZP was replaced by DPB must therefore result from a much more effective production of initiating radicals with this novel photoinitiator. The reasons for such a large effect of the phenoxy substituents on the photolysis mechanism of benzophenone are now being investigated by laser spectroscopy.

An additional advantage of this new photoinitiator is to provide a better through-cure of the coating and to reduce the amount of acrylate functions which have not polymerized. A 36 μm tack-free coating photocured with DPB was found to contain only 12 % residual unsaturation, as compared to 25 % with benzophenone, thus improving its long-term stability. As a consequence of the enhanced polymerization at the coating-substrate interface, the adhesion of the polymer film onto the support was markedly increased ; this parameter is of prime importance with respect to the practical applications of curable systems as protective coatings and adhesives.

Another interest of this new photoinitiator is to improve some of the mechanical properties of the cured polymer such as the hardness and scratch resistance. Polyurethane-acrylate coatings, UV cured with the DPB (2 %) + MDEA (5 %) photoinitiator system, appeared to be substantially harder (Persoz hardness = 200 s) than BZP + MDEA based coatings (120 s), as expected from the enhanced through-cure provided by DPB. Lowering the amount of tertiary amine, which acts both as plasticizer and chain transfer agent, was shown to further increase the coating hardness, but at the expense of the cure speed. Extremely hard and glassy polymers were obtained by using a bisphenol A-epoxy-acrylate as prepolymer instead of a polyurethane-acrylate, the Persoz hardness reaching then values beyond 300 s (Figure 2).

It should finally be emphasized that the use of DPB is not restricted to acrylic compounds since it proved to be also a very efficient photoinitiator for the polymerization of vinyl monomers, like N-vinylpyrrolidone (NVP). In addition, DPB appeared to be particularly well-suited to photo-cure systems that need hydrogen abstraction type photoinitiators, like the thiol-polyenes resins (7), since it is then to be compared to the poor-performing benzophenone.

NEW REACTIVE ACRYLIC MONOMERS

The actual trends for the development of new diluents usually follow a number of important guide-lines. Besides decreasing the formulation viscosity, all tend to :
- lower the volatility and the toxicity ;
- increase the reactivity, especially for soft coatings ;
- decrease the amount of unreacted functional groups ;
- improve the final product properties.

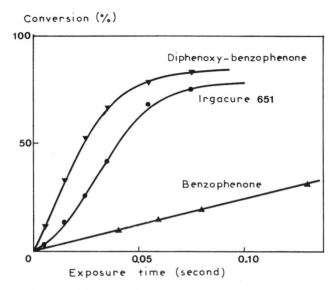

Figure 1. Influence of the photoinitiator on the kinetics of photopolymerization of a polyurethane-acrylate + HDDA photoresist. [Photoinitiator] = 2 %; [MDEA] = 5 % ; Fluence rate : $1,5 \times 10^{-6}$ einstein s^{-1} cm^{-2}.

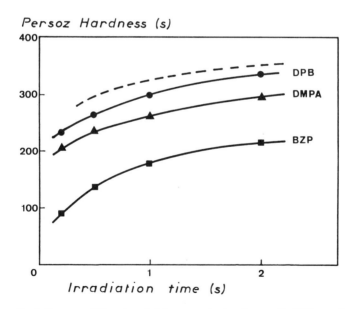

Figure 2. Influence of the photoinitiator on the hardness of a UV cured epoxy-acrylate coating. [Photoinitiator] = 2 % ; [MDEA] = 5 % ; Film thickness = 36 µm ; ----- : 2 % DPB + 1 % MDEA.

For economic reasons, such performance should be achieved at the lowest possible cost to be compatible with large scale industrial production.

Some novel monomers, recently developed by SNPE[8], proved to satisfactorily fulfill these requirements. They consist of monoacrylates containing either carbamate or oxazolidone structures :

Acticryl CL 959

$$CH_2 = CH - \underset{\underset{O}{\|}}{C} - O - CH_2 - CH_2 - N \begin{array}{c} -CH_2 \\ | \\ C \\ \nearrow \; \diagdown \\ O \quad\; O \end{array} \begin{array}{c} \\ CH_2 \\ / \\ CH \\ | \\ CH_3 \end{array}$$

Acticryl CL 960

$$CH_2 = CH - \underset{\underset{O}{\|}}{C} - O - CH_2 - CH_2 - NH - \underset{\underset{O}{\|}}{C} - O - \underset{\underset{CH_3}{|}}{CH}$$

Besides their low volatility, low draize index and high diluting power, these monomers were found to polymerize much faster than typical monoacrylates, like ethyldiethyleneglycolacrylate (EDGA), and even faster than diacrylates, like tripropyleneglycoldiacrylate (TPGDA) or HDDA (Figure 3). Owing to the high molecular mobility characteristic of monofunctional monomers, close to 100 % conversion can be reached with these compounds, by contrast to di- or triacrylates where the formation of a tight network limits the extent of the polymerization.

Similar results were obtained with a resin formulation containing these new monomers as diluent and a polyurethane-diacrylate or an epoxy-diacrylate oligomer (Figure 4). The rates of polymerization were significantly increased compared to formulations based on neat mono- or diacrylate monomers and the photosensitivity (S), i.e., the energy required to reach 50 % conversion, dropped as low as 0.4 mJ cm^{-2} (Table I).

Table I. Performance analysis of UV-curable systems
Photoinitiator : [Irgacure 651] = 5 % ; Film thickness : 30 µm

Formulation 50/50	[Acrylate]$_0$ mol l^{-1}	f^a	R$_p^b$ mol l^{-1}s^{-1}	S^c mJ cm^{-2}	τ^d %	δ^e s	l^f m	m
Polyurethane-acrylate								
EDGA	3.4	1	30	10	2	40	0	
TPGDA	4.0	2	110	4	10	170	0	
HDDA	5.2	2	100	4,3	15	200	1	
CL 960	3.2	1	300	0,6	4	80	0	
CL 959	3.4	1	450	0,4	3	90	0	
Epoxy-acrylate								
EDGA	4.6	1	35	9	5	60	0	
TPGDA	5.3	2	120	4	17	280	7	
CL 960	4.5	1	200	1,3	9	300	0,5	
CL 959	4.7	1	500	0,6	10	320	1,5	

a) functionality
b) maximum rate of polymerization
c) photosensitivity 50 % conversion
d) residual unsaturation ⎫
e) Persoz hardness   ⎬ of the tack-free coating
f) mandrel T bend   ⎭

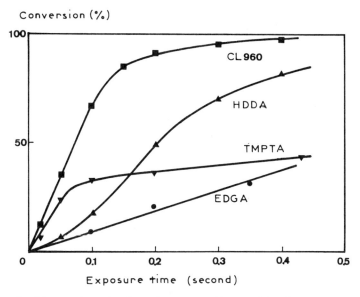

Figure 3. Photopolymerization of monoacrylic (EDGA, CL 960), diacrylic (HDDA) and triacrylic (TMPTA) monomers in the presence of air. [Irgacure 651] = 5 % ; Fluence rate = $1.5.10^{-6}$ einstein s^{-1} cm^{-2}.

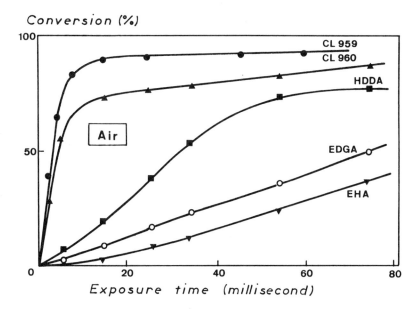

Figure 4. Influence of the reactive diluent on the photopolymerization of a polyurethane-acrylate coating irradiated in the presence of air. [Irgacure 651] = 5% ; [Diluent] = 47 % ; Film thickness : 24 μm.

Another remarkable feature is that the formulation containing the monomer CL 960 (or CL 959) polymerized almost as fast in the presence of air as the conventional TPGDA-based formulation, photo-cured in a nitrogen atmosphere (Figure 5). These novel monomers are thus helping to solve the crucial oxygen inhibition effect (9,10), one of the major factors which limit the speed of the production line in the industrial applications of UV-curing. Because of their monofunctional structure and the resulting increase of the chain reptation mobility, the polymerization of photoresists containing these monomers proceeds more extensively and reaches higher degrees of conversion. The photocured polymer thus contains less residual unsaturation than with di- or triacrylate diluents (Table I), with the expected beneficial consequences on its long-term performance. The reasons for the great reactivity of these monoacrylate monomers are still unclear but they are expected to result, to some extent, from an enhanced chain transfer reaction. Indeed, the fact that the fully cured homopolymer is essentially insoluble in the organic solvents, by contrast to the others monoacrylates polymers, indicates the occurence of a crosslinking process and therefore implies the contribution of the carbamate or oxazolidone function in the polymerization reaction.

A final advantage of these new acrylic monomers is that they improve as well the mechanical properties of the UV-cured polymer, leading either to highly reactive soft coatings in the case of polyurethane oligomers, or to very hard but still flexible coatings in the case of epoxy-acrylates (Table I). The latter coatings, based on carbamate - or oxazolidone-acrylate monomers, thus present the distinct advantage of being scratch resistant and almost as hard as glass, while they proved at the same time to be highly flexible, with break elongation values above 50 %.

For photocured polymer materials, the Persoz hardness was shown to depend mainly on the chemical structure of the prepolymer chain, on the functionality of the reactive diluent, on the photoinitiator system and on the duration of the UV exposure. The hardness of tack-free coatings thus continues to increase with further irradiation, as shown by Figure 6 for polyurethane-acrylates. This slow hardening process is likely to result from some further polymerization of the unreacted acrylic functions since their concentration was found to decrease concomitantly.

LASER-INDUCED CURING

The effectiveness of lasers for inducing the UV curing of acrylic systems has been first demonstrated only a few years ago (10-13). Owing to their very large light intensity, lasers appear today as the ultimate light source to reach extremely high speeds of cure. The laser-induced radical production in photosensitive systems offers several advantages over conventional UV initiation, in particular a large power output, the narrow bandwith of the emission and the spatial coherence of the laser beam which can be finely focused.

The UV emission of both continuous wave (12) (Ar^+) and pulsed (13) (N_2) lasers was shown to induce an instant cure of multifunctional acrylated monomers, in the presence of adequate photoinitiators, with polymerization quantum yields reaching values as high as 10,000 mol / photon (14). High resolution polymer relief images were thus directly drawn onto negative photoresists with a sharply focused CW laser beam by a maskless technology (15). In the case of pulsed lasers, the repetition rate was recently found to have a marked effect on both the polymer yield and the molecular weight distribution (16,17), the polymerization developing essentially in the dark period between the 10 ns wide laser shots.

The most promising potential applications of laser curing are expected to profit from the distinct advantages provided by these powerful light sources, e.g. the large speed of cure (coating of optical fibers), the high selectivity affording micronic resolution (photoimaging) and the great penetration into organic materials (18) (3 D modelling, curing of lacquers, composites, ...).

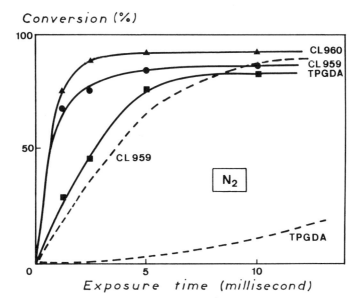

Figure 5. Influence of the reactive diluent on the photopolymerization of a polyurethane-acrylate coating irradiated in a nitrogen-saturated atmosphere. [Irgacure 651] = 5 % ; [Diluent] = 47 % ; Film thickness : 24 µm ; (----- : in the presence of air).

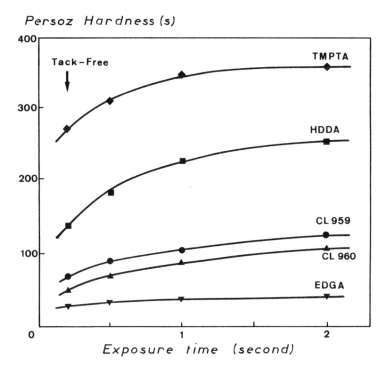

Figure 6. Influence of the reactive diluent on the hardness of a UV-cured polyurethane-acrylate coating of 36 μm thickness. [Diluent] = 50%.

A critical and most important problem that still remained to be solved was to find an adequate analytical method that would enable us to follow in real time the kinetics of such ultra-fast polymerizations. Different techniques have already been used to study the laser-induced curing of acrylic monomers (DSC, IR spectroscopy, interferometry, ...) but none of them provide conversion *versus* exposure time curves for polymerization that occur in a fraction of a second. The results obtained by some of these techniques are now briefly reported for comparison with those obtained by using real-time infra-red (RTIR) spectroscopy, a powerful method newly developed in our laboratory to investigate the kinetics of ultrafast photopolymerizations ([6](#)).

Gel fraction ([14](#)). The cure extent is evaluated here from the amount of crosslinked polymer formed upon laser irradiation. Figure 7 shows some typical kinetic profiles where the normalized thickness of the insoluble polymer film is plotted against the log of the exposure time for a UV and laser-irradiated epoxy-diacrylate. The slope of these curves corresponds to the contrast which characterizes the sharpness of the relief image obtained after solvent development. By using a highly focused Ar^+ laser beam, tuned to its emission lines in the UV range, high resolution polymer patterns can thus be directly drawn onto the photoresist coating, at writing speeds up to 10 m/s ([15](#)).

- **Infrared spectroscopy** ([5,13](#)). By following the disapearance of the IR absorption of the acrylic double bond one can quantitatively evaluate both the maximum rate of polymerization and the amount of residual unsaturation in the laser-cured polymer. Polymerization profiles are shown on Figure 8 for the curing of a polyester-acrylate photoresist under pulsed laser irradiation at 337.1 nm in the presence of air or pure nitrogen. The main disadvantage of IR spectroscopy is that, like gel fraction determination, it requires discrete measurements at various exposures and thus becomes a tedious, time consuming technique.

- **Differential Photocalorimetry (DPC)** ([19,20](#)). The polymerization being an exothermal process, the reaction can be monitored in real time by differential scanning calorimetry (DSC). From the recorded thermogram which shows the variation of the heat flow with the irradiation time, the rate of polymerization can be directly calculated, provided the standard heat of polymerization (ΔH_0) is known. For acrylic monomers, ΔH_0 values are usually in the range of 78 to 86 kJ mol^{-1}, depending on the monomer considered.

While uncertainties about the value of ΔH_0 affect only slightly the calculated value of the rate of polymerization, they have a drastic effect on the evaluation of the residual unsaturation content of the cured polymer which can therefore not be determined accurately by DPC. Another serious limitation of this technique is the slow response time of the detection that prevents monitoring polymerizations which occur in a fraction of a second. This is the reason why DPC investigations are usually carried out at low light-intensities and in the absence of oxygen.

- **Laser-interferometry** (18). This method is based on the variation of the refractive index of the resin as it undergoes polymerization. The interference pattern of two intersecting laser beams can thus be impressed onto the photoresist, leading to the formation of a grating. By monitoring the intensity of the diffracted beam, which accurately reflects the extent of the polymerization, we were able to record in real time the kinetics of curing reactions which develop in the millisecond time scale (Figure 9). This elegant and fast method proved unable to give quantitative information about the actual rate of polymerization or about the crosslink density of the cured material. Furthermore it is limited to laser-induced polymerization and cannot be applied to conventional UV curing.

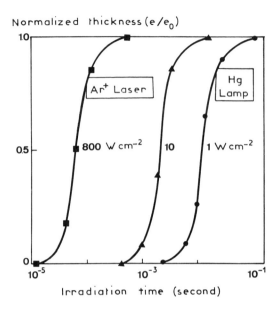

Figure 7. Insolubilization profile of an epoxy-acrylate photoresist exposed to high intensity UV radiation or to an Ar$^+$ laser beam ($\lambda = 363.8$ nm) in the presence of air. Film thickness : $e_0 = 10$ μm.

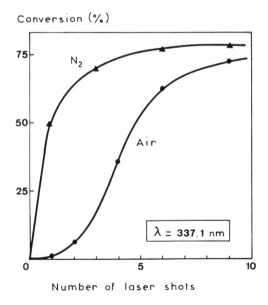

Figure 8. Photopolymerization of a polyester-acrylate photoresist exposed to a pulsed nitrogen laser ($\lambda = 337.1$ nm) in the presence of air or pure nitrogen. Pulse width = 8 ns ; pulse frequency = 5 Hz ; pulse energy = 5 mJ.

Real-time infrared spectroscopy (RTIR) (21). The basic principle of this analytical technique consists of exposing the sample simultaneously to the polymerizing UV beam and to the analyzing IR beam, and monitoring on a high speed recorder the sharp decrease of the acrylic absorbance at 812 cm^{-1}. Conversion *versus* time curves have thus been recorded for the first time for photopolymerizations that develop extensively in a fraction of a second (21). If the reaction time drops into the millisecond range, a transient memory recorder (22) or an oscilloscope with storage function can be used to shorter the time resolution further.

This technique proved particularly well suited to follow ultra-fast polymerizations induced by laser beams and thus to obtain directly the kinetic profile of the reaction, as shown by Figure 10. One of the unique advantages of this method is that it now becomes possible, based on these kinetic curves, to determine instantly and at any moment, both the true rate of polymerization and the amount of residual unsaturation in the cured polymer, i.e., its crosslink density.

Compared to the other techniques of real-time investigation (DPC, dilatometry), RTIR spectroscopy permits the reduction by a 50-fold factor of the time scale, owing to the short response of the IR detector (30 ms), and thus to, operate at high light intensities and in the presence of air, i.e. under experimental conditions comparable to those used in practical applications. It was also found very valuable to evaluate precisely the extent of the dark polymerization which occurs just after the laser has been switched off, as shown by Figure 11. This post-effect was found to represent a significant part (up to 80 %) of the overall process (21), especially in the absence of oxygen where the living polymer chains continue to grow for a few seconds before they terminate by radical trapping (23) or interaction with another radical.

Similar polymerization profiles were obtained when such acrylic photoresists were exposed to the UV emission at 337.1 nm of a pulsed nitrogen laser. Owing to the extremely short duration of the pulse, 8 ns which is about the time needed to generate the initiating radicals, polymerizations develops here only in the dark after the laser shot. When the laser is operated on a multiple pulse mode, the time interval between successive pulses can be easily changed by acting on the laser repetition rate. Figure 12 shows the kinetic curves recorded at three frequencies 2,5, 10 and 40 Hz, corresponding to a pulse interval of 400, 100 and 25 ms, respectively. As the repetition rate is increased, the multiple jump profile progressively smoothes down into a regular and monotonous variation.

Comparison of the various techniques. Table II compares the performance analysis for the various analytical methods most commonly used to monitor curing reactions.

Table II. Performance analysis of various methods of kinetic investigation of UV radiation or laser-induced polymerization

Techniques	Gel fraction	IR spectroscopy	DSC	Interferometry	RTIR
Real-time Analysis	no	no	yes	yes	yes
Response time (s)	-	-	> 1	0.001	0.03
Light-intensity	high	high	low	high	high
Atmosphere	air	air	N_2	air	air
Quantitative data	no	yes	yes	no	yes
Properties evaluation	no	yes	no	no	yes

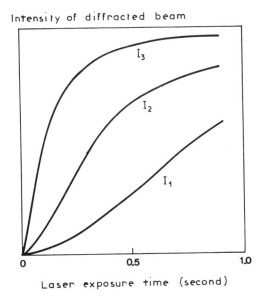

Figure 9. Polymerization profiles recorded by laser-interferometry for a polyester-acrylate photoresist exposed to an Ar^+ laser beam ($\lambda = 363.8$ nm) at various light intensities in the presence of air.

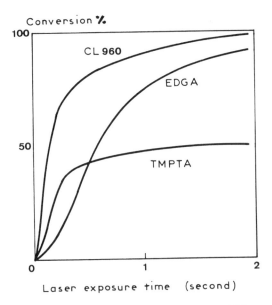

Figure 10. Conversion *versus* time curves recorded by RTIR spectroscopy for a polyurethane-acrylate photoresist, containing different types of diluent, exposed to the UV emission lines of an Ar^+ laser in the presence of air. Power output : 100 mW ; Film thickness : 24 μm.

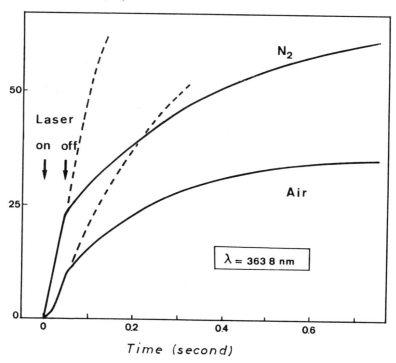

Figure 11. Post-polymerization profiles recorded by RTIR spectroscopy for a polyurethane-acrylate + CL 960 photoresist irradiated for 50 ms with an Ar+ laser beam (l = 363.8 nm) in the presence of air or pure nitrogen. Power output = 100 mW.

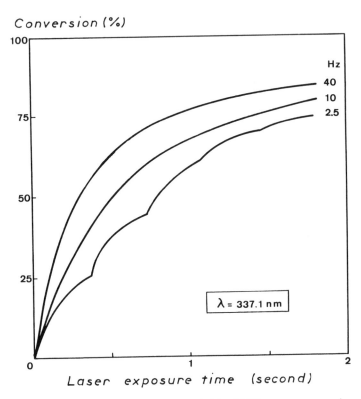

Figure 12. Polymerization profiles recorded by RTIR spectroscopy for a polyurethane-acrylate + CL 960 photoresist exposed to a pulsed nitrogen laser (λ = 337 nm) operated at various repetition rates.

RTIR spectroscopy appears as the only technique presently available that permits determination, quantitatively and in real time, of the kinetics of ultrafast polymerizations induced by high-intensity radiation in particular CW and pulsed lasers. The great sensitivity of IR spectroscopy allows very small changes in the monomer concentration to be detected, e.g. less than 1 % conversion in a 1 µm thick coating, a performance unchallenged so far. This analytical method proved extremely valuable to determine rapidly the efficiency of new photoinitiators and reactive monomers, as well as to study the effect of some physical parameters (light intensity, film thickness, O_2 concentration) on the speed of cure (21). An additional advantage is that this technique is not restricted to the polymerization of acrylic monomers but can be applied to any fast polymerizing system, provided it exhibits a distinct IR absorption characteristic of the reactive functional group.

CONCLUSION

Recent advances in the radiation curing technology have been presented, with special emphasis being directed towards the enhancement of the speed of cure by means of a high performance photoinitiator, newly developed acrylic monomers and the use of UV laser beams. Diphenoxybenzophenone associated to a tertiary amine revealed to be much more efficient than the widely used benzophenone for initiating the photopolymerization of acrylic monomers, while providing at the same time a better through-cure. A similar improvement was obtained with some novel monomers containing one acrylate function and carbamate or oxazolidone structures ; they proved to be much more reactive than difunctional acrylates, leading to close to 100 % conversion. Both the hardness and the flexibility of UV-cured polyurethane-acrylates and epoxy-acrylates were improved when such monomers were used as reactive diluents.

The rate of polymerization and the cure penetration can be further increased by employing powerful lasers as radiation source, the exposure time dropping then into the millisecond range. Polymer relief images were thus obtained at micronic resolution by simply scanning the photosensitive plate with a highly focused laser beam. The kinetic profiles of such ultrafast reactions were directly recorded by using time-resolved infrared spectroscopy which allows instantaneous evaluation at any moment of the actual rate of polymerization and the precise amount of residual unsaturation in the cured polymer.

REFERENCES :

1. Roffey C., *"Photopolymerization of Surface Coatings"* John Willey, New York 1982
2. Tazuke S. in *"Developments in Photochemistry - 3"* Ed. Allen N.S., Appl. Sci. Publ. London 1982 p 53
3. Pappas S.P., *"UV Curing - Science and Technology - Vol 2"* Technology Marketing Corp. Norwalk, Conn. 1985
4. Decker C., J. Coat. Techn. 1987, 59(751), 97
5. Decker C., Bendaikha T., Europ. Polym. J., 1984, 20, 753
6. Decker C., Moussa K., Polym. Preprints 1988, 29(1), 516
7. Morgan C.R., Magnolta F., Ketley A.D., J. Polym. Sci., Polym. Chem. Ed., 1977, 15, 627
8. Chevallier F., Chevalier S., Decker C., Moussa K., SME Techn. Paper 1987, FC 87, 9
9. Decker C., Fizet M., Faure J., Org. Coat. Plast. Chem., 1980, 42, 710
10. Decker C., Jenkins A., Macromolecules 1985, 18, 124
11. Decker C., Proc. Int. Conf. Microcircuit Engineering, Grenoble, 1982 p. 299
12. Decker C., Polym. Photochem.1983, 3, 131

13. Decker C., J. Polym. Sci., Poly. Chem. Ed. 1983, 21, 2451
14. Decker C., ACS Symposium Series 1984, 266, 207
15. Decker C., J. Coat. Tech. 1984, 56(713), 29
16. Hoyle C.E., Trapp M.A., Chang C.H., Polym. Mater. Sci. Eng. 1987, 57, 579
17. Hoyle C.E., Trapp M.A., Chang C.H., Latham D.D., Mc Laughin K.W., Macromolecules (in press)
18. Decker C. in "Radiation Curing of Polymers" Ed. D. Randell, Royal Society of Chemistry 1987, 64, 16
19. Moore J.E. in "UV Curing, Science and Technology, Vol 1", Ed. Pappas S.P., Technology Marketing Corp., Stamford, Conn., 1978, p 133
20. Tryson G.R., Shultz A.R., J. Polym. Sci. Polym. Phys. Ed., 1979, 17, 2059
21. Decker C., Moussa K., Makrom. Chemie, 1988, 189, 2831.
22. Decker C., Moussa K., Macromolecules (in press).
23. Decker C., Moussa K., J. Appl. Polym. Sci., 1987, 34, 1603

RECEIVED September 13, 1989

HIGH-ENERGY RADIATION CURING

High-Energy Radiation Curing

Curing of polymer systems with electron beams and other high energy radiation sources such as gamma rays is different in many respects from curing with light despite the fact that both involve the use of highly energetic sources to effect an immobilization of a coating or plastic material. However, there are distinct similarities with respect to the types of components comprising the electron beam curable formulations. And, of course, both electron beam curing and photocuring involve the same basic polymerization or mechanisms.

The first entry in this section (Chapter 32) describes the use of a gamma ray source to initiate reduction of onium salts with free radicals or solvated electrons. The resultant Brönsted acid or carbenium ion generated is responsible for the ensuing cationic curing of divinyl ethers. This paper leads to speculation of the potential use of such high energy sources for specialized commercial applications. The next paper in this section (Chapter 33) presents a fascinating study of the importance of the morphology of an electron beam induced crosslinking of a copolymer of caprolactone and allyl glycidyl ether. A convincing argument is set forth for the need to consider the crystalline state of a radiation curable material at the time of exposure and its effect on the final physical properties of the crosslinked polymer. Chapter 34 describes an investigation of the function of both inert and reactive solvents on electron beam radiation chemistry. In addition, perfluorinated acrylates mixed with non-fluorinated difunctional acrylate were exposed to electron beams to yield films whose surface free energies could be correlated directly with the concentration of the perfluorinated acrylates. This paper demonstrates the latitude provided by using perfluorinated monomers in radiation curable materials to control the surface properties of the cured coating. Chapter 35 deals with the dependence of dissolution and performance of negative and positive electron beam resists on polymer molecular weight. The paper offers an excellent mix of radiation induced scission and crosslinking theory and measurement of the lithographic properties of films exposed to electron beams. The final paper in this section (Chapter 36) describes the use of electron beams to effect grafting of vinyl silane primers and other common monomers onto the surface of polyethylene and polypropylene films. This chapter points out the benefits obtained when using electron beam initiated grafting to alter surface properties of bulk films for specific applications.

Chapter 32

High-Energy-Radiation-Induced Cationic Polymerization of Vinyl Ethers in the Presence of Onium Salt Initiators

Per-Erik Sundell[1], Sonny Jönsson[2], and Anders Hult[1]

[1]Department of Polymer Technology, The Royal Institute of Technology, S–100 44 Stockholm, Sweden
[2]AB Wilhelm Becker, Box 2041, S–19502, Märsta, Sweden

Cationic polymerization of diethyleneglycol divinyl ether and butanediol divinyl ether in the presence of onium salts was induced by γ-irradiation. The mechanism for the initiation process involves the reduction of onium salts either by organic free radicals or solvated electrons depending on the reduction potentials of the onium salts. The reduction potentials of sulfonium salts was determined by polarography at the dropping mercury electrode. Only solvated electrons were capable of reducing the salts with reduction potentials lower than approximately -100 kJ/mol. The redox process liberates the non-nucleophilic anion from the reduced onium salt and leads to the formation of a Brönsted acid or a stabilized carbenium ion. These species are the true initiators of cationic polymerization in this system. The γ-induced decomposition of onium salts in 2-ethoxyethyl ether was also followed by measuring the formation of protons. An ESR study of the structure of the radicals formed in the γ-radiolysis of butanediol divinyl ether showed that only α-ether radicals were formed.

During the last decade there has been a growing interest in industrial applications of radiation induced polymerizations. Photoimaging (1), photocuring of coatings (2) and printing inks (3) are examples of industrial applications depending on photo-crosslinking of polymers. These processes are based on the photogeneration of radical or cationic species which initiate polymerization and crosslinking. Photoinitiated free radical polymerization of organic coatings is a well established technology in the coating industry today. Although there are a few industrial processes based on photoinitiated cationic polymerization, this field is in an early state of development. Only acrylate systems are used for industrial electron beam curing today. No initiator is required for electron beam curing since free radicals are formed directly upon electron beam irradiation of the coating.

The cationic photoinitiators that are most effective for coating applications are various onium salts. The existence of these onium salts has been known for nearly a century (4, 5). However, the utility of these substances as cationic initiators was not realized until seventy years after their discovery. Now, a wide range of thermally stable photoinitiators based on diaryliodonium (6-8) and triarylsulfonium (9, 10) salts of non-nucleophilic, complex metal halides has been synthesized and evaluated. UV-irradiation of these compounds results in cleavage of a carbon-iodine (11) or carbon-sulfur (12) bond to generate a reactive cation radical which abstracts hydrogen from the surrounding medium to ultimately form a stable, long-lived, strong Brönsted acid. This photogenerated acid can be employed to initiate a variety of cationic polymerization processes as well as other acid catalyzed reactions.

Diaryliodonium and some sulfonium salts can also produce active cationic species via a redox reaction with a photosensitizer (13). An electron is transferred from the excited photosensitizer to the onium salt resulting in generation of a photosensitizer cation radical capable of initiating polymerization. The reduced onium salt decomposes to a radical, aryliodide or a sulfide, and the counterion is liberated. Electron transfer to onium salts also occurs from electron donating organic free radicals (14), which are oxidized to highly reactive carbenium ions. The radical may be generated thermally (14), photochemically (15), or by ionizing radiation (16, 17). Solvated electrons are also capable of reducing diphenyliodonium (18) and triphenylsulfonium (19) salts. This process has been found to promote high energy radiation-induced cationic polymerization of tetrahydrofuran (20). In these cases, reduction of the onium salts does not directly generate cationic species. Instead, the effect is due to the scavenging of solvated electrons by the onium salt and stabilization of radiolytically produced cations by the non-nucleophilic counterion which is liberated when the onium salt decomposes.

The main reason for the limited industrial application of radiation-induced cationic polymerization is that oligomers that offer high cure rates are not commercially available. So far, due to their excellent combination of chemical, physical and electrical properties, only multifunctional epoxy oligomers have been used, but these have poor cure speeds compared to acrylate based systems. This is a serious limitation. Interesting alternatives to the epoxys are oligomers based on highly reactive vinyl ethers (21, 22) and the recently developed distyrene ethers (23), which are as reactive as vinyl ethers. Multifunctional vinyl ethers in the presence of onium salts have also been shown to be an alternative to acrylate systems for electron beam curing (17, 22). Three major advantages of vinyl ethers compared to acrylates are 1) the low toxicity (24), 2) the low dosage required for EB curing and 3) reduced susceptibility to hydrolysis of the cured coating since there are no ester groups in crosslinks.

The present paper reports a study of the initiation mechanism for high energy radiation-induced cationic polymerization of divinyl ethers in the presence of various onium salts. Although there is a great difference in the dose rates of γ-radiation and electron beam, the radiation chemistry is essentially the same.

Experimental Section

Materials. Diethylene glycol divinyl ether, DEGDVE (GAF), butanediol divinyl ether, BDDVE (GAF) and 2-ethoxyethyl ether (Aldrich) were distilled from calcium hydride prior to gamma experiments. All starting materials for the synthesis of onium salts were obtained from Aldrich and used without further purification.

Synthesis. Diphenyliodonium hexafluorophosphate, Ph_2IPF_6, and triphenylsulfonium hexafluoroantimonate, Ph_3SSbF_6, were synthesized using literature procedures (25). Tetra-n-butylammonium hexafluorophosphate, $TBAPF_6$, was obtained by direct metathesis between tetra-n-butylammonium bromide (99%) and potassium hexafluorophosphate (tech.), KPF_6.

The following general procedure has been employed for the synthesis of sulfonium salts listed below. The synthesis is accomplished by the condensation of a halide and a sulfide followed by metathesis to give the desired anion. The yields ranged from 35 to 70%. Equivalent amounts (0.04 mol) of the halide and sulfide were stirred in 15 ml acetonitrile (99%) at room temperature. After five days the solvent was evaporated and the residue was dissolved in 30 ml water. Sodium hexafluoroantimonate (tech.), $NaSbF_6$ (s) (0.04 mol), or KPF_6 (0.04 mol), in 15 ml water was added to give the corresponding SbF_6^- and PF_6^- salts. The salts usually precipitated immediately and were filtered, washed with cold diethyl ether and dried. All salts were recrystallized twice from ethanol (99.5%).

Phenacyltetramethylenesulfonium salts, $PTSPF_6$ and $PTSSbF_6$, were synthesized from 2-chloroacetophenone (99%), tetrahydrothiophene (99%) and KPF_6 or $NaSbF_6$. Benzyltetramethylenesulfonium hexafluoroantimonate, $BTSSbF_6$, was synthesized from benzyl bromide (97%), tetrahydrothiophene and $NaSbF$. 4-Methoxybenzyltetramethylenesulfonium hexafluoroantimonate, $MBTSSbF_6$, was synthesized from 4-methoxybenzylchloride (97%), tetrahydrothiophene and NaSbF. Di-n-butylmethylsulfonium hexafluorophosphate, BMSPF$_6$, was synthesized from di-n-butylsulfide (96%), iodomethane (99%) and KPF$_6$. Satisfactory analytical values were found for all compounds, Table I.

Table I. Characteristics and elemental analysis data of synthesized onium salts. Elemental analysis shows found and (calculated) values

Onium salt	MW g/mol	mp °C	% C	% H	% P	% S	% Sb
Ph$_2$IPF$_6$	426.1	143	33.6 (33.8)	2.0 (2.4)	7.3 (7.3)		
Ph$_3$SSbF$_6$	499.1	177	42.5 (43.3)	2.7 (3.0)		7.3 (6.4)	25.3±1.3 (24.4)
PTSSbF$_6$	443.1	186	32.3 (32.5)	3.1 (3.4)		7.3 (7.2)	29.3±1.5 (27.5)
BTSSbF$_6$	415.0	122	31.8 (31.8)	3.4 (3.6)		7.9 (7.7)	30.7±1.5 (29.3)
MBTSSbF$_6$	429.1	62	32.2 (33.6)	3.7 (4.0)		7.0 (7.5)	28.3±1.4 (28.4)
BMSPF$_6$	306.3	120	35.2 (35.3)	6.9 (6.9)	10.4 (10.5)	10.5 (10.5)	

Procedures. Polymerizations were carried out in polypropylene tubes (1 cm diameter). Solutions (2 ml) of oligomers with an initiator concentration of 10^{-3} M were bubbled with either argon, Ar, nitrous oxide, N$_2$O, or oxygen, O$_2$, for 20 minutes and then the tubes were sealed. The reactions were induced by exposure to gamma, γ, rays from an AECL$_{220}$Co^{60} γ-cell giving a dose rate of 1500 Gy/h. Samples were checked at intervals of 1 minute (25 Gy) until 12 minutes (300 Gy) had passed and thereafter at 2 minute intervals (50 Gy). The dosage required for inducing polymerization was taken as the gelation point of the sample. The γ-radiation-induced proton formation from 10^{-3} M onium salt in 2-ethoxyethyl ether was followed spectroscopically at 540 nm using α-naphtyl red as indicator. A calibration curve with methanesulfonic acid in 2-ethoxyethyl ether was used to determine the H$^+$ concentration of irradiated samples. UV measurements were conducted on a Hewlett Packard 8451A Diode Array Spectrophotometer. BDDVE and 2-methyl-2-nitrosopropane as a spin trap were used for ESR experiments. ESR samples were placed in 3 mm Suprasil quartz ESR tubes, evacuated with the freeze-thaw method, sealed and exposed to a dose of 1500 Gy. The ESR instrument used was a Bruker EPR 420. The polaro-graphic apparatus was a Princeton Applied Research Model 174A Polarographic Analyzer. The aqueous pH 7.6 buffer for polarography was a solution of 0.042 M potassium dihydrogen phosphate (p.a.) and 0.015 M sodium tetraborate decahydrate (p.a.). The buffer solution containing 10^{-3} M onium salt was bubbled with nitrogen for 15 minutes to remove oxygen before measurement and the half wave potential, $E_{1/2}$, of the onium salt was determined at the dropping mercury cathode with a saturated calomel electrode as anode. A three second drop time was employed. Gelatin (0.002 %) was used as a maximum supressor.

Results and Discussion

When an organic coating is exposed to γ-rays, fast electrons are generated in the sample by essentially three processes: photoelectric absorption, Compton scattering and pair production. Fast electrons dissipate most of their energy in matter causing ionization, Reaction 1 and excitation of molecules, Reaction 2. When a molecule is excited above its ionization potential (10-12 eV for most organic molecules) it may lose energy by ionization or dissociation, Reaction 5, 6. The yield of ions in a liquid depends on the distance the ejected electron travels before reaching thermal energy levels (about 0.025 eV at room temperature). This distance depends to a large extent on the electron density of the medium. For liquids with low dielectric constants, ε, the G-value for free ion production very low, indicating that most electrons rapidly recombine with their geminate positive ions, Reaction 3. The recombination produces an excited molecule with an energy excess lower than the ionization potential but often high enough to cause radical fragmentation. As ionized and excited species undergo further reactions, neutral free radicals are produced by several processes, Reaction 4-10. The yield of free radicals is much higher than that of free ions. For example, in the radiolysis of liquid hydrocarbons, $\varepsilon \approx 2$, the stationary state concentration of free radicals is about two orders of magnitude higher than that of free ions (26) due to lower G-value (about 0.15 (27)) and higher recombination rates for ions relative to radicals (G_r=5.5 (28)). In ethers, having $\varepsilon \approx 4$, the yield of ions is higher, G_i(diethyl ether)=0.35 (27), reportedly due to the possibility of electrons becoming solvated (29). Whether electrons really are solvated or not in DEGDVE and BDDVE is not clear but electrons will be considered to be solvated in the following discussion. Thus the main product from radiolysis of organic liquids is free radicals, which, are consequently responsible for a majority of the subsequent chemical reactions.

$$\text{=\!\!O}\!\!\frown\!\!R \xrightarrow{\gamma} \text{=\!\!\overset{+\cdot}{O}}\!\!\frown\!\!R + e^- \quad (1)$$

$$\text{=\!\!O}\!\!\frown\!\!R \xrightarrow{\gamma} \left(\text{=\!\!O}\!\!\frown\!\!R\right)^* \quad (2)$$

$$\text{=\!\!\overset{+\cdot}{O}}\!\!\frown\!\!R + e^- \longrightarrow \left(\text{=\!\!O}\!\!\frown\!\!R\right)^* \quad (3)$$

$$\text{=\!\!\overset{+\cdot}{O}}\!\!\frown\!\!R \longrightarrow \text{=\!\!O}\!\!\frown\!\!\overset{\cdot}{R} + H^+ \quad (4)$$

$$\left(\text{=\!\!O}\!\!\frown\!\!R\right)^* \longrightarrow \text{=\!\!O}\!\!\frown\!\!\overset{\cdot}{R} + H\cdot \quad (5)$$

$$\left(\text{=\!\!O}\!\!\frown\!\!R\right)^* \longrightarrow \text{=\!\!O}\cdot + R\cdot \quad (6)$$

$$H^+ + e^- \longrightarrow H\cdot \quad (7)$$

$$H\cdot + \text{=\!\!O}\!\!\frown\!\!R \longrightarrow \text{=\!\!O}\!\!\frown\!\!\overset{\cdot}{R} + H_2 \quad (8)$$

$$R\cdot + \text{=\!\!O}\!\!\frown\!\!R \longrightarrow \text{=\!\!O}\!\!\frown\!\!\overset{\cdot}{R} + RH \quad (9)$$

$$R\cdot + \text{=\!\!O}\!\!\frown\!\!R \longrightarrow R\!\!\smile\!\!\overset{\cdot}{\underset{O}{}}\!\!\frown\!\!R \quad (10)$$

$$e_s^- + N_2O \longrightarrow N_2 + O^- \quad (11)$$

$$O^- + H^+ \longrightarrow \cdot OH \quad (12)$$

Reaction 1

Vinyl ethers are among the most reactive monomers for cationic polymerization but they can not be homopolymerized by a radical mechanism. This was illustrated by the fact that no polymerization occurred when BDDVE and DEGDVE were exposed to γ-rays in the absence of onium salt initiators even at a dosage of 15 kGy. When BDDVE is γ-irradiated one would expect several radicals to be formed. However, the ESR spectrum of γ-irradiated BDDVE, using 2-methyl-2-nitrosopropane as a spin trap, exhibited only one detectable radical, Figure 1. The signal was very sharp and strong. Comparing the hyperfine splitting constants with known radicals shows that the radical formed from γ-irradiation of BDDVE is an α-ether radical. Since this is the only radical found, it seems that mainly C-H bonds was cleaved when excited molecules of BDDVE dissociate, Reaction 5. The adduct arising from C-C bond cleavage, Reaction 6, followed by radical addition to BDDVE, Reaction 10, would certainly exhibit a different ESR spectrum. The radical is α to an ether and therefore stable and is not likely to add to the vinyl ether double bond since this would be analogous to radical homopolymerization of the vinyl ether monomer which is known not to occur.

Iodonium and sulfonium salts undergo electron reductions by solvated electrons and electron donating free radicals, both products from the γ-radiolysis. Scheme 1 shows this processes for diphenyliodonium (14, 18) and triphenylsulfonium salts (14, 19). In the first process, Route 1, reduction of the onium salt liberates the counterion which promotes polymerization either by forming a strong Brönsted acid with radiolytically produced protons or by stabilizing other positive ions that are formed during the radiolysis. Route 2 yields a stabilized carbenium ion directly by an electron transfer reaction between onium salt and an organic free radical. The reduced onium salt decomposes immediately to phenyliodide or diphenylsulfide and a phenyl radical which abstracts hydrogen from its environment, producing another free radical which might also be capable of reducing the onium salt and thus initiating a chain reaction. The Brönsted acid and ion pair formed are the true initiators of cationic polymerization in this system.

In order to distinguish between the solvated electron (Route 1) and the radical mechanism (Route 2), samples were bubbled with N_2O or O_2 and the dosages for curing were compared with samples bubbled with Ar. Ar-saturated samples required smaller dosages. In the presence of nitrous oxide electrons are scavenged producing nitrogen, Reaction 11, and OH· radicals, Reaction 12, which either abstract hydrogen or add to double bonds. Both abstraction and addition results in an α-ether radical thus insuring that the concentration of oxidizable radicals is maintained. Therefore, only samples initiated according to Route 1 should require higher dosages in the presence of N_2O.

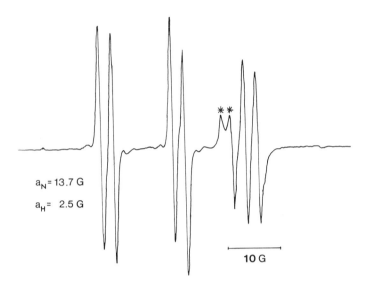

Figure 1. The ESR spectrum of γ-irradiated BBDVE using 2-methyl-2-nitrosopropane as a spin trap. * = Quartz signal.

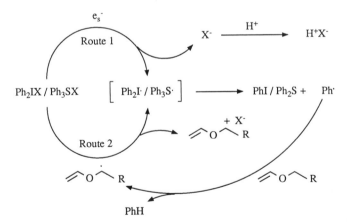

Scheme 1. Electron reductions of Ph_2IPF_6 and Ph_3SSbF_6 by solvated electrons, Route 1, and by electron donating free radicals, Route 2.

Oxygen-saturation of the samples results in scavenging of radicals, but also some electrons. Therefore, oxygen saturation increases the dosages for samples that are initiated according to both mechanisms but the effect is most pronounced for radical induced polymerization (Route 2). The mechanism that dominates depends on the redox potential of the onium salt and on the radiation yield of oxidizable radicals and solvated electrons. Since the stationary state concentration of radicals is much higher than that of solvated electrons, one would expect that radical reduction of the salt is the dominating process. This was indeed true for some onium salts, but two thermodynamic prerequisites have to be fulfilled. First, the energy required to oxidize the radical, E_r^{ox}, must be greater than the energy required for reduction of the onium salt, E_o^{red}. Second, the free energy, ΔG, for the reaction must be negative (exothermic), Equation 1.

$$\Delta G = E_r^{ox} - E_o^{red} < 0 \qquad \text{Eq. 1}$$

α-Ether radicals are easily oxidized due to the possibility of forming a double bond (30). The oxidation potential for α-oxygen radicals derived from diethyl ether and tetrahydrofuran is -0.8 eV (-78 kJ/mol) (30). Since the radicals derived from BDDVE and DEGDVE have similar structures, their oxidation potentials should be of about the same, i.e. $E_r^{ox} \approx -78$ kJ/mol. The reduction potential, E^{red} for all onium salts except for TBAPF$_6$ was determined polarographically and the results are given in Table II together with calculations of ΔG-values for electron transfer reduction of onium salt by α-oxygen radicals and the results of the polymerization studies.

For salts with reduction potentials lower than approximately -100 kJ/mol, N$_2$O had a huge effect on required dosages, typically a 400-500% increase in dose was required for polymerization compared to Ar-saturated samples. This implies that these salts are at least mainly reduced by solvated electrons. The effect of oxygen was most pronounced for samples containing salts with reduction potentials higher than approximately -65 kJ/mol, indicating that these salts are preferentially reduced by free radicals. Also, the fact that salts with lower reduction potentials were also affected by O$_2$, can be attributed to some scavenging of solvated electrons by oxygen. The exact value of the reduction potential at which electron transfer from radicals becomes less efficient and reduction by solvated electrons begins to dominate must lie in the vicinity of the reduction potential for α-ether radicals, i.e. -78 kJ/mol. For salts with more negative reduction potentials, the electron transfer process becomes unfavorable due to positive ΔG-values. Thus, the iodonium and phenacylsulfonium salts are essentially reduced by free radicals (Route 1) while Ph$_3$SSbF$_6$, BTSSbF$_6$ and BMSPF$_6$ are reduced by solvated electrons (Route 2).

Table II. Results of polymerization studies. Cationic polymerization of DEGDVE induced by γ-rays in the presence of onium salt initiators. Dose rate = 1500 Gy/h. The reduction potential, E^{red}, for the salts and the free energy, ΔG, for electron transfer reduction of onium salts by α-ether radicals are given

Initiator	E^{red} (kJ/mol)	ΔG (kJ/mol)	Dose (Gy) Ar	N_2O	O_2
Ph_2IPF_6	-14	-59	75	100	600
$PTSSbF_6$	-63	-15	250	275	775
$MBTSSbF_6$	-88	+10	300	600	625
Ph_3SSbF_6	-112	+34	400	1900	1000
$BTSSbF_6$	-100	+22	500	2150	1150
$BMSPF_6$	-146	+68	600	2650	1625
$TBAPF_6$	-249[a]	+171	no polymerization		

[a] Reference (31)

The substituted benzylsulfonium salt, $MBTSSbF_6$, was equally affected by N_2O and O_2, indicating that both free radicals and solvated electrons played important roles in the redox process which was a bit surprising since ΔG is positive for Route 1. This indicates that there was no abrupt change in the reduction mechanism rather there seems to be a gradually changing process as the reduction potential changes. In the presence of $TBAPF_6$ no polymerization was observed, even at 15 kGy. The very high reduction potential of quartenary ammonium salts appears to make these substances useless as redox initiators for cationic polymerization. A discrepancy between E^{red} and dose for Ph_3SSbF_6 and $BTSSbF_6$ was also found. Although having a higher E^{red} value, Ph_3SSbF_6 required a lower dose than $BTSSbF_6$. One reason for this observation could be that the sulfides produced by decomposition of the sulfonium salts have different nucleophilicity. Thus, tetrahydrothiophene is much more basic than diphenylsulfide and might be able to terminate polymerization by forming a new alkylsulfonium salt. This effect has previously been shown to occur in polymerization photoinitiated with Ph_2IPF_6 in which phenyliodide terminates the growing chain end by forming a new iodonium salt[32]. The addition of 5 mM THT to a sample led to a fourfold increase in the dose, while addition of the same amount of Ph_2S caused only a moderate increase. The same effect was observed for photoinitiated polymerization of samples containing Ph_3SSbF_6 and THT or Ph_2S. THT requires a threefold and Ph_2S a twofold increase in dose.

Proof that $BTSSbF_6$ was really reduced at a lower dose than Ph_3SSbF_6 was demonstrated when the decomposition of onium salts was studied by measuring the H^+ concentration in 2-ethoxyethyl ether at different doses. An H^+ concentration of 10 mM was obtained when $BTSSbF_6$ was exposed to 250 Gy while Ph_3SSbF_6 required 300 Gy to reach the same value. Figure 2 shows the results for $PTSSbF_6$, Ph_3SSbF_6 and $BMSPF_6$. The dose required for inducing polymerization of DEGDVE, i.e. 250, 500 and 600 Gy respectively, correspond to a proton concentration of about 0.15-0.20 mM in 2-ethoxyethyl ether. Figure 3 shows the effect of Ar, O_2 and N_2O on H^+ formation in γ-radiolysis of Ph_3SSbF_6 in 2-ethoxyethyl ether. The results are in good agreement with the results from the polymerization study and the suggested reduction mechanism for Ph_3SSbF_6 (Route 2).

No difference in initiation efficiency was observed for sulfonium ions with identical structure containing different counterions. Phenacyltetramethyl-enesulfonium salts with PF_6^- and SbF_6^- as counterions required the same dose for inducing polymerization. This indicates that the counterion does not take part in the redox process, which is consistent with the above outlined mechanisms. On the other hand, there was a remarkable difference between salts with different anions in how the polymerization proceeded. Salts containing SbF_6^- usually gave violent polymerizations, evolving so much heat that the PP-tubes melted and resulting in strongly discolored samples. PF_6^- salts produced a much smoother polymerization giving clear, transparent and uncolored polymer samples.

Figure 4 shows the effect of Ph_3SSbF_6 and $PTSSbF_6$ concentration on the dose required for inducing polymerization of DEGDVE. The salts exhibit similar behavior. There is a dramatic increase in the dose when the concentration is dropped below 1 mM while at concentrations higher than 10 mM there was very little gain in dosage with increasing concentration. This same behavior is observed in the case of electron beam induced polymerization of DEGDVE in the presence of triphenylsulfonium salt (17), although the curve is shifted an order of magnitude to the right. The dosages are also about thirty times lower for γ-induced polymerization than for electron beam induced polymerization. Some of the difference is explained by the use of purified monomers in this work, but the dose rate is also an important factor to consider. At low dose rates, the stationary state

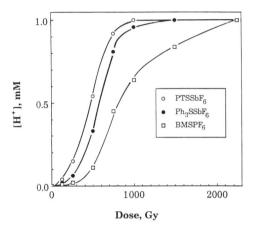

Figure 2. The concentration of protons produced by γ-irradiation of PTSSbF$_6$, Ph$_3$SSbF$_6$ and BMSPF$_6$ in Ar-saturated 2-ethoxyethyl ether.

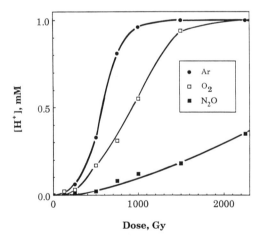

Figure 3. The concentration of protons produced by γ-irradiation of Ph$_3$SSbF$_6$ in 2-ethoxyethylether as a function of dose. Differences between Ar, O$_2$ and N$_2$O-saturated samples.

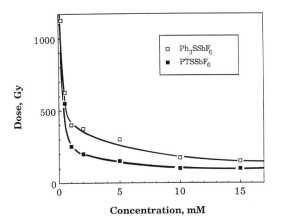

Figure 4. Effect of Ph$_3$SSbF$_6$ and PTSSbF$_6$ concentration on the dosages required for inducing polymerization of Ar-saturated DEGDVE.

concentration of active species is low and most of the reactions are diffusion controlled. The lifetime of active species is proportional to the square root of the dose. As the dose rate in EB-irradiation is about a thousand times higher than that in γ-irradiation, a difference of a factor of about thirty is expected. So, at sufficiently high dose rates, radical-radical reactions predominate and other radical reactions become unimportant (28). Thus a very high dose rate is unfavorable for the efficiency of the electron transfer reduction of onium salts since the product yields are dependent of the dose rate. A consequence of this dose rate dependence is that higher initiator concentrations are required in EB applications.

Conclusions

In conclusion, the main features of high energy radiation-induced cationic polymerization of vinyl ethers in the presence of onium salts are: First, the generation of solvated electrons and oxidizable radicals by high energy radiation. Second, the reduction of onium salts, resulting in the formation of stable cationic species and finally, the initiation of cationic polymerization by these stable cationic species. In the irradiation process, the G-values for radical and ion formation, and hence the dose rate, influence the efficiency in the following processes, in which the reduction potential of the onium salt and the oxidation potential for radicals are the main factors to consider when choosing initiators.

References and Notes

(1) Thompson, L.F.; Willson, C.G.; Bowden, M.J. (eds.): In Introduction to Microlithography; ACS Symposium Series No.219, American Chemical Society, Washington D.C., 1983.
(2) Pappas, S.P. (ed.): UV Curing Science and Technology; Technology Marketing Corp., Norwalk, Conn., 1978.
(3) Wentik, S.G.; Kock, S.D. (eds.): UV Curing in Screen Printing for Printed Circuits and the Graphic Arts; Technology Marketing Corp., Norwalk, Conn., 1981.
(4) Hartmann, C.; Meyer, V., Ber. Dtsch. Chem. Ges. 1894, 27, 502; 1592
(5) von Halban, H., Z. Phys. Chem., Stoechiom. Verwandtschaltsl. 1909, 67, 129
(6) Smith, G.H. Belgium Patent 828 841, 1975.
(7) Imperial Chemical Industries, Belgium Patent 837 782, 1970.
(8) Crivello, J.V. U.S. Patent 3 981 897, 1976.
(9) Crivello, J.V. Belgium Patent 828 670, 1974.
(10) Imperial Chemical Industries, Belgium Patent 833 472, 1976.
(11) Crivello, J.V.; Lam, J.H.W. Macromolecules 1977, 10, 1307.

(12) Crivello, J.V.; Lam, J.H.W. J. Polym. Sci., Polym. Chem. Ed. 1979, 17, 977.
(13) Pappas, S.P.; Jilek, J.H. Photogr. Sci. Eng. 1979, 23, 140.
(14) Ledwith, A. Polymer 1978, 19, 1217.
(15) Abdul-Rasoul, F.A.M.; Ledwith, A.; Yagci, Y. Polymer 1978, 19, 1219.
(16) Mah, S.; Yamamoto, Y.; Hayashi, K. Macromolecules 1983, 16, 681.
(17) Lapin, S.C.: Radcure '86: Conference Proceedings, Association for Finishing Processes, Baltimore, 1986.
(18) Bachofner, H.E.; Beringer, F.M.; Meites, L. J. Am. Chem. Soc. 1958, 80, 4269.
(19) Mckinney, P.S.; Rosenthal, S. J. Electroanal. Chem. 1968, 16, 261.
(20) Ma, X-H.; Yamamoto, Y.; Hayashi, K. Macromolecules 1987, 20, 2703.
(21) Dougherty, J.A.; Vara, F.J.; Anderson, L.R.: Radcure '86: Conference Proceedings, Association for Finishing Processes, Baltimore, 1986.
(22) Lapin, S.C.: Radtech '88: Conference Papers, New Orleans, 1988, 395.
(23) Ericsson, J.; Hult, A. Polym. Bull. 1987, 18, 295.
(24) Dougherty, J.A.; Vara, F.J.: Radcure Europe '87: Conference Proceedings, Association for Finishing Processes, Munich, 1987.
(25) Crivello, J.V.; Lam, J.H.W. J. Org. Chem. 1978, 43, 3055.
(26) Williams, F.: In Fundamental Processes in Radiation Chemistry, Ausloos, P. (ed.), Interscience, New York, 1968, 515.
(27) Schmidt, W.F.; Allen, A.O. J. Phys. Chem. 1968, 72, 3730.
(28) Holroyd, R.A.: In Fundamental Processes in Radiation Chemistry, Ausloos, P. (ed.), Interscience, New York, 1968, 413.
(29) Arai, S.; Sauer Jr., M.C. J. Chem. Phys. 1966, 44, 2297.
(30) Henglein, A. J. Electroanal. Chem. 1976, 9, 163.
(31) van Rysselberghe, P.; McGee, J.M. J. Am. Chem. Soc. 1945, 67, 1039.
(32) Yagci, Y.; Ledwith, A. J. Polym. Sci. Polym. Chem. Ed. 1988, 26, 1911.

RECEIVED September 13, 1989

Chapter 33

Structure—Property Behavior of Caprolactone—Allyl Glycidyl Ether Copolymers Cross-Linked by Electron-Beam Irradiation

Ha-Chul Kim[1,3], Abdel M. El-Naggar[1,4], Garth L. Wilkes[1,5], Youngtai Yoo[2], and James E. McGrath[2]

[1]Department of Chemical Engineering and Polymer Materials and Interfaces Laboratory, Virginia Polytechnic Institute and State University, Blacksburg, VA 24061
[2]Department of Chemistry and Polymer Materials and Interfaces Laboratory, Virginia Polytechnic Institute and State University, Blacksburg, VA 24061

> Semicrystalline copolymers of caprolactone and allyl glycidyl ether(CL-AGE) have been crosslinked at various temperatures both above and below the crystalline melting point by electron beam irradiation. Gel fraction increases with irradiation dose up to 40 Mrad and shows slightly higher values when irradiated above the melting point compared to irradiation below the melting point. A stable spherulitic crossed polarizer optical microscopy pattern of the CL-AGE copolymer cross-linked in the semicrystalline state persisted over a wide range of temperature even in the melt. In contrast, random introduction of crosslinks into this material in the melt state resulted in some restriction on the crystallization process limiting the development of ordered superstructure upon solidification. Measurements of thermal and mechanical properties support the gel fraction results and the microscopy observations. It was concluded that the final solid state properties of the EB-irradiated CL-AGE copolymers are strongly dependent on the morphological state of the material at the time of the EB irradiation process.

[3]Current address: Research and Development Department, Technical Center, Mobil Chemical Company, Macedon, NY 14502
[4]Current address: National Center for Radiation Research and Technology, Cairo, Egypt
[5]Address correspondence to this author.

Crosslinking and chain scission of polymers induced by high energy radiation are generally simultaneous phenomena and the ratio of these two events determines the net effect of the process. Depending on the value of this ratio, most polymers are often categorized into two distinct classes- those which crosslink and those which degrade. Polyethylene which usually crosslinks and polyisobutylene which primarily degrades by radiation are two opposing examples with polypropylene often being considered as intermediate with regard to the chemical structure and the response to radiation(1,2).

Crosslinkable semicrystalline polymers such as polyethylene can be irradiated by an electron beam source at any desired temperatures to produce a network by crosslinking, while many other network forming processes such as peroxide crosslinking are typically undertaken with best results at temperatures above the melting or softening point. Polyethylene or other semicrystalline polymers crosslinked in the melt state generally result in changes in crystallinity as well as in mechanical response when compared to the same system crosslinked in the semicrystalline state. Therefore, it can be deduced that radiation effects on semicrystalline polymers are strongly dependent on sample morphology(and/or chain mobility) at the time of irradiation(3).

The overall net effect of radiation on saturated linear aliphatic polyesters has been known to depend on the ratio of the methylene group concentration to that of the ester group. D'alelio et al.(4) showed that the increased methylene to ester ratio resulted in a higher gel content for a given dose. Poly(glycolic acid), which possesses a one-to-one concentration of methylene groups to ester groups, degrades at a relatively low radiation dose while high density polyethylene (no ester group) predominantly crosslinks representing extreme cases in this regard. Radiation effects on polycaprolactone, a semicrystalline polymer with the methylene to ester group ratio of 5, has been reported to show an initial gel formation in a critical dose range of 20-30 Mrad when irradiated at room temperature(5). It is this latter polymer that will be of interest within the present study.

In order to achieve lower gelation doses for saturated linear aliphatic polyesters such as polycaprolactone, certain functional groups including unsaturation can be incorporated. In this regard, and as will be discussed here, ε-caprolactone has been copolymerized with a relatively small amount of allyl glycidyl ether. Therefore, the randomly distributed allyl glycidyl ether units(2-6 mole percent) in the synthesized caprolactone-allyl glycidyl ether(CL-AGE) copolymers(semicrystalline at room temperature) provide crosslinkable sites by electron beam irradiation in a relatively low dose range thereby limiting damage to the backbone. It has been of interest to investigate whether the radiation-induced crosslinking of the CL-AGE copolymers in the semicrystalline state will result in a network which differs considerably in properties from that produced by crosslinking in the amorphous or melt state.

Experimental

Materials

Caprolactone-allyl glycidyl ether(CL-AGE) copolymers were synthesized as shown in Figure 1. First, the aluminum-porphyrin catalyst, which is an effective initiator of alkylene oxide and lactone polymerization, was prepared following the general procedure reported by Aida and Inoue(6). Secondly, after the catalyst had been synthesized, a mixture of purified ε-caprolactone and allyl glycidyl ether(Aldrich Chemical Co.) was added to a reactor. The monomer feed which was composed of primarily ε-caprolactone with 5, 10, 15 mole % of allyl glycidyl ether was brought into reaction by stirring under a nitrogen atmosphere at 60°C. After an appropriate period of time, the reaction mixture was cooled down to room temperature and the final product was isolated by precipitation in methanol. Although, 5, 10, 15 mole percent of allyl glycidyl ether was included initially, approximately 2, 4, 6 mole percentages of allyl glycidyl ether were found in the final respective copolymer product when subjected to proton-NMR analysis(7). The number average molecular weight was found to be nearly 30,000 as determined by GPC analysis calibrated with polystyrene and polymethyl methacrylate standards. The glass transition temperature and the crystalline melting temperature of the prepared CL-AGE copolymer(6 mol% AGE) were found to be ca. -45°C and 56°C, respectively. A melting point depression of ca. 4°C by the introduction of AGE units(6 mol%) was observed in comparison to pure polycaprolactone of same molecular weight(Tm=60°C)-this behavior being discussed in more detail later.

Preparation of Samples for Irradiation

The synthesized CL-AGE copolymers were dissolved in tetrahydrofuran(THF) at room temperature to give a homogeneous solution. After film casting and drying at an atmospheric condition for 24 hours, the

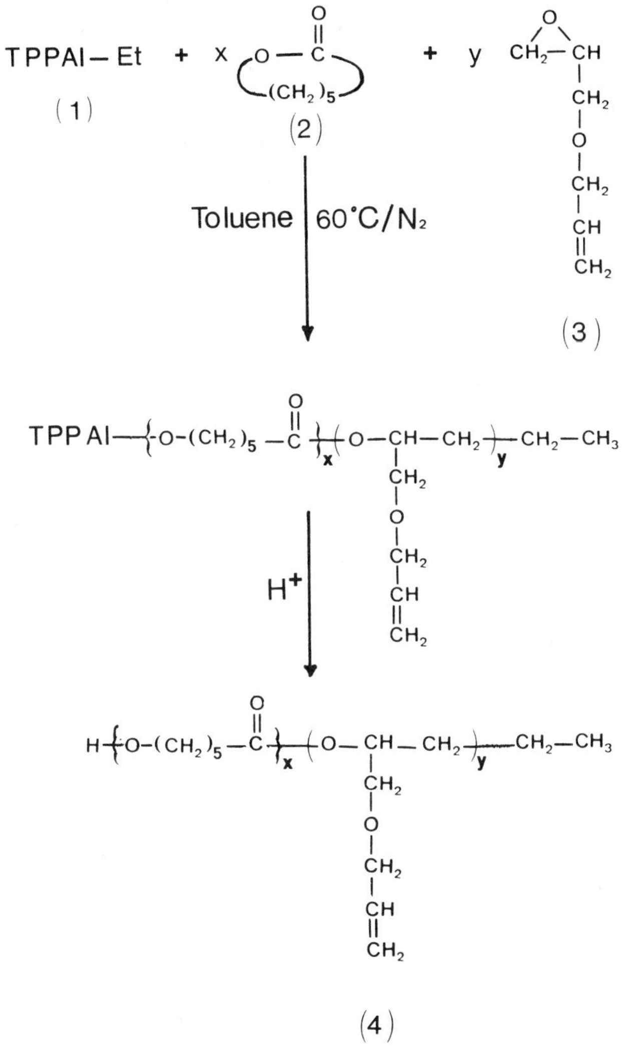

Figure 1. Synthetic scheme for the CL-AGE copolymers: (1)aluminum-porphyrin catalyst; (2)ε-caprolactone; (3)allyl glycidyl ether; (4)caprolactone and allyl glycidyl ether copolymer.

CL-AGE copolymer films of ca. 3-mil-thickness were additionally vacuum-dried for 48 hours. Two sample temperatures, 25°C(semi-crystalline state) and 80°C(melt state), were utilized for irradiation. For the higher temperature of 80°C, a heated steel plate was used as the substrate for the copolymer. The film sample was placed on this temperature-controlled steel plate and passed through the electrocurtain system described below. For the irradiation process which required two passes, following the first pass the sample was again immediately placed on another 80°C steel plate and passed through the electrocurtain system a second time.

Irradiation and EB Apparatus

Electron beam irradiation has been carried out with an electrocurtain accelerator manufactured by Energy Sciences, Inc.(model CB/150/15/180). The samples were placed on steel plates in aluminum trays and passed through the conveyor system of the electron beam apparatus. The maximum available dose per pass was 20 Mrad, hence, for the highest dose used in this study(40 Mrad), two passes were utilized. In light of the depth-dose profile at 175 kilovolts electron energy level of the EB system, the radiation dose will be nearly unifiorm throughout the sample thickness(3 mil).

Gel Percent Measurement

The respective gel fraction of the irradiated CL-AGE copolymer materials was evaluated as follows. First, a 10 to 40 milligram sample of the irradiated CL-AGE copolymer free film was immersed and stirred in 200 ml of THF at room temperature for 72 hours. Then, the sample was removed from THF and dried in a vacuum oven at room temperature for 96 hours and the weight of the dried sample was measured. The gel content was given by the percent ratio of the final dried gel weight to the initial weight.

Optical and Scanning Electron Microscopy Analysis

A Zeiss polarizing microscope equipped with a Leitz 350 heating stage and a 35 mm camera was utilized to investigate the change in the crystalline structure by changing temperature. The temperature calibration of the heating stage was performed with naphthalene, indium, anthraquine and sodium nitrate. A scanning electron microscope(SEM, Cambridge Stereoscan 200) was also utilized for morphological investigations.

Thermal Test

Measurements of thermal properties were performed using a differential scanning calorimeter(Perkin-Elmer DSC-4). The samples were run from 15 °C up to 75°C with a heating rate of 10°C per minute and the data obtained were labeled as "first run." After cooling to 10°C with a cooling rate of 10 °C per minute, samples were again heated up to 75°C at the same scan rate. These data were labeled as "second run." The crystalline melting point of the CL-AGE material was determined as functions of the radiation dose and the morphological state at the time of irradiation.

Mechanical Analysis

The irradiated samples were tested for mechanical properties such as Young's modulus at room temperature using the ASTM D638 specification. Samples were prepared in a dog-bone shape of initial dimensions corresponding to gauge length of 10 mm and width of 2.8 mm. An Instron tensile tester(model 1122) was utilized throughout this study with an extension ratio of 50 % per minute based on the initial sample length.

Results and Discussion

EB Crosslinking of CL-AGE Copolymers

CL-AGE copolymers having different contents(2, 4, 6 mole percent) of allyl glycidyl ether were EB-irradiated at room temperature by changing the irradiation dose from 0 to 40 Mrad. Figure 2 shows the gel fraction of these CL-AGE copolymers as a function of the irradiation dose. Up to 20 Mrad, the gel fraction increases as the content of AGE increases suggesting an improved crosslinking reactivity of the copolymers upon EB irradiation. However, at 40 Mrads the differences in the gel fraction values between the samples of different AGE contents were somewhat reduced and, in fact, the order was partly reversed. It was reported by Narkis et al.(5) that the critical gelation dose level for pure polycaprolactone(MW 35,000), which is the major backbone of the CL-AGE copolymers in this study, is 26 Mrads and, above this irradiation dose, measurable gel contents are formed with about 15% gel at 40 Mrads. Therefore, in the data shown in Figure 2, it can be deduced that, up to 20 Mrads, the crosslinking of the CL-AGE copolymers is principally attributed to the AGE content although the CL content likely affects the gel fraction data particularly at the higher dose level. To support this discussion, samples of pure polycaprolactone(MW 30,000) were EB-irradiated with a dose of up to 20 Mrads. Figure 3 shows typical

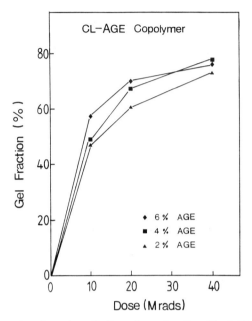

Figure 2. Percent gel vs. irradiation dose for the CL-AGE copolymers of different AGE mole percent.

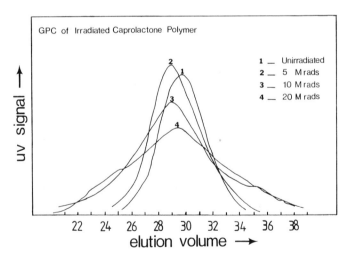

Figure 3. GPC chromatograms of irradiated pure polycaprolactone in the pregelation stage.

chromatograms obtained by gel permeation chromatography(GPC) as a function of irradiation dose in the pregelation stage of pure polycaprolactone. By irradiation with 5 Mrad, the average molecular weight appears to be shifted slightly to a higher value suggesting an increased molecular weight by EB irradiation. Then, the narrower molecular weight region of the distribution gradually widens with increasing dose and, at 20 Mrad, low and high molecular weight tails become clearly evident. It would be expected that this high molecular weight fraction shown in the tail would be transformed to the gel fraction upon further irradiation. Therefore, it can be deduced that when the CL-AGE copolymers are EB irradiated in the dose range up to 20 Mrad, the randomly distributed allyl glycidyl ether along the poly(ε-caprolactone) backbone provides readily crosslinkable sites without any major gelling contribution to the gel fraction from the caprolactone chain component. On the basis of these data, the CL-AGE copolymer with a 6 mole percentage of AGE units was chosen as the system for further analysis of radiation effects as will now be discussed below.

Shown in Figure 4 are the gel fractions measured for the CL-AGE(6 mol% AGE) copolymer irradiated at room temperature(semicrystalline state) as well as in the melt state at at 80°C(T_m = 56°C) with increasing irradiation dose up to 40 Mrad. Although irradiation in the melt state results in higher gel fractions, the data suggest that a considerably high degree of crosslinking also occurred for irradiation in the semicrystalline state(room temperature irradiation). As discussed later in this paper, when the copolymer is irradiated in the semicrystalline state, a certain degree of the ordered crystalline state which is present at the time of irradiation is maintained in the process of the network formation. In addition, this ordered state is shown to persist beyond the crystalline melting transition.

Optical and Scanning Electron Microscopy Analysis

Shown in Figure 5 are the cross-polarized optical micrographs of the CL-AGE copolymers crosslinked by EB irradiation with 20 Mrad at room temperature(semicrystalline state) and at 80°C(amorphous melt state) along with the control sample(no EB treatment). In order to study the morphological structure at different temperatures, a microscope heating stage was used for the irradiated sample. The sample temperature was raised to 120°C and then cooled down to room temperature in a step-by-step comparison with the control sample given the same thermal history in the hot stage.

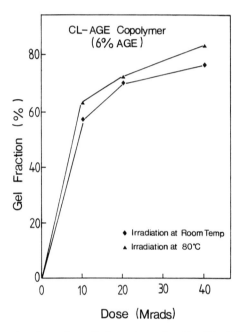

Figure 4. Percent gel vs. irradiation dose for the CL-AGE(6 mol% AGE) copolymers irradiated at room temperature and 80°C.

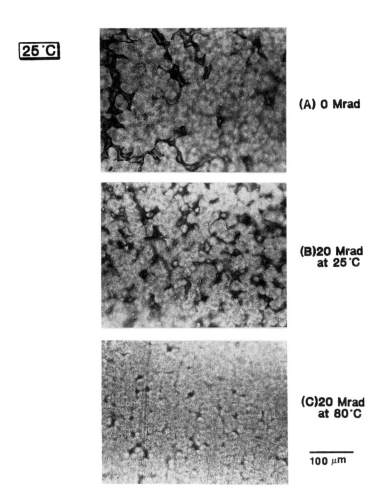

Figure 5a. Cross polarizer optical micrographs taken for the CL-AGE copolymer control sample and those irradiated at 25 °C and 80 °C with changing temperature of the heating stage: 25 °C.

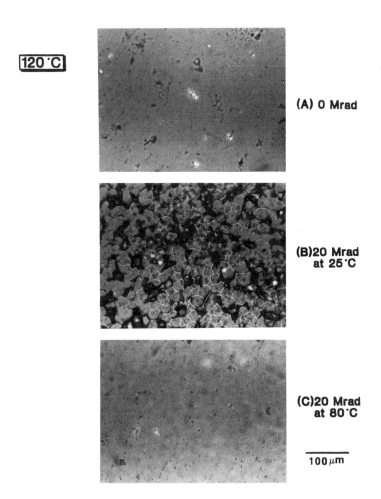

Figure 5b. Cross polarizer optical micrographs taken for the CL-AGE copolymer control sample and those irradiated at 25 °C and 80 °C with changing temperature of the heating stage: 120 °C.

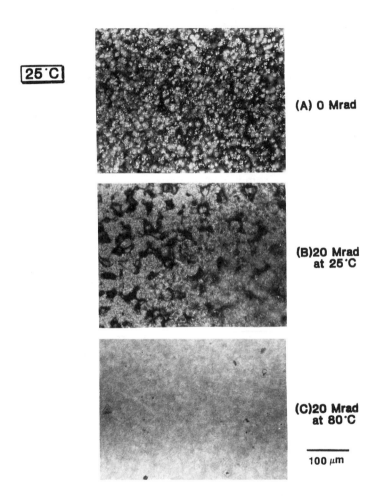

Figure 5c. Cross polarizer optical micrographs taken for the CL-AGE copolymer control sample and those irradiated at 25 °C and 80 °C with changing temperature of the heating stage: 25 °C.

At 25°C as shown in Figure 5(a), the micrograph of the CL-AGE sample EB-irradiated in the semicrystalline state shows a strong birefringent spherulitic pattern although the irradiated sample is highly crosslinked. It is very similar to that found for the control sample. As previously indicated in Figure 4, the gel fraction of this room-temperature-irradiated(20 Mrad) sample is approximately 70 percent. On the other hand, the sample, which was first irradiated in the amorphous melt state at 80°C and subsequently cooled down to room temperature, shows a relatively lower order and somewhat disrupted spherulitic pattern. The random introduction of crosslinks into the molten CL-AGE copolymer distinctly plays an important role in the restriction on the crystallization process since these crosslink units would be excluded from the crystalline regions and hence interfere with the kinetics of the crystallization process.

Micrographs of the samples which have been heated up to 120°C and which well exceed the crystalline melting temeperature(ca. 55°C) are illustrated in Figure 5(b). The birefringent spherulitic pattern of the room-temperature-irradiated polymer continues to persist beyond 120°C, while the birefringent pattern of the control sample and the copolymer cross-linked in the amorphous melt state were shown to disappear after exceeding the crystalline melting temperature(sample becomes black between crossed polarizer). Therefore, it is clear that irradiation of the sample in the spherulitic semicrystalline state results in a persisting optical anisotropy or order that exists well beyond the melting point although clearly the spherulites will have undergone the crystalline melting transition as shown in the micrographs or by later DSC results that will be presented shortly.

Upon isothermal cooling at 25°C from the melt state(120°C) as shown in Figure 5(c), some recrystallization of the room-temperature-irradiated sample was observed to occur within the former persisting spherulitic pattern. On the other hand, the unirradiated control sample underwent the usual transformation from an isotropic liquid to a spherulitic semicrystalline state. In a striking contrast to the samples irradiated at a temperature below the melting point, the formation of the spherulitic order was observed to be hindered for the case of 80°C-irradiated sample. Therefore, these microscopy results clearly emphasize the importance of the morphological state at the time of EB irradiation.

Scanning electron micrographs of the unirradiated copolymer and the copolymers irradiated with 20 Mrad at room temperature and 80°C are shown in Figure 6. All micrographs show spherulitic structure although the

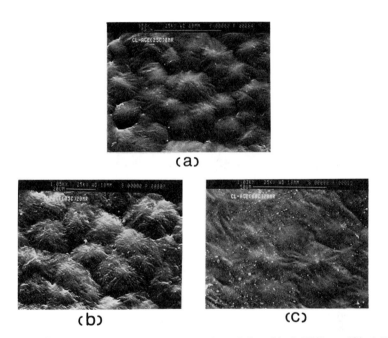

Figure 6. Scanning electron micrographs of the CL-AGE(6 mol% AGE) copolymers: (a)unirradiated;(b)irradiated with 20 Mrads at room temperature;(c)irradiated with 20 Mrads at 80°C.

surface textures are clearly different. The initial smooth surface texture of the unirradiated copolymer sample was significantly roughened when irradiated with a single pass of 20 Mrad at room temperature, while the overall shape and size of the spherulites were maintained. The increased roughness for this room-temperature-irradiated sample is considered to be due to melting **during** the irradiation process with 20 Mrad. In contrast, the initial spherulitic structure was almost destroyed for the case of the sample irradiated with 20 Mrad at 80°C indicating that the formation of the spherulite was greatly hindered due to the random introduction of cross-linking in the melt state.

As indicated earlier, irradiation with 20 Mrad can induce some heating of the samples. An approximate calculation of the temperature rise due to electron beam irradiation process should be useful in this regard(8). In brief, an energy balance based on a unit volume element of the medium provides the following equation:

$$\rho Cp(\frac{dT}{dt}) = k\nabla^2 T + S$$

where ρ is density, Cp is the heat capacity, k represents the heat conductivity, T is the temperature, S is the heat generation term. If the adiabatic conditions are assumed and if energy deposition is only by penetrating electrons i.e. no consideration of the exothermic reaction, the above equation can be reduced to the following simple relation:

$$\Delta T = \frac{E't}{Cp}$$

where E' is dose rate in Mrad per second and t is time in seconds. For the case of a 20 Mrad-irradiation which is equivalent to the process with 0.4 [sec] of exposure time and 50 [Mrad/sec] of dose rate assuming 0.5 [cal/gK] as heat capacity, a maximum temperature increase of about 96°C can be calculated for the irradiation process. Although this is a rough estimation with some simplifications, it implies that the sample temperature could be rapidly and easily increased above the melting point in the irradiation process. Thus, this temperature rise can induce, at least, melting and subsequent recrystallization when the sample is cooled down to room temperature after the irradiation process.

Thermal Properties

The crystalline melting transition behavior of the first and second DSC scans of the irradiated copolymers are shown in Figure 7 and 8, respec-

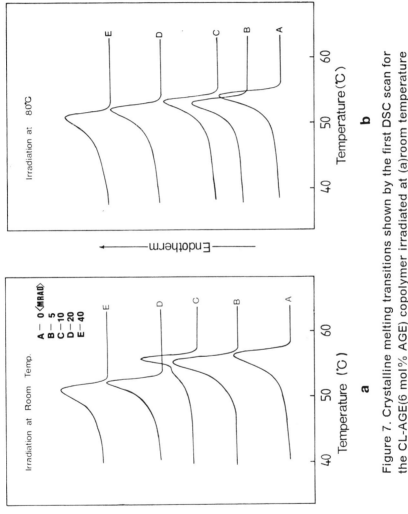

Figure 7. Crystalline melting transitions shown by the first DSC scan for the CL-AGE(6 mol% AGE) copolymer irradiated at (a)room temperature and (b)80°C: (A) 0 Mrad; (B) 5 Mrad; (C) 10 Mrad; (D) 20 Mrad; (E) 40 Mrad.

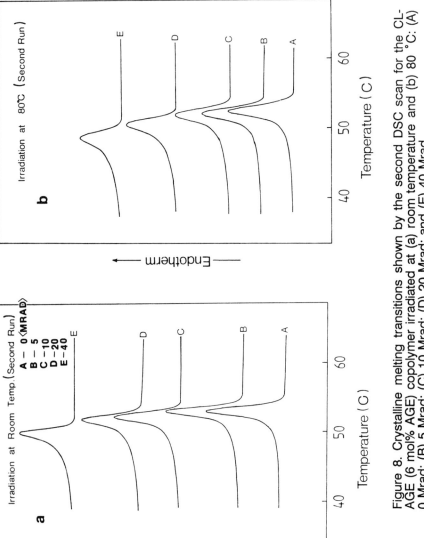

Figure 8. Crystalline melting transitions shown by the second DSC scan for the CL-AGE (6 mol% AGE) copolymer irradiated at (a) room temperature and (b) 80 °C: (A) 0 Mrad; (B) 5 Mrad; (C) 10 Mrad; (D) 20 Mrad; and (E) 40 Mrad.

tively. The crystalline melting point and the heat of fusion of a pure polycaprolactone having a number average molecular weight of 30,000 have been reported to be 60°C and 33[cal/g], respectively. However, due to the random introduction of the 6 mole percent of allyl glycidyl ether units which generate steric hindrances for crystallization, a melting point depression is expected for the unirradiated CL-AGE copolymer comparing to the pure polycaprolactone of comparable molecular weight. As shown in the first DSC scan of the **unirradiated** sample also having 6 mole percent AGE units, a melting temperature of 56.3°C, which is equivalent to approximately a depression of 4°C, was measured as shown in Figure 7. For a truly random copolymer, a melting point depression can be approximated by using the well-known Flory relation(8) as shown by:

$$\frac{1}{Tm} - \frac{1}{Tm°} = (\frac{-R}{\Delta Hu}) \ln(Xa)$$

where $Tm°$ is the melting point without non-crystallizable units in [K] under the same melting procedure, R is the gas constant, ΔHu is the enthalpy of fusion per repeating unit in [cal/mol] and Xa is the mole fraction of the crystallizable units. A Tm of 56°C was calculated from the above relation showing good agreement between theory and experiment for the unirradiated CL-AGE copolymer.

In the first DSC scans as shown in Figure 7, the melting points of the samples irradiated at room temperature do not appear to be greatly affected by radiation-induced crosslinking up to 10 Mrad. However there is a relatively abrupt melting point depression observed for the samples irradiated with 20 and 40 Mrad. The abrupt melting temperature depression for the samples irradiated at room temperature with 20 and 40 Mrad is again believed to be due to the heating effect of the EB irradiation process as discussed previously. In contrast, there is a distinct gradual melting point depression shown for the samples irradiated at 80°C over the entire dose range. As can be seen in the DSC data, the melting points are depressed further and the peaks become broader and less steep with increasing irradiation dose for both room temperature and 80°C irradiation suggesting that irradiation introduces imperfections, and crosslinks which induce a molecular weight increase. The general trend of decreasing sharpness of the melting peaks with increasing irradiation dose is in good agreement with increasing gel contents with dose as previously shown in Figure 4.

The second DSC scans, as shown in Figure 8, indicate a crystalline melting behavior which is similar to that of the first scans except the disappearance of the abrupt depression of the melting point for the samples irradiated with 20 and 40 Mrad at room temperature. This can be expected since all the samples of room-temperature-irradiation are recrystallized by cooling down to room temperature after the first DSC scan(heating). That is, the abrupt melting point depression shown in the first scans induced by the EB heating for irradiation with 20 and 40 Mrad may well be nullified in the second scan.

The melting point depression behaviors with increasing irradiation dose for the first and second DSC scans are summarized as shown in Figure 9. The effects of EB heating and thermal "history" of the CL-AGE copolymers on the crystalline melting transition are well demonstrated in these plots. In the first DSC scan, the melting point of the room-temperature-irradiated sample shows abrupt depressions for the doses of 20 and 40 Mrad. In contrast, irradiation in the melt state resulted in a gradual melting point decrease showing Tm values lower than that for the sample irradiated in the semicrystalline state up to 10 Mrad but similar melting points for the 20 and 40 Mrad dosages. It should be emphasized that the melting point differences up to 10 Mrad in the first DSC run up to 10 Mrad are likely attributed to the difference in the thermal "history." Specifically, the material irradiated in the melt state was first heated to 80°C followed by irradiation and, subsequently cooled down to room temperature whereas the samples irradiated at 25°C had not undergone any significant thermal heat treatment except by EB heating effect for 20 and 40 Mrad irradiation. After the first DSC run, all the materials were recrystallized at room temperature and, hence, the effects of the thermal "history" which includes EB heating effects during the irradiation process were expected to be minimized in the following thermal test. In the second scan as shown in Figure 9, the differences in the melting point between the samples irradiated at 25°C and 80°C for doses up to 10 Mrad were decreased in contrast to the case of the first run thereby supporting our discussion. Hence, the difference in the melting point depression behavior in the second run is somewhat widened with increasing irradiation dose showing only the effects of the difference in the degree of crosslinking.

The crystallization processes, as shown by the DSC peaks in Figure 10, appear to be more sensitive to the degree of crosslinking and the morphological state at the time of irradiation. These data show that the crystallization is delayed to a higher degree for the samples irradiated at a temperature above the melting point. In addition, this effect becomes more noticeable as the irradiation dose is increased as well be expected.

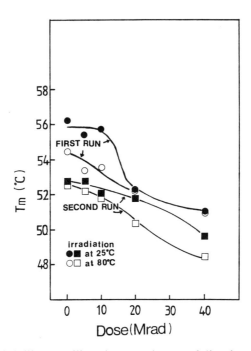

Figure 9. Crystalline melting temperatures of the irradiated CL-AGE copolymer(6 mol% AGE) with increasing irradiation dose for the first and second DSC scans.

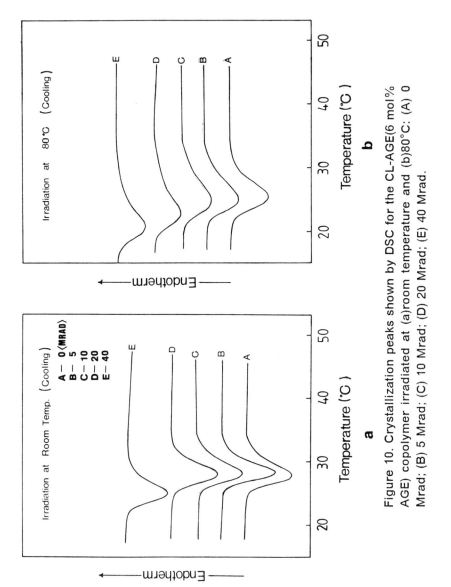

Figure 10. Crystallization peaks shown by DSC for the CL-AGE(6 mol% AGE) copolymer irradiated at (a) room temperature and (b) 80°C: (A) 0 Mrad; (B) 5 Mrad; (C) 10 Mrad; (D) 20 Mrad; (E) 40 Mrad.

Mechanical Properties

Limited mechanical testing was performed at room temperature on the EB-irradiated CL-AGE copolymers. As shown in Figure 11, the modulus of the room-temperature-irradiated sample increases slightly up to 10 Mrad and, upon further irradiation, it decreases abruptly. The increase in modulus of the room-temperature-irradiated sample up to 10 Mrad is believed to be due to the increase in crosslinking mainly occurring at the AGE units which are excluded from the crystallites. The decrease in modulus of the room-temperature-irradiated sample above 10 Mrads is caused by a slightly lower crystalline content due to the EB heating effect with crosslinking-this remark being indirectly supported by the previous DSC and microscopy data.

In summary, the irradiation process carried out below the melting point should be differentiated from that above the melting point for several reasons. First, radiation- induced crosslinking in the semicrystalline state results in a relatively high gel percent and a locked-in molecular anisotropy resulting which will persist the melt state. Second, crosslinking events induced by irradiation below the melting temperature may not be expected to occur in a random fashion as may well be the case in the irradiation process above melting point. Third, for irradiation below the melting point, the crystalline melting transition can occur if sufficient EB heating occurs with crosslinking reactions occurring at the same time. Finally, for irradiation above the melting point, the crosslinking reaction proceeds in the liquid state above the melting temperature and likely in a random fashion thereby more distinctly influencing the recrystallization behavior upon cooling.

Conclusions

A number of conclusions can be made from this study on the effect of electron beam irradiation on the CL-AGE copolymers. Gel fraction of the CL-AGE copolymers increases with irradiation dose up to 40 Mrad. Crosslinking in the melt state results in slightly higher gel contents than irradiated in the semicrystalline state, although both values are quite high. Direct microscopic observations show that the polymer crosslinked in the semicrystalline state produces a stable spherulitic pattern which persists through the molten state. However, for the unirradiated sample the birefringent spherulitic pattern disappears at the temperature of the crystalline melting transition. In contrast, introduction of crosslinks into the CL-AGE copolymer in the molten state

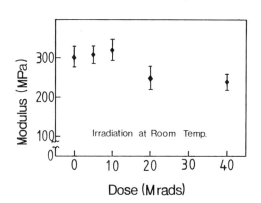

Figure 11. Modulus vs. irradiation dose for the CL-AGE(6 mol% AGE) copolymer irradiated at room temperature.

leads to restriction on the development of ordered spherulitic superstructure. Thermal and mechanical behavior was found to indirectly support the microscopic observations. It was concluded that the final solid state properties of the EB- irradiated CL-AGE copolymers are strongly dependent on the morphological state of the material at the time of irradiation process.

Acknowledgement: The financial support from 3M Corp. for this study is gratefully acknowledged.

Literature Cited

1. A. Chapiro, Radiation Chemistry of Polymer Systems, Interscience, New York, 1962.

2. A. Charlesby, Atomic Radiation and Polymers, Pergamon Press, New York, 1960.

3. G. N. Patel, and A. Keller, J. Polym. Sci. Polym. Phys. Ed., 13, 303(1975).

4. G. F. D'Alelio, R. Haberli, and G. F. Pezdirtz, J. Macromol. Sci. Chem., A2(3), 501(1968).

5. M. Narkis, S. Sibony-Chaouat, A. Siegmann, S. Shkolnik, and J. P. Bell, Polymer, 26, 50(1985).

6. Y. Yoo, Ph.D. Thesis, VPI&SU, 1988.

7. H. Kim, and G. Wilkes, in preparation.

8. L. Mandelkern, Crystallization of Polymers, McGraw-Hill, 1964.

RECEIVED October 27, 1989

Chapter 34

Electron-Beam Exposure of Organic Materials

Radiation Curing of Perfluorinated Acrylates

J. Pacansky and R. J. Waltman

IBM Almaden Research Center, 650 Harry Road, San Jose, CA 95120-6099

This study is divided into two parts: the first involves electron beam exposures on CO_2 isolated in rare gas solids to investigate the role of an inert solvent on radiation chemistry; further experiments were conducted on diazoketones mixed into polymers to investigate the role of reactive solvents. The second part of the study deals with electron beam exposure of fluorinated acrylates and perfluorinated ethers. These reveal that highly fluorinated acrylates when mixed with non-fluorinated acrylic monomers may be used to produce radiation cured films with a broad range of contact angles.

Processes utilizing electron beams for curing of coatings, sterilization, etc., almost always involve exposure of a mixture of substances. A pertinent example is the curing of coatings via an electron beam induced free radical polymerization. Here, the rate of the polymerization depends on the efficient use of the absorbed dose to form free radicals for the initiation of the polymerization. Since the material that initiates the polymerization (whose identity in most cases is unknown) is only one of the components in the mixture absorbing energy from the incident electron beam, energy not directed to it is wasted. In the experiments presented herein, it is shown that the electron beam sensitivity of a component in a mixture may be altered by changing its concentration. We investigate two component mixtures: one consists of a molecular system in a chemically inert solvent; the second case relaxes the chemical reactivity by using a polymer as a matrix. Specifically, the former is carbon dioxide isolated in solid argon while the latter is a diazoketone blended into a phenolic polymeric matrix. An understanding of the radiation-induced effects in the more "simple" CO_2/Ar system will provide a framework from which the irradiation of more complicated systems such as of guest-polymer (organic) formulations may be understood.

Finally, since perfluorinated materials are widely used as coatings for a number of applications, we investigated the electron beam curing of mixtures of perfluorinated and non-perfluorinated acrylates to produce formulations for radiation curing with a wide degree of contact angles (with water). Additionally, the feasibility of directly converting liquid perfluorinated ethers to solid films using electron beams was investigated.

CO_2 IN SOLID ARGON

The rare gas matrix isolation technique has been successfully used to study reactive intermediates generated by both chemical and physical methods[1,2]. In this report we extend the utility of the matrix isolation technique by demonstrating that chemistry may be initiated in molecular systems isolated in rare gas matrices using electron beams with an accelerating voltage of 25 kV.

Experimental Apparatus. The experimental equipment was described in detail in another report[3], hence, only those parts special to the matrix application are presented herein. The apparatus, shown in Fig. 1, consisted of a 55 liter stainless steel vacuum chamber (Model 202 Displex) with a 8 inch cold surface. Pressure in the chamber was typically in the 10^{-8} torr range; pressures in the 10^{-9} to 10^{-10} torr range were reached with a nominal amount of baking. Connected to the vacuum chamber was an electron beam gun, a quadrupole mass spectrometer (UTI Model 100C), and a He/Ne laser interferometer.

The electron beam gun could be operated at energies up to 30 keV. Its main features were first, the electron beam could be focussed using magnetic lenses, and second, the beam could be positioned and raster scanned using a dual ramp generator with a Celco blanking amplifier. During the course of an experiment the electron beam current was measured with a Faraday cup connected to a picoammeter (Keithly 480). The current measurement was made by positioning the electron beam onto the Faraday cup. The sample was mounted onto a rotable sample holder that could be adjusted from outside of the chamber. Thus, a matrix could be deposited and simultaneously interrogated by a He/Ne laser interferometer; in this manner the thickness of the matrix was determined.

Also shown in Fig. 1 is the location of the reflectance infrared spectrometer relative to the vacuum chamber. The spectrometer consists of a double beam goniometer which accurately controls the angle of incidence on the reflectance sample and reference, and a Spex Industries Incorporated Model 1701, 3/4 meter monochromator.

The gas handling system used for sample preparation was described previously[3]. An Air Products low temperature closed cycle refrigerator was used to maintain the sample temperature. The refrigerator was mounted in a rotable flange at a position above the point marked RS in Fig. 1 such that a 180° rotation allowed a matrix to be deposited, irradiated or its infrared spectrum recorded.

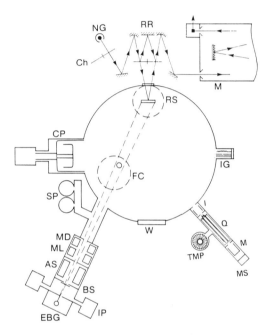

Figure 1. Experimental apparatus for spectrometric detection of electron beam induced reactions. Upper portion of figure describes infrared spectrometer while lower portion describes electron beam gun and vacuum chamber. NG = Nernst glower; Ch = chopper; RR = reflectance reference; RS = reflectance sample; M = monochrometer; CP = cryopump; FC = Faraday cup; IG = ionization gauge; SP = sorption pumps; W = window; MD = magnetic deflection; ML = magnetic lens; IP = ion pumps; EBG = electron beam gun; TMP = turbomolecular pump; MS = mass spectrometer; BS = blanking section; and AS = alignment section.

Argon Matrix Exposures. We have chosen the system CO_2 in argon for its simplicity. The chemical changes induced by the electron beam were determined by following the changes in infrared spectra as a function of incident charge density. The net effect of the electron beam exposure is the conversion of CO_2 to CO as shown below:

$$CO_2 \rightarrow CO + O \qquad (1)$$

Qualitative evidence for this reaction is obtained by noting the synchronous decrease and increase in the CO_2 and CO absorptions at 2340 and 2138 cm^{-1} as a function of dose. The reaction was quantitatively related by measuring the current density in the electron beam with a Faraday cup and by measuring the thickness of the matrix using an interferometer. Fig. 2a shows a plot for the log of N/N_o (normalized decay of CO_2) as a function of incident charge density (current density = 0.26 μamps/cm²); note that the concentration of CO_2 in the argon matrix decreases as the rate of decay for CO_2 increases; concomitantly, the rate of formation of CO also increases with CO_2 dilution in argon. This latter observation is clearly shown in Fig. 2b for the formation of CO as a function of incident charge density (For the matrix thickness and beam energy employed here, a charge density of 1 μC/cm², is equivalent to an absorbed dose of 2.23 Mrads.).

The mechanism for the enhanced rate of decay for CO_2 shown in Fig. 2 is most likely related to migration and transfer of electronic energy from the argon matrix to CO_2. An analysis of the stopping power[4] for particulate radiation incident upon a solid shows that it is proportional to the number density of the materials in the matrix. Consequently, the rate of decomposition for CO_2, based upon stopping power considerations alone, should decrease with dilution in solid argon. However, since it is observed that the rate increases with dilution, an energy transfer mechanism is required that efficiently directs the energy deposited by the electron beam to the matrix isolated CO_2 molecules. As a further consideration, the G values for decomposition of CO_2 in argon (concentration: 1/100) were determined to be 3.6 based on the energy absorbed by the matrix, and 150 if only the energy absorbed by CO_2 is used (The latter value was obtained by using the electron fraction of CO_2). Clearly, this G value is very high and another mechanism other than direct electron hits on CO_2 is required. In the discussion that follows a kinetic analysis is presented for a quantitative evaluation of the process.

The electron beam excitation of the matrix is given by

$$M \rightarrow M^x \quad k_1 \qquad (2)$$

Here M^x may be an electronically excited or charged entity in the matrix, for example an exciton or hole, which may dissipate its energy back to the matrix or transfer it to CO_2

$$M^x + M \rightarrow M + M^x \quad k_2 \qquad (3)$$

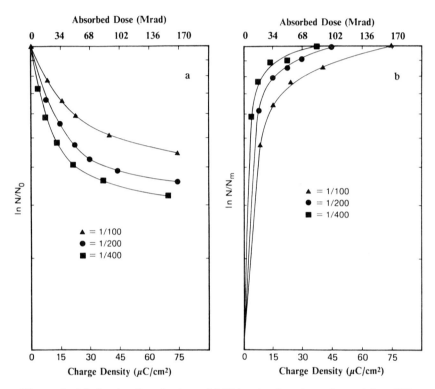

Figure 2. (a) A plot for the log of N/N_o, the fraction of surviving CO_2 molecules, as a function of incident charge density. The N/N_o was determined by following the changes in the infrared absorption of the 2340 cm^{-1} band. The film thicknesses were 2.9 μm for all three samples. The concentration of CO_2 in matrix is given as mole fraction. (b) Growth of the CO band at 2138 cm^{-1} as a function of incident charge density. N_m = maximum number of CO molecules on the growth curve.

$$M^x + CO_2 \rightarrow M + CO_2^x \quad k_3 \tag{4}$$

CO_2^x may dissociate to give CO or return to the ground state (or some other entity)

$$CO_2^x \rightarrow CO + O \quad k_4 \tag{5}$$

$$CO_2^x \rightarrow CO_2 \quad \text{or other products} \quad k_5 \tag{6}$$

The rate for the formation of CO by direct irradiation of CO_2 is

$$CO_2 \rightarrow CO + O \quad k_6 \tag{7}$$

If we use the steady state approximation then the concentration of CO formed in unit time is

$$[CO] = \frac{k_1 k_3 k_4}{k_4 + k_5} \frac{[M][CO_2]}{k_2[M] + k_3[CO_2]} + k_6[CO_2] \tag{8}$$

The rate for reaction 7 may be neglected because of the dilute concentration mandating infrequent direct hits on CO_2; therefore,

$$\frac{1}{[CO]} = \frac{k_2(k_4 + k_5)}{k_1 k_3 k_4 [CO_2]} + \frac{k_4 + k_5}{k_1 k_4 [M]} \tag{9}$$

A plot of $\dfrac{1}{[CO]}$ versus $\dfrac{1}{[CO_2]}$ at a particular dose should be linear with

$$\frac{\text{intercept}}{\text{slope}} = \frac{k_3}{k_2[M]} \tag{10}$$

Thus, we find that the ratio of k_3/k_2 = 140, that is, the rate constant for energy transfer of M^x to CO_2 is 140 times greater than for transfer back to the matrix.

The identity of M^x is consistent with an excited state of the matrix. For example, when CO was irradiated under the same conditions as CO_2, the G value for decomposition was 0.09 (based on energy absorbed by the matrix) and 0.006 for CO decomposition in Xe (concentration: 1/100); the value for argon is consistent with inefficient or no energy transfer, for xenon decomposition by direct hits on CO_2 appears to be the only viable process. In Fig. 3, an energy level diagram is shown summarizing the energetics for decomposition of CO_2 to CO + O, and CO to C + O, along with the energies for the $E_{n=1}$ excitons[5]. Note that the experimental results discussed above fit the

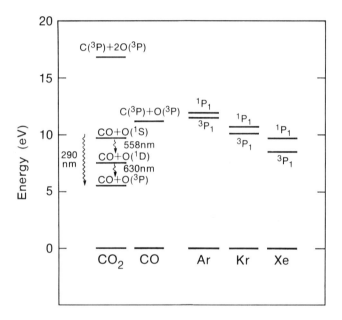

Figure 3. Energy level diagram for CO_2 and CO, and exciton levels for Ar, Kr and Xe.

required energetics that M^x is an exciton. Electron beam excitation of solid Ar, Kr and Xe, would thus produce an exciton with sufficient energy to produce CO + O from CO_2; For CO only exposures in solid argon may be aided by energy transfer, a similar process in Kr and Xe is not energetically possible.

MIXTURE OF DIAZOKETONE IN A PHENOLIC POLYMERIC MATRIX

The system chosen to study consisted of the diazoketone 1 in the resin 2. The diazoketone plays the role of a guest mixed into the host resin. Films were prepared by spin coating onto a silicon substrate using diglyme as a solvent.

Electron beam exposure was performed on a CD 150 Electron Beam Processor (Energy Sciences, Woburn, Mass.). The decay of the diazoketone was monitored by recording the transmittance of the diazo-stretching at 2110 cm^{-1} as a function of absorbed dose; these results are summarized in the form of a N/N_o versus dose plot in Fig. 4. The salient result of the plot is that the efficiency for the decomposition of the diazoketone increases as its concentration in the phenolic resin decreases as shown in Fig. 4 by the gradual increase in slope with decreasing diazoketone concentration. As noted above for the CO_2/argon experiments, a reasonable explanation for the diazoketone/resin results is migration of energy deposited by the electron beam in the resin followed by energy transfer to the diazoketone. More experiments are presently in progress to study this phenomenon.

THE DIAZOKETONE/RESIN SYSTEM: VACUUM ELECTRON BEAM EXPOSURE

The photochemical Wolff rearrangement is the mechanism by which a diazoketone after excitation loses N_2 and converts to a ketene. All of this occurs on a reaction path that presumably does not have a minimum between the diazoketone and ketene. As shown in the scheme below, reasonable evidence for the ketene is in fact the two trapping experiments, i.e., in air, the

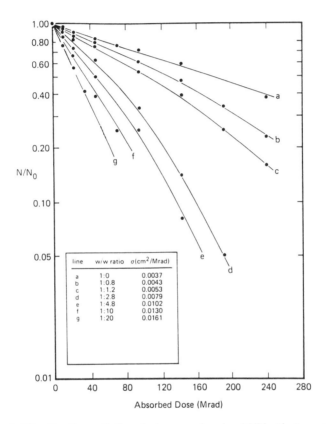

Figure 4. The fraction of diazoketone molecules N/N_o that survive an absorbed dose (Mrad) in a diazoketone/resin mixture. The legend labels each line with concentration by weight of diazoketone 1 in resin 2.

thermal reaction of the ketene with water to form the indene carboxylic acid and, in vacuum, the thermal reaction with phenolic −OH groups to form the carboxylic acid ester.

For the case of the vacuum electron beam exposures we have two observations that apparently lead to a paradox. The first is electron beam exposure produces a ketene but with a much lower yield at low temperatures; this is shown in Fig. 5 by the characteristic C=C=O stretching frequency at $\simeq 2130$ cm^{-1} which appears upon irradiation. The second is that the carboxylic acid ester does not appear to be formed when the vacuum electron beam exposures are performed at room temperature. A comparison of Figs. 6 & 7 supports this claim; for example, Fig. 6 summarizes the results for a UV exposure of the system to demonstrate the major IR absorption for the ester; Fig. 7 contains the results for the electron beam exposure under identical conditions and shows that the C=O band for the ester at $\simeq 1750$ is not evident.

If the ketene was produced via a Wolff rearrangement at low temperature then the carboxylic acid ester should certainly form as a result of the room temperature exposure. In view of this we conclude that the mechanism for the vacuum electron beam exposure does not involve a Wolff rearrangement. Instead we propose that a carbene formed by loss of N_2 from the diazoketone plays a central role in the chemistry that ensues after electron beam exposure. The proposed mechanism for the electron beam induced chemistry discussed thus far is summarized in the scheme below.

Electron beam exposure at low temperatures produces a carbene 7, which upon further electron beam excitation rearranges to a ketene 5. The carbene 7 may exist for relatively long periods of time at $T = 10K$ because of the frozen, immobile environment at this temperature. In effect, this dramatically retards carbene reactions with the resin and opens a channel for further excitation of the carbene to ketene. At room temperature, however, the electron beam generated carbene rapidly reacts with the resin and hence has too short a lifetime for further excitation to ketene to be an issue.

EVIDENCE FOR CARBENE FORMATION

Mixtures of diazoketone 1, with benzene were exposed to the 175 kV electron beam and infrared spectra recorded as a function of absorbed dose. Fig. 8 contains spectra taken after D = 90, 180 and 300 Mrads were administered to the sample. In this case the spectral region where −OH groups characteristically absorb is not initially obscured and the increase in absorption in this region as a result of the electron beam induced chemistry is clearly evident. Furthermore, a number of the absorptions observed as a result of the electron

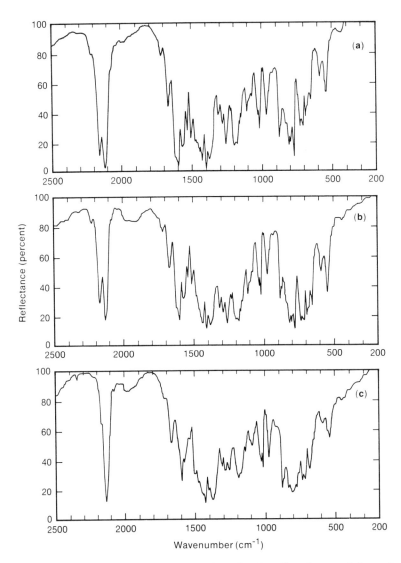

Figure 5. IR spectra of 5.2 μm film of equimolar diazoketone 3 in resin 2. The sample was exposed to an electron beam (25 kV, I = 1 nA/cm^2, $\tilde{T}$ = 10 K). Incident charge density Q equals: (a) 0; (b) 40; (c) 100 μC/cm^2.

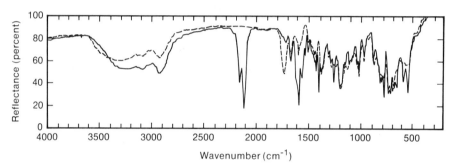

Figure 6. IR spectra of a thin film of equimolar diazoketone 3 in resin 2. The sample was exposed to UV light ($\lambda > 300$ nm) in vacuum at room temperature using a 150 W high pressure Xe lamp: (———) before exposure; (– – –) after exposure.

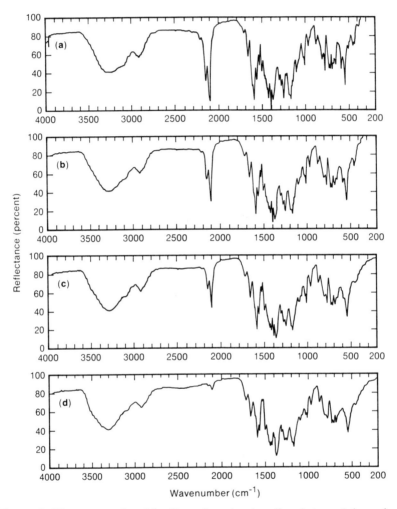

Figure 7. IR spectra of a thin film of equimolar diazoketone 3 in resin 2. The sample was exposed to an electron beam (25 kV, $\tilde{I}$ = 0.5 nA/cm^2) in vacuum at room temperature. The incident charge density Q equals: (a) 0; (b) 22; (c) 33; and (d) 70 μC/cm^2.

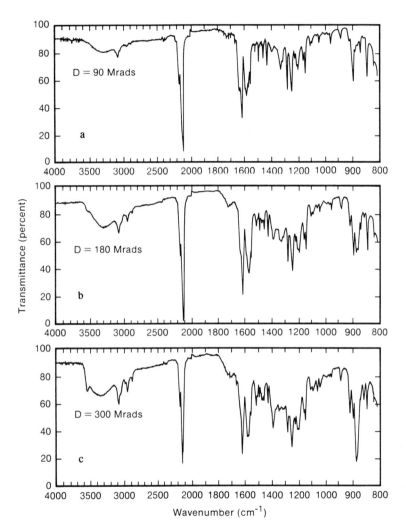

Figure 8. IR spectra of diazoketone 1 in benzene (film). The sample was exposed to an electron beam (175 kV) at room temperature under N_2. The absorbed dose equals: (a) 90; (b) 180; and (c) 300 Mrad.

beam exposure of the diazoketone 1, and resin 2, and diazoketone 3, and meta-cresol are in common with those observed for the benzene irradiations; in particular, these are the bands in the high frequency region of the IR where OH and CH groups appear. The respective band centers for those common to all of the exposures studied herein are 3550, 3140, 1460, 1380, 1080, 1040, 810, and 780 cm^{-1}. Another broad absorption appears at $\simeq$ 1700 cm^{-1} but it is clearly due to a secondary product of the irradiation because the intensity of this band continues to increase with exposure after the diazoketones have been consumed by the radiation.

SUMMARY OF THE ELECTRON BEAM EXPOSURES OF DIAZOKETONES 1, 3

The experimental results reported herein reveal electron beam excitation at room temperature does not proceed through a Wolff rearrangement. The major reaction path appears to mitigate an intermediate which reacts with aromatic CH bonds and perhaps aromatic CC bonds. The intermediate appears to also react with OH bonds; irradiations of mixtures of the diazoketones in tertiary butyl alcohol[7] are very similar to the resin, meta-cresol, and benzene results presented above.

Our contention therefore, is that electron beam excitation of the diazoketones at room temperature produces a carbene which in the presence of benzene reacts with CH and CC bonds to ultimately produce substituted naphthols for example as outlined in the scheme below. At low temperatures, the carbene is confined by the rigid environment and when excited by another electron forms a ketene. In summary, these results not only uncover a divergence between the excitation by photons and charged particulate radiation but also provide a rationale for the positive working nature of electron beam resists using diazoketones. The substituted naphthols produced by the electron beam induced chemistry are soluble in aqueous KOH developers and hence the exposed regions of the resists may be imaged by the high energy electron beam.

ELECTRON BEAM CURING OF PERFLUORINATED ACRYLATES: A PRACTICAL EXAMPLE

The surface characteristics of metallic and other substrates frequently require modification to meet particular criteria; often, it is important to alter the surface energy of a substrate for a particular application. Here, contact angles control a variety of important surface properties, such as wettability, detergency and waterproofing, etc. The contact angle is defined simply as the angle formed between the surface of a solid and that of a liquid; it relates the surface tensions at the solid, liquid and liquid-solid interfaces.[6]

A facile method by which organic thin films may be readily coated upon substrates employs the electron beam curing of materials such as of monomeric and/or oligomeric acrylates and methacrylates. The perfluorinated acrylates 2-(N-butylperfluorooctanesulfonamido)ethyl acrylate (BFOSA), 2-(N-ethylperfluorooctanesulfonamido)ethyl acrylate (EFOSA), and perfluoropolyether diacrylate (PFEDA) were obtained from the 3M corporation, St. Paul, Minnesota; 1,6-Hexanediol diacrylate (HDODA) was obtained from Interez, Inc., Louisville, Kentucky.

Thin films were spin-coated onto virgin, optically flat NaCl disks or flat, polished Si wafers and electron beam cured at 5-10Mrad under N_2. The oxygen content in the irradiation chamber was $\leq \simeq 175$ ppm. The equipment used for electron beam curing was a CB 150 Electrocurtain Processor (Energy Sciences, Inc., Woburn, Mass.) that permits the exposure of samples in the presence of air or an inert atmosphere like N_2. A series of experiments were performed to determine the optimal absorbed dose for curing; this was determined to be about 10 Mrads.

Contact Angle Measurements. The surface energies of the perfluorinated acrylates may be characterized by measurement of the contact angle θ of water (74.5 erg/cm²) with the various radiation cured perfluorinated films. For practical purposes, if $\theta < 90°$, the liquid is said not to wet the surface, while wetting occurs when $\theta = 0°$. Once the contact angles are determined, the spreading coefficients may be calculated. One mixture is considered here for illustrative purposes. The fluorinated acrylate BFOSA is added to 1,6-hexanediol diacrylate (HDODA) to investigate the variance of water wettability on surfaces which vary in fluorine content (mole fraction) from 0 to 0.52. The contact angle measurements of the radiation cured films, shown in Fig. 9a, reveal that the contact angle of water on the films increase (or $\cos\theta$ decrease) as a function of increasing fluorine content. The spreading coefficient of liquid on solid, defined as the difference between adhesion work and cohesion work, $S_{l/s}$, is given by:[7]

$$S_{l/s} = \gamma_{l/v}(\cos\theta - 1) \tag{11}$$

where $\gamma_{l/v} = 74.5 \text{erg/cm}^2$ for H_2O, and θ is the contact angle. Thus, if $S > 0$, spreading is accompanied by a decrease in free energy. $S_{l/s}$ as a function of fluorine content for the above formulations are shown in Fig. 9b. These data reveal that H_2O does not spread on the perfluorinated films. Indeed, the spreading coefficient of water on EFOSA, BFOSA and PFEDA are quite comparable to Teflon, which has $S_{l/s} \simeq -102$ for water. Note also for the composite film HDODA/BFOSA, that addition of up to $\simeq 20\%$ by weight of HDODA, a non-fluorinated diacrylate, in BFOSA (mole fraction of F = 0.33 in Fig. 9) does not dramatically increase the spreading coefficient of water on the film. This has important consequences for controlling or tailoring film mechanical properties and their surface energies. For the latter, addition of acrylates with suitable chemical structure and in appropriate concentrations allows a range of surface energies to be achieved. As regards mechanical

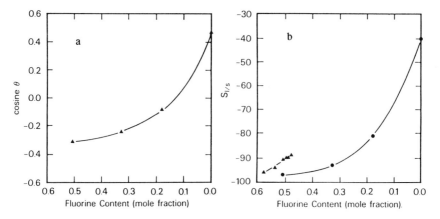

Figure 9. (a) Cosine of contact angle with water as a function of fluorine mole fraction in HDODA/BFOSA mixtures. The composite films were electron beam cured at 10 Mrad. (b) The spreading coefficients $S_{l/s}$ of water on (electron beam) cured films of HDODA/BFOSA (●) and PFEDA/BFOSA (▲) mixtures as a function of fluorine mole fraction.

properties, up to ≃20% by weight of tri- or penta-acrylates may be added instead of a diacrylate (depending upon its chemical structure and molecular weight) without increasing surface energies significantly. The addition of these multifunctional acrylates allow the formation of 3-dimensional networks (crosslinking) upon radiation curing which makes possible films with less porous structure, or of vastly different mechanical properties such as, for example, higher tensile strength. Thus, perfluorinated acrylates represent viable materials for which the surface energies of substrates may be lowered.

LITERATURE CITED

1. Jacox, M. Chem. Phys. 1981, 59, 213 - 230.
2. Andrews, L.; Keelan, B.W. J. Am. Chem. Soc. 1981, 103, 99 - 103.
3. Pacansky, J.; Coufal, H. J. Am. Chem. Soc. 1980, 102, 410 -412.
4. Pacansky, J.; Waltman, R.J. J. Rad. Curing 1986, 13, 24 - 28.
5. Z. Ophir, B. Raz, and J. Jortner, J. Chem. Phys., 1975, 62, 650 - 665.
6. Fox, H.W.; Zisman, W.A. J. Colloid. Chem. 1952, 7, 109 - 121.
7. Adamson, A.W. Physical Chemistry of Surfaces 3rd ed., John Wiley & Sons, New York, 1976; Chapter 12.

RECEIVED October 27, 1989

Chapter 35

Molecular Weight Dependence of Electron-Beam Resist Sensitivity

F. Rodriguez[1], B. C. Dems[1], A. A. Krasnopoler[1], Y. M. N. Namaste[2], and S. K. Obendorf[2]

[1]School of Chemical Engineering, Olin Hall, Cornell University, Ithaca, NY 14853
[2]Fiber Science Program, MVR Hall, Cornell University, Ithaca, NY 14853

Any measure of lithographic sensitivity must take into account both the events during exposure (latent image formation) and their subsequent conversion into an etch-resistant stencil (real image formation) by selective dissolution. Poly(methylmethacrylate), a popular "positive" chain-scissioning resist, is developed by dissolving exposed areas. Poly(chloromethyl styrene), a "negative" resist, crosslinks on exposure and is developed by dissolving the unexposed polymer. In each case the equations relating primary exposure events to parameters such as molecular weight need to be coupled with the dissolution characteristics to define sensitivity. The reconciliation of lithographic sensitivity (and resolution) with conventional radiation measurements also requires an understanding of important secondary phenomena. Some, like microporosity and gel-swelling, may also depend on the initial molecular weight and distribution of the polymer.

Microlithography is the process of producing microscopic patterns on a surface. The most prominent commercial application of the process is in the electronics industry. Patterns forming transistors, capacitors, wires, and other features are produced on silicon single-crystal substrates. Gallium-arsenide is another semiconductor surface used. The same features may be required on a quartz, glass or other substrate for use as a mask in replicating patterns. The feature dimensions shrink as more devices are crowded onto a central processing unit (CPU) chip or random-access memory (RAM) chip for computers. Photolithography with UV light has been used commercially for features down to about 0.8 µm. Both the masks used to produce these patterns and the patterns themselves when dimensions below about 1 µm are needed, often make use of computer-controlled, focussed electron beams.

In practice, a thin film of polymer (the "resist") is coated on the surface, typically in a thickness of 0.5 to 2 µm. Selected areas are altered by exposure to the electron beam. The most common reactions induced by the beam (Fig. 1) are chain scission, crosslinking, and polymerization. Chain scissioning increases the solubility of the exposed resist which can then be washed away leaving behind the unexposed polymer to act as a mask during etching, doping, metallizing, etc. A polymer that responds to radiation by increased solubility is termed a "positive" resist. Crosslinking and polymerization of polyfunctional monomers both insolubilize the exposed area. In this case, a solvent is used to wash away the unexposed material. Such a polymer system (a "negative" resist) leaves behind a mask made of insolubilized polymer.

The sensitivity of a polymer for lithographic purposes differs somewhat from the classical criteria of radiolytic yield. The usual measure of sensitivity is expressed in terms of chain scissions resulting from 100 electron volts of absorbed energy, G(s), or crosslinks resulting from the same amount of absorbed energy, G(x). The absorbed energy can also be expressed as rads or Grays (1 Gray = 100 rad). Other yields (H_2, monomer, other fragments) can be measured, but G(s) and G(x) often suffice to characterize the response of a polymer to radiation.

From the lithographic viewpoint, scissioning and crosslinking are only important insofar as they permit the production of a mask or stencil - a cohesive, adherent, dimensionally stable polymer film with open areas. This means that the latent image produced by radiation has to be converted to a 3-dimensional image by dissolution.

For a positive resist, there is seldom a perfect developer which will dissolve exposed polymer and not affect the unexposed material. The approach most often used to express lithographic response combines the effects of exposure and development in a "contrast curve" (Fig. 2). Experimentally, contrast curves are made using exposures at various dose levels. For a given developing time, the film thickness remaining, d, may be measured for each dose and normalized in terms of the original thickness, d_o. Occasionally one sees curves in which the thickness is normalized by dividing by the thickness of unexposed polymer remaining, d_u, after development of the exposed areas. This latter method can be deceptive since it will not reveal even drastic thinning of the original polymer mask. Also, this can result in a different apparent contrast. Good resolution of fine lines in the polymer film requires a high contrast (a steep slope). The time of development can be varied to produce a series of contrast curves, and, with forced developing, any exposure can be developed.

The contrast, γ, is defined as (-1/slope) of a plot of normalized thickness remaining versus log D, where D is the incident dose of electrons. The same slope should be obtained when the abcissa is any quantity proportional to the absorbed dose. In practice, thickness does not decrease linearly with dose but shows some curvature. In this case, contrast is calculated from the slope of the portion of the curve near zero thickness. Lithographic sensitivity can be defined in several ways. Since any dose can be developed, a different criterion is needed. One rather subjective definition of sensitivity is the minimum dose which will give a "satisfactory" pattern. Somewhat more objective is to select the minimum dose which will give vertical wall patterns under specified conditions. An easier and more commonly used technique is to determine the minimum dose required for complete development of an exposed area while removing no more than 10% of the unexposed film. To determine this dose, a "thinning" curve (Fig. 3) can be constructed from a family of contrast curves.

For a negative resist, a contrast curve also can be constructed (Fig. 2). In a crosslinked system, the time of development should not alter the curve since extraction of sol from the thin film network is very rapid. Contrast is defined now as (1/slope). A problem with all polymer resists, but especially acute with negative ones, is that of distortion of the remaining pattern by solvent swelling during development.

Negative (Crosslinking) Resists

Theoretical Models for Insolubilization. The equations relating gel formation to crosslinking have been presented by Flory[1] and Charlesby[2]. At the point of gelation,

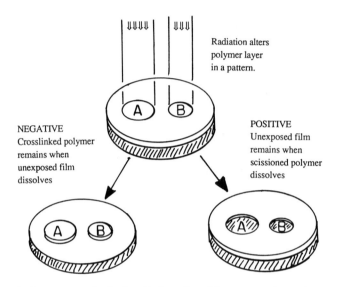

Figure 1. Pattern transfer by irradiation of positive or negative polymer resist.

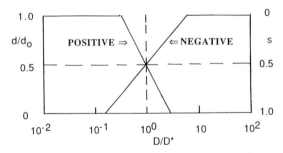

Figure 2. Idealized contrast curves. D* is the dose that allows 1/2 of the original film to remain after development.

the crosslink density, $v/2$, (mols/g) for any distribution of molecular weights with weight-average molecular weight M_w is[1]

$$v/2 = 1/2M_w \tag{1}$$

where each crosslink connects (on the average) two molecules of molecular weight M_w. Since the crosslink yield, $G(x)$ is in crosslinks per 100 e.v. absorbed energy per gram, the dose to gel, $D(g)$, is related to M_w by

$$G(x)D(g)/(100N) = 1/2M_w \tag{2}$$

where N is Avogadro's number, 6.023×10^{23} molecules/mol and $D(g)$ is in e.v./g.

For a monodisperse polymer, the soluble fraction s (sol) is given by[1]

$$-\ln(s) = \delta(1-s) \tag{3}$$

where d is the ratio of crosslinks at any point to the crosslinks needed to cause incipient gelation. Assuming that intermolecular crosslinks are introduced in proportion to radiation dose, d is also $D/D(g)$.

The expressions for a polymer with the "most probable" distribution** undergoing both scissioning and crosslinking were presented by Charlesy and Pinner[2]. The expression for dose to gel, $D(g)$, is modified to be

$$G(x)D(g)/(100N) = (1/2M_w)/\{1 - (G(s)/4G(x))\} \tag{4}$$

When $G(s)$ equals or exceeds $4G(x)$, no gel forms. The expression for the sol fraction s when gel does form is[2]:

$$s + s^{1/2} = [G(s)/2G(x)] + 100N/[M_w DG(x)] \tag{5}$$

Equation 5 should be used with some caution, since it has been successfully applied usually only when s is less than 0.5 which is, of course, no longer a situation of great lithographic interest. Examples of radiation plots of $(s + s^{1/2})$ versus $(1/D)$ for high-molecular weight polyethylene[2] or poly(ethylene terephthalate)[4] show distinct concave downward curvature.

To be useful in lithography, a polymer should be capable of being almost completely insolubilized by radiation. That is, s should approach something less than 0.05 as D becomes large. In terms of equation 5, this is a condition of

$$s + s^{1/2} = G(s)/2(Gx) < 0.2736 \tag{6}$$

<u>Simulation of Contrast Curves.</u> To convert the various predictions of these theoretical equations to contrast curves, it is convenient to use as a reference condition the point at which a dose of D^* yields a gel fraction = "sol" fraction, s, of 0.5. Then, if $a = G(s)/2G(x)$:

$$s + s^{1/2} - a = 100N/[M_w DG(x)] \tag{7}$$

and

$$D/D^* = (1.2071 - a)/(s + s^{1/2} - a) \tag{8}$$

**In the most probable distribution, the weight fraction of x-mer, $w(x)$ is given by*
$$w(x) = x(1-p)^2 p^{x-1}$$
where x is the degree of polymerization (number of repeat units of molecular weight M_m), the number average degree of polymerization is
$$x_n = 1/(1-p),$$
and the weight average degree of polymerization is
$$x_w = (1+p)/(1-p).$$
Neglecting end groups in polymer chains, molecular weight $M = xM_m$.

The contrast, γ, can be obtained by differentiating Eq. 7 and combining it with Eq. 8 to give:

$$ds/dD + (0.5s^{-1/2})ds/dD = (1.2071 - a)D^*/(-D^2) \quad (9)$$

$$ds/d\ln D = Dds/dD = -(1.2071 - a)(D^*/D)/(1 + 0.5s^{-1/2}) \quad (10)$$

and $\quad \gamma = -ds/d\log D = -2.303 ds/d\ln D \quad (11)$

Combining these with equation 8 gives a general equation for contrast as a function of s:

$$\gamma = 4.606\, s^{1/2}(s + s^{1/2} - a)/(1 + 2\,s^{1/2}) \quad (12)$$

At D^*, $s = 1/2$, the contrast is

$$\gamma^* = 1.349(1.2071 - a) \quad (13)$$

At $s = 1$, it is

$$\gamma = 1.535(2 - a) \quad (14)$$

In similar fashion, contrast can be derived from equation 3 for the monodisperse polymer (only crosslinking, no chain scission).

$$(\ln s)/(1-s) = -D/D(g) = -1.3863\,(D/D^*) \quad (15)$$

$$\{(1-s)^{-1}s^{-1} - (1-s)^{-2}\ln s\}\,ds/dD = -1.3863/D^* \quad (16)$$

Multiplying by D and substituting back for D/D^* from equation 15 gives:

$$\{s^{-1} + (1-s)^{-1}\ln s\}\,ds/d\ln D = \ln s \quad (17)$$

and the contrast as a function of s becomes:

$$\gamma = -2.303 s(1-s)\ln s/(1-s + s\ln s) \quad (18)$$

At $s = 1/2$, $\gamma^* = 2.601$, and as s goes to 1, $\gamma = 2(2.303) = 4.606$.

Simulated contrast curves (Fig. 4) for equation 8 with a = 0, 0.25 and 0.5 and for equation 15 illustrate several effects. Polydispersity decreases contrast as does chain scissioning. Unlike the idealized contrast curve of Fig. 2, the slopes of these plots are not constant. That is to say, the contrast varies with the thickness remaining. What is important to the lithographer is that attempts to overcome, say 5% thinning, are likely to run into trouble because the doses needed become disproportionately large, even in a well-ordered system where little scissioning occurs. Actually, most lithographers are more concerned with the contrast and dose to achieve an s of 0.5. It has to be emphasized that equation 8 is not very reliable for s greater than 0.5. In some cases, plots of $(s + s^{1/2})$ seem to consist of two branches with a lower value of a at higher doses.

Comparison with Experimental Data. Contrast curves for two samples of poly(chloromethylstyrene) (Fig. 5) confirm the prediction of equation 1. The dose at $s = 0.5$, D^*, is inversely proportional to M_w. The same can be said for the data of other workers who examined three molecular weights of the same polymer[5]. Moreover, the contrast curve predicted by equation 15 (monodisperse) gives a very good fit to the lower molecular weight sample. The higher molecular weight sample has a slope which does not correspond to the ideal. On the other hand, this behavior could be expected from the non-linear plots of $(s + s^{1/2})$ versus $1/D$ that have been reported on gamma radiation of several polymers. In fact, the data can be fitted approximately with a lower half of a= 0.5 and an upper half of a = 0.25. This should not be taken as a suggestion that chain scissioning is occurring. It does confirm a correlation between the two types of experiments. Some of the same reasons advanced by Charlesby and others to explain the non-linearity of gamma radiation

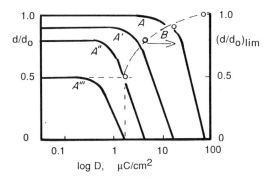

Figure 3. The thinning curve, B, is obtained from a series of contrast curves (development time increasing from A to A'''). The limiting value of normalized thickness is plotted versus the corresponding dose needed to remove polymer completely.

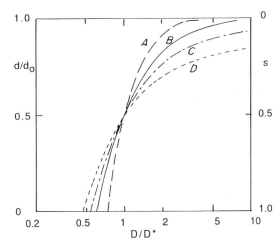

Figure 4. Plots of equation 15, (A); and equation 8 with a = 0, (B); a = 0.25, (C); and a = 0.5, (D).

results can be invoked here. Extraction of non-network, highly-branched polymer chains may be inefficient. Impurities may play a larger role with higher molecular weight polymers because the doses involved are lower and competitive reactions more important.

A further example (Fig. 6) illustrates the importance of molecular weight distribution. An unfractionated vinyl chloride terpolymer exhibits very low contrast. However, a low molecular weight fraction of that terpolymer yields a curve which approximates equation 15 for the monodisperse polymer. Unfortunately, a curve for fractionated polymer of the same weight-average molecular weight as the unfractionated polymer is not available.

Positive (Scissioning) Resists

Theoretical Models for Scission. Random chain scission characterized by a yield of $G(s)$ increases the number of polymer chains in a system while crosslinking decreases the number. A plot[2] of the number of molecules per unit mass, $1/M_n$, should be linear with dose, D:

$$1/M_n - 1/M_{no} = [G(s) - G(x)]D/(100N) \quad (19)$$

where M_{no} is the number-average molecular weight at $D = 0$. Any initial molecular weight distribution will tend to approach the "most probable" as scissioning proceeds in the absence of crosslinking. If the initial distribution is the "most probable", the weight average molecular weight, M_w, will follow a similar pattern:

$$1/M_w - 1/M_{wo} = [G(s) - 4G(x)]D/(200N) \quad (20)$$

As noted earlier in equation 4, no gel should form when $G(s) > 4 G(x)$. Since both equations are linear in D, the ratio of the slope for equation 19 to that for equation 20, defined as $K(r)$, will give the value of $G(x)/G(s)$, that is:

$$[G(s) - G(x)]/[G(s)/2 - 2G(x)] = K(r) \quad (21)$$

And thus:

$$G(s)/G(x) = [4K(r) - 2]/[K(r) - 2] \quad (22)$$

In the case where $G(x)$ is nil, $K(r)$ is 2, and so on.

Models for Dissolution Rate. Unlike the sensitivity of typical negative resists, lithographic sensitivity for a positive resist depends on the relative dissolution rates of exposed and unexposed areas. A number of studies have been reported on the effect of molecular weight, M, on rate of dissolution, R.

A 3-parameter model was proposed by Greeneich[6]:

$$R = R° + ß/M^a \quad (23)$$

where $R°$, ß, and a are fitted constants with no particular physical significance. Others[7-9] have used log-log plots of R versus M with an assumption that the slope will be relatively constant over a restricted range. This agrees with equation (23) when $R \gg R°$.

A further complication arises from the observation that R for an exposed polymer may be greater than that for an unexposed polymer of the same molecular weight. That is, the process of irradiation induces a change in addition to lowering molecular weight which increases dissolution rate by a factor of up to two. "Microporosity", probably caused by gas evolution during exposure, has been invoked for PMMA[7] and for a copolymer of alphamethylstyrene and maleic anhydride[10]. Stillwagon[8] showed that for the case of a polysulfone, the change could be erased by annealing an exposed polymer without changing its molecular weight.

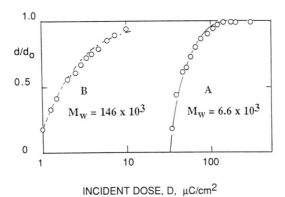

Figure 5. Experimental contrast curves (symbols) for poly(chloromethylstyrene) fitted (lines) with equation 15, (A); and equation 8, (B) in which the lower half uses a value of a = 0.5 and the upper half uses a = 0.25.

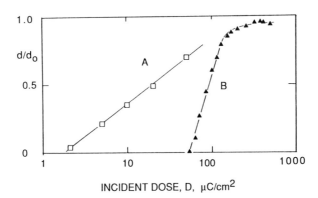

Figure 6. Experimental contrast curves for A, and unfractionated vinyl chloride terpolymer, $M_n = 48 \times 10^3$, $M_w = 85 \times 10^3$, and B, a fraction isolated from the same polymer with $M_n = 30 \times 10^3$ and $M_w = 42 \times 10^3$.

Cooper[11] measured the dissolution rates of unexposed PMMA films with a variety of molecular weights, R(unexp). These were compared with films from a high molecular weight PMMA which had been reduced in molecular weight by electron beam radiation, R(exp). She found that the ratio of R(exp) to R(unexp) for a given molecular weight was relatively constant (Fig. 7). The Cooper data for unexposed PMMA can be replotted (Fig. 8) to be consistent with

$$\log(8.00\,R) = 1.09 \times 10^5 / M_n \qquad (24)$$

If the rate is referred to a condition of R^* corresponding to a molecular weight of M^*, we can write:

$$\log(R/R^*) = (1/M_n - 1/M_n^*) 1.09 \times 10^5 \qquad (25)$$

Since most positive resists will approach the most probable distribution on scissioning, it may seem immaterial whether M_n or M_w is used to correlate R. However, a simple test can be made by combining two molecular weights of the same polymer in various proportions to give a range of values of M_n and M_w. When this is done (Fig. 9) it becomes obvious that M_n is the proper parameter. This result is not intuitively evident. One might expect R to scale with other transport-related properties such as melt viscosity (which correlates well with M_w) or the radius-of-gyration (which correlates well with M_v). One can conclude that the dissolution process is more sensitive to the low molecular-weight end of a distribution than it is to the high end.

Simulation of Contrast Curves. In order to simulate contast curves, two models for dissolution rate dependence on molecular weight can be used. The Gamma model is based on equation 25 and the Exponential model is based on a modification of equation 23 with two parameters.

A. The Gamma Model. Equation (19) with $G(x) = 0$ and equation (25) can be combined to give a contrast curve by eliminating M_n. The reference condition of D^* (related to resist sensitivity) where the dissolution rate is R^* is used. R^* is selected to be the rate corresponding to dissolution of half the original film thickness, $d°$, in an arbitrary development time t_d. We assume that R is uniform so that it is simply related to the thickness removed $(d°-d)$:

$$R = (d° - d)/t_d = (d°/t_d)(1 - d/d°) \qquad (26)$$

If we let $y = R/R^*$, then:

$$y = 2(1 - d/d°) \qquad (27)$$

We can rewrite equations 19 and 25 as

$$1/M_n - 1/M_n^* = G(s)(D - D^*)/(100N) \qquad (28)$$

$$1/M_n - 1/M_n^* = (1/K)\log y \qquad (29)$$

where K is the proportionality constant for rate of dissolution (eg., 1.09×10^5 in equation 25).
Then, letting $x = D/D^*$ and $K_L = \{KG(s)D^*/100N\}$,

$$\log y = \{KG(s)D^*/100N\}(D/D^* - 1) = K_L(x - 1) \qquad (30)$$

The contrast curve is a function of only one lumped parameter, K_L. The contrast γ is (-1/slope) of the contrast plot or

$$-\gamma = d(d/d°)d\log D = -(dy/d\log D)/2 \qquad (31)$$

Differentiation of equation 30 gives

$$(1/y)(dy/dD) = 2.303\, K_L/D^* \qquad (32)$$

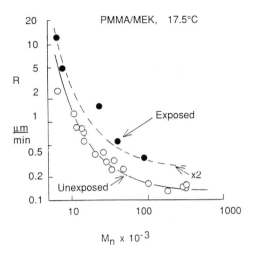

Figure 7. Dissolution rates for PMMA in methyl ethyl ketone. Data are from ref. 11.

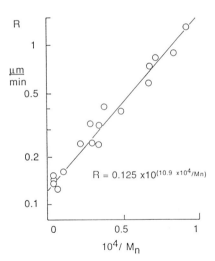

$R = 0.125 \times 10^{(10.9 \times 10^4/M_n)}$

Figure 8. Data of Figure 7 replotted.

Rearrangment results in
$$Ddy/dD = 2.303 \, yK_L(D/D^*) \tag{33}$$
And, using equation 31,
$$dy/d\log D = (2.303)^2 yK_L x = 2\gamma \tag{34}$$
Equation 30 gives
$$x = (\log y)/K_L + 1 \tag{35}$$
and
$$\gamma = (2.303)^2 \, yK_L\{(\log y)/K_L + 1\}/2 \tag{36}$$
so that
$$\gamma = (2.303)^2 \, y\{(\log y) + K_L\}/2 \tag{37}$$
At D^*, $y = 1$, and the contast reduces to
$$\gamma^* = (2.303)^2 \, K_L/2 \tag{38}$$
Thus the lumped parameter K_L essentially determines the contrast at D^*. Which is given by
$$\gamma^* = 2.652KD^*G(s)/100N \tag{39}$$

This predicts that the contrast will be highest for resists of low sensitivity (high D^*), high chain scissioning yields (high $G(s)$) and high molecular weight dissolution sensitivity (high K). All of these predictions have been observed with experimental data.

Eliminating K_L between equations 30 and 38 gives
$$\log y = 0.3772\gamma^* (x - 1) \tag{40}$$
This expression can be used to plot theoretical contrast curves for various values of γ^* (Figure 10). These plots indicate that considerable thinning of unexposed film is predicted when contrast is less than 4. This is not always observed in experimental results.

B. The Exponential Model. A contrast curve can be constructed from a two-parameter modification of equation 23:
$$R/R° = (M°/M)^a \tag{41}$$
Equation 19 with $G(x) = 0$ gives
$$M°/M = M°G(s)D + 1 \tag{42}$$
Combining the two and using the reference condition of R^* again:
$$R^*/R° = b = (1 + M°G(s)D^*)^a \tag{43}$$
where $(1/2b)$ is the fractional thinning of unexposed polymer when polymer exposed to dose D^* has been developed to half its original thickness. For example, when $b = 10$, then $d/d° = 0.95$ for the unexposed (but developed) polymer and $d/d° = 0.5$ at a dose of D^*.

Now with $y = R/R^*$ and $x = D/D^*$ again, dividing equation 41 by equation 43,
$$y = (1 + M°G(s)D)^a/b = (1 + M°G(s)D^*x)^a/b \tag{44}$$
But, also using equations 41 and 43,
$$M°G(s)D^* = (M°/M^*) - 1 = (b^{1/a} - 1) \tag{45}$$
Thus we can have an equation for the contrast curve in terms of two parameters corresponding to the slope of the dissolution rate-molecular weight plot and the thinning of unexposed polymer.
$$y = \{1 + (b^{1/a} - 1)x\}^a/b \tag{46}$$

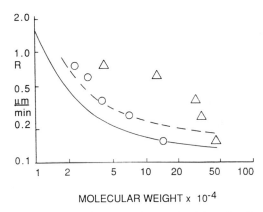

Figure 9. Dissolution rates for mixtures of PMMA with molecular weights calculated as weight-average (Δ) or number-average (o). Solvent was methyl ethyl ketone at 23°C. Cooper's data[11] at 17.5°C (solid line) for unexposed polymer is adjusted to 23°C (dashed line) for comparison.

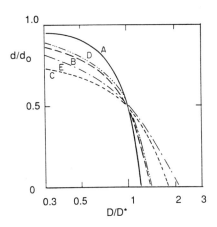

Figure 10. Contrast curves drawn according to equation 40 with $\gamma^* = 4$, (A); 2, (B); and 1, (C); and according to equation 46 with $b = 10$ and $a = 4$, (D); and $a = 1$, (E).

An expression now can be derived for contrast based on equation 31.

$$dy/dx = (a/b)\{1 + (b^{1/a} - 1)x\}^{a-1}(b^{1/a} - 1) \qquad (47)$$

And $\qquad \gamma = (2.303/2)x\,dy/dx \qquad (48)$

At $x = 1$, $\gamma = \gamma^*$,

$$\gamma^* = (2.303/2)a\{1 - b^{-1/a}\} \qquad (49)$$

Representative plots for equation 46 are included in Figure 10. Curves D and E with a = 4 and 1, respectively, are not very different in the region where d/d_0 is less than 0.5.

Comparison with Experimental Data. For PMMA in MIBK, contrast data can be fitted, as one might expect, by the Gamma model (Fig. 11). Workers usually select the slope at zero thickness to define the contrast. In this case, the contrast would range from 4.4 to 5.8 going from left to right.

Sensitivity does not vary greatly with molecular weight. Indeed, there is little difference observable for three molecular weights all developed with the same solvent (Fig. 12). It should be noted that the test on the highest molecular weight was run several years before the other two. The reason for the insensitivity lies in the fact that dissolution rate is not a linear function of molecular weight. The same dose (same number of chain scissions) that changes a molecular weight from 1,000,000 to 100,000 will change a molecular weight of 100,000 only to 50,000. However, the ratio of dissolution rates (R/R*) is the same in either case according to equation 25. For a given polymer, according to equation 32, the contrast should decrease and the thinning of unexposed areas should increase as development is pushed to obtain lower values of D*.

As mentioned earlier, an objective way of defining sensitivity is to plot the thinning of unexposed film when a film exposed to dose D is developed completely. The "sensitivity" can be defined as the dose at which thinning is 10%. This treatment ignores the differences in contrast which may result. With that caveat in mind, curves for PMMA and a copolymer of methyl methacrylate with monomethyl itaconate illustrate the point (Fig. 13)[12].

Sensitivity Relationships for Positive Resists - Nomograph. Specialists in radiation degradation sometimes have trouble communicating with specialists in microlithography. When dealing with a chain-scissioning polymer, the first group prefers G(s) as a measure of sensitivity, while the second group prefers the incident dose of electrons D_i needed to achieve a given difference in dissolution rate, contrast, or actual pattern transfer. D_i is a function of polymer thickness, polymer density, and the accelerating voltage (see below, equation 54).

A nomographic solution to the various equations involved is useful for several reasons:

1. It permits rapid comparison between the sensitivity criteria used by the two groups of workers,
2. It gives a picture of the relative importance of each parameter in arriving at the measure of sensitivity,
3. It allows rapid estimation of the effect of changing one variable on the consequent value of some other variable.

The complicating feature in joining the two measures of sensitivity is the variable dissipation of energy depending on the thickness penetrated by an electron beam. Gamma rays are scarcely attenuated at all when traversing as much as a few cm of an organic material. Thus, a polymer film only several μm thick receives the same gamma radiation dose throughout its cross section. In contrast, electrons in the range

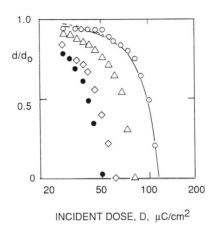

Figure 11. Contrast curves for PMMA developed in methyl isobutyl ketone for 1, 3, 5, and 10 minutes (right to left). Solid line is equation 30 with $\gamma^* = 4$. Dotted line (almost superimposed except at low doses) is equation 46 with $a = 8$ and $b = 80$.

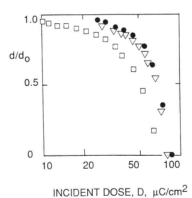

Figure 12. Contrast curves for PMMA developed in methyl isobutyl ketone. $M_w = 38 \times 10^3$, developed 0.5 min., (□); $M_w = 496 \times 10^3$, developed 1 min. (●); and $M_w = 950 \times 10^3$, developed 3 min. (▽).

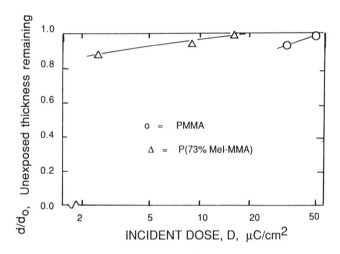

Figure 13. Comparison of thinning behavior[12] for a copolymer of monomethyl itaconate (73 wt.%) with methyl methacrylate (Δ) and PMMA (o).

of 10 to 50 kV are much more limited in penetrating effectiveness. As the electrons slow down, they are capable of depositing more energy per unit thickness. For example, virtually all the energy of 5 kV electrons is deposited in the first μm of an organic resist that is encountered. For a polymer film of density ρ and thickness z irradiated with electrons of energy V_a, the ratio of energy absorbed, D_a, to incident dose, D_i, is given by[13]

$$D_a/D_i = (k_1 V_a/\rho z) \int \Lambda(f) df \quad (50)$$

where $\quad f = z/R_g \quad (51)$

$$R_g = (0.046/\rho)(V_a)^{1.75} \text{ (The Grün range)} \quad (52)$$

$$\Lambda(f) = 0.74 + 4.7f - 8.9f^2 + 3.5f^3 \text{ (The depth-dose function)} \quad (53)$$

This set of equations is further complicated by the units employed in common practice. Conversion factors used in arriving at a value of the constant $k_1 = 1.00$ are:

1 electron = 1.602×10^{-19} C
(C = coulomb)
1 rad = 100 erg/g
1 erg = 6.242×10^{11} electron volt

The equations as written are consistent with the following units:
D_a, Mrad; D_i, μC/cm^2; ρ, g/cm^3; z, μm; and V_a, kV

The scissioning yield is given by equation 19 rewritten as

$$G(s) = k_2 \{ 1/M_n - 1/M_{no}\}/D_a \quad (54)$$

where $k_2 = 0.965 \times 10^6$ when the units used are:

G(s), scissions/100 electron volt; M_n, g/mol, and D_a, Mrad.

The present nomograph (Figure 14) has been constructed by conventional methods[14] and is relatively compact. Compromises have been made in the choice of ranges for each variable and in the accuracy with which the depth dose function can be represented. All of the equations are used with the following assumptions:
1. There is no cross-linking by radiation.
2. There is no accounting for electrons scattered from the substrate and re-entering the film.

General description:
There are four sets of axes labelled I, II, III, and IV. A straight line connecting any two axes (with the same roman numeral) will intersect the third axis of the set to solve a single equation. The systems are:

I. Multiplication of D_a times G(s).

II. Division of D_a by D_i.

III. Equation 54, the product of G(s) and D_a given as a function of initial and final molecular weights.

IV. Equation 50, D_a/D_i given as a function of V_a and the product of ρ and z.

Example
  Under conditions where an absorbed dose of 15 Mrad changes the molecular weight from an original value of 300,000 to a final value of 40,000,
  a) What is (G(s)?
  b) What equivalent D_i is required at a V_a of 20 kV?

Additional data: z = 1.10 μm, ρ = 1.20 g/cm^3.

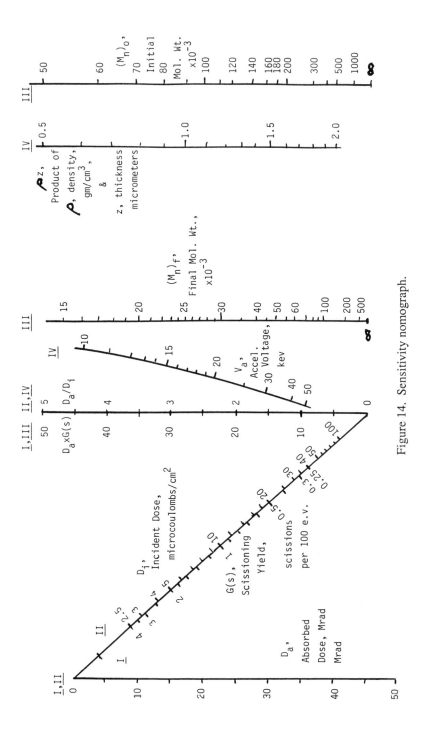

Figure 14. Sensitivity nomograph.

Nomographic solutions are:
Use System III to get the product of $G(s) \times D_a = 20.7$ from the initial and final molecular weights.
Use System I to get $G(s) = 1.39$ from $D_a = 15$ and $G(s) \times D_a = 20.7$
Use System IV to get $D_a/D_i = 2.37$ from ρz and V_a.
Use System II to get $D_i = 6.33$ from D_a and D_a/D_i.

The nomograph, of course, can only be read to two significant figures rather than the three indicated above.

Conclusions

The theoretical equations for scissioning and crosslinking can be used to correlate the results of lithographic behavior of thin films on irradiation. The same deviations can be expected that are found when the equations are used for bulk samples exposed to gamma radiation. Positive resists of the type considered here depend on a change in dissolution rate, R, with polymer molecular weight, M. The form of the relationship between R and M is still a matter of some controversy.

The idealized contrast curve consisting of a linear thickness dependency on logarithm of dose is not likely to be realized in practice. Indeed, such a curve would contradict the clear predictions of theory for both positive and negative resists.

Acknowledgments

This work is supported in part by the Office of Naval Research. The cooperation of the National Nanofabrication Facility at Cornell is gratefully acknowledged.

Literature Cited

1. Flory, P. J., "Principles of Polymer Chemistry," Cornell Press, Ithaca, NY, 1953, Chapter 9.
2. Charlesby, A., and S. H. Pinner, Proc. Roy. Soc., Ser. A 249, 367 (1959).
3. Flory, P. J., op. cit., Chapter 8.
4. Turner, D. T., in "The Radiation Chemistry of Macromolecules," Vol. 2, M. Dole (ed), Academic Press, New York, 1973, Chap. 8.
5. H. S. Choong and F. J. Kahn, J. Vac. Sci. Tech., 19 (4), 1121 (1981).
6. Greeneich, J. S., J. Electrochem. Soc., 121, 1669 (1974).
7. Ouano, A. C., Polym. Eng. Sci., 18, 306 (1978); 20, 160 (1980).
8. Stillwagon, L., in "Polymer Materials for Electronic Applications," E. D. Feit and C. Williams, Jr. (eds), ACS, Washington, DC, 1982, p. 19.
9. Manjkow, J., J. S. Papanu, D. S. Soong, D. S. Hess, and A. T. Bell, J. Appl. Phys., 62, 2 (1987).
10. Rodriguez, F., R. J. Groele, and P. D. Krasicky, SPIE Vol. 539 Advances in Resist Technology and Processing (1985), p. 14.
11. Cooper, W. J., P. D. Krasicky, and F. Rodriguez, Polymer, 26, 1069 (1985).
12. Anderson, C.C., PhD Thesis, Cornell University, 1985.
13. Thompson, L. F., C. G. Willson, and M. J. Bowden, "Introduction to Microlithography, ACS, Washington, DC, 1983, Chapter 2
14. Levens, A. S., "Nomography," 2nd ed., Wiley, New York, 1959.

RECEIVED September 29, 1989

Chapter 36

Electron-Initiated Graft Modification of Polyolefins

Sam V. Nablo[1], I. J. Rangwalla[1], and John E. Wyman[2]

[1]Energy Sciences Inc., 8 Gill Street, Woburn, MA 01801
[2]Consultant, 17 Monadnock Drive, Westford, MA 01886

Modern electron processors offer high speed (high dose-rate) curing of low viscosity liquid coatings so that surface modification of films becomes practicable. A process has been developed for grafting vinyl monomers to polyolefin film surfaces with the aid of functional silane primers using electron initiated polymerization. A one step process is described using a pre-dried vinylbenzylamine siloxane primer, overcoated with the functional monomer, in this case acrylonitrile. Infrared analysis, tape tests, and gravimetric analysis after washing and flexing have shown the electron initiated cure grafts the primer to the film surface and simultaneously causes the copolymerization of the monomer-primer system.

Monomers which have also been demonstrated to have been grafted to polyethylene and polypropylene film by this process include styrene and various acrylated esters, mixtures of acrylated esters, and acrylonitrile. The application of this technology to improving the barrier properties of films is discussed. Some initial data on the O_2 barrier, fragrance/aroma, and haze performance of these graft modified films indicate substantial improvement and place them in the high barrier range where dry O_2 permeabilities of less than 1 cc/100 in^2/24 hrs. are obtained.

This work was directed to a demonstration of a new process developed in our laboratories(1) for graft polymerization and copolymerization initiated by energetic electrons.
    In the past few years several radiation-chemical methods have been developed for the preparation of graft polymers and copolymers. These radiation chemical methods are much easier to handle than

conventional chemical processes, and the costs are comparable. Most of this work used low dose rate gamma rays from a Co^{60} source. Very little has been published concerning radiation-chemical grafting using high energy electrons from an accelerator which delivers those electrons at a 10^6-10^9 rads/sec dose rate. Such high dose rates make the radiation chemical process commercially attractive. We have found that by using the new process(1) and high energy electrons from an electron beam processor the technique can be extended to graft polymerize and copolymerize a wide range of polymers in any desired combination and achieve excellent adhesion to the surface of (biaxially oriented) polyolefin films. For example, several coatings based upon acrylated urethane oligomers and monomers such as vinyl pyrrolidone and 1,6, hexanediol-diacrylate, have been coated on primed polyolefin films. These coatings were cured on the dried primer in a "one step" process. Adhesion tests conducted with adhesive tape(2) and flexing showed that surface anchoring to the primed and electron grafted film was superior to that demonstrated with the unprimed film.

A brief review of the physics of electron energy absorption in matter, and of the equipment used to accomplish electron initiation of these reactions at high dose rates (speeds) is now presented.

The Electron Energy Source

Practical application of electron induced polymerization has only become possible with high dose rate machinery. Modern electron processors conveniently provide dose rates compatible with the curing of products at speeds of production machinery. These speeds are typically 1-300 m/min for coating, printing and laminating with doses to cure of 1-10 Mrads. Such a modern curtain type, Selfshielded electron processor is outlined in Figure 1. As the process speed requirements increase, multiple curtains or strip electron beams are employed to increase the power output of the curing station. In this way very high processing speeds are achievable from a single module. Some of these relations for a typical 1050 millimeter wide unit are shown in Figure 2.

These processors are designed so that they provide a uniform treatment of a product cross-web (i.e. across the 1050 millimeter process width). However, in the longitudinal direction the processor's electron flux has a Gaussion distribution due to window and air path scattering. Some knowledge of this dose rate variation in the processor is desirable in that many of the processes of interest are rate dependent, as indeed are most free radical initiated reactions.(3) In seeking to achieve a "low" average dose rate at the product plane for high power processors, a large "effective window area" is desirable. This increases the residence time of the product in the process zone.

An example of the relatively protracted process zone achieved for web treatment in an industrial unit, employing 2 modules each capable of delivering one megarad at 380 meters/minute, is illustrated in Figure 3. Although the unit will deliver energy at high rates (this corresponds to 300 mA of 200 kilovolt electrons or 60

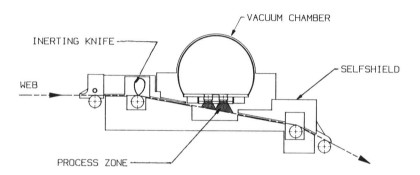

Figure 1. Schematic view of a curtain-type selfshielded electron processor. The product path typically involves a modest angle change to facilitate radiation shielding. An inert gas knife coupled with a small process volume eliminates oxygen inhibition and ozone generation problems. Process speeds to 300 m/minute are typical.

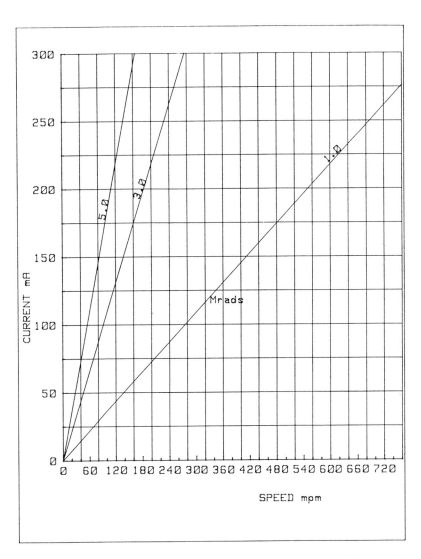

Figure 2. Typical isodose performance curves for a 1050 mm wide processor, of the type shown in Figure 1.

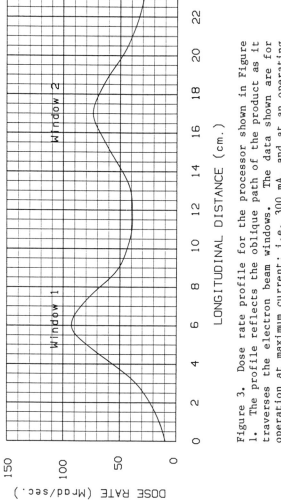

Figure 3. Dose rate profile for the processor shown in Figure 1. The profile reflects the oblique path of the product as it traverses the electron beam windows. The data shown are for operation at maximum current; i.e. 300 mA, and at an operating voltage of 165 kV.

kilowatts of beam power) it does so over a protracted distance (approximately 20 cm) so that dose rates well under 100 Mrads/second are experienced by the product. Figure 3 shows the dose rate profile along the actual path of the product as it obliquely traverses the process zone, as shown in Figure 1.

Because the free radical initiated surface chemistry of interest here is normally quenched by the presence of oxygen, units of the type shown in Figure 1 are designed to provide inerting in the process zone, typically at levels under 0.1% of atmospheric concentration; i.e. < 210 ppm. Very modest air volumes are affected by the electron beam in these Selfshielded designs, so that small N_2 flow is adequate to provide this level of purging e.g. 1.5 m^3/min for a 1 meter wide system. In addition, the possible generation and dispersion of ozone into the work area external to the processor is eliminated by providing this level of inerting in the highly ionized process zone.

Surface Treatment of Matter

Many of the surface property modifications of commercial interest can be adequately accomplished by the use of controlled depth of penetration of the electron beam into the substrate. This penetration depth is controlled by the high voltage of the processor which in turn controls the kinetic energy of the electrons and their penetration. For industrial use, preservation or perhaps even enhancement of the bulk physical properties of the substrate are often desired while altering the surface properties. These modifications may be further delineated by the controlled diffusion or penetration of the added modifying agent into the substrate. For electron initiated surface modification, these processes usually consist of crosslinking of the beam affected zone in the substrate,[4] or substrate copolymerization with a surface deposited/-coated polymer. For most materials, electrons should penetrate well beyond any surface connected pore structure, in order to obtain the desired modification. This must be accomplished after penetration of the modifying coating itself. In general, this necessitates ranges of a few tens of micrometers (or tens of g/m^2) in polymeric materials. Figure 4 shows a Monte Carlo calculated depth dose profile in polypropylene for electrons from a 70 kilovolt processor;[5] taking into account energy losses in the processor window foil and air path. It shows that some 70% of the beam energy can be delivered to the top 25 micrometers of material. In this manner, a few grams/m^2 of a monomer can be grafted to (or otherwise cured on) a labile substrate without significant alteration of the bulk physical properties of the base material. It is known from previous work that insignificant radiolytic products or mechanical property changes occur in the bulk polymer under these conditions.[6] One should recall that the energy deposited in the beam-affected layer is too small (<10 cal/gm) to cause such changes.

An important feature of this application of low energy electrons lies in the rapid increase in electron stopping power (dE/dx) with decreasing energy.[7] If such low energy spectra can be produced

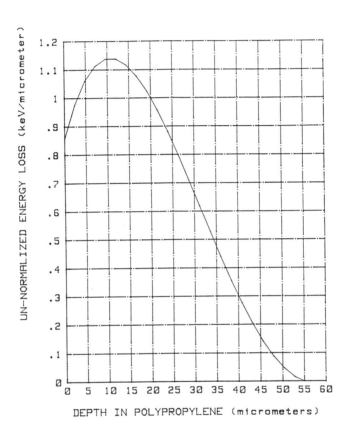

Figure 4. A calculated depth-dose profile in polypropylene for 70 kilovolt operation. A zero air path to the product is assumed.

practically, they will deliver dose to a surface with much higher efficiency (e.g. in keV/micrometer) than their more conventional, higher energy counterparts.

The product being irradiated may be supported on a cooled (or heated) drum in order to control the substrate temperature(8-9) during and immediately subsequent to electron treatment. Although the electron energy investment in the product may be relatively small in order to initiate polymerization, e.g. 1 Mrad or 10 kGy equals 2.4 cal/gm, it is often necessary to limit temperature excursions which could result from a high heat of polymerization on a thermally sensitive substrate.

It is known that most radiation initiated polymerization processes are initiated by the free radicals created by radiolysis of the monomers. If a monomer or a mixture of monomers is irradiated in the presence of a polymer, a graft copolymer is formed which has different physical properties. For example, if a second polymer like polyacrylonitrile or polyvinylidine chloride, which possess superior barrier properties against permeation by oxygen, carbon dioxide, water vapor etc., is grafted to a polyolefin film like polyethylene, polypropylene, etc., the barrier properties of the composite film are greatly enhanced compared with the polyolefin film.

Good adhesion of a second polymer to the base polyolefin film is often difficult to achieve. Conventional methods used to achieve this, for example extrusion coating with tie coatings are expensive. Another way to achieve improved adhesion is to surface treat the polyolefin film to raise its surface energy. This is done by a treatment with sodium dichromate/sulfuric acid solution, flame, or corona treatment. This allows the surface of the treated polyolefin film to be wetted by a liquid adhesive or polymer which would not otherwise spread on untreated polyolefin film so that an adhesive bond can be formed between the treated polyolefin film and the second polymer. Silane materials have also been used to modify surfaces to improve adhesion. Dow Corning Corp. supplies a variety of organofunctional silanes which are designed to improve adhesion between a mineral filler and a polymer.(10-11) Certain of these organofunctional silane coupling agents will also promote adhesion between dissimilar polymers which are otherwise difficult to bond. We have found that oligomers of a hydrolyzed cationic vinyl benzyl amine silane deposited from a alcoholic solution on the untreated surface of a polyolefin film, then dried and treated by electron beam radiation, form a uniform grafted surface layer of a crosslinked organofunctional polysiloxane. This polysiloxane has a critical surface tension higher than that of the polyolefin film and consequently will be wetted by a variety of liquid monomers, polymer, and liquid dispersions of various kinds. Vinyl groups of the organofunctional siloxane are grafted onto the polyolefin film by the influence of the electron beam radiation. Grafting of the siloxane to the polyolefin substrate produces the excellent adhesion obtained by the process. Also, a compatible monomer can be added to the hydrolyzed vinyl benzyl amino silane primer in methanol. Following evaporation of the methanol the polysiloxane/monomer mixture can be polymerized and grafted onto the polyolefin film. Both these processes are described in detail in Reference 1.

The monomer chosen for this demonstration was acrylonitrile. It is compatible with the hydrolyzed vinyl benzyl amine silane primer in methanol, and will polymerize and copolymerize with moderate doses of electron beam energy. This study was designated to demonstrate the effectiveness of the new process in radiation grafting of acrylonitrile onto polyolefin films.

Experimental Procedure

The surface primer was prepared as a 40% methanol solution of a vinyl benzyl amine functional silane (Dow Corning Corporation Z-6032) by slowly adding methanol and then water. The mixture was allowed to stand for 24 hours to hydrolyze the silane and allow the low molecular weight oligomers to form. Samples of low density polyethylene (LDPE) film .005" thick, supplied by U.S. Industrial Chemical Company (Resin #NA140-00 containing no antioxidants) were prepared for use as the olefinic substrate.

Solutions of the hydrolyzed silane primer in methanol and acrylonitrile were then prepared at acrylonitrile:primer ratios of 1, 2, and 3:1. Individual samples of the LDPE film were immersed in each solution, as well as in the primer above, and irradiated on both sides (i.e. bilaterally) on the processor at 5 megarads at a 200 kV operating level at O_2 concentrations of less than 100 ppm in the process zone. Prior to the runs, the processor was characterized using film dosimeters.(12-13)

Air drying was used only for the sample treated with primer alone and was omitted for the three samples containing primer and acrylonitrile since acrylonitrile has a higher vapor pressure then methanol.(14) As a result, the electron treatment was performed on wet samples. The fifth LDPE sample was used as a virgin polyethylene control. After irradiation, all samples were washed with luke warm water to remove any residual monomer. Because of the highly toxic nature of acrylonitrile, these coating procedures were conducted under carefully controlled conditions. All five samples (i.e. primed, 1:1, 2:1, 3:1 and control) were examined by Fourier Transform Infrared Spectroscopy (FTIR).

Results

The physical appearances of all the chemically treated and irradiated samples indicated that a polymer layer was deposited on the LDPE film surface. The first two samples which had been treated with 1) the vinyl benzylamine silane primer and 2) the 1:1 volume ratio of acrylonitrile and the vinyl benzyl amine silane primer provided clear samples with a rough surface. The 2:1 and the 3:1 acrylonitrile/vinyl benzyl amine silane treated samples were mottled in appearance, suggesting the formation of regions of polyacrylonitrile homopolymer.

Transmission FTIR spectra were recorded for all five samples using an open beam reference. The spectra were taken on a Digilab FTS 45 FTIR with HgCd Te detector with an infrared response from 4000 to 750 cm^{-1}.

Figure 5 shows an FTIR spectrum of virgin LDPE. Figure 6 shows the spectrum of LDPE and primer and acrylonitrile with the LDPE spectrum of Figure 5 subtracted, and exhibits peaks at 1100 cm^{-1} and at 2241 cm^{-1}, corresponding to Si-O-C/Si-O-Si bonds and to the C≡N bond from the nitrile group respectively. The presence of this latter bond is further illustrated in Figure 7. Here the <u>primed</u> polyethylene spectrum was subtracted from the grafted LDPE and primer and acylonitrile spectrum, again showing the peak at 2241 cm^{-1}.

Unexpectedly, the radiated 2:1 acrylonitrile:primer sample showed a greater quality of aliphatic nitrile than either the 3:1 or 1:1 samples. Grafting of the acrylonitrile to the substrate was demonstrated by repeated washing and flexing of the samples with no measurable loss of the grafted polymer on the substrate as indicated by gravimetric analysis of the cured samples. Tape tests were also performed on the grafted samples which confirmed the excellent film anchoring.

A preliminary integration of the peaks obtained indicated a grafted layer of 1-2 micrometers thickness of the silane/acrylonitrile on polyethylene. A gravimetric analysis of the 2:1 sample indicated a weight gain of the treated polyethylene of about 43 mg. This weight gain for a 20 cm x 12 cm sample corresponds to a 1.5 micrometer thick layer.

Application

The objective of this study was to demonstrate that the adhesion of polymers which are known to have excellent gas and oil barrier properties to a polyolefin surface could be greatly improved by grafting them to the polyolefin surface with the aid of a siloxane primer. Permeability tests showed that the grafted siloxane primer itself has excellent gas and oil barrier properties. Some of these results are shown in Figure 8 in which the dry oxygen permeability of the primer is shown as a function of thickness. These tests utilized three different base films, namely 13 micrometer polyester, 38 micrometer oriented polypropylene and 135 micrometer low density polyethylene. As shown in figure 8, all three untreated films have comparable dry oxygen permeabilities. The results show that oxygen permeabilities below 1 cc/100 in^2/24 hours at room temperature (23°C) can be achieved with primer thicknesses of approximately 10 micrometers. Thicknesses of only a few micrometers provide a dramatic improvement in barrier properties. Equally impressive improvement in oil resistance of the polyolefins is achieved by the use of the grafted siloxane polymer. Some haze measurements performed on LDPE film are presented in Table I which was tested in accordance with ASTM test method D-1003.

Comparison of the dependence of dry oxygen permeability on temperature with some conventional barrier polymers (after Watanabe)(<u>15</u>) are presented in Figure 9 in normalized units of cc/m^2/24 hours for a one mil (25.4 micrometers) film thickness. The temperature coefficient of the grafted siloxane polymer appears to be comparable to that of BAREX 210 and ethylene vinyl alcohol. The EVAL shown (EVAL-E) is 44% by weight ethylene. The silane is considerably better than that of the polyvinylidene chloride (Saran 468) sample shown in the figure. The data are plotted so that the temperature increases to the right on the abscissa.

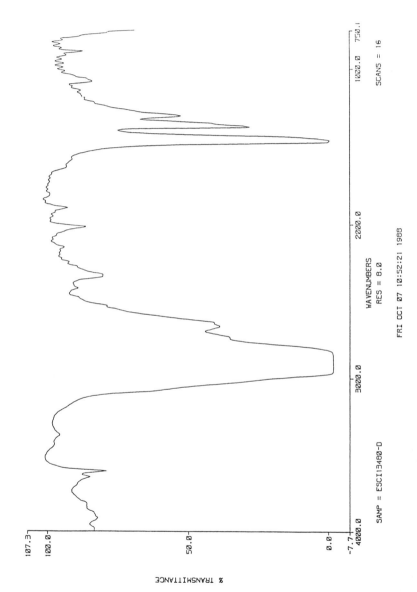

Figure 5. FTIR spectrum for low density polyethylene. These spectra were recorded on a Digilab FTS 45 FTIR with HgCdTe detector.

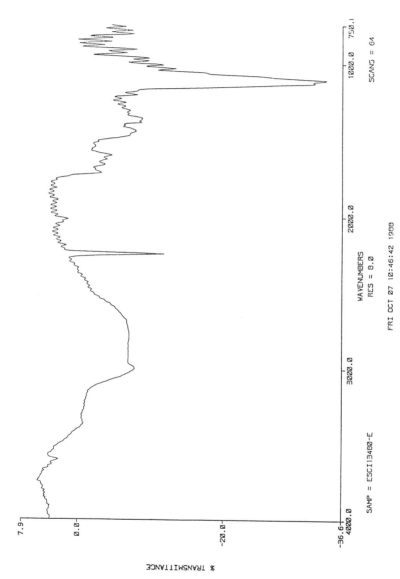

Figure 6. FTIR spectrum for the primed and acrylonitrile modified low density polyethylene with the substrate spectrum of Figure 5 subtracted. The C≡N bonds at 2241 cm^{-1} and the Si-O-Si at 1100 cm^{-1} are evident.

546 RADIATION CURING OF POLYMERIC MATERIALS

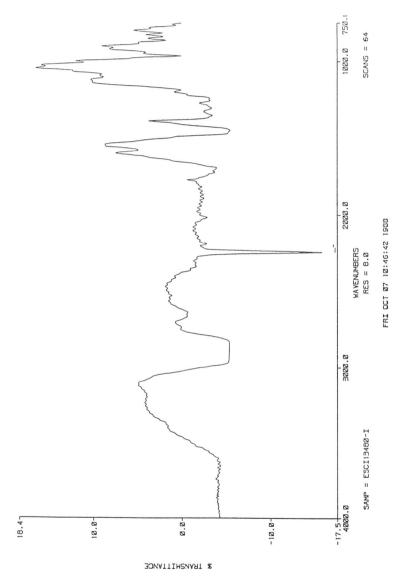

Figure 7. FTIR spectrum for the primed and acrylonitrile modified low density polyethylene with the primer-substrate spectrum subtracted. The C≡N bonds for the acrylonitrile graft are clearly shown.

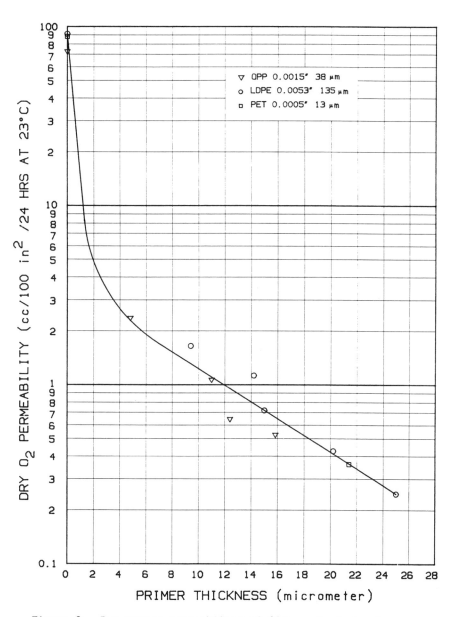

Figure 8. Dry oxygen permeability of 38 μm oriented polypropylene, 135 μm low density polyethylene and 13 μm polyester, treated with varying thicknesses of the silane primer. These data were recorded with a Modern Controls Inc. Oxtran 1000 permeability analyzer.

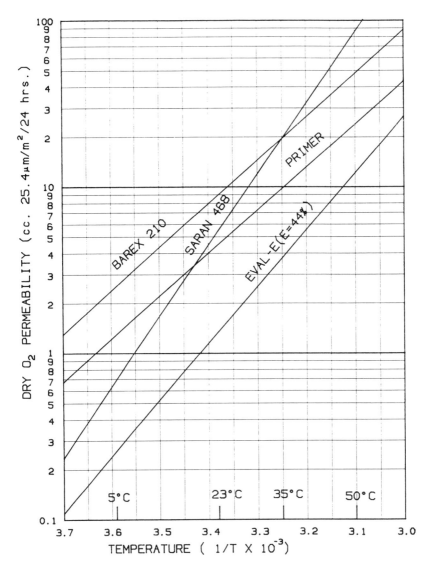

Figure 9. Dry oxygen permeability as a function of temperature for the electron grafted organofunctional silane, Saran 468. Barex 210 and EVAL-E (E 44% ethylene). The data are presented as a function of increasing temperature (after Watanabe (15)).

TABLE I. HAZE MEASUREMENTS - LDPE
135 MICROMETER FILM:5 MICROMETER
PRIMER BILATERAL

ASTM TEST METHOD D-1003

SAMPLE	TRANSMITTANCE %	HAZE %
VIRGIN FILM	88.8	14.9
PRIMED FILM	88.9	2.5

Tests on the fragrance/flavor barrier properties of siloxane coated polyethylene and polypropylene film, using the ASTM test method D-1653 are given in Table II, which show excellent performance for the primed film.

TABLE II. LIMONENE TRANSMISSION - LDPE/OPP

ASTM TEST METHOD D-1653

(75°F - 35% R.H.)

SAMPLE	THICKNESS (MICROMETERS)	LOSS (MG/100 IN^2/24 HOURS)	NORMALIZED
VIRGIN LDPE	135	8014.9	100.0
PRIMED LDPE	APPROX. 145	6.0	.07
VIRGIN OPP	50	3359.6	100.0
PRIMED OPP	APPROX. 60	31.8	.94

The grafted siloxane produces a very high gloss (low haze) surface which greatly enhances the appearance of these films. Electron micrographs at approximately ten thousand magnification are shown in Figure 10(a) and reveal the smooth surface produced by a five micrometer thick grafted siloxane coating on a 125 micrometer thick LDPE film. The untreated film surface is shown in 10(b) for comparison.

Sections of these coated films were examined by x-ray dispersive spectroscopy (EDX) which showed that the siloxane was confined strictly at the film surface with no significant migration into the polyolefin. These results illustrate that the exceptional adhesion which has been achieved with the use of the siloxane primer is most likely the result of grafting of the primer to the surface.

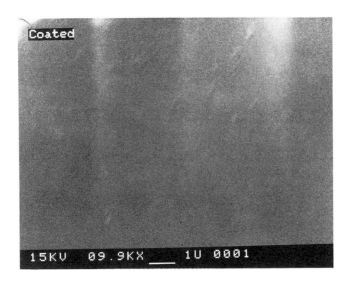

Figure 10(a). Electron micrograph of the primed low density polyethylene film at 9.9 X 10^3 magnification. Five μm of organofunctional silane were grafted to the 125 μm LDPE film (USI-NA-140-00).

Figure 10(b). Electron micrograph of the untreated LDPE film surface at 9.7 X 10^3 magnification.

Conclusion:

This preliminary study has shown that radiation initiated graft polymerization and co-polymerization on a polyolefin film surface can be achieved using the above described process with electron initiation. This process can be employed with a wide range of monomers.

One of the siloxane polymers used as a primer on the surface of polyolefin films has demonstrated barrier properties toward oxygen, oil, and flavors comparable with the best known barrier polymers.

Acknowledgments

The authors appreciate the continuing assistance and encouragement of Gerald Venn and Dr. Peter Pape of Dow Corning Corporation.

Literature Cited

(1) Wyman, J.E. U.S. Patent 4,803,126, 7 February 1989. Additional applications are on file.
(2) Standard Test Methods for Measuring Adhesion by Tape Test, ASTM 3359-87, American Society for Testing and Materials, 1988.
(3) Chapiro, A. Radiation Chemistry of Polymeric Systems, John Wiley and Sons: New York, NY, 1961; Vol, XV, p 598.
(4) Sax, J.E.; Thalacker, V.P.; Larson, E.G. J. Rad. Phys. Chem., 1988, 31, p 887-896.
(5) Colbert, H.M., "SANDYL - A Computer Program for Calculating Combined Photon - Electron Transport in Complex Systems", Sandia Laboratories SLL-74-0012, Livermore, California, May 1974.
(6) Josephson, E.S.; Peterson, M.S., Ed.; Preservation of Food by Ionizing Radiation; CRC Press Inc.: Boca Raton, FL, 1983; Vol. 2, Chapter 8, p 319.
(7) Stopping Powers and Ranges of Electrons and Positrons, U.S. Department of Commerce, NBS IR 82-2550-A: Washington, DC, 1982.
(8) Fletcher, P.M.; Williams, K.E. J. Rad. Phys. Chem. 1988, 31, 337-348.
(9) Nablo, S.V.; Tripp, E.P. U.S. Patent 4,521,445, 1985.
(10) Handbook of Adhesives; Skeist, I., Ed.; Van Nostrand Reinhold Company: New York, NY, 1977; p 640-841.
(11) Plueddeman E.R. Silane Coupling Agents; Plenum Press: New York, NY, 1982; p 1-204.
(12) Nablo, S.V.; Frutiger, W.A. J. Rad. Phys. Chem. 1980, 18, 1023.
(13) McLaughlin, W.L. et al J. Rad. Phys. Chem., 1983, 22, 325.
(14) Handbook of Chemistry and Physics; Weast, R.C., Ed.; CRC Press Inc.: Boca Raton, FL, 1986-87; p C58.
(15) Watanabe, H. In Future-Pak '88, Sixth International Ryder Conference on Packaging Innovations, 1988, p 442.

RECEIVED September 13, 1989

INDEXES

Author Index

Abdulaziz, Mahmoud, 220
Allen, G. M., 242
Allen, Norman S., 72, 346
Bayer, H., 412
Bett, Stephen J., 128
Bishop, T. E., 258
Cage, M. R., 308
Catalina, F., 72
Chen, W., 72
Clancy, Roy, 346
Crivello, J. V., 398
Decker, C., 439
Dektar, John L., 82
Dems, B. C., 516
Desobry, V., 92
Dietliker, K., 92
Dirlikov, Stoil K., 194
Drain, K. F., 242
Dworjanyn, Paul A., 128
Eckberg, Richard P., 382
El-Naggar, Abdel M., 474
Feldman, L., 297, 308
Feng, Yan, 194
Fields, Barry A., 128
Fouassier, Jean-Pierre, 59
Gaber, Dennis J., 325
Garnett, John L., 128
Gatechair, Leslie R., 27
Glaser, David M., 160
Gloskey, David J., 210
Graham, John C., 210, 325
Green, P. N., 72
Green, W. A., 72
Gummeson, Joel J., 176
Hacker, Nigel P., 82
Hoyle, Charles E., 1, 429
Hult, Anders, 459
Hüsler, R., 92
Jacobine, Anthony F., 160
Jankiewicz, Stan V., 128
Jönsson, Sonny, 459
Kim, Ha-Chul, 474
Kinstle, James F., 17
Kiser, T. K., 308
Koehler, M., 106
Krajewski, J. J., 258
Krasnopoler, A. A., 516
Kukkala, Pravin K., 325

Lapin, Stephen C., 363
Lee, J. L., 398
Lehner, B., 412
Liang, R. C., 308
Liu, Yifang, 325
Lougnot, Daniel-Joseph, 59
Mateo, J. L., 72
McConnell, JoAnn A., 272
McGrath, James E., 474
Meador, Mary Ann B., 220
Meador, Michael A., 220
Misev, L., 92
Moussa, K., 439
Murphy, E. J., 151
Nablo, Sam V., 534
Nakos, Steven T., 160
Namaste, Y. M. N., 516
Nobbs, James H., 43
Noren, G. K., 151, 258
Obendorf, S. K., 516
Ohngemach, J., 106
Oldring, Peter K. T., 43
Pacansky, J., 498
Rånby, Bengt, 140
Rangwalla, I. J., 534
Rembold, M., 92
Riding, Karen D., 382
Rist, G., 92
Robinson, Peter J., 346
Rodriguez, F., 516
Russell, Geoffrey A., 284
Rutsch, W., 92
Sangster, David F., 128
Sastre, R., 72
Shi, D. J., 308
Skiens, W. Eugene, 284
Sundell, Per-Erik, 459
Tiefenthaler, Ann M., 27
Trapp, Martin A., 429
Waltman, R. J., 498
Ward, T. C., 297
White, Nicholas J., 346
Wilkes, Garth L., 474
Willard, F. Kurt, 272
Wu, Xiaosong, 194
Wyman, John E., 534
Yoo, Youngtai, 474
Zimmerman, J. M., 258

Affiliation Index

AB Wilhelm Becker, 459
Academia Sinica, 72
Allied–Signal Engineered Materials Research Center, 363
Ciba–Geigy Corporation, 27,242
Ciba–Geigy Ltd., 92
Commonwealth Scientific and Industrial Research Organisation, 128
Cornell University, 516
DeSoto Inc., 151,258
E. Merck, 106
Eastern Michigan University, 194,210,325
Ecole Nationale Supérieure de Chimie, 439
Energy Sciences Inc., 534
General Electric Company, 382,398
Harcros Chemicals (UK) Ltd., 346
Howson-Algraphy Ltd., 346
IBM Almaden Research Center, 82,498
Institute of Photographic Sciences, Academia Sinica, 72
Instituto Plasticos y Caucho, CSIC, 72
James River Corporation, 17
Loctite Corporation, 160
Manchester Polytechnic, 72,346
Mead Imaging, 297,308
Monsanto Chemical Company, 176
NASA Lewis Research Center, 220
Optical Data, Inc., 284
Polycure Pty. Ltd., 128
Radcure Specialties Inc., 272
The Royal Institute of Technology, 140,459
Siemens AG, 412
Thomas Swan & Company Ltd., 43
Unité Associée au Centre National de Recherche Scientifique, 59
University of Leeds, 43
University of New South Wales, 128
University of Southern Mississippi, 1,429
Virginia Polytechnic Institute and State University, 297,474
Ward-Blenkinsop and Company Ltd., 72

Subject Index

A

Abstraction-type photoinitiators
 examples, 7
 mechanism, 7
Acrylate(s), disadvantages, 363
Acrylate reactive diluents
 advantages, 206
 solubility, 206t,207
Acrylate resin systems, properties, 151
Acrylated epoxy oligomer
 advantages, 9
 structure, 9
Acrylated melamines in UV-curable coatings
 accelerated weathering, 181,185f
 cure procedures
  thermal cure, 179
  UV cure, 179
 demonstration of dual cure, 186,188f,189,190–191f
 description, 176,177t
 dual-curve procedure, 179
 dynamic–mechanical analytical temperature scan, 186,188f
 effect of mustard staining, 192f
 film properties, 181,183–184,186–187

Acrylated melamines in UV-curable coatings—Continued
 isothermal scan, 186,189,190–191f
 overprint varnish formulation viscosities, 181,182t
 physical properties, 176,177t
 preparation, 179
 properties of dual-cured films, 189,191t,192f
 shrinkage, 181,183t
 thermal curve, 186,187f,t
 Tukon hardness, 181,184f,186,187f
 type A structure, 176,178f
 type B structure, 176,178f
 viscosity, 179,180f,181,182f
 viscosity in N-vinylpyrrolidone, 176,180f
Acrylated oligomers, classes, 272
Acrylated polyester
 properties, 10
 structure, 10
Acrylated polyether
 properties, 10
 structure, 10
Acrylated urethane oligomer
 properties, 10
 structure, 10

Acrylic monomers
 characteristics, 442,444
 examples, 444
 laser-induced curing, 446–445
 polymerization, 444,445f
 UV-radiation- and laser-induced
  polymerization, 439–455
Additives
 advantages and disadvantages, 11–12
 properties, 5,6t,11
Adhesives, characterization in terms of
  degree of cure, 242–256
Alcohols, effect on cure speed, 364
4-(2-Azidoethoxy)phenyl 2-hydroxy-2-propyl
  ketone, synthesis, 109

B

2-Benzyl-2-(dimethylamino)-1-(4-morpholino-
  phenyl)butanone
 absorption characteristics, 93,94f
 chemically induced dynamic nuclear
  polarization spectrum, 96,97f
 decomposition products, 96,97f
 effect of initiator concentration on
  optical density, 101,103f
 effects
  on flexographic printing plate, 101,103f
  on offset printing inks, 101,102f
  on resist formulations, 101,102f
  on white pigmented lacquer, 98,100f,101
 emission spectrum of Nyloprint bulb vs.
  absorption, 101,103f
 photochemical reaction pathways, 93,95f
 photochemistry, 93,95f,96
 photolysis with H-donor trapping,
  96,98,99f
 photolysis with TMPO trapping, 98,99f
 preparative photochemistry, 96,98,99f
 synthesis, 93,94f
 trapping reactions with 2-tert-
  butylacrylic acid methyl ester, 98,100f
1,4-Bis(2-benzyloxybenzoyl)benzene
 characterization, 226
 preparation, 226
Bis(dodecylphenyl)iodonium
  hexafluoroantimonate
 synthesis, 384
 use as catalyst for epoxysilicones,
  384–385
1,4-Bis(2-hydroxybenzoyl)benzene
 characterization, 225
 preparation, 225
Bis(2-hydroxypropyl ether) of bisphenol A
 GC after photolysis, 335,336f
 MS analysis after photolysis,
  335,337t,339

Bis(2-hydroxypropyl ether) of bisphenol A—
  Continued
 photodegradation pathways, 335,338–339
 photolysis, 335–339
 preparation, 328–329
Bis(methyl ester) of bisphenol A
 GC after photolysis, 339,340f,343
 MS analysis after photolysis,
  339,342t,343
 photodegradation pattern, 339,341,343
 photolysis, 339–343
 preparation, 329
2,2-Bis{4-[2-(norborn-2-ene-5-carboxy)-
  ethoxy]phenyl}propane, preparation,
  163–164
Bisphenol-A-based epoxy coatings, effect of
  diluent on surface softening, 326,327f
Bisphenol A diglycidyl ether,
  photodegradation products, 326
Bisphenol A–epichlorohydrin condensation
  polymer(s)
 instrumentation, 329–330
 photolysis procedure, 329
 preparation of model compounds, 328–329
 structure, 328
Bisphenol A–epichlorohydrin condensation
  polymer coatings, degradation by UV light,
  325–326
Bis(n-propyl ether) of bisphenol A
 GC after photolysis, 330,331f
 MS analysis after photolysis,
  330,333t,334–335
 photodegradation pathways, 330,332
 photolysis, 330–335
 preparation, 328

C

Caprolactone–allyl glycidyl ether
  copolymers, synthetic scheme, 476,477f
Caprolactone–allyl glycidyl ether copolymers
  cross-linked by electron-beam
  irradiation
 calculation of temperature rise due to
  electron-beam irradiation, 488
 cross polarizer optical micrographs,
  481,483–485f,486
 crystalline melting temperature vs.
  irradiation dose, 492,493f
 crystalline melting transitions,
  488,489–490f,491–492
 crystallization peaks, 492,494f
 electron-beam irradiation and
  apparatus, 478
 experimental materials, 476,477f
 gel permeation chromatograms,
  479,480f,481

INDEX

Caprolactone–allyl glycidyl ether copolymers cross-linked by electron-beam irradiation—*Continued*
  mechanical analysis, 479
  mechanical properties, 495,496f
  melting point depression, 491–492
  modulus vs. irradiation dose, 495,496f
  optical and scanning electron microscopic analysis, 478,481,483–488
  percent gel measurement, 478
  percent gel vs. irradiation dose, 479,480f,481,482f
  preparation of samples for irradiation, 476,478
  scanning electron micrographs, 486,487f,488
  thermal properties, 488–494
  thermal test, 479
Cationic photocurable systems
  advantages, 5
  mechanism, 4
  scheme, 2–3
Cationic photoinitiators
  advantages, 9
  mechanism, 8
  rearrangement reactions from direct irradiation of diarylhalonium salts, 82–90
Cationic polymerization of vinyl ethers, cure rates, 364
Chain scissioning, description, 516
Chlorendic anhydride, use in synthesis of UV-curable coatings, 210
Chlorendic-anhydride-containing polymeric coatings
  experimental materials, 211
  instrumentation and equipment, 212
  MS analysis of photoproducts, 215,218
  MS fragmentation pattern, 215,217
  photolysis conditions, 211–212
  photolysis products, 212,213t,215,216t
  photolytic decomposition pathways of diethoxyacetophenone, 212,214–215
Chlorendic-based reactive diluents
  effect on UV cure rate, 210–211
  structures, 210–211
2-(Chlorodimethylsilyl)propenylnorborn-2-ene-5-carboxylate, preparation, 164
Cleavage-type photoinitiators
  examples, 8
  mechanism, 7–8
CO_2 in solid argon
  argon matrix exposures, 501–505
  conversion of CO_2 to CO, 501
  electron-beam excitation of matrix, 501,503
  energy level diagram, 503,504f,505
  experimental apparatus, 499,500f

CO_2 in solid argon—*Continued*
  fraction of surviving CO_2 molecules vs. incident charge density, 501,502f
Concurrent grafting during radiation curing, effect of monomer structure, 128–138
Contact angle
  definition, 512–513
  measurements, 513,514f,515
Contrast, definition, 517
Coreactive photoinitiators for surface polymerization
  coupling to silica, 110–114,117
  examples, 106–107
  interaction of phenols with silica, 113,117t
  materials and methods, 107–108
  modified silica(s)
    thermal gravimetric analysis, 113,114f
    transmission spectra, 110,112f
  photoinitiator loading of modified silicas, 110,111t,113
  photoreactivity tests, 109–110,111t
  surface polymerization on photoinitiator-modified silica, 115–123
  synthesis, 108–109
Corrosion protection, long-term performance requirements, 194
Cross-linking, description, 516–517
Cure profiling of UV-cured coatings
  absorbance and transmittance
    vs. thickness, 34,35t
    vs. wavelength, 31,34,35t
  comparison of photoinitiators, 37t
  curing conditions, 30
  effect of photoinitiator concentration on cure response, 34,37t
  effects of concentration, 34,36f
  experimental procedure, 30
  extinction coefficient spectrum
    developmental photoinitiator, 31,32f
    spectra of photoinitiators, 37,40f
  importance, 27
  lamp emission spectra, 31,33f
  optimum density of coating, 28–29
  percent incident radiation absorbed
    calculations, 29–30
    for photoinitiators, 37,40f,41
    vs. total optical density, 31,33f
  photoinitiator structures, 30,32,37,39
  surface cure and through cure vs. photoinitiator concentration, 34,36f
  surface cure vs. percent incident radiation absorbed, 37,38f
  theory, 28–30
  through cure vs. percent incident radiation absorbed, 37,38f

D

Decomposition products of chlorendic
  anhydride, analysis, 211–218
Diacetylene reactive diluents
  preparation, 207
  structure, 207
Diaryliodonium ion
  structure, 399
  use as photoinitiators, 399
Diaryliodonium salts, use as cationic
  photoinitiator, 460
Diazoketone in phenolic polymeric matrix
  electron-beam exposures, 512
  evidence for carbene formation,
    507,511f,512
  fraction of molecules that survive
    absorbed dose, 505,506f
  structures, 505
Diazoketone–resin system
  IR spectra, 507,508–510f
  mechanism for electron-beam-induced
    chemistry, 507
  photochemical Wolff rearrangement,
    505,507
  vacuum-electron-beam exposure,
    505,507–510
Diepoxysiloxane monomers and oligomers
  cure response, 391,392t
  glass transition temperatures, 391,392t
  preparation, 389,391
  structures and properties, 389,391t
  UV flux, 391,393f
Diethoxyacetophenone, photolytic
  decomposition pathways, 212,214–215
Diphenoxybenzophenone
  advantages, 442
  effect on hardness of coating, 442,443f
  effect on polymerization kinetics,
    441–442,443f
Dual free radical–cationic photocurable
  systems, advantages, 12
Dual photocurable–thermal-curable coatings
  development, 12
  properties, 12

E

Electron(s)
  considerations for utilization, 18–19
  electromagnetic spectrum, 17,18t
  irradiation, 21–22
  isolation, 17–18
  polymerization, 19–21
  reactions with organic molecules, 19
Electron-beam-cured materials,
  applications, 258
Electron-beam curing of coatings,
  examples, 498
Electron-beam curing of perfluorinated
  acrylates
  measurements of contact angles, 513,514f,515
  modification of parameters, 512–513
Electron-beam curing of polymeric materials
  advantages of penetration ability of
    electrons, 20–21
  influencing factors, 20–21
  irradiation, 21–22
  rate expressions
    initiation, 19–20
    propagation, 20
    termination, 20
Electron beam resist, molecular weight
  dependence of sensitivity, 516–533
Electron-induced reactions, commercial
  importance, 17
Electron-initiated graft modification,
  electron micrographs, 549,550f,551
Electron-initiated graft modification of
  polyolefins
  adhesion, 541
  advantages, 534–535
  application, 543,547–551
  depth–dose profile, 539,540f
  dose–rate profile for electron processor,
    535,538f,539
  dry-oxygen permeability, 543,547–548f
  electron energy source, 535–539
  electron stopping power, 539,541
  experimental procedure, 542
  haze measurements, 543,549t
  limonene transmission, 549t
  schematic view of curtain-type
    self-shielded electron processor,
    535,536f
  surface treatment of matter,
    539,540f,541–542
  transmission Fourier transform IR spectra,
    542–543,544–546f
  typical isodose performance curves,
    535,537f
Electron irradiation
  applications, 22
  mechanism, 21–22
Enthalpy relaxation in bulk polymeric
  glasses
  absorption spectrum of photoinitiator,
    285,286f
  basis for structural relaxation, 285
  determination of degree of cure, 292
  differential scanning calorimetry
    results for cured films, 289t
  thermograms, 287,288f,289
  use, 285
  excess enthalpy vs. annealing time, 289,290f

INDEX

Enthalpy relaxation in bulk polymeric glasses—*Continued*
experimental materials, 285,286f,287t
experimental procedures, 285,287
formulation parameters, 285,287t
Fourier transform IR difference spectrum, 292,293f
percent conversion vs. UV dose, 292,294f,295
percent extractable materials vs. cure dose, 295t
schematic representation showing effects of structural relaxation, 285,286f
thermomechanical analytical thermograms, 289,291f,292
Epoxies, advantages, 364
Epoxy acrylate polymeric films, effect of diluent on surface softening, 326,327f
Epoxy coatings
effect of alcohol on cure speed, 364
effect of vinyl ether addition, 364
properties, 364
α,ω-Epoxy-functionalized poly(dimethylsiloxane) oligomers, preparation, 403
Epoxy resins, UV-induced polymerization, 412–424
Epoxysilicone(s)
effect of alcohols on cure response
hexanediol, 394,395f
monofunctional alcohols, 394,396t
triol, 394t
iodonium catalysts, 384–385,386f
polymerization, 383–384
properties of cured film, 396t
reaction with polyols, 394
requirements for radiation-cured release coatings, 383
synthesis, 383
UV absorbance spectra, 389,390f
UV cure of modified compounds, 388t,389
Epoxysilicone monomers, UV cure, 398–410
Epoxysilicone polymer modifications
comparison of properties of silicone hydride and silicone epoxy fluids, 387t
hydroxy ester synthesis, 387–388t
targeting of reactions, 385,387
Epoxysiloxane monomers, differential scanning calorimetric characterization, 403,405,406–408f
Ethylene–propylene–diene elastomers
development of rapid cross-linking method, 140–141
experimental procedures, 141,142f
photoinitiated cross-linking, 146–149
2-Ethyl-2-(hydroxymethyl)-1,3-propanediol-trinorborn-2-ene-5-carboxylate, preparation, 163

Excitation transfer between thioxanthones and morpholinoketone
excited-state processes in thioxanthones, 63,64t,65
interaction between thioxanthones and morpholinoketone, 63,65,66f,67
primary processes in morpholinoketone, 63–64
spectra, 62,64f
Exotherm measurements of photocuring, usefulness, 15

F

Free radical-photocurable system
mechanism, 2–4
scheme, 2–3
Free radical photoinitiators
computer modeling of absorption, 27–28
effect of concentration on photopolymerization rate, 28
optimum concentration, 28–29

H

High-energy-radiation-induced cationic polymerization of vinyl ethers with onium salt initiators
characteristics and elemental analysis of synthesized onium salts, 462t
concentration of protons produced by γ irradiation, 469,470f
effect of counterion(s)
on dosages, 469,471f,472
on initiation efficiency, 469
effect of oxygen saturation, 467
electron reductions, 465–467
electron spin resonance spectrum, 465,466f
experimental materials, 461
procedures, 462–463
reaction, 463–464
reduction potentials, 467,468t,469
synthetic procedures, 461,462t
Hydrophobic coatings, use as low-emission-radiation-curable coatings, 194–209
2-Hydroxy-2-methylpropiophenone, use as photoinitiator, 106–107

I

Intermolecular hydrogen abstraction, photoinitiation of polymerization, 151–152,156

Iodonium catalysts for epoxysilicones
bis(dodecylphenyl)iodonium
hexafluoroantimonate, 384–385
(4-octyloxyphenyl)phenyliodonium
hexafluoroantimonate, 385
UV absorbance spectra, 385,386f
IR spectroscopy, limitations for analysis of
UV-curable adhesives, 242
IR spectroscopy of radiation-cured adhesives
comparison with mechanical spectroscopy,
247,252f,253
experimental procedure, 243,246t
polymerization results, 247t,253
spectra, 247,249f
Irradiation, electrons, 21–22

K

Kubelka–Munk optical treatment of photon
flux
boundary effects, 47–48
calculated down flux, 49,50f,51
calculated up flux, 49,50f,51
comparison with results of N101 cure tester,
53,55,56–57f
comparison with single-flux model, 48–49
determination of absorbed light, 45,47
determination of reflected light, 45,47
distribution of light within a layer,
49t,50f,51
energy flux vs. distance from irradiated
surface of coating, 44–45,46f
factors
concentration, 51,52f
film thickness, 53,54f
substrate reflectance, 51,53,54f
light absorbed within layer, 51,52f
method of parameter calculation, 48
parameter values vs. depths within
layer, 49
sections within coating layer, 45,46f
theory of two-flux model, 44–45,46f,47
transmittance determination, 45,47

L

Laser-induced curing of acrylic systems
advantages, 446
comparison of techniques, 451t,455
conversion vs. time, 451,452f
differential photocalorimetry, 449
gel fraction, 449,450f
IR spectroscopy, 449,450f
laser interferometry, 449,452f
polymerization profiles, 451,454f
postpolymerization profiles, 451,453f

Laser-induced curing of acrylic systems—
Continued
real-time IR spectroscopy, 451,452–454f
see also Laser-initiated polymerization of
multifunctionalized acrylates
Laser-initiated polymerization of
multifunctionalized acrylates
development, 429
exotherm rate vs. time, 430,431f
experimental materials and procedure,
429–430
factors affecting percent conversion vs.
laser repetition rate
number of pulses, 433,434f
photoinitiator concentration, 433,435f,436
photoinitiator cross-linking, 436,437f
percent conversion vs. laser repetition
rate, 430,432f,433
see also Laser-induced curing of
acrylic systems
Light sources, use in UV curing processes,
13,14t
Linear unsaturated polyesters, development of
rapid cross-linking method, 140–141
Low-emission-radiation-curable coatings
adhesion, 202
characterization, 202,203t,204–206
corrosion resistance, 204–205
flexibility, 204
hardness, 202,204
impact strength, 204
IR spectra, 205–206
IR spectra of dipropargyl ether,
198,199–201f
low-volatile-organic-compound coatings,
206t,207
monomer preparation, 195–196,197t
polymer structures, 195,205
polymerization of propargyl monomers,
198,199–201f
preparation, 202
properties vs. cure time, 202,203t
solvent resistance, 204
water absorption, 204
Low-volatile-organic-compound coatings
acrylate and methacrylate reactive diluents,
206t,207
diacetylene reactive diluents, 207
procedures
coating preparation, 208
monomer polymerization, 208
monomer preparation, 207–208

M

Mechanical spectroscopy of radiation-cured
adhesives
comparison with IR spectroscopy,
247,252f,253

INDEX

Mechanical spectroscopy of radiation-cured adhesives—*Continued*
  determination of coefficient of thermal expansion, 253t
  determination of glass transition temperature, 253t
  experimental procedure, 246t
  frequency-dependent spectra
    vs. conversion, 247,250–251f
    vs. temperature, 253,254f,255
  polymerization results, 247t,253
  temperature-dependent spectrum, 247,250f
Methacrylate reactive diluents
  advantages, 206
  solubility, 206t,207
Methyl(mercaptopropylsiloxy)dimethylsiloxy silicone copolymers, preparation, 164
Microcolumn imaging
  characterization procedure, 312
  critical parameters, 309
  density vs. sensitivity, 313,316f
  diagrams of storage vs. loss modulus, 319,320f
  fractional dye control vs. fractional conversion of monomer, 319,320f
  image density and modulus change vs. log intensity, 319,321f
  photographic sensitivity vs. initiator concentration, 313,315f
  photographic shoulder speed and final image color density vs. pore diameter, 312–313,315f
  photographic shoulder speed vs. membrane thickness, 313,316f
  photomicrographs of curing stages, 313,317f
  process description, 308
  reciprocity behavior, 312,314f
  sample preparation, 309,312
  scanning electron micrographic images, 313,317f
  schematic representation, 309,311f
  T_g and fractional density change vs. exposure, 319,321f
  typical sensitometric curve, 309,310f
  vinyl group conversion vs. monomer conversion, 313,318f
Microencapsulated imaging system
  color density vs. photographic step number for exposure temperatures, 300,302f
  experimental procedures, 298
  exposure and image development, 298,299f
  factors influencing color density, 298
  phase diagrams of color density levels, 300,301f
  phase transformation profiles, 300,302–305f,306
  schematic representation, 297–298,299f

Microencapsulated imaging system—*Continued*
  sensitometric curves, 300,303f
  sensitometric response, 298,301f
Microlithography
  commercial application, 516
  contrast curve, 517,518f
  description, 516
  pattern transfer by irradiation of polymer resist, 516–517,518f
  thinning curve, 517,521f
Molecular weight, effect on sensitivity of electron-beam resist, 516–533
Monomer(s)
  properties, 5,6t,11
  types, 11
Monomer structural effects on concurrent grafting during radiation curing
  acid enhancement, 132–133t
  additive effect in ionizing-radiation-initiated grafting, 132–134t
  additive effect in UV-initiated grafting, 130–132t
  asymptotic approach to equilibrium, 135t
  combined effect of acids and salts with multifunctionalized acrylates, 137
  effect of various substances
    electrolytes, 135,136t,137
    lithium salts, 133t
    organic additives, 132t
    trimethylolpropane triacrylate, 130t
  experimental materials and methods, 129–130
  initial swelling behavior, 134,135t
  mechanism of acid and salt effect in UV and radiation grafting, 134,135–136t,137
  mechanism of additive effect in UV and radiation grafting, 134
  significance of multifunctional grafting work in curing applications, 137–138
  synergistic effects
    di- and monofunctionalized monomers, 131t
    trifunctionalized monomers, 130t
    trimethylolpropane triacrylate, 131t
    trimethylolpropane triacrylate and acid, 133,134t
Multifunctionalized acrylates
  examples, 128–129
  laser-initiated polymerization of multifunctionalized acrylates, 429–437
  role in concurrent grafting during radiation curing, 128–138
Multifunctionalized amine-terminated diacrylates
  absorption spectroscopy, 349
  alkyl band index vs. Microscal irradiation, 355–356t
  amine acrylate synthesis, 348
  C–H and carbonyl stretching regions, 355

Multifunctionalized amine-terminated
diacrylates—*Continued*
complexity of oxidation mechanism for
amine functionality, 356
dependence of photoyellowing on amine
structure, 354
effect of Microscal irradiation
on growth of UV band at 275 nm,
356,357f,358
on hydroxyl index for electron-beam-cured
acrylates, 350,352t
on second-derivative UV absorption band
for electron-beam-cured acrylates, 352,353f
for UV-cured acrylates, 350,351f
effect of polymer backbone on
photooxidative stability, 358t,359
electron-beam-curing procedure, 348
experimental materials, 347
hydroxyl index
effect of structure, 358t
effect of terminal amine functionality on
profile, 349–350
values, 349t
IR measurements, 347
oxidation site, 354–355
photodegradation vs. terminal amine
functionality, 356
photooxidation procedure, 349
photooxidative stability of electron-beam-
cured acrylates, 352
photoyellowing order for electron-beam-
cured acrylates, 352,354
problems with studying photooxidation and
photoyellowing, 346–347
total amine value, 348
UV-curing equipment, 347–348
Multifunctionalized epoxysilicone monomers
structure, 401
synthesis, 399,400t,401
Multifunctionalized urethane acrylate oligomers
gel permeation chromatographic curve,
260,262f,263
synthesis, 260–261
Multifunctionalized vinyl-ether-terminated
urethane oligomers
effect of humidity on UV cure speed,
378,380t
effect of monomer on viscosity, 375,377f
effect of reagent stoichiometry, 375t
GC, 375,376f
maximum tack-free cure speeds, 378,379f
preparation, 374
properties of cured coatings, 375t,378

N

N101 cure tester
effect of film weight, 55,56f,57
effect of initiator concentration, 57f

N101 cure tester—*Continued*
function, 53
procedure, 53
simple kinetic model, 55
trace of force vs. irradiation time,
53,56f
Negative (cross-linking) resists
comparison with experimental data,
520,522,523f
contrast curve, 520,522,523f
importance of molecular weight
distribution, 522,523f
simulation of contrast curves,
519–520,521f
theoretical models for insolubilization,
517,519
Non-acrylate curing mechanisms
effect of photoinitiator on cure, 152t
factors
amine structure, 154t,155
benzophenone level and UV dosage, 152,153t
ratio of unsaturation equivalent to amine
equivalent, 155,156–157f
structure of unsaturation source, 153t,154
oligomers used for amine structure study,
154t
photoinitiator studies, 152–153t
simplex experimental design study,
155,157f
thermal behavior of cured films, 155,158f
thermal mechanical analysis, 155,158f
Norborn-2-ene-5-carbonyl chloride,
preparation, 163
Norbornene-functionalized siloxanes,
preparation, 173,174f
Norbornene resin, preparation, 173f
Norbornene-terminated poly(dimethylsiloxane)
prepolymers, preparation, 164
Novel cyclic epoxy-functionalized siloxanes
preparation, 401–403,404t
structures, 403,404t

O

Olefins, determination of relative reaction
ratios, 164–165
Oligomers
properties, 5,6t
structure, 9
types, 9–10
Onium salt(s)
catalysis of cationic polymerization
reactions, 363
cationic photoinitiator mechanism, 460
initiators, high-energy radiation-induced
cationic polymerization of vinyl ethers,
459–472
photoinitiators, applications, 82

INDEX

Onium salt(s)—*Continued*
  use in coating applications, 460
Ordered aromatic and heteroaromatic polymers
  characteristics, 220
  photochemical approaches, 222–237
  preparation, 220–221
  structures, 220–221
Oxidation index, definition, 349
4-(Oxiranylmethoxy)phenyl 2-hydroxy-2-propyl ketone, synthesis, 109
Oxygen inhibition in photopolymerization processes, measurement of steady-state concentration of oxygen, 12–13

P

Percent incident radiation absorbed
  calculations, 29–30
  importance, 27
  theory, 28–29
Perfluorinated acrylates, electron-beam curing, 512–513,514f,515
Phase transformation of UV-curable systems, temperature effect, 297–306
Photochemical approaches to ordered polymers
  1,4-bis(2-hydroxybenzoyl)benzene, 225
  ^{13}C cross-polarization–magic angle spinning spectra, 232,236f,237
  characterization of polymers, 226–231
  experimental procedures, 222,225
  ^1H and ^{13}C NMR spectra, 229–232,234–235
  IR spectra, 226,228f
  photochemistry, 229,232–237
  photocyclization of o-benzyloxyphenyl ketones, 222–223
  preparation of 1,4-bis(2-benzyloxybenzoyl)benzene, 226
  route based on phenyl-substituted benzofuran repeat unit, 222,224
  side reactions, 232–233
  synthesis of polymers, 226,229
  UV spectra, 226,227f
Photochemistry, water-soluble benzophenone initiators, 72–80
Photo-cross-linking of ethylene–propylene–diene elastomers
  diene effectiveness, 148,149f
  experimental procedure, 146–147
  IR spectra of copolymer, 147,149f
  reaction, 146
  spin trapping, 147–148
Photo-cross-linking of norbornene resins with thiols
  calculations with cross-linked network model, 167,168–169f
  competitive thiol addition, 165f
  effect of monofunctional oligomer, 168,169f

Photo-cross-linking of norbornene resins with thiols—*Continued*
  effect of ring strain on reactivity, 166
  enthalpy of cross-linking reaction vs. temperature, 170f
  experimental procedures, 162–165
  gel point calculations, 168–169f
  model studies, 165f,t,166t,167f
  olefin heats of hydrogenation, 166t
  physical properties of norbornene-terminated silicone fluids, 172t
  reaction, 161,162f
  relative olefin addition rates, 165t,166
  testing of bulk physical properties, 171–172t
  thermal analytical studies, 170f,171
  thiol–olefin cooxidation, 166,167f
Photo-cross-linking of polyethylenes
  effect of cross-linker additive on cross-linking homogeneity, 143,145f
  effect of irradiation temperature, 141,142f
  mechanism, 141,143
  rate of cross-linking measured as gel content vs. light intensity, 143,145f,146
  rate vs. chain length, 143,144f
  rate vs. cross-linker addition, 143,144f
Photocuring mechanisms, discussion, 2–5
Photodifferential scanning calorimetry of radiation-cured adhesives
  apparatus, 243,245f
  polymerization results, 247t
  thermograms, 247,248f
Photoinitiated cationic polymerizations
  advantages, 398–399
  photoinitiators, 399
  preparation of α,ω-epoxy-functionalized poly(dimethylsiloxane) oligomers, 403
  preparation of novel cyclic epoxy-functionalized siloxanes, 401–403,404t
  synthesis of di-, tri-, and tetrafunctionalized epoxysilicone monomers, 399,400t,401
Photoinitiators
  abstraction type, 7
  cationic type, 8–9
  cleavage type, 7–8
  properties, 5,6t
  synergistic reaction, 60
Photooxidative stability of multifunctionalized amine-terminated diacrylates, 346–359
Photopolymerization activity of water-soluble benzophenone initiators, 72–80
Photopolymerization processes, oxygen inhibition, 12–13
Photopolymerization rate
  effect of intensity of absorbed radiation, 28
  influencing factors, 28

Photopolymerized epoxysilicone monomers
  film properties, 405,409t
  thermogravimetric analytical curves,
    405,410f
Photosensitizer, synergistic reaction, 60
Photoyellowing of multifunctionalized
  amine-terminated diacrylates, 346–359
Poly(dimethylsiloxane)s, applications, 382
Polyester acrylate films, effect of diluent
  on surface softening, 326,327f
Polyester resin–styrene systems,
  properties, 151
Polyethylenes
  development of rapid cross-linking method,
    140–141
  experimental procedures, 141,142f
  photoinitiated cross-linking, 140–146
Polymer(s), cross-linking classes, 475
Polymerization
  description, 516–517
  electron-beam curing, 19–21
Polymerization of acrylic monomers
  analysis, 441
  effect of reactive diluent
    on hardness, 446,448f
    on photopolymerization, 444,445f,446,447f
  experimental materials, 440
  irradiation procedure, 440–441
  performance analysis of UV-curable
    systems, 444t
  polymerization in air, 444,445f
  use of diphenoxybenzophenone as
    photoinitiator, 441–442,443f
Polyolefins, electron-initiated graft
  modification, 534–551
Positive (scissioning) resists
  comparison of contrast data with
    experimental data, 528,529–530f
  exponential model for dependence of
    dissolution rate on molecular weight,
    526,528
  γ model for dependence of dissolution rate
    on molecular weight, 524,526,527f
  models for dissolution rate,
    522,524,525f,527f
  sensitivity relationships,
    528,531,532f,533
  simulation of contrast curves,
    524,526,527f,528
  theoretical models for scissioning, 522
  thinning behavior, 528,530f
Propargyl monomers
  advantages of using propargyl chloride in
    preparation, 196
  formula, 197
  melting points of crystalline compounds,
    197t
  polymerization, 198,199–201f

Propargyl monomers—*Continued*
  preparation methods, 195–196
  properties, 197
Pulsed xenon lamp source, advantages,
  13,14t

R

Radiation-curable coatings, types, 363
Radiation-curable formulations, effect of
  adding multifunctionalized acrylate, 128
Radiation-cured adhesives
  activation energy, 255t
  comparison of IR spectroscopy, mechanical
    spectroscopy, and photodifferential
    scanning calorimetry, 247–256
  comparison of thermal, mechanical, and
    spectroscopic techniques, 242–256
  frequency-dependent master curve,
    255,256f
  IR spectroscopic procedure, 243,246
  mechanical spectroscopic procedure, 246t
  photodifferential scanning calorimetric
    apparatus, 243,245f
  sample preparation, 243,244f
  temperature dependence of shift factor,
    255,256f
  thermomechanical analysis, 243,246t
  WLF constants, 255t
Radiation-cured materials, adhesive
  applications, 242
Radiation curing
  applications, 439
  effect of monomer structure on concurrent
    grafting, 128–138
Radiation-curing technology, microcolumn
  imaging, 308–321
Radiation-induced cationic curing,
  development, 363–364
Radiation-induced cationic curing of
  vinyl-ether-terminated urethane oligomers
  effect of humidity on cure speed,
    378,380t
  effect of oxygen, 378
  effect of sulfonium salt concentration
    on electron-beam cure speed, 370,372f
    on UV cure speed, 370,371f
  IR analysis, 370,373f
  oligomer structure, 368
  physical properties of electron-beam-cured
    coating, 369t,370
  physical properties of UV-cured coating,
    369t,370
  properties, 370,374–375t
  radiation-curing procedure, 368
  synthesis of compounds, 366,368
Radiation-induced polymerization(s)
  examples of industrial applications, 459–460

INDEX

Radiation-induced polymerization(s)—
Continued
  limitations of cationic polymerization, 460–461
Rearrangement reactions from direct irradiation of diarylhalonium salts
  acid generation from formation of rearrangement product, 85,87,89f
  experimental materials and procedures, 82–83
  mechanism of diarylhalonium salt photodecomposition, 88,89f,90
  photolysis products of diphenylhalonium salts, 83,85,86f
  photolysis products of 4,4′-ditolylhalonium salts, 85,86f
  product distribution, 88t
  product quantum yields, 83,85,87t
  UV absorption of diphenyliodonium triflate and photoproducts, 83,84f
  UV absorption spectra monitoring $Ph_2I^+CF_3SO_3^-$ consumption, 83,84f

S

Saturated linear aliphatic polyesters, radiation dependence on ratio of methylene to ester, 475
Semicrystalline polymers, effect of temperature on cross-linking, 475
Silicone(s), *See* Poly(dimethylsiloxane)s
Silicone polymer structures, shorthand system of abbreviation, 382–383
Single-pass dose, definition, 348
Sulfonium salts, use as cationic photoinitiator, 460
Surface polymerization on photoinitiator-modified silica
  effect of initiator silica type and irradiation time, 116,117t
  effect of reduced initiator loading, 122
  initiation by radical transfer, 119
  IR spectra, 116,118f,119,121f
  monomer(s), 115
  monomer dilution, 119–123
  reaction, 115
Synergism, description, 60–62
Synergistic processes in photoinitiators of polymerization
  excitation transfer between thioxanthones and morpholinoketone, 62–67
  role of addition of photosensitizer of thioxanthone series, 67–68,69t
  triplet energy levels of compounds, 68,69t,70
  use of time-resolved laser spectroscopy, 59–70

T

Thermomechanical properties of filled resins
  coefficient of thermal expansion, 416,419–420,422t
  displacement of sample ends, 419,420f
  modulus, 416,418f
  properties of resins with different filler contents, 419,422t
  volume shrinkage, 419
Thiol-initiated acrylic polymerization, reaction, 161f
Thiol–olefin polymerizations, development, 160
Thiolene addition polymerizations, reaction, 160,161f
Thiolene resin systems, properties, 151
Time-resolved laser spectroscopy, study of synergistic processes in photoinitiators of polymerization, 59–70
Time–temperature–transformation state diagram, application to thermosets, 297
Traditional photocurable-system components
  additives, 6t,11–12
  monomers, 6t,11
  oligomers, 6t,9–10
  photoinitiators, 5–9
Triarylsulfonium ion, structure, 399
Triethoxysilyl-substituted (2-hydroxy-2-propyl)phenones, synthesis, 109
Triethylene glycol monovinyl ether, synthesis, 366
Trimethylolpropane triacrylate, application in radiation grafting reactions, 128–129

U

Urethane acrylate(s)
  applications, 272
  compositional variables vs. end properties, 273t
  effects of acrylate functionality
    on coating properties, 281,282t
    on stress–strain properties, 279,280t
    on viscosity and cure speed, 275,276t,278
  effects of composition
    on coating properties, 281–283t
    on cure speed, 275,276–278t
    on properties, 275–283
    on stress–strain properties, 278,279–280t
    on viscosity, 275,276–278t
  effects of diisocyanate
    on coating properties, 281t
    on stress–strain properties, 278,279t
    on viscosity and cure speed, 275,276t,278

Urethane acrylate(s)—*Continued*
  effects of molecular weight
    on coating properties, 281,283t
    on stress–strain properties, 279,280t
    on viscosity and cure speed, 275,277t,278
  effects of polyol type
    on coating properties, 281,282t
    on stress–strain properties, 278,279t
    on viscosity and cure speed, 277,278t
  effects of urethane content
    on coating properties, 281,283t
    on stress–stain properties, 279,280t
    on viscosity and cure speed, 275,277t,278
  experimental procedures, 275
  experimental reaction products of urethane formation, 273,274f
  formation, 272–273
  properties, 259
  structure, 272,274f
  synthesis, 259
Urethane acrylate films, effect of diluent on surface softening, 326,327f
UV-curable coating
  factors affecting light absorption, 43–44
  Kubelka–Munk optical treatment of photon flux, 44–57
UV-curable resins, classifications, 151
UV-curable systems
  application areas, 2,6t
  general characteristics, 1–2
  performance improvement, 439
  temperature effect on phase transformation, 297–306
  types, 2–5
UV cure, epoxysilicone monomers, 398–410
UV-cured coatings
  advantages, 398
  depth of cure profiling, 27–41
UV-cured coatings containing multifunctionalized acrylates
  coating formulations and physical properties of films, 260,265t,267,269t
  composition and properties of oligomers, 260,263,264t,266t
  experimental materials and procedures, 259
  gel permeation chromatographic curve, 260,262t,263
  influencing factors
    diol molecular weight, 263,266t,267,269t
    diol structure, 263,264–265t
    functional group on trifunctional branching agent, 267
    oligomer functionality, 268,269–270t
    overall composition, 267–268,269–270t
    reactive diluent, 268,270t
  synthesis, 260–261

UV-cured epoxy coatings, enthalpy relaxation, 284–295
UV-cured materials
  advantages, 258
  applications, 258
  curing mechanism, 258
  effect of components on performance, 258
  properties vs. time after cure, 284
UV curing
  components, 5–12
  development of specialized photoinitiators, 92
  exotherm measurements, 15
  light sources, 13,14t
  oxygen inhibition in photopolymerization processes, 12–13
UV-curing industry, development, 1
UV curing of epoxy resins, advantages for globe top applications, 412–413
UV curing of filled epoxy resins
  inorganic filler, 414
  light source spectrum, 414,415f,416
  optical density, 416,417f
  resin matrix, 414
  resistance and capacitance monitoring, 416,417f
  spectrometry with transmitted light, 414,415f,416,417f
  thermal postcure, 416
  transmitted-light spectra, 414,415f,416
UV-induced polymerization of highly filled epoxy resins
  dynamic–mechanical testing, 414
  experimental materials, 414
  flexibilization and toughening, 419,422t,423
  preparation of resin formulations, 414
  test performance, 423,424t
  thermomechanical properties of filled resins, 416–422
  UV curing, 414,415f,416,417f
  UV curing and light monitoring, 414
  viscosity, 419,421f
  viscosity measurements, 413–414

V

Vinyl ether(s)
  high-energy-radiation-induced cationic polymerization, 459–472
  synthesis, 365
Vinyl-ether-terminated urethane oligomers
  preparation, 370,374
  properties of cured coatings, 374t
  radiation-induced cationic curing, 366–380
  relationship with hydroxyvinyl ether synthesis, 366–367

INDEX

Vinyl-ether-terminated urethane oligomers—*Continued*
  synthesis, 365–366,368

W

Water-soluble benzophenone initiators
  effect of amines
    on polymerization, 76,77t
    on transient absorption, 79,80t
  effect of pH on transient absorption, 79,80t
  end-of-pulse absorption spectra, 78–79t

Water-soluble benzophenone initiators—*Continued*
  experimental materials, 73
  flash photolysis, 74,78–80t
  photopolymerization, 74
  photopolymerization activity, 75,76–77t,78
  photoreduction quantum yields, 74,75t
  quantitative photocalorimetric data, 75,76t
  spectroscopic measurements, 73
  spectroscopic properties, 77t,78
  structures, 73,75t

*Production: Becki K. Weiss and Paula M. Bérard
Indexing: Deborah H. Steiner
Acquisition: Robin Giroux*

*Elements typeset by Hot Type Ltd., Washington, DC
Printed and bound by Maple Press, York, PA*

Paper meets minimum requirements of American National Standard for Information Sciences—Permanence of Paper for Printed Library Materials, ANSI Z39.48–1984 ∞

Other ACS Books

Chemical Structure Software for Personal Computers
Edited by Daniel E. Meyer, Wendy A. Warr, and Richard A. Love
ACS Professional Reference Book; 107 pp;
clothbound, ISBN 0–8412–1538–3; paperback, ISBN 0–8412–1539–1

Personal Computers for Scientists: A Byte at a Time
By Glenn I. Ouchi
276 pp; clothbound, ISBN 0–8412–1000–4; paperback, ISBN 0–8412–1001–2

Biotechnology and Materials Science: Chemistry for the Future
Edited by Mary L. Good
160 pp; clothbound, ISBN 0–8412–1472–7; paperback, ISBN 0–8412–1473–5

Polymeric Materials: Chemistry for the Future
By Joseph Alper and Gordon L. Nelson
110 pp; clothbound, ISBN 0–8412–1622–3; paperback, ISBN 0–8412–1613–4

The Language of Biotechnology: A Dictionary of Terms
By John M. Walker and Michael Cox
ACS Professional Reference Book; 256 pp;
clothbound, ISBN 0–8412–1489–1; paperback, ISBN 0–8412–1490–5

Cancer: The Outlaw Cell, Second Edition
Edited by Richard E. LaFond
274 pp; clothbound, ISBN 0–8412–1419–0; paperback, ISBN 0–8412–1420–4

Practical Statistics for the Physical Sciences
By Larry L. Havlicek
ACS Professional Reference Book; 198 pp; clothbound; ISBN 0–8412–1453–0

The Basics of Technical Communicating
By B. Edward Cain
ACS Professional Reference Book; 198 pp;
clothbound, ISBN 0–8412–1451–4; paperback, ISBN 0–8412–1452–2

The ACS Style Guide: A Manual for Authors and Editors
Edited by Janet S. Dodd
264 pp; clothbound, ISBN 0–8412–0917–0; paperback, ISBN 0–8412–0943–X

Chemistry and Crime: From Sherlock Holmes to Today's Courtroom
Edited by Samuel M. Gerber
135 pp; clothbound, ISBN 0–8412–0784–4; paperback, ISBN 0–8412–0785–2

For further information and a free catalog of ACS books, contact:
American Chemical Society
Distribution Office, Department 225
1155 16th Street, NW, Washington, DC 20036
Telephone 800–227–5558